An Introduction to Medicinal Chemistry

An Introduction to
Medicinal Chemistry

Second Edition

Graham L. Patrick

Department of Chemistry
Paisley University

OXFORD
UNIVERSITY PRESS

OXFORD

UNIVERSITY PRESS

Great Clarendon Street, Oxford OX2 6DP

Oxford University Press is a department of the University of Oxford.
It furthers the University's objective of excellence in research, scholarship,
and education by publishing worldwide in

Oxford New York

Athens Auckland Bangkok Bogotá Buenos Aires Calcutta
Cape Town Chennai Dar es Salaam Delhi Florence Hong Kong Istanbul
Karachi Kuala Lumpur Madrid Melbourne Mexico City Mumbai
Nairobi Paris São Paulo Singapore Taipei Tokyo Toronto Warsaw

with associated companies in Berlin Ibadan

Published in the United States
by Oxford University Press Inc., New York

A catalogue record for this book is available from the British Library

Library of Congress Cataloguing in Publication Data
(data available)

ISBN 0 19 850533 7 (Pbk)

Typeset by Florence Production Ltd, Stoodleigh, Devon
Printed in Great Britain on acid-free paper by Bath Press Ltd., Bath, Avon

Preface

This text is aimed at undergraduates and graduates who have a basic grounding in chemistry, and who are interested in learning about drug design and the molecular mechanisms by which drugs act in the body. Consequently, the book is of particular interest to students who might be considering a future career in the pharmaceutical industry. It attempts to convey something of the fascination of working in a field which overlaps the disciplines of chemistry, biochemistry, cell biology, and pharmacology.

The first edition of this textbook was published in 1995 and proved extremely successful. Since then medicinal chemistry has developed rapidly and continues to do so. There are clearly exciting times ahead, and this second edition includes fresh material which reflects many of these developments.

The book is divided into two main sections.

Section A covers the basic principles and techniques of medicinal chemistry. The first seven chapters cover the basics of cell structure, proteins, and nucleic acids as applied to drug design, and also include a new chapter (Chapter 6) which gives a detailed description of receptors and the mechanism by which they 'switch on' effects within cells. Chapters 8–13 describe the tactics and tools which are used in developing an effective drug and also the difficulties faced by the medicinal chemist in this task. New chapters on combinatorial synthesis, and the use of computers in medicinal chemistry reflect the importance of these techniques in modern day drug design.

Section B covers a selection of specific topic areas within medicinal chemistry. The chapters on antibacterial agents, cholinergics and anticholinesterases, opiate analgesics and cimetidine have been updated from the first edition, and a new chapter covering adrenergic agents has been included.

As the title indicates, this textbook is an introduction to the rapidly expanding scientific discipline of medicinal chemistry, and has been published at an affordable price for students. The book is intended to whet the appetite and interested readers are encouraged to read further by consulting the more specialised texts described under 'Further Reading'.

Paisley   G.L.P.
January 2001

Acknowledgements

The various cartoons in the textbook were designed and drawn by Mr Gary Learie, to whom the author is indebted. The author also wishes to express his gratitude to Dr Stephen Bromidge of SmithKline Beecham for permitting the description of his work on selective 5-HT$_{2C}$ antagonists (Chapter 13), and for providing many of the diagrams used in that Case Study. Thanks are also due to Cambridge Scientific, Oxford Molecular, and Tripos for their advice and assistance in the writing of Chapter 13.

Finally, the author acknowledges the many helpful and constructive comments which were made in this text by Professor John Mann, Dr Stephen Gorham and other reviewers. Comments and suggestions from readers of the first edition were particularly welcome and the author will be delighted to receive any similar feedback on this edition. Any errors in the text are the author's responsibility alone.

Contents

Classification of drugs

The way drugs are classified or grouped can be confusing. Different textbooks group drugs in different ways.

1 By pharmacological effect

Drugs are grouped depending on the biological effect they have, e.g. analgesics, antipsychotics, antihypertensives, anti-asthmatics, antibiotics, etc.

This is useful if one wishes to know the full scope of drugs available for a certain ailment. However, it should be emphasized that such groupings contain a large and extremely varied assortment of drugs. This is because there is very rarely one single way of dealing with a problem such as pain or heart disease. There are many biological mechanisms which the medicinal chemist can target to get the desired results, and so to expect all painkillers to look alike or to have some common thread running through them is not realistic.

A further point is that many drugs do not fit purely into one category or another and some drugs may have more uses than just one. For example, a sedative might also have uses as an anticonvulsant. To pigeon-hole a drug into one particular field and ignore other possible fields of action is folly.

The antibacterial agents (Chapter 14) are a group of drugs that are classified according to their pharmacological effect.

2 By chemical structure

Many drugs which have a common skeleton are grouped together, e.g. penicillins, barbiturates, opiates, steroids, catecholamines, etc.

In some cases (e.g. penicillins) this is a useful classification since the biological activity (e.g. antibiotic activity for penicillins) is the same. However, there is a danger that one could be confused into thinking that all compounds of a certain chemical group have the same biological action. For example, barbiturates may look much alike and yet have completely different uses in medicine. The same holds true for steroids.

It is also important to consider that most drugs can act at several sites of the body and have several pharmacological effects.

The opiates (Chapter 17) are a group of drugs with similar chemical structures.

3 By target system

These are compounds which are classed according to whether they affect a certain target system in the body – usually involving a chemical messenger, e.g. antihistamines, cholinergics, etc.

This classification is a bit more specific than the first, since it is identifying a system with which the drugs interact. It is, however, still a system with several stages and so the same point can be made as before – for example, one would not expect all anti-histamines to be similar compounds since the system by which histamine is synthesized, released, interacts with its receptor, and is finally removed, can be attacked at all these stages.

4 By site of action

These are compounds which are grouped according to the enzyme or receptor with which they interact. For example, anticholinesterases (Chapter 15) are a group of drugs which act through inhibition of the enzyme acetylcholinesterase.

This is a more specific classification of drugs since we have now identified the precise target at which the drugs act. In this situation, we might expect some common ground between the agents included, since a common mechanism of action is a reasonable though not inviolable assumption.

It is easy, however, to lose the wood for the trees and to lose sight of why it is useful to have drugs which switch off a particular enzyme or receptor site. For example, it is not intuitively obvious why an anticholinergic agent should paralyse muscle and why that should be useful.

1 Drugs and the medicinal chemist

In medicinal chemistry, the chemist attempts to design and synthesize a medicine or a pharmaceutical agent which will benefit humanity. Such a compound could also be called a 'drug', but this is a word which many scientists dislike since society views the term with suspicion.

With media headlines such as 'Drugs Menace', or 'Drug Addiction Sweeps City Streets', this is hardly surprising. However, it suggests that a distinction can be drawn between drugs which are used in medicine and drugs which are abused. Is this really true? Can we draw a neat line between 'good drugs' like penicillin and 'bad drugs' like heroin? If so, how do we define what is meant by a good or a bad drug in the first place? Where would we place a so-called social drug like cannabis in this divide? What about nicotine, or alcohol?

HEY MAN
DRUGS ARE
COOL

The answers we get would almost certainly depend on who we were to ask. As far as the law is concerned, the dividing line is defined in black and white. As far as the party-going teenager is concerned, the law is an ass. As far as we are concerned, the questions are irrelevant. To try and divide drugs into two categories – safe or unsafe, good or bad – is futile, and could even be dangerous.

First of all, let us consider the so-called 'good' drugs – the medicines. How 'good' are they? If a medicine is to be truly 'good' it would have to satisfy the following criteria. It would have to do what it is meant to do, have no toxic or unwanted side-effects, and be easy to take.

How many medicines fit these criteria?

The short answer is 'none'. There is no pharmaceutical compound on the market today which can completely satisfy all these conditions. Admittedly, some come quite close to the ideal. Penicillin, for example, has been one of the most effective anti-bacterial agents ever discovered and has also been one of the safest. Yet it too has drawbacks. It has never been able to kill all known bacteria and, as the years have gone by, more and more bacterial strains have become resistant. Nor is penicillin totally safe. There are many examples of patients who show an allergic reaction to penicillin and are required to take alternative antibacterial agents.

Whilst penicillin is a relatively safe drug, there are some medicines which are distinctly dangerous. Morphine is one such example. It is an excellent analgesic, yet it suffers from serious side-effects such as tolerance, respiratory depression, and addiction. It can even kill if taken in excess.

Barbiturates are also known to be dangerous. At Pearl Harbor, American casualties were given barbiturates as general anaesthetics before surgery. However, because of a poor understanding about how barbiturates are stored in the body, many patients received sudden and fatal overdoses. In fact, it is reputed that more casualties died at the hands of the anaesthetists at Pearl Harbor than died of their wounds.

To conclude, the 'good' drugs are not as perfect as one might think.

What about the 'bad' drugs then? Is there anything good that can be said about them? Surely there is nothing we can say in defence of the highly addictive drug heroin?

Well, let us look at the facts about heroin. Heroin is one of the best painkillers known to man. In fact, it was named heroin at the end of the nineteenth century because it was thought to be the 'heroic' drug which would banish pain for good. The drug went on the market in 1898, but 5 years later the true nature of heroin's addictive properties became evident and the drug was speedily withdrawn from general distribution.

However, heroin is still used in medicine today – under strict control of course.

The drug is called diamorphine and it is the drug of choice when treating patients dying of cancer. Not only does diamorphine reduce pain to acceptable levels, it also produces a euphoric effect which helps to counter the depression faced by patients close to death. Can we really condemn a drug which can do that as being all 'bad'?

By now it should be evident that the division between good drugs and bad drugs is a woolly one and is not really relevant to our discussion of medicinal chemistry. All

drugs have their good points and their bad points. Some have more good points than bad and *vice versa* but, like people, they all have their own individual characteristics. So how are we to define a drug in general?

One definition could be to classify drugs as 'compounds which interact with a biological system to produce a biological response'.

This definition covers all the drugs we have discussed so far, but it goes further. There are chemicals which we take every day and which have a biological effect on us. What are these everyday drugs?

One is contained in all the cups of tea, coffee, and cocoa which we consume. All of these beverages contain thè stimulant caffeine. Whenever you take a cup of coffee, you are a drug user. We could go further. Whenever you crave a cup of coffee, you are a drug addict. Even kids are not immune. They get their caffeine 'shot' from coke or pepsi. Whether you like it or not, caffeine is a drug. When you take it, you experience a change of mood or feeling.

So too, if you are a worshipper of the 'nicotine stick'. The biological effect is different. In this case you crave sedation or a calming influence, and it is the nicotine in the cigarette smoke which induces that effect.

There can be little doubt that alcohol is a drug and as such causes society more problems than all other drugs put together. One only has to study road accident statistics to appreciate that fact. It has been stated that if alcohol was discovered today, it would be restricted in exactly the same way as cocaine or cannabis. If one considers alcohol in a purely scientific way, it is a most unsatisfactory drug. As many will testify, it is notoriously difficult to judge the correct dose of alcohol required to gain the beneficial effect of 'happiness' without drifting into the higher dose levels which produce

Caffeine

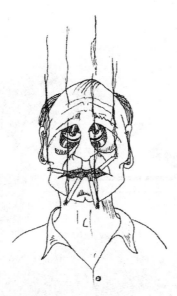

unwanted side-effects. Alcohol is also unpredictable in its biological effects. Happiness or depression may result depending on the user's state of mind. On a more serious note, addiction and tolerance in certain individuals have ruined the lives of addicts and relatives alike.

Our definition of a drug can also be used to include compounds which at first sight we might not consider to be drugs.

Consider how the following examples fit our definition.

- Morphine – reacts with the body to bring pain relief.
- Snake venom – reacts with the body to cause death!

- Strychnine – reacts with the body to cause death!
- LSD – reacts with the body to produce hallucinations.
- Coffee – reacts with the body to wake you up.
- Penicillin – reacts with bacterial cells to kill them.
- Sugar – reacts with the tongue to produce a sense of taste.

All of these compounds fit our definition of drugs. It may seem strange to include poisons and snake venoms as drugs[1], but they too react with a biological system and produce a biological response – a bit extreme perhaps, but a response all the same.

The idea of poisons and venoms acting as drugs may not appear so strange if we consider penicillin. We have no problem in thinking of penicillin as a drug, but if we were to look closely at how penicillin works, then it is really a poison. It interacts with bacteria (the biological system) and kills them (the biological response). Fortunately for us, penicillin has no effect on human cells.

Even those medicinal drugs which do not act as poisons have the potential to become poisons – usually if they are taken in excess. We have already seen this with morphine. At low doses it is a painkiller. At high doses, it is a poison which kills by suffocation. Therefore, it is important that we treat all medicines as potential poisons and keep them well protected from children searching the house for concealed sweets.

There is a term used in medicinal chemistry known as the **therapeutic index** which indicates how safe a particular drug is. The therapeutic index is a measure of the drug's beneficial effects at a low dose *versus* its harmful effects at a high dose. To be more precise, the therapeutic index compares the drug dose levels which lead to toxic effects in 50% of cases studied, with respect to the dose levels leading to maximum therapeutic effects in 50% of cases studied. A high therapeutic index means that there is a large safety margin between beneficial and toxic doses. The values for cannabis and alcohol are 1000 and 10 respectively, which might imply that cannabis is safer and more predictable than alcohol. Indeed, it has been proposed that cannabis would be a useful medicine for the relief of chronic pain which is untreatable by normal analgesics (e.g. morphine). This may well be the case, but legalizing cannabis for medicinal use does not suddenly make it a safe drug. For example, the favourable therapeutic index of cannabis does not indicate its potential toxicity if it is taken over a long period (chronic use). For example, the various side-effects of cannabis include dizziness, sedation, dry mouth, blurred vision, mental clouding, disconnected thought, muscle twitching, and impaired memory. More seriously, it can lead to panic attacks, paranoid delusions, and hallucinations. Clearly, the safety of drugs is a complex matter and it is not helped by media sensationalism.

If useful drugs can be poisons at high doses or over long periods of use, does the opposite hold true? Can a poison be a medicine at low doses? In certain cases, this is found to be the case.

---

[1] Substances which are foreign to the particular biological system under study are known as xenobiotics. The word is derived from the Greek word *xenos*, meaning foreign, and *bios*, meaning life.

Arsenic is well known as a poison, but arsenic-based compounds were used at the beginning of the century as antiprotozoal agents.

Curare is a deadly poison which was used by the Incas to tip their arrows such that a minor arrow wound would be fatal, yet compounds based on the tubocurarine structure (the active principle of curare) have been used in surgical operations to relax muscles. Under proper control and in the correct dosage, a lethal poison may well have an important medical role.

Since our definition covers any chemical which interacts with any biological system, we could include all pesticides and herbicides as drugs. They interact with bacteria, fungi, and insects, killing them and thus protecting plants.

Sugar (or any sweet item for that matter) can be classed as a drug. It reacts with a biological system (the taste buds on the tongue) to produce a biological response (sense of sweetness).

Even food can be a drug. Junk foods and fizzy drinks have been blamed for causing hyperactivity in children. It is believed that junk foods have high concentrations of certain amino acids which can be converted in the body to neurotransmitters. These are chemicals which pass messages between nerves. If an excess of these chemical messengers should accumulate, then too many messages are transmitted in the brain, leading to the disruptive behaviour observed in susceptible individuals. Allergies due to food additives and preservatives are also well recorded.

Some foods even contain toxic chemicals! Broccoli, cabbage and cauliflower all contain high levels of a chemical which can cause reproductive abnormalities in rats. Peanuts and maize sometimes contain fungal toxins, and it is thought that fungal toxins in food were responsible for the biblical plagues. Basil contains over 50 compounds which are potentially carcinogenic, while other herbs contain some of the most potent carcinogens known! Carcinogenic compounds have also been identified in radishes, brown mustard, apricots, cherries, and plums. Such unpalatable facts might put you off your dinner, but take comfort! These chemicals are present in such

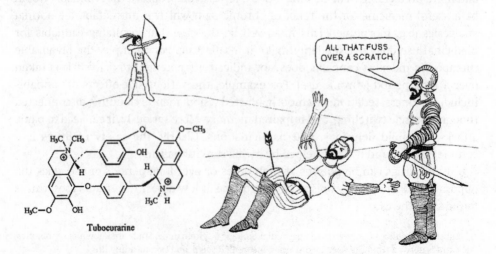

Tubocurarine

small quantities that the risk is insignificant. Therein lies a great truth which was recognized as long ago as the 15th century when it was stated that 'Everything is a poison, nothing is a poison. It is the dose that makes the poison'.

Almost anything taken in excess will be toxic. You can make yourself seriously ill by taking 100 aspirin tablets or a bottle of whisky or 9 kg of spinach. The choice is yours!

Having discussed what drugs are, we shall now consider why, where, and how drugs act.

2 The why and the wherefore

2.1 Why should drugs work?

Why indeed? We take it for granted that they do, but why should chemicals, some of which have remarkably simple structures, have such an important effect on such a complicated and large structure as a human being? The answer lies in the way in which the human body operates. If we could see inside our bodies to the molecular level, we would no doubt get a nasty shock, but we would also see a magnificent array of chemical reactions taking place, keeping the body healthy and functioning.

Drugs may be mere chemicals, but they are entering a world of chemical reactions with which they interact. Therefore, there should be nothing odd in the fact that they can have an effect. The surprise might be that they can have such specific effects. This is more a result of *where* they react in the body.

2.2 Where do drugs work?

2.2.1 Cell structure

Since life is made up of cells, then quite clearly drugs must act on cells. The structure of a typical cell is shown in Fig. 2.1.

All cells in the human body contain a boundary wall called the cell membrane. This encloses the contents of the cell – the cytoplasm. The cell membrane seen under the electron microscope consists of two identifiable layers. Each layer is made up of an ordered row of phosphoglyceride molecules such as phosphatidylcholine (lecithin).[1] Each phosphoglyceride molecule consists of a small polar head group, and two long hydrophobic chains (Fig. 2.2).

---

[1] The outer layer of the membrane is made up of phosphatidylcholine whereas the inner layer is made up of phosphatidylethanolamine, phosphatidylserine, and phosphatidylinositol.

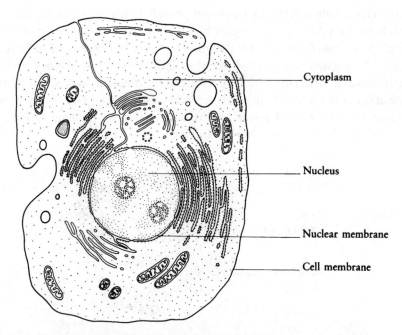

Fig. 2.1 A typical cell. (Taken from J. Mann, *Murder, magic, and medicine,*
Oxford University Press (1992), with permission.)

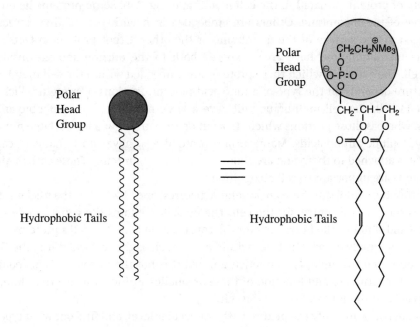

Fig. 2.2 Phosphoglyceride structure.

In the cell membrane, the two layers of phospholipids are arranged such that the hydrophobic tails point to each other and form a fatty, hydrophobic centre, while the ionic head groups are placed at the inner and outer surfaces of the cell membrane (Fig. 2.3). This is a stable structure since the ionic, hydrophilic head groups can interact with the aqueous media inside and outside the cell, while the hydrophobic tails maximize van der Waals interactions with each other and are kept away from the aqueous environments. The overall result of this structure is to construct a fatty barrier between the cell's interior and its surroundings.

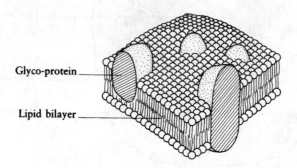

Glyco-protein

Lipid bilayer

Fig. 2.3 Cell membrane. (Taken from J. Mann, *Murder, magic, and medicine*, Oxford University Press (1992), with permission.)

The membrane is not just made up of phospholipids, however. There is a large variety of proteins situated in the cell membrane (Fig. 2.4). Some proteins lie on the surface of the membrane. Others are embedded in it with part of their structure exposed to one surface of the membrane or the other. Other proteins traverse the whole membrane and have areas exposed both to the outside and the inside of the cell. The extent to which these proteins are embedded within the cell membrane structure depends on the type of amino acid present. Portions of protein which are embedded in the cell membrane will have a large number of hydrophobic amino acids, whereas those portions which stick out on the surface will have a large number of hydrophilic amino acids. Many surface proteins also have short chains of carbohydrates attached to them and are thus classed as glycoproteins. These carbohydrate segments are important to cell recognition.

Within the cytoplasm there are several structures, one of which is the nucleus. This acts as the 'control centre' for the cell. The nucleus contains the genetic code – the DNA – and contains the blueprints for the construction of all the cell's proteins.

There are many other structures within a cell, such as the mitochondria, the Golgi apparatus, and the endoplasmic reticulum, but it is not the purpose of this book to look at the structure and function of these organelles. Suffice it to say that different drugs act at different locations in the cell.

We shall now magnify the picture to the molecular level, and find out what types of molecules in the cell are affected by drugs. There are four main molecular targets:

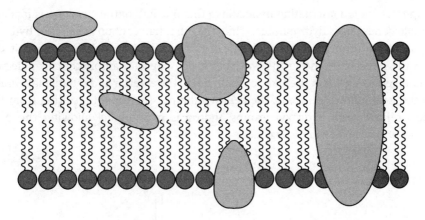

Fig. 2.4 The position of proteins associated with the cell membrane.

(1) lipids

(2) carbohydrates

(3) proteins

(4) nucleic acids

2.2.2 Lipids as drug targets

The number of drugs which interact with lipids is relatively small and, in general, they all act in the same way – by disrupting the lipid structure of cell membranes. For example, it has been proposed that general anaesthetics work by interacting with the lipids of cell membranes to alter the structure and conducting properties of the cell membrane.

The antifungal agent amphotericin B (Fig. 2.5) (used against athlete's foot) interacts with the lipids of fungal cell membranes to build 'tunnels' through the membrane. Once in place, the contents of the cell are drained away and the cell is killed.

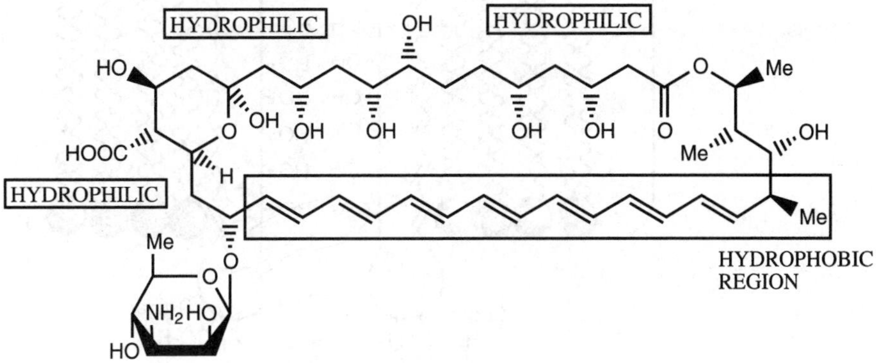

Fig. 2.5 Amphotericin B.

Amphotericin is a fascinating molecule in that one half of the structure is made up of double bonds and is hydrophobic, while the other half contains a series of hydroxyl groups and is hydrophilic. It is a molecule of extremes and, as such, is ideally suited to act on the cell membrane in the way that it does. Several amphotericin molecules cluster together such that the alkene chains are to the exterior and interact favourably with the hydrophobic centre of the cell membrane. The tunnel resulting from this cluster is lined with hydroxyl groups and so is hydrophilic, allowing the polar contents of the cell to escape (Fig. 2.6).

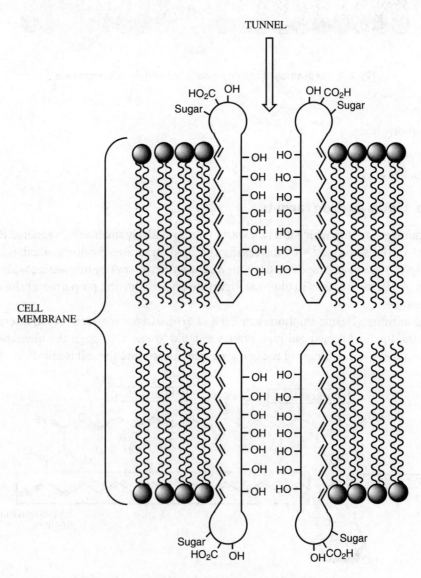

Fig. 2.6 Amphotericin-formed channel through the cell membrane.

The antibiotics valinomycin and gramicidin A operate by acting within the cell membrane as ion carriers and ion channels, respectively (see Chapter 14).

2.2.3 Carbohydrates as drug targets

Carbohydrates are polyhydroxy structures, many of which have the general formula $C_nH_{2n}O_n$. Examples of some simple carbohydrate structures include glucose, fructose, and ribose (Fig. 2.7). These are called monosaccharides since they can be viewed as the monomers required to make more complex polymeric carbohydrates. For example, glucose monomers link together to form the natural polymers glycogen, cellulose (Fig. 2.8), and starch.

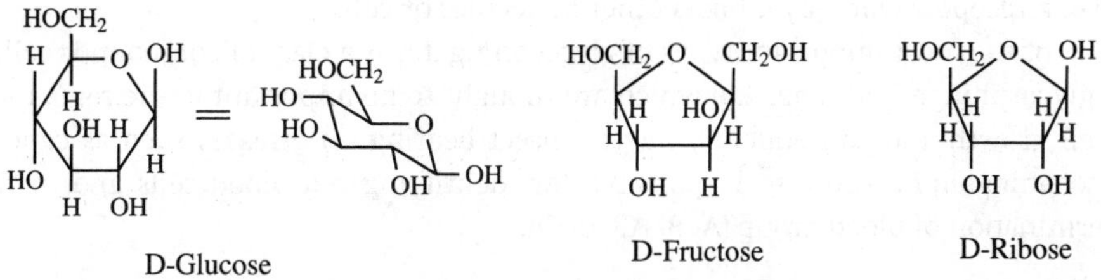

D-Glucose   D-Fructose   D-Ribose

Fig. 2.7 Examples of monosaccharides.

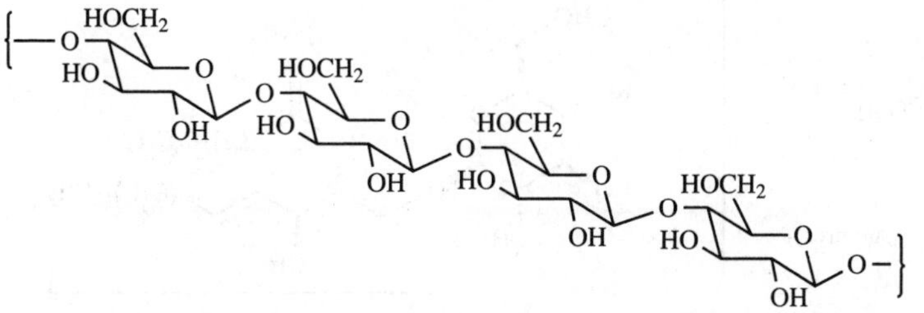

Fig. 2.8 Cellulose.

Until relatively recently, carbohydrates were not seen as useful targets for drugs. The main roles for carbohydrates in the cell were seen as being those of energy storage (e.g. glycogen) or structural (e.g. starch and cellulose). As a result, there are very few clinically useful drugs which act on carbohydrate targets. However, this may change in the future. It is now known that carbohydrates have important roles to play in various cellular processes such as cell recognition, regulation, and growth. Various disease states are associated with these cellular processes.

For example, bacteria and viruses have to recognize host cells before they can infect them and so the carbohydrate molecules involved in this cell recognition are crucial to the process. Designing drugs to bind to these carbohydrates may well block the ability of bacteria and viruses to invade host cells. Alternatively, vaccines may be developed based on the structure of carbohydrates which are distinctive to the invading cells.

It has also been observed that autoimmune diseases and cancers are associated with changes in the structure of cell surface carbohydrates. Understanding how carbohydrates are involved in cell recognition and regulation may well allow the design of novel drugs to treat these diseases.

Many of the important cell recognition roles played by carbohydrates are not acted out by pure carbohydrates, but by carbohydrates linked to proteins (glycoproteins or proteoglycans) or lipids (glycolipids). Such molecules are called glycoconjugates. Usually, the lipid or protein portion of the molecule is embedded within a cell membrane with the carbohydrate portion 'hanging free' on the outside of the membrane like the streamer of a kite. This allows the carbohydrate portion to serve the role of a molecular 'tag' which labels and identifies the cell. The tag may also take the role of a receptor whereby it binds other molecules or cells.

One of the most important types of glycoconjugates is a class of compounds called the glycosphingolipids (Fig. 2.9) which are thought to be important in the regulation of cell growth and consequently have a direct bearing on diseases such as cancer. Glycosphingolipids are also responsible for identifying red blood cells and in the determination of blood group (A, B, AB, or O).

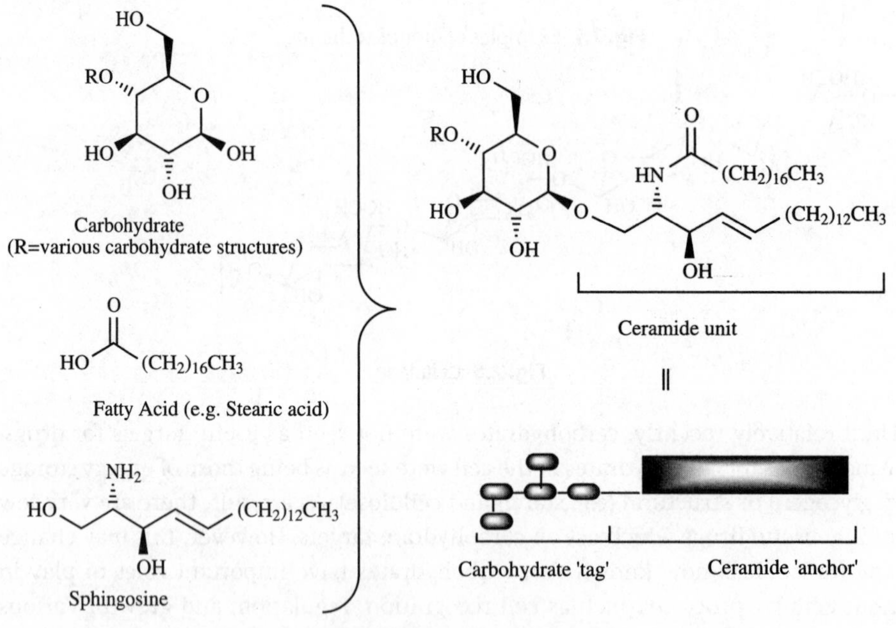

Fig. 2.9 Glycosphingolipids.

The glycosphingolipids are made up of three components – a carbohydrate structure which can be highly variable and complex, a structure called sphingosine which consists of a 2-amino-1,3-diol unit linked to a long chain hydrocarbon, and a fatty acid (e.g. stearic acid). The portion of the molecule consisting of the sphingosine and the fatty acid is called a ceramide.

The ceramide portion of the molecule is hydrophobic and is embedded within the cell membrane, thus acting as an anchor for the highly polar carbohydrate section. This portion lies outside the cell membrane and acts as the molecular 'tag' for the cell (Fig. 2.10).

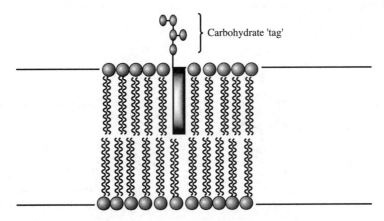

Fig. 2.10 Glycosphingolipids as molecular 'tags'.

There is actually good sense in having a carbohydrate as a molecular tag rather than a peptide or a nucleic acid. This is because there are far more variations in possible structure for carbohydrates than there are for these other types of structure. For example, there is only one possible dipeptide which can be formed between two molecules of alanine since there is only one way in which they can be linked (Fig. 2.11). However, because of the different hydroxyl groups on a carbohydrate, there are 11 possible disaccharides which can be formed from two glucose molecules. This allows the creation of an almost infinite number of molecular tags based on different numbers and types of sugar units.

There are several examples of drugs which are carbohydrates or contain carbohydrates as part of their structure. For example, several important antibiotics, e.g. streptomycin (see Chapter 14), contain carbohydrate units within their structure. Frequently, these carbohydrate units have rare and unusual structures and have been found to be important to the compound's activity as well as lowering the chances of antibacterial resistance developing. An understanding of why these sugar units are important at the molecular level would be extremely useful to the development of further antibacterial drugs.

The anti-HIV drug AZT and the antiherpes drug acyclovir contain unusual sugar units which are important to their mechanism of action. These are discussed in Chapter 7.

The glycosides digoxin and digitoxin (Fig. 2.12) are used in cardiovascular medicine, and contain three carbohydrate rings attached to a steroid nucleus.

Carbohydrate structures offer tremendous potential for novel drugs in the future. However, the amount of research carried out on carbohydrate-based drugs is

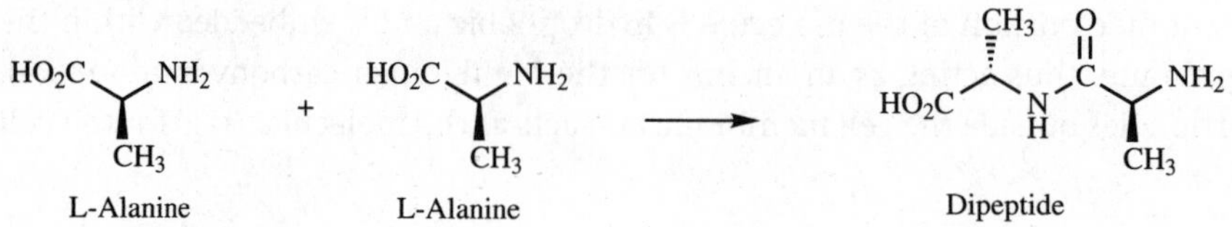

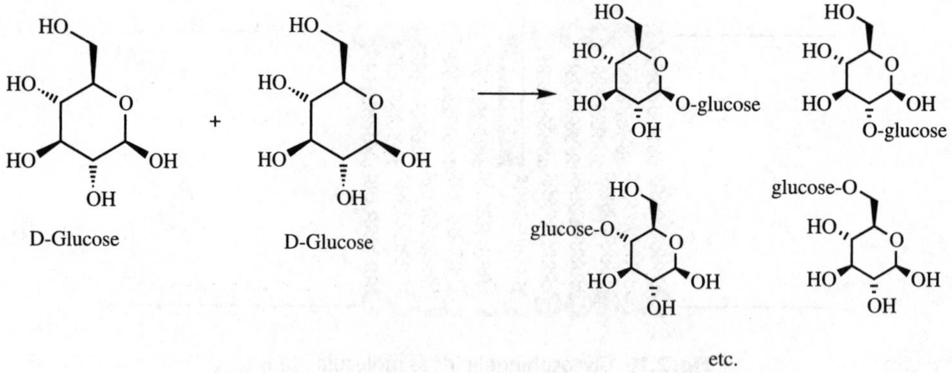

Fig. 2.11 Variability in carbohydrate structures.

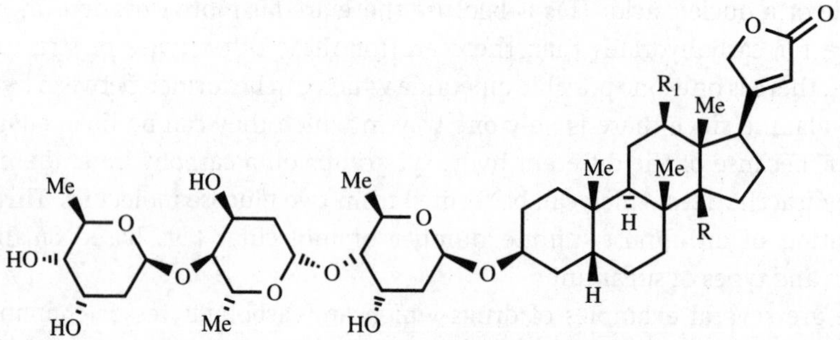

Fig. 2.12 Digoxin: R=OH, R₁=OH; Digitoxin: R=OH. R₁=H.

relatively small compared with, for example, peptide-based drugs. This is partly because of the greater complexity involved in synthesizing or modifying carbohydrates. Carbohydrates are 'loaded' with hydroxyl groups and asymmetric centres. For example, D-glucose has a cyclic structure containing five hydroxyl groups and five asymmetric centres. As a result, carbohydrate synthesis is a demanding 'sport', where the chemist is often required to carry out a reaction at a specific hydroxyl group whilst leaving the remaining hydroxyl groups unaffected. Moreover, reactions have to be performed under mild conditions to avoid racemization. The situation is made more complicated by the fact that many carbohydrates can easily form two different isomers called **anomers** in solution. For example, when D-glucose is dissolved in water, the cyclic structure opens up to form a linear structure containing an aldehyde functional group. The ring can then reform, but can do so in two ways such that the original ring structure is formed or its anomer is formed (Fig. 2.13). An equilibrium is set up favouring the more stable anomer, resulting in partial epimerization of an asymmetric centre. Therefore, controlling a synthesis to produce a single anomer becomes extremely difficult.

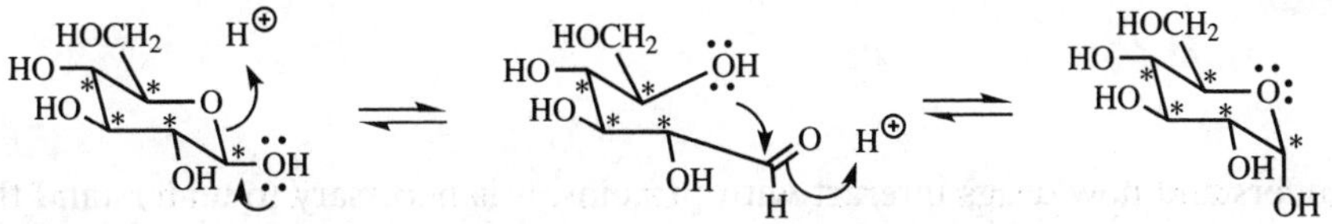

Fig. 2.13 Reversible ring opening of D-glucose.

As a result of these difficulties, the solid phase synthesis of carbohydrates is not yet possible, thus ruling out the use of combinatorial synthesis (see Chapter 12) for the generation of large numbers of carbohydrate structures. Should such techniques ever become possible, there are far more carbohydrate structures which could be synthesized from linking carbohydrate monomers together than there are for peptide structures synthesized from linking amino acids together.

2.2.4 Proteins and nucleic acids as drug targets

We have seen some examples of drugs which interact with the lipids of cell membranes. However, the vast majority of drugs used in medicine are targetted on proteins and nucleic acids, and these targets will be the subject of the next few chapters.

3 | Protein structure

To understand how drugs interact with proteins, it is necessary to understand their structure. Proteins have four levels of structure – primary, secondary, tertiary, and quaternary.

3.1 The primary structure of proteins

The primary structure is quite simply the order in which the individual amino acids making up the protein are linked together through peptide bonds (Fig. 3.1). The 20 common amino acids found in man are listed in Table 3.1 and the structures are shown in Appendix 1. Three letter or one letter codes are frequently used to represent the individual amino acids. For example, Ala or A is the code for alanine.

The primary structure of Met-enkephalin (one of the body's own painkillers) is shown in Fig. 3.2.

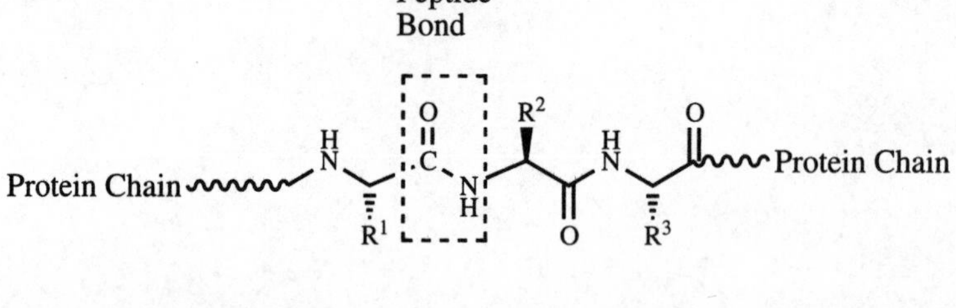

Fig. 3.1 Primary structure (R^1, R^2 and R^3 are amino acid residues).

3.2 The secondary structure of proteins

The secondary structure of proteins consists of regions of ordered structures taken up by the protein chain. There are two main structures – the alpha helix and the beta-pleated sheet.

Table 3.1 Amino acids found in man.

Synthesized in the human body			Essential to the diet		
Amino acid	Codes		Amino acid	Codes	
Alanine	Ala	or A	Histidine	His	or H
Arginine	Arg	or R	Isoleucine	Ile	or I
Asparagine	Asn	or N	Leucine	Leu	or L
Aspartic acid	Asp	or D	Lysine	Lys	or K
Cysteine	Cys	or C	Methionine	Met	or M
Glutamic acid	Glu	or E	Phenylalanine	Phe	or F
Glutamine	Gln	or Q	Threonine	Thr	or T
Glycine	Gly	or G	Tryptophan	Trp	or W
Proline	Pro	or P	Valine	Val	or V
Serine	Ser	or S			
Tyrosine	Tyr	or Y			

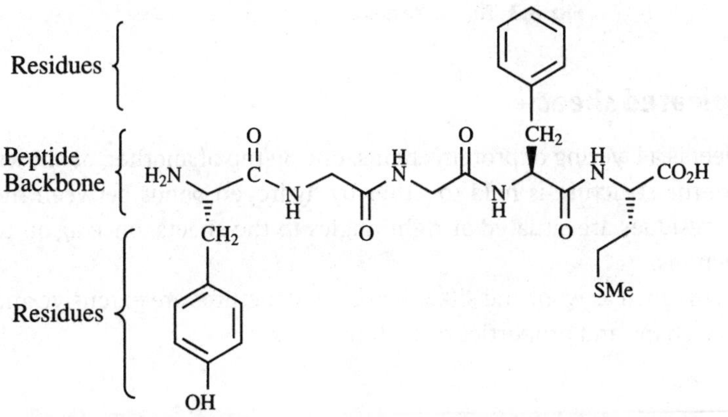

SHORTHAND NOTATION: H-TYR-GLY-GLY-PHE-MET-OH or YGGFM

Fig. 3.2 Met-enkephalin.

3.2.1 The alpha helix

The alpha helix results from coiling of the protein chain such that the peptide bonds making up the backbone are able to form hydrogen bonds between each other. These hydrogen bonds are directed along the axis of the helix as shown in Fig. 3.3. The residues of the component amino acids 'stick out' at right angles from the helix, thus minimizing steric interactions and further stabilizing the structure.

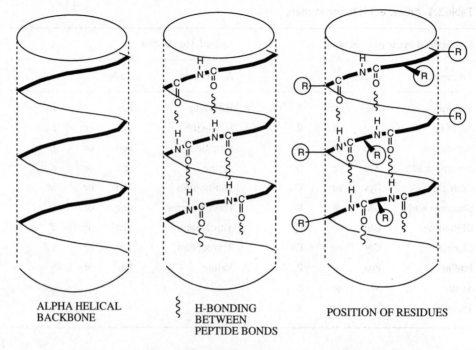

ALPHA HELICAL
BACKBONE

H-BONDING
BETWEEN
PEPTIDE BONDS

POSITION OF RESIDUES

Fig. 3.3 The alpha helix.

3.2.2 The beta-pleated sheet

The beta-pleated sheet is a layering of protein chains, one on top of another, as shown in Fig. 3.4. Here too, the structure is held together by hydrogen bonds between the peptide chains. The residues are situated at right angles to the sheets, once again to reduce steric interactions.

In structural proteins such as wool and silk, secondary structures are extensive and determine the overall shape and properties of such proteins.

3.3 The tertiary structure of proteins

The tertiary structure is the overall 3D shape of a protein. Structural proteins are quite ordered in shape, whereas other proteins such as enzymes and receptors fold up on themselves to form more complex structures. The tertiary structure of enzymes and receptors is crucial to their function and also to their interaction with drugs. Therefore, it is important to appreciate the forces which control their tertiary structure.

Enzymes and receptors have a great variety of amino acids arranged in what appears to be a random fashion – the primary structure of the protein. Regions of secondary structure may also be present, but the extent varies from protein to protein.

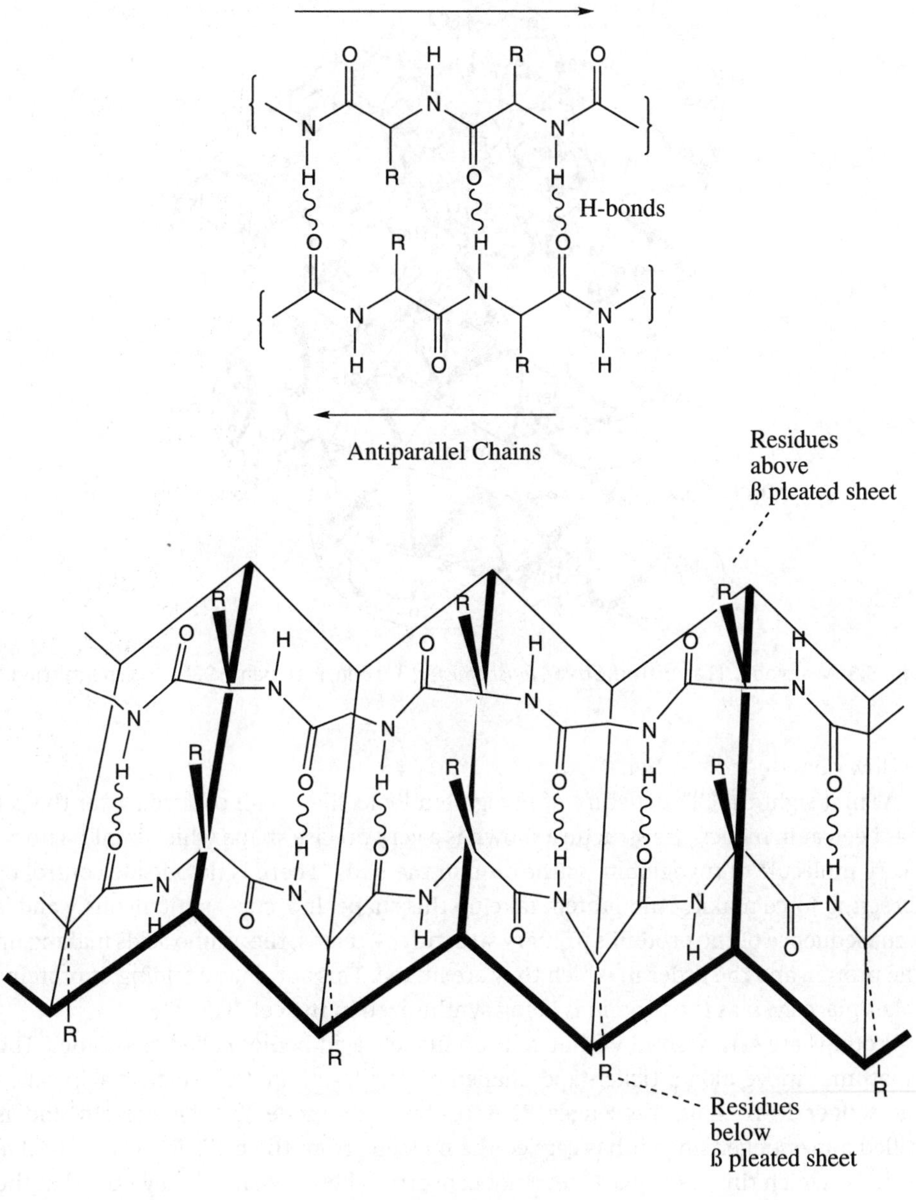

Fig. 3.4 β-pleated sheet.

For example, myoglobin (a protein which carries oxygen molecules) has extensive regions of alpha helical secondary structure (Fig. 3.5), whereas the digestive enzyme chymotrypsin has very little secondary structure. Nevertheless, the protein chains in both myoglobin and chymotrypsin fold up upon themselves to form a complex globular shape such as that shown for myoglobin (Fig. 3.5).

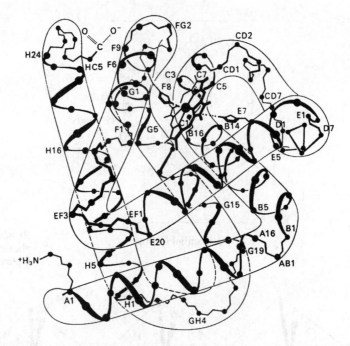

Fig. 3.5 Myoglobin. (Taken from Stryer, *Biochemistry*, 3rd edn, Freeman (1988), with permission.)

How does this come about?

At first sight the 3D structure of myoglobin looks like a ball of string after the cat has been at it. In fact, the structure shown is a very precise shape which is taken up by every molecule of myoglobin synthesized in the body. There is no outside control or directing force making the protein take up this shape. It occurs spontaneously and is a consequence of the protein's primary structure[1] – that is, the amino acids making up the protein and the order in which they are linked. This automatic folding of proteins takes place even as the protein is being synthesized in the cell (Fig. 3.6).

Proteins are synthesized within cells on nucleic acid bodies called ribosomes. The ribosomes move along 'ticker-tape'-shaped molecules of another nucleic acid called messenger RNA. This messenger RNA contains the code for the protein and is called a messenger since it has carried the message from the cell's DNA. The mechanism by which this takes place need not concern us here. We need only note that the ribosome holds on to the growing protein chain as the amino acids are added on one by one. As the chain grows, it automatically folds into its 3D shape so that by the time the protein is fully synthesized and released from the ribosome, the 3D shape is already adopted. The 3D shape is identical for every molecule of the particular protein synthesized.

---

[1] Some proteins contain species known as cofactors (e.g. metal ions or small organic molecules) which also have an effect on tertiary structure.

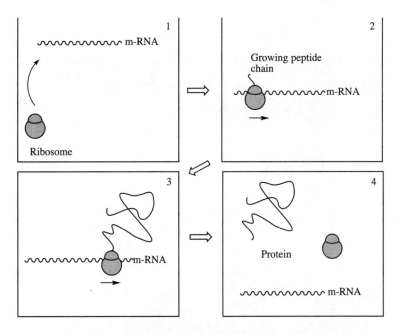

Fig. 3.6 Protein synthesis.

This poses a problem. Why should a chain of amino acids take up such a precise 3D shape? At first sight, it does not make sense. If a length of string is placed on the table, it does not fold itself into a precise complex shape. So why should a 'string' of amino acids do such a thing?

The answer lies in the fact that a protein is not just a bland length of string. It has a whole range of chemical functional groups attached along the length of its chain. These are, of course, the residues of each amino acid making up the chain. These residues can interact with each other. Some will attract each other. Others will repel. Thus the protein will twist and turn to minimize the unfavourable interactions and to maximize the favourable ones until the most favourable shape (conformation) is found – the tertiary structure (Fig. 3.7).

What then are these important interactions? Let us consider the bonding interactions first. There are four to consider: covalent bonds, ionic bonds, hydrogen bonds, and van der Waals interactions.

3.3.1 Covalent bonds

Covalent bonds are the strongest bonding force available.

$$\text{Bond strength (S–S)} = 250 \text{ kJ mol}^{-1}$$

When two cysteine residues are close together, a covalent disulphide bond can be formed between them as a result of oxidation. A covalent bonded bridge is thus formed between two different parts of the protein chain (Fig. 3.8).

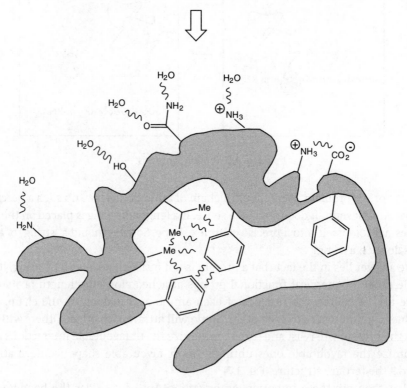

Fig. 3.7 Tertiary structure.

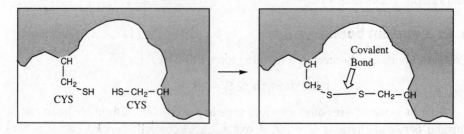

Fig. 3.8 Covalent bond.

3.3.2 Ionic bonds

Ionic bonds are a strong bonding force between groups having opposite charges.

$$\text{Bond strength} = 20 \text{ kJ mol}^{-1}$$

An ionic bond can be formed between the carboxylate ion of an acidic residue such as aspartic acid and the ammonium ion of a basic residue such as lysine (Fig. 3.9).

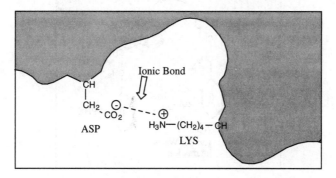

Fig. 3.9 Ionic bond.

3.3.3 Hydrogen bonds

$$\text{Bond strength} = 7\text{–}40 \text{ kJ mol}^{-1}$$

Hydrogen bonds are formed between electronegative atoms, such as oxygen, and protons attached to electronegative atoms (Fig. 3.10).

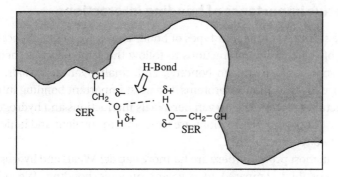

Fig. 3.10 Hydrogen bond.

3.3.4 Van der Waals interactions

$$\text{Bond strength} = 1.9 \text{ kJ mol}^{-1}$$

Van der Waals interactions take place between hydrophobic molecules (for example, between two aromatic residues such as phenylalanine (Fig. 3.11) or between

two aliphatic residues such as valine). It arises from the fact that the electronic distribution in these neutral, non-polar residues is never totally even or symmetrical. As a
result, there are always transient areas of high electron density and low electron
density, such that an area of high electron density on one residue can have an attrac-
tion for an area of low electron density on another molecule.

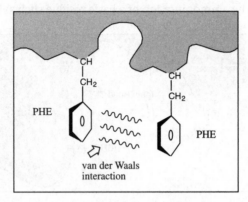

Fig. 3.11 Van der Waals interaction.

3.3.5 Repulsive interactions

Repulsive interactions arise if a hydrophilic group such as an amino function is too
close to a hydrophobic group such as an aromatic ring. Alternatively, two groups with
identical charge are repelled.

3.3.6 Relative importance of bonding interactions

Based on the strengths of the four types of bonds above, we might expect the relative
importance of the bonding interactions to follow the same order as their strengths,
that is, covalent, ionic, hydrogen bonding, and, finally, van der Waals. In fact, the
opposite is usually true. In most proteins, the most important bonding interactions in
tertiary structure are those due to van der Waals interactions and hydrogen bonding,
while the least important interactions are those due to covalent and ionic bonding.

There are two reasons for this.

First of all, in most proteins there are far more van der Waals and hydrogen bonding
interactions possible, compared to covalent or ionic bonding. We only need to
consider the number and types of amino acids in any typical globular protein to see
why. The only covalent bond which can contribute to tertiary structure is a disulphide
bond. Only one amino acid out of our list can form such a bond – cysteine. However,
there are eight amino acids which can interact with each other through van der Waals
interactions: Gly, Ala, Val, Leu, Ile, Phe, Pro, and Met.

Having said that, there are examples of proteins with a large number of disulphide
bridges in which the relative importance of the covalent link to tertiary structure is

more significant. Disulphide links are also more significant in small polypeptides such as the peptide hormones vasopressin (Fig. 3.12) and oxytocin (Fig. 3.13). However, in the majority of proteins, disulphide links play a minor role in controlling tertiary structure.

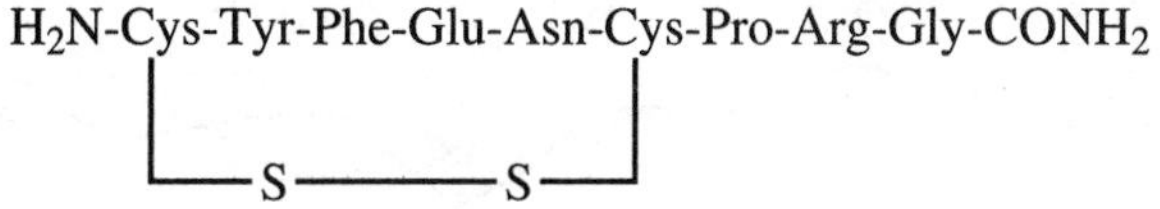

$$H_2N\text{-Cys-Tyr-Phe-Glu-Asn-Cys-Pro-Arg-Gly-}CONH_2$$

Fig. 3.12 Vasopressin.

$$H_2N\text{-Cys-Tyr-Ile-Glu-Asn-Cys-Pro-Leu-Gly-}CONH_2$$

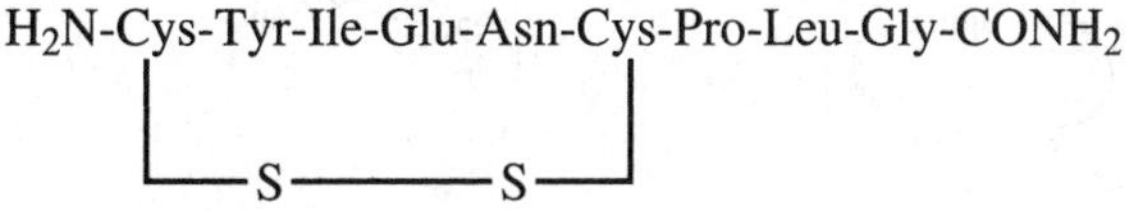

Fig. 3.13 Oxytocin.

As far as ionic bonding is concerned, only four amino acids (Asp, Glu, Lys, Arg) are involved, whereas eight amino acids can interact through hydrogen bonding (Ser, Thr, Cys, Asn, Gln, His, Tyr, Trp). Clearly, the number of possible ionic and covalent bonds is greatly outnumbered by the number of hydrogen bonds or van der Waals interactions.

There is a second reason why van der Waals interactions, in particular, can be the most important form of bonding in tertiary structure. Proteins do not exist in a vacuum. The body is mostly water and as a result all proteins will be surrounded by this medium. Therefore, amino acid residues at the surface of proteins must interact with water molecules. Water is a highly polar compound which forms strong hydrogen bonds. Thus, water would be expected to form strong hydrogen bonds to the hydrogen-bonding amino acids previously mentioned.

Water can also accept a proton to become positively charged and can form ionic bonds to aspartic and glutamic acids (Fig. 3.14).

Therefore, water is capable of forming hydrogen bonds or ionic bonds to the following amino acids: Ser, Thr, Cys, Asn, Gln, His, Tyr, Trp, Asp, Glu, Lys, and Arg. These polar amino acids are termed hydrophilic amino acids. The remainder (Gly, Ala, Val, Leu, Ile, Phe, Pro, and Met) are all non-polar amino acids which are hydrophobic or lipophilic. As a result, they are repelled by water.

Therefore, the most stable tertiary structure of a protein will be the one in which most of the hydrophilic groups are on the surface so that they can interact with water, and most of the hydrophobic groups are in the centre so that they can avoid the water and interact with each other.

Since hydrophilic amino acids can form ionic/hydrogen bonds with water, the number of ionic and hydrogen bonds contributing to the tertiary structure is reduced. The hydrophobic amino acids in the centre have no choice in the matter and must

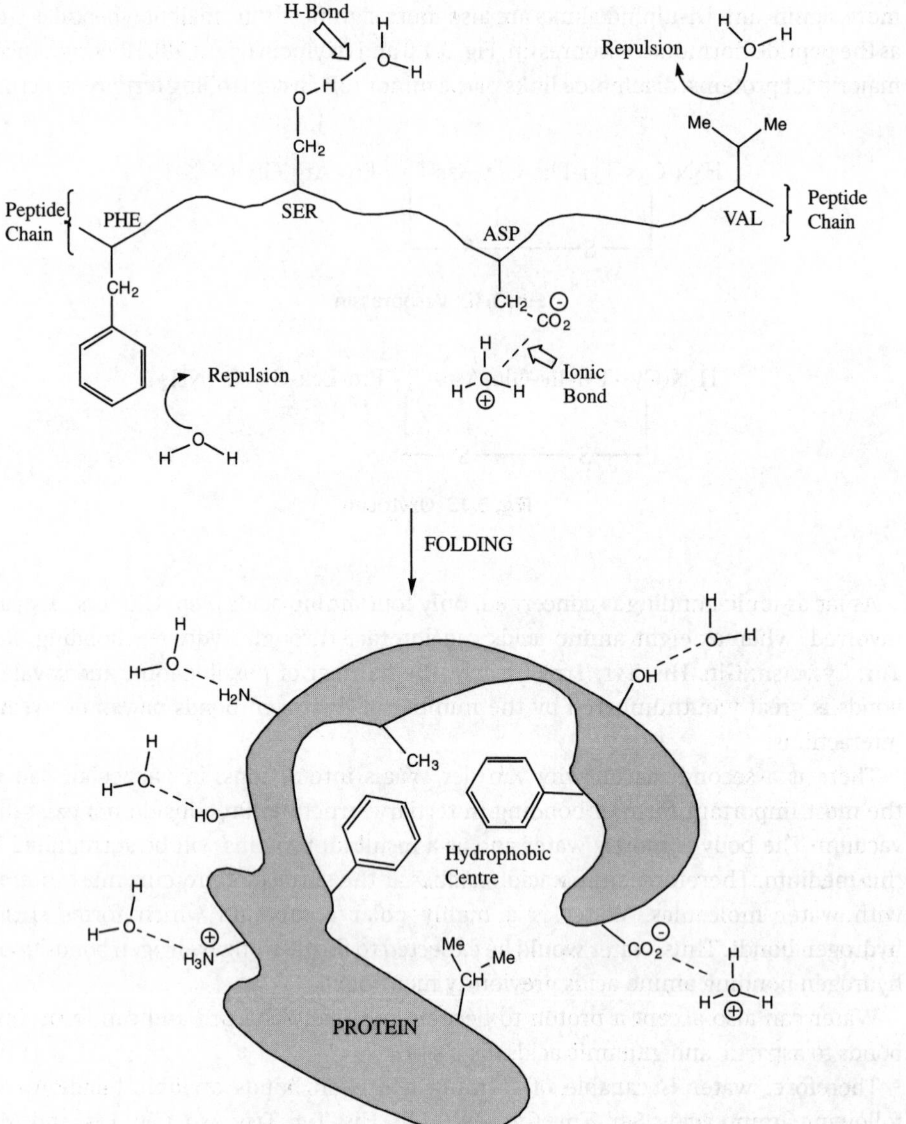

Fig. 3.14 Bonding interactions with water.

interact with each other. Thus, the hydrophobic interactions within the structure outweigh the hydrophilic interactions and control the shape taken up by the protein.

One important feature of this tertiary structure is that the centre of the protein is hydrophobic and non-polar. This has important consequences as far as the action of enzymes is concerned and helps to explain why reactions, which should be impossible in an aqueous environment, can take place in the presence of enzymes. The enzyme can provide a non-aqueous environment in which the reaction can take place.

3.4 The quaternary structure of proteins

Quaternary structure is confined to those proteins which are made up of a number of protein subunits (Fig. 3.15). For example, haemoglobin is made up of four protein molecules – two identical alpha subunits and two identical beta subunits (not to be confused with the alpha and beta terminology used in secondary structure). The quaternary structure of haemoglobin is the way in which these four protein units associate with each other.

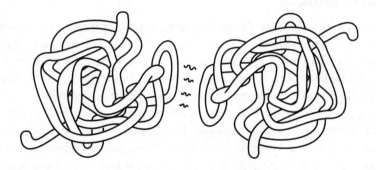

Fig. 3.15 Quaternary structure.

Since this must inevitably involve interactions between the exterior surfaces of proteins, ionic bonding can be more important to quaternary structure than it is to tertiary structure. Nevertheless, hydrophobic (van der Waals) interactions have a role to play. It is not possible for a protein to fold up such that all hydrophobic groups are placed to the centre. Some such groups may be stranded on the surface. If they form a small hydrophobic area on the protein surface, there is a distinct advantage in two protein molecules forming a dimer such that the two hydrophobic areas face each other rather than be exposed to an aqueous environment.

3.5 Conclusion

We have now discussed the four types of structure. The tertiary structure is the most important feature as far as drug action is concerned, although it must be emphasized again that tertiary structure is a consequence of primary structure. Tertiary structure is a result of interactions between different amino acid residues and interactions between amino acids and water.

3.6 Drug action at proteins

We are now ready to discuss the various types of protein with which drugs interact – enzymes, receptors, carrier proteins, and structural proteins. Enzymes are the body's catalysts and will be discussed in Chapter 4. Receptors are the body's 'letter boxes' and are crucial to the communication between cells. These are discussed in Chapters 5 and 6.

3.6.1 Carrier proteins

Carrier proteins are present in the cell membrane and act as the cell's 'smugglers' – smuggling the important chemical building blocks (e.g. amino acids, sugars, and metal ions) that we obtain from food into the cell such that the cell can construct proteins, carbohydrates, and nucleic acids. But why is this smuggling operation necessary? Why can't these molecules pass through the membrane by themselves?

Quite simply, molecules such as amino acids are polar structures and cannot pass through the hydrophobic cell membrane. The carrier proteins are hydrophobic on the outside and can float freely through the cell membrane. When they reach the outer surface, they bind the polar molecule (e.g. an amino acid), 'wrap it up' into a hydrophilic pocket and ferry it across the membrane to release it on the other side (Fig. 3.16).

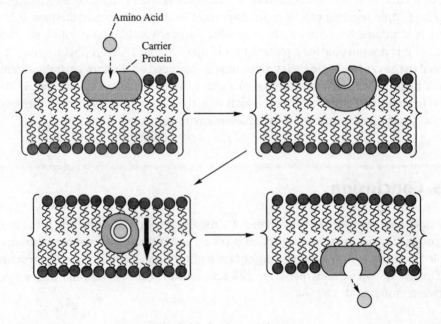

Fig. 3.16 Carrier proteins.

Carrier proteins are not all identical and there are specific carrier proteins for the different molecules which need to be smuggled across the membrane. These carrier proteins have a recognition site which allows them to bind and encapsulate their specific 'guest', but some drugs fool the carrier protein into thinking that they are the usual guest molecule. The carrier protein accepts the drug and carries it across the cell membrane, delivering the drug into the cell. This is one way of delivering highly polar drugs into cells or across cell membranes (see Chapter 10).

In some cases, drugs are tightly bound to the carrier protein once they are 'wrapped up' and remain as permanent lodgers. The carrier protein is then no longer able to carry its usual guest across the cell membrane. Important drugs such as cocaine and the tricyclic antidepressants act in this way; they prevent the neurotransmitter noradrenaline from re-entering nerve cells (see Chapter 16). This results in an increased level of noradrenaline in nerve synapses and has the same effect as adding drugs which mimic noradrenaline. The hugely successful antidepressant drug fluoxetine (Prozac) (Fig. 3.17) is a selective inhibitor of the carrier protein responsible for the uptake of another neurotransmitter called serotonin.

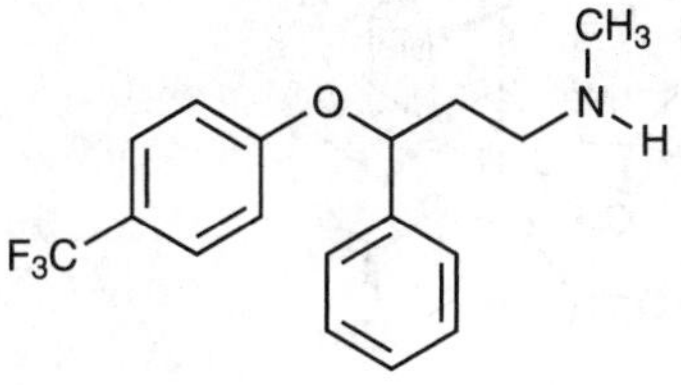

Fig. 3.17 Prozac.

3.6.2 Structural proteins

Structural proteins do not normally act as drug targets. However, the structural protein tubulin is an exception. Tubulin molecules polymerize to form small tubes called microtubules in the cell's cytoplasm (Fig. 3.18). These microtubules have various roles within the cell, including the maintenance of shape, exocytosis, and release of neurotransmitters. They are also involved in the mobility of cells. For example, inflammatory cells called neutrophils are mobile cells which normally protect the body against infection. However, they can also enter joints, leading to inflammation and arthritis. One way of treating this disease has been to administer

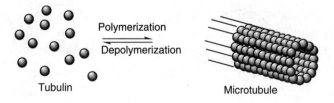

Fig.3.18 Polymerization of tubulin.

colchicine (Fig. 3.19) which binds to tubulin and causes the microtubules to depoly-merize. Once the microtubules have been broken down, the neutrophils lose their mobility and can no longer migrate into the joints. Unfortunately, colchicine has many side-effects and is not an ideal drug for this treatment.

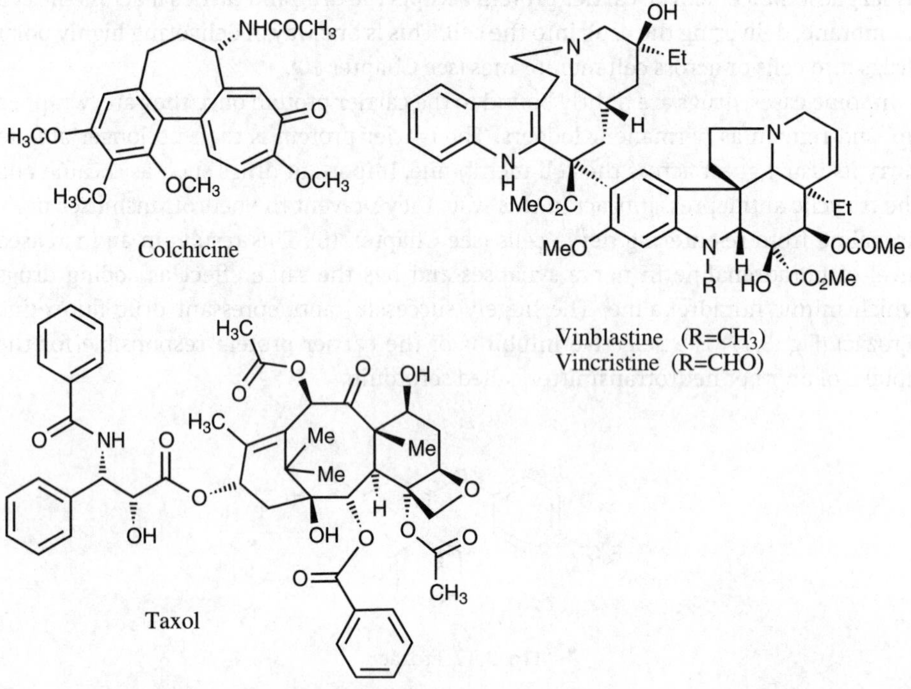

Colchicine

Vinblastine (R=CH$_3$)
Vincristine (R=CHO)

Taxol

Fig. 3.19 Drugs acting against tubulin.

Tubulin is also crucial to cell division. When a cell is about to divide, its micro-tubules depolymerize to give tubulin. The tubulin is then repolymerized to form a structure called a spindle which serves to push apart the two new cells and to act as a framework on which the chromosomes of the original cell are transferred to the nuclei of the daughter cells (Fig. 3.20). Several important anti-cancer drugs work by

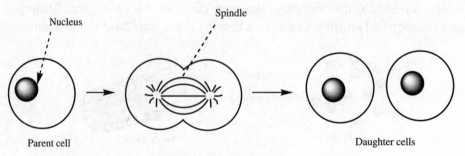

Nucleus

Spindle

Parent cell

Daughter cells

Fig. 3.20 Cell division.

binding to tubulin to prevent this polymerization/depolymerization cycle and thus inhibit the growth of tumours. For example, vincristine (Fig. 3.19) prevents polymerization of tubulin to form spindles, whereas taxol (Fig. 3.19) stimulates the formation of the microtubules and thus prevents the depolymerization process.

3.7 Peptides and proteins as drugs

Many of the body's important hormones are peptides and proteins, and in certain diseases the body fails to produce sufficient quantities. Therefore, it is useful to provide such peptides or hormones as drugs to make up for any deficiency. Several such hormones have been introduced to the market (e.g. insulin, human growth factor, interferon, and erythropoietin) and the availability of such proteins owes a great deal to genetic engineering. It is extremely tedious and expensive to obtain substantial quantities of these chemicals by any other means. For example, isolating and purifying a hormone from blood samples is doomed to failure because of the low quantities of hormone present. It is far more practical to use recombinant DNA techniques, whereby the human genes for the protein are cloned and then incorporated into the DNA of fast growing bacterial, yeast or mammalian cells. These cells then produce sufficient quantities of the protein.

The above examples are success stories. However, in general, it has to be stated that peptides and proteins make poor drugs since they are difficult to administer, are rapidly cleared from the body and can potentially cause an immune response. As a result, there is often a reluctance to consider peptide drugs as potential medicines. We will consider this problem in more detail in Chapter 10.

3.8 Antibodies in medicinal chemistry

Before leaving this chapter, it is worth mentioning another important group of proteins which are present in the body – antibodies (Fig. 3.21). Antibodies are crucial to the recognition and destruction of foreign cells by the immune response. They are Y-shaped molecules which are made up of two heavy and two light peptide chains. At the amino terminals of these chains there is a highly variable region of amino acids which differs from antibody to antibody. It is this region which recognizes and binds to particular chemical groupings on the invader. These groupings are called antigens and are normally the glycoconjugates on the outside of the cell (see section 2.2.3). Once an antigen is recognized, the antibody binds to it and recruits the body's immune response to destroy the foreign cell (Fig. 3.22). All cells (including our own) have antigens on their outer surface. They act as a molecular signature for different cells, allowing the body to distinguish between its own cells and 'foreigners'.

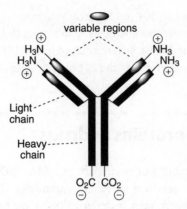

Fig. 3.21 Structure of an antibody.

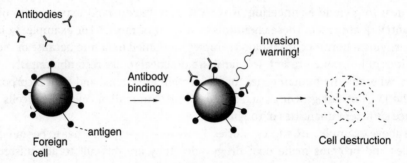

Fig. 3.22 Role of antibodies in cell destruction.

Fortunately, the body does not produce antibodies against its own cells and so our own cells are safe from attack.

Since antibodies can recognize the chemical signature of a particular cell, they have great potential in targetting drugs to specific cells in the body. Antibodies which recognize a particular human cell are generated by exposing another organism to that cell. A drug can then be attached to the antibody and the antibody used to carry the drug to the target cell (see Chapter 10).

The first antibody to be used clinically was specific for an antigen on lymphocytes. The antibody was used to tag lymphocytes which had turned cancerous and encouraged the body to destroy them.

4 Drug action at enzymes

4.1 Enzymes as catalysts

Enzymes are the body's catalysts. Without them, the cell's chemical reactions would be too slow to be useful. A catalyst is an agent which speeds up a chemical reaction without being consumed. An example of a catalyst used frequently in the laboratory is palladium on activated charcoal (Fig. 4.1).

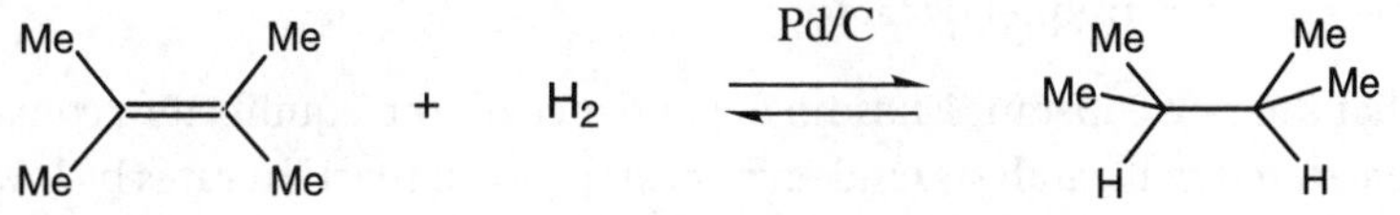

Fig. 4.1 An example of a catalyst.

Note that the above reaction is shown as an equilibrium. It is therefore more correct to describe a catalyst as an agent which can speed up the approach to equilibrium. In an equilibrium reaction, a catalyst will speed up the reverse reaction just as efficiently as the forward reaction. Consequently, the final equilibrium concentrations of the starting materials and products are unaffected by a catalyst.

How do catalysts affect the rate of a reaction without affecting the equilibrium? The answer lies in the existence of a high-energy intermediate or transition state which must be formed before the starting material can be converted to the product. The difference in energy between the transition state and the starting material is the activation energy, and it is the size of this activation energy which determines the rate of a reaction rather than the difference in energy between the starting material and the product (Fig. 4.2).

A catalyst acts to lower the activation energy by helping to stabilize the transition state. The energies of the starting material and products are unaffected, and therefore the equilibrium ratio of starting material to product is unaffected.

We can relate energy, and the rate and equilibrium constants with the following equations:

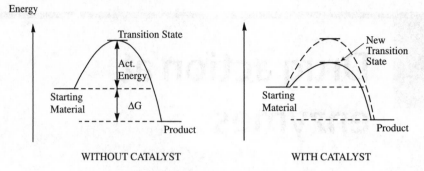

Fig. 4.2 Activation energy.

Energy difference $= \Delta G = -RT \ln K$ where
K = equilibrium constant = [products]/[reactants]
$R = 8.314\,\text{J}\,\text{mol}^{-1}\,\text{K}^{-1}$
T = temperature

Rate constant $= k = Ae^{-E/RT}$ where

E = activation energy
A = frequency factor

Note that the rate constant k has no dependence on the equilibrium constant K.

We have stated that catalysts (and enzymes) speed up reaction rates by lowering the activation energy, but we have still to explain how.

4.2 How do catalysts lower activation energies?

There are several factors at work.

- Catalysts provide a reaction surface or environment.
- Catalysts bring reactants together.
- Catalysts position reactants correctly so that they easily attain their transition state configurations.
- Catalysts weaken bonds.
- Catalysts may participate in the mechanism.

We can see these factors at work in our example of hydrogenation with a palladium–charcoal catalyst (Fig. 4.3). In this reaction, the catalyst surface interacts with the hydrogen molecule and in doing so weakens the H–H bond. The bond is broken and the hydrogen atoms are bound to the catalyst. The catalyst can then interact with the alkene molecule, so weakening the π bond of the double bond. The hydrogen

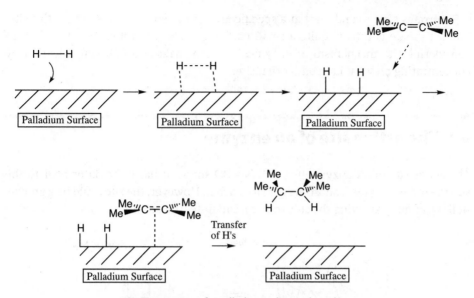

Fig. 4.3 Action of a palladium–charcoal catalyst.

atoms and the alkene molecule are positioned on the catalyst conveniently close to each other to allow easy transfer of the hydrogens from catalyst to alkene. The alkane product then departs, leaving the catalyst as it was before the reaction.

Therefore, the catalyst helps the reaction by providing a surface to bring the two substrates together. It participates in the reaction by binding the substrates and breaking high-energy bonds, and it then holds the reagents close together to increase the chance of them reacting with each other.

Enzymes may have more complicated structures than, say, a palladium surface, but they catalyse reactions in the same way. They act as a surface or focus for the reaction, bringing the substrate or substrates together and holding them in the best position for reaction. The reaction takes place, aided by the enzyme, to give products which are then released (Fig. 4.4). Note again that it is a reversible process. Enzymes can catalyse both forward and backward reactions. The final equilibrium mixture will, however, be the same, regardless of whether we supply the enzyme with substrate or product.

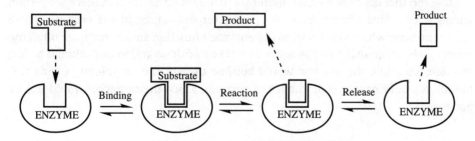

Fig. 4.4 Enzyme catalyst.

Substrates bind to and react at a specific area of the enzyme called the active site. The active site is usually quite a small part of the overall protein structure but, in considering the mechanism of enzymes, we can make a useful simplification by concentrating on what happens at that site.

4.3 The active site of an enzyme

The active site of an enzyme (Fig. 4.5) is a 3D shape. It has to be on or near to the surface of the enzyme if substrates are to reach it. However, the site could be a groove, hollow, or gully allowing the substrate to 'sink into' the enzyme.

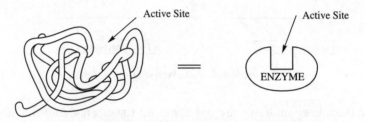

Fig. 4.5 The active site of an enzyme.

Because of the overall folding of the enzyme, the amino acid residues which are close together in the active site may be extremely far apart in the primary structure. For example, the important amino acids at the active site of lactate dehydrogenase are shown in Fig. 4.6. The numbers refer to their positions in the primary structure of the enzyme.

The amino acids present in the active site play an important role in enzyme function and this can be demonstrated by comparing the primary structures of the same enzyme from different organisms. In such a study, we find that the primary structure differs from species to species as a result of mutations over millions of years. The variability is proportional to how far apart the organisms are on the evolutionary ladder and this is one method of determining such a relationship.

However, that does not concern us here. What does is the fact that there are certain amino acids which remain constant, no matter the source of the enzyme. These are amino acids which are crucial to the enzyme's function and, as such, are often the amino acids which make up the active site. If one of these amino acids should be lost through mutation, the enzyme would become useless and an animal bearing this mutation would have a poor chance of survival. Thus, the mutation would not be preserved[1].

---

[1] The only exception to this would be if the mutation either introduced an amino acid which could perform the same task as the original amino acid, or improved substrate binding.

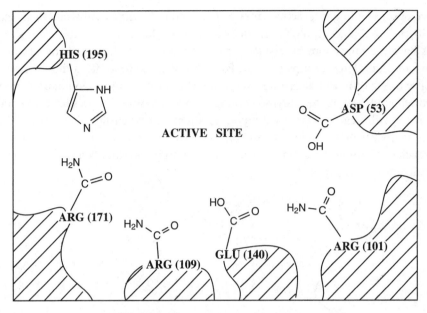

Fig. 4.6 The active site of lactate dehydrogenase.

This consistency of active site amino acids can often help scientists determine which amino acids are present in an active site if this is not known already.

Amino acids present in the active site can have one of two roles.

1. Binding – the amino acid residue is involved in binding the substrate to the active site.

2. Catalytic – the amino acid is involved in the mechanism of the reaction.

We shall study these in turn.

4.4 Substrate binding at an active site

4.4.1 The binding interactions

The interactions which bind substrates to the active sites of enzymes include ionic and hydrogen bonding, as well as van der Waals interactions. These bonding interactions are the same bonding interactions responsible for the tertiary structure of proteins and have already been described in Chapter 3. However, although the bonding interactions are the same, the relative importance of the interactions differs. Ionic bonding plays a relatively minor role in protein tertiary structure compared to hydrogen bonding or van der Waals interactions, but it can play a crucial role in the binding of a substrate to an active site – not too surprising since active sites are located on or near the surface of the enzyme.

Two other bonding interactions may be involved in binding a substrate to an active site. These are dipole–dipole interactions or induced dipole interactions. Many molecules have a dipole moment resulting from the different electronegativities of the atoms and functional groups present. For example, a ketone has a dipole moment because of the different electronegativities of the carbon and oxygen making up the carbonyl bond. The active site also contains functional groups and so it is inevitable that it too will have various local dipole moments on its surface. It is possible that dipole moments on the substrate and on the active site will interact as the substrate approaches, aligning the substrate such that the dipole moments are parallel and in opposite directions (Fig. 4.7).

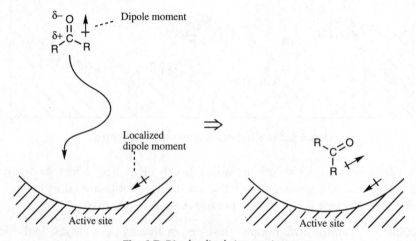

Fig. 4.7 Dipole–dipole interactions.

Interactions involving an induced dipole moment have been proposed. There is evidence that an aromatic ring can interact with an ionic group such as a quaternary ammonium ion. Such an interaction is feasible if the positive charge of the quaternary ammonium group distorts the π electron cloud of the aromatic ring to produce a dipole moment such that the face of the aromatic ring is electron-rich and the edges are electron-deficient (Fig. 4.8).

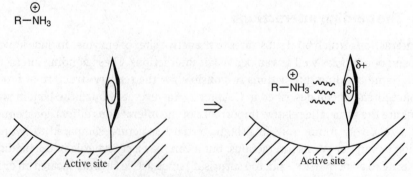

Fig. 4.8 Induced dipole interaction.

Since we know the bonding forces involved in substrate binding, it is possible to look at the structure of a substrate and postulate the probable interactions that it will have with its active site. As an example, let us consider the substrate for lactate dehydrogenase – an enzyme which catalyses the reduction of pyruvic acid to lactic acid (Fig. 4.9).

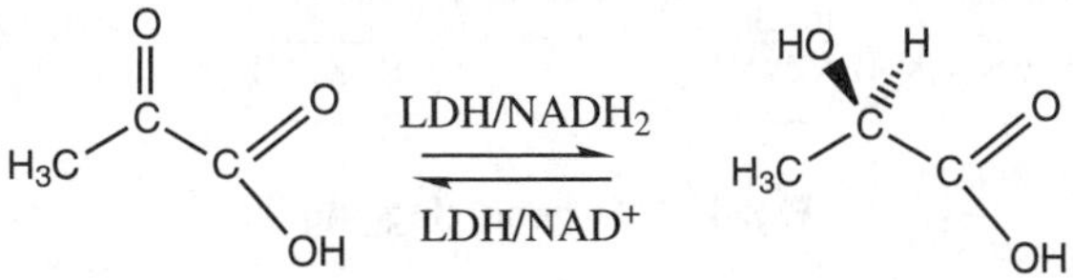

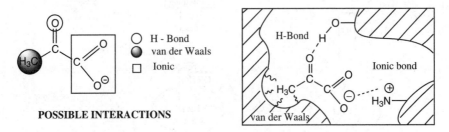

Pyruvic acid   Lactic acid

Fig. 4.9 Reduction of pyruvic acid to lactic acid.

If we look at the structure of pyruvic acid, we can propose three possible interactions with which it might bind to its active site – an ionic interaction involving the ionized carboxylate group, a hydrogen bond involving the ketonic oxygen, and a van der Waals interaction involving the methyl group (Fig. 4.10). If these postulates are correct, then there must be suitable amino acids at the active site to take part in these bonds. Lysine, serine, and phenylalanine residues would fit the bill respectively.

○ H - Bond
● van der Waals
□ Ionic

H-Bond

Ionic bond

van der Waals

POSSIBLE INTERACTIONS

Fig. 4.10 Interactions between pyruvic acid and lactate dehydrogenase.

4.4.2 Competitive (reversible) inhibitors

Binding interactions between substrate and enzyme are clearly important. If there were no interactions holding the substrate to the active site, then the substrate would drift in and drift out again before there was a chance for it to react.

Therefore, the more binding interactions there are, the better the substrate will be bound, and the better the chance of reaction. But there is a catch! What would happen if a substrate bound so strongly to the active site that it was not released again (Fig. 4.11)?

The answer, of course, is that the enzyme would become 'clogged up' and would be unable to accept any more substrate. Therefore, the bonding interactions between substrate and enzyme have to be properly balanced so that they are strong enough to

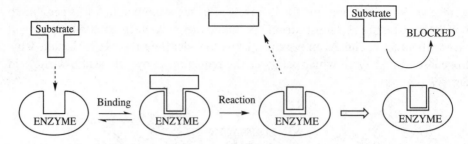

Fig. 4.11 Enzyme 'clogging'.

keep the substrate(s) at the active site to allow reaction, but weak enough to allow the product(s) to depart. This bonding balancing act can be turned to great advantage if the medicinal chemist wishes to inhibit a particular enzyme or to switch it off altogether. A molecule can be designed which is similar to the natural substrate and can fit the active site, but which binds more strongly. It may not undergo any reaction when it is in the active site, but as long as it is there, it blocks access to the natural substrate and the enzymatic reaction stops (Fig. 4.12). This is known as competitive inhibition since the drug is competing with the natural substrate for the active site.

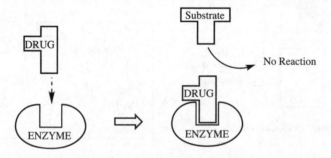

Fig. 4.12 Competitive inhibition.

The longer the inhibitor is present in the active site, the greater the inhibition. Therefore, if the medicinal chemist has a good idea which binding groups are present in an active site and where they are, a range of molecules can be designed with different inhibitory strengths.

Competitive inhibitors can generally be displaced by increasing the level of natural substrate. This feature has been useful in the treatment of accidental poisoning by antifreeze. The main constituent of antifreeze is ethylene glycol which is oxidized in a series of enzymatic reactions to oxalic acid (Fig. 4.13). It is the oxalic acid which is responsible for toxicity and if its synthesis can be blocked, recovery is possible.

The first step in this enzymatic process is the oxidation of ethylene glycol by alcohol dehydrogenase. Ethylene glycol is acting here as a substrate, but we can view it as a

CH$_2$–OH   ADH   C—H   Enzymes   COOH
| |
CH$_2$–OH   CH$_2$–OH   COOH

Ethylene   Oxalic Acid
Glycol

Fig. 4.13 Formation of oxalic acid from ethylene glycol.

competitive inhibitor since it is competing with the natural substrate for the enzyme. If the levels of natural substrate are increased, it will compete far better with ethylene glycol and prevent it from reacting. Toxic oxalic acid will no longer be formed and the unreacted ethylene glycol is eventually excreted from the body (Fig. 4.14). The cure then is to administer high doses of the natural substrate – alcohol! Perhaps one of medicine's more acceptable cures?

OXIDATION OF ETHYLENE GLYCOL

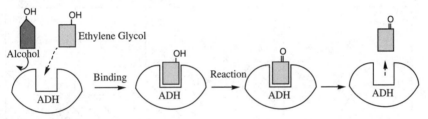

BLOCKING WITH EXCESS ALCOHOL

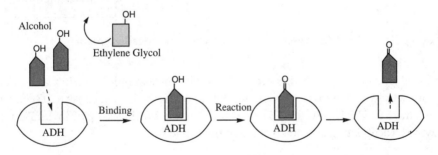

Fig. 4.14 Oxidation of ethylene glycol and blocking with excess alcohol.

There are many examples of useful drugs which act as competitive inhibitors. For example, the sulphonamides inhibit bacterial enzymes (see Chapter 14), while anti-cholinesterases inhibit the mammalian enzyme acetylcholinesterase (see Chapter 15). Many diuretics used to control blood pressure are competitive inhibitors, as are many antidepressive agents.

4.4.3 **Non-competitive (irreversible) inhibitors**

To stop an enzyme altogether, the chemist can design a drug which binds irreversibly to the active site and blocks it permanently. This would be a non-competitive form of inhibition since increased levels of natural substrate would not be able to budge the unwanted squatter. The most effective irreversible inhibitors are those which can react with an amino acid at the active site to form a covalent bond. Amino acids such as serine and cysteine which bear nucleophilic residues (OH and SH respectively) are commonly present in enzyme active sites since they are frequently involved in the mechanism of the enzyme reaction (see later). By designing an electrophilic drug which fits the active site, it is possible to alkylate these particular groups and hence permanently clog up the active site (Fig. 4.15). The nerve gases (see Chapter 15) are irreversible inhibitors of mammalian enzymes and are therefore highly toxic. In the same way, penicillins (see Chapter 14) are highly toxic to bacteria by irreversibly inhibiting an enzyme crucial to cell wall synthesis.

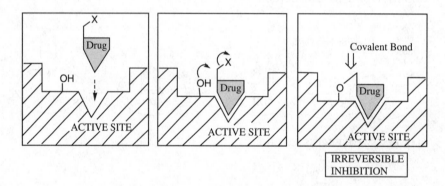

Fig. 4.15 Irreversible inhibition (X = halogen leaving group).

Not all irreversible inhibitors are highly toxic. For example, aspirin's anti-inflammatory activity is due to irreversible inhibition of the cyclo-oxygenase (COX) enzyme – an enzyme which is required for prostaglandin synthesis. Aspirin inhibits the enzyme by acetylating a serine residue in the active site.

4.4.4 **Non-competitive, reversible (allosteric) inhibitors**

So far we have discussed inhibitors which bind to the active site and prevent the natural substrate from binding. We would therefore expect these inhibitors to have some sort of structural similarity to the natural substrate. We would also expect reversible inhibitors to be displaced by increased levels of natural substrate.

However, there are many enzyme inhibitors which appear to have no structural similarity to the natural substrate. Furthermore, increasing the amount of natural substrate has no effect on the inhibition. Such inhibitors are therefore non-competitive inhibitors but, unlike the non-competitive inhibitors mentioned above, the inhibition can be reversible or irreversible.

Non-competitive or allosteric inhibitors bind to a different region of the enzyme and therefore do not compete with the substrate for the active site. However, since binding is occurring, a moulding process takes place (Fig. 4.16) which causes the enzyme to change shape. If that change in shape hides the active site, then the substrate can no longer react.

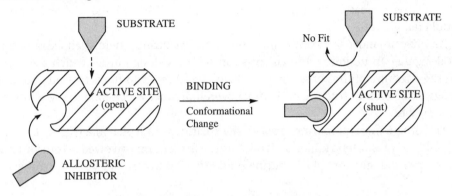

Fig. 4.16 Non-competitive, reversible (allosteric) inhibition.

Adding more substrate will not reverse the situation, but that does not necessarily mean that the inhibition is irreversible. If the inhibitor uses non-covalent bonds to bind to the allosteric binding site, it will eventually depart in its own good time. If the inhibitor uses covalent bonds, it will bind irreversibly.

But why should there be this other binding site?

The answer is that allosteric binding sites are important in the control of enzymes. A biosynthetic pathway to a particular product involves a series of enzymes, all working efficiently to convert raw materials into final product. Eventually, the cell will have enough of the required material and will want to stop production. Therefore, there has to be some sort of control which says enough is enough. The most common control mechanism is one known as feedback control, where the final product controls its own synthesis. It can do this by inhibiting the first enzyme in the biochemical pathway (Fig. 4.17). Therefore, when there are low levels of final product in the cell, the first enzyme in the pathway is not inhibited and works normally. As the levels of final product increase, more and more of the enzyme is blocked and the rate of synthesis drops off in a graded fashion.

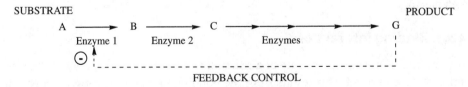

Fig. 4.17 Control of enzymes.

We might wonder why the final product inhibits the enzyme at an allosteric site and not the active site itself. There are two explanations for this.

First, the final product has undergone many transformations since the original starting material and is no longer 'recognized' by the active site. It must therefore bind elsewhere on the enzyme.

Second, binding to the active site itself would not be a very efficient method of feedback control, since it would then have to compete with the starting material. If levels of the latter increased, then the inhibitor would be displaced and feedback control would fail.

An enzyme under feedback control offers the medicinal chemist an extra option in designing an inhibitor. The chemist can not only design drugs which are based on the structure of the substrate and which bind directly to the active site, but also drugs based on the structure of the final overall product which bind to the allosteric binding site.

The drug 6-mercaptopurine, used in the treatment of leukaemia (Fig. 4.18), is an example of an allosteric inhibitor. It inhibits the first enzyme involved in the synthesis of purines and therefore blocks purine synthesis. This in turn blocks DNA synthesis.

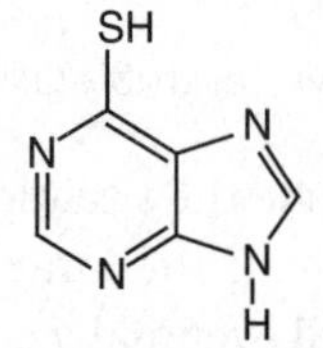

Fig. 4.18 6-Mercaptopurine.

4.5 The catalytic role of enzymes

We now move on to consider the mechanism of enzymes, and how they catalyse reactions. In general, enzymes catalyse reactions by providing the following:

(1) binding interactions

(2) acid/base catalysis

(3) nucleophilic groups

(4) cofactors.

4.5.1 Binding interactions

As mentioned previously, the rate of a reaction is increased if the energy of the transition state is lowered. This results from the bonding interactions between substrate and enzyme.

In the past, it was thought that a substrate fitted its active site in a similar way to a key fitting a lock. Both the enzyme and the substrate were seen as rigid structures with the substrate (the key) fitting perfectly into the active site (the lock) (Fig. 4.19). However, such a scenario does not explain how some enzymes can catalyse a reaction on a range of different substrates. It implies instead that an enzyme has an optimum substrate which fits it perfectly and which can be catalysed very efficiently, whereas all other substrates are catalysed less efficiently. Since this is not the case, the lock and key analogy is invalid.

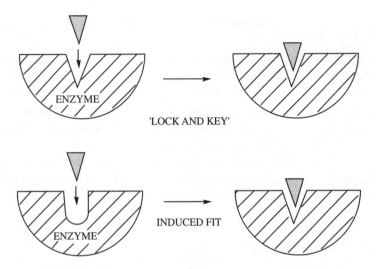

Fig. 4.19 Bonding interactions between substrate and enzyme.

It is now proposed that the substrate is not quite the ideal shape for the active site and, when it enters the active site, it forces the latter to change shape – a kind of moulding process. This theory is known as **Koshland's Theory of Induced Fit** since the substrate induces the active site to take up the ideal shape to accommodate it (Fig. 4.19).

For example, a substrate such as pyruvic acid might interact with its active site via one hydrogen bond, one ionic bond, and one van der Waals interaction (Fig. 4.20). However, if the fit is not perfect, the three bonding interactions are not ideal either (e.g. the binding groups are a bit too far apart). In order to maximize the strength of these bonds, the enzyme changes shape so that the amino acid residues involved in the binding move closer to the substrate.

This theory of induced fit helps to explain why enzymes can catalyse a wide range of substrates. Each substrate induces the active site into a shape which is ideal for it and, as long as the moulding process does not distort the active site too much such that the reaction mechanism proves impossible, the reaction can proceed.

But note this. The substrate is not just a passive spectator to the moulding process going on around it. As the enzyme changes shape to maximize bonding interactions,

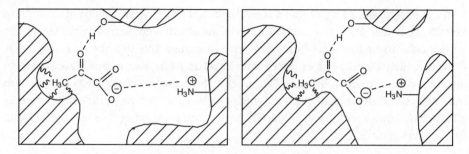

Fig. 4.20 Example of induced fit.

the same thing is going to happen to the substrate. It too will alter shape. Bond rotation may occur to fix the substrate in a particular conformation – not necessarily the most stable conformation. Bonds may even be stretched and weakened. Consequently, this moulding process, designed to maximize binding interactions, may force the substrate into the ideal conformation for the reaction to follow and may also weaken the very bonds which have to be broken.

Once bound to an active site, the substrate is now held ready for the reaction to follow. Binding has fixed the 'victim' so that it cannot evade attack and this same binding has weakened its defences (bonds) so that reaction is easier (lower activation energy).

4.5.2 Acid/base catalysis

Usually acid/base catalysis is provided by the amino acid histidine. Histidine is a weak base and can easily equilibrate between its protonated form and its free base form (Fig. 4.21). In doing so, it can act as a proton 'bank'; that is, it has the capability to accept and donate protons in the reaction mechanism. This is important since active sites are frequently hydrophobic and will therefore have a low concentration of water and an even lower concentration of protons.

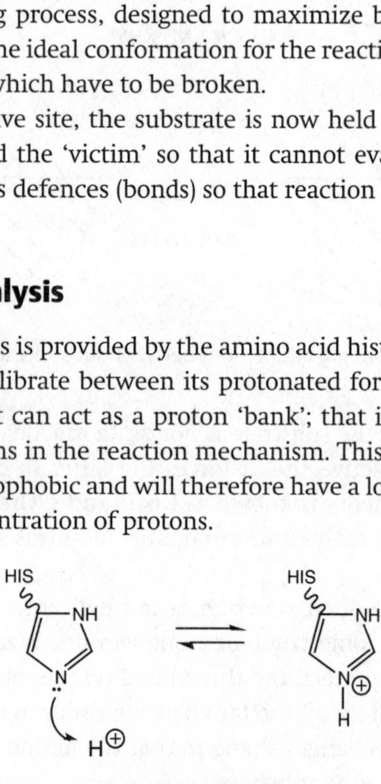

Fig. 4.21 Histidine.

4.5.3 Nucleophilic groups

The amino acids serine and cysteine are commonly present in active sites. These amino acids have nucleophilic residues (OH and SH respectively) which are able to participate in the reaction mechanism. They do this by reacting with the substrate to form intermediates which would not be formed in the uncatalysed reaction. These intermediates offer an alternative reaction pathway which may avoid a high-energy transition state and hence increase the rate of the reaction.

Normally, an alcoholic group such as that on serine is not a good nucleophile. However, there is usually a histidine residue close by to catalyse the reaction. For example, the mechanism by which chymotrypsin hydrolyses peptide bonds is shown in Fig. 4.22.

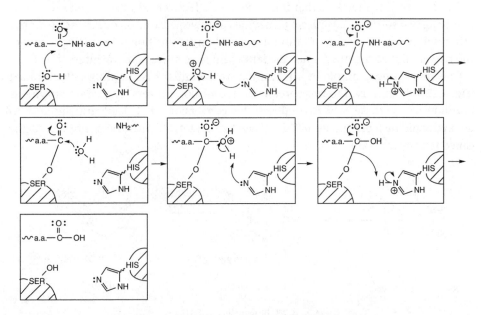

Fig. 4.22 Hydrolysis of peptide bonds by chymotrypsin.

The presence of a nucleophilic serine residue means that water is not required in the initial stages of the mechanism. This is important since water is a poor nucleophile and may also find it difficult to penetrate a hydrophobic active site. Secondly, a water molecule would have to drift into the active site, and search out the carboxyl group before it could attack it. This would be something similar to a game of blind man's buff. The enzyme, on the other hand, can provide a serine OH group positioned in exactly the right spot to react with the substrate. Therefore, the nucleophile has no need to search for its substrate. The substrate has been delivered to it.

Water is required eventually to hydrolyse the acyl group attached to the serine residue. However, this is a much easier step than the hydrolysis of a peptide link since esters are more reactive than amides. Furthermore, the hydrolysis of the peptide link means that one half of the peptide can drift away from the active site and leave room for a water molecule to enter. A similar enzymic mechanism is involved in the action of the enzyme acetylcholinesterase described in Chapter 15.

An understanding of an enzyme mechanism can help medicinal chemists design more powerful inhibitors. For example, it is possible to design inhibitors which bind so strongly to the active site (using non-covalent forces) that they are effectively irreversible inhibitors – a bit like inviting the mother-in-law for dinner and finding that she has moved in on a permanent basis. One way of doing this is to design a drug which resembles the transition state for the catalysed reaction. Such a drug should bind more strongly to the enzyme than either the substrate or the product. If this compound is designed to remain intact, then the enzyme becomes strongly inhibited. Such compounds are known as **transition-state analogues**.

The beauty of the tactic is that it can be used effectively against enzyme reactions involving two substrates. With such enzymes, inhibitors based on one substrate or the other will not be as good as an inhibitor based on a transition-state analogue where the two are linked together, since the latter will have more bonding interactions.

One interesting example of a transition-state inhibitor is the drug 5-fluorouracil (Fig. 4.23) which is used to treat breast, liver, and skin cancers. The target enzyme is thymidylate synthase which catalyses the conversion of 2′-deoxyuridylic acid to deoxythymidine-5-phosphate (dTMP) (Figs 4.23 and 4.24) with the aid of a cofactor called tetrahydrofolate.

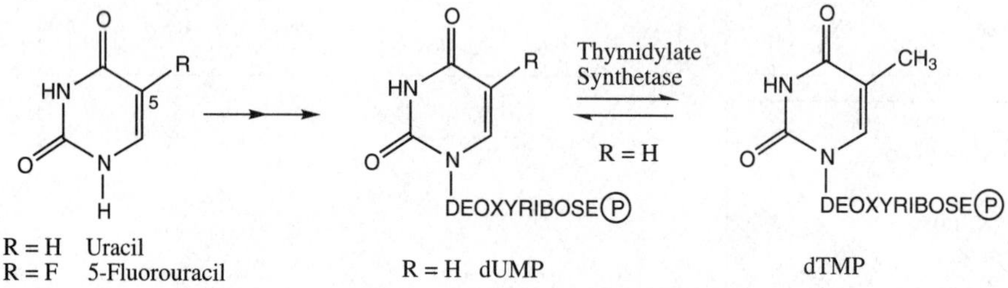

R = H   Uracil
R = F   5-Fluorouracil

R = H   dUMP

dTMP

Fig. 4.23 Biosynthesis of dTMP.

5-Fluorouracil is not the transition-state analogue itself. It is converted in the body to the fluorinated analogue of 2′-deoxyuridylic acid which then combines with the enzyme and the cofactor to form a transition-state analogue *in situ*. Up until this point, nothing unusual has happened and the reaction mechanism has been proceeding normally. The tetrahydrofolate has formed a covalent bond to the uracil skeleton via a methylene unit (the unit which is usually transferred to uracil). However, now things start to go wrong. At this stage, a proton is usually lost from the 5-position of uracil. However, 5-fluorouracil has a fluorine atom at that position instead of a hydrogen. Further reaction is impossible since it would require fluorine to leave as a positive ion. As a result, the fluorouracil skeleton remains covalently and irreversibly bound to the active site. The synthesis of thymidine is now terminated, which in turn stops the synthesis of DNA. As a result, replication and cell division are blocked. 5-Fluorouracil is a particularly useful drug for the treatment of skin cancer since it shows a high level of selectivity for cancer cells over normal skin cells.

Transition-state analogues have also been successful in inhibiting protease enzymes. These are enzymes which cleave peptides (see Chapter 10).

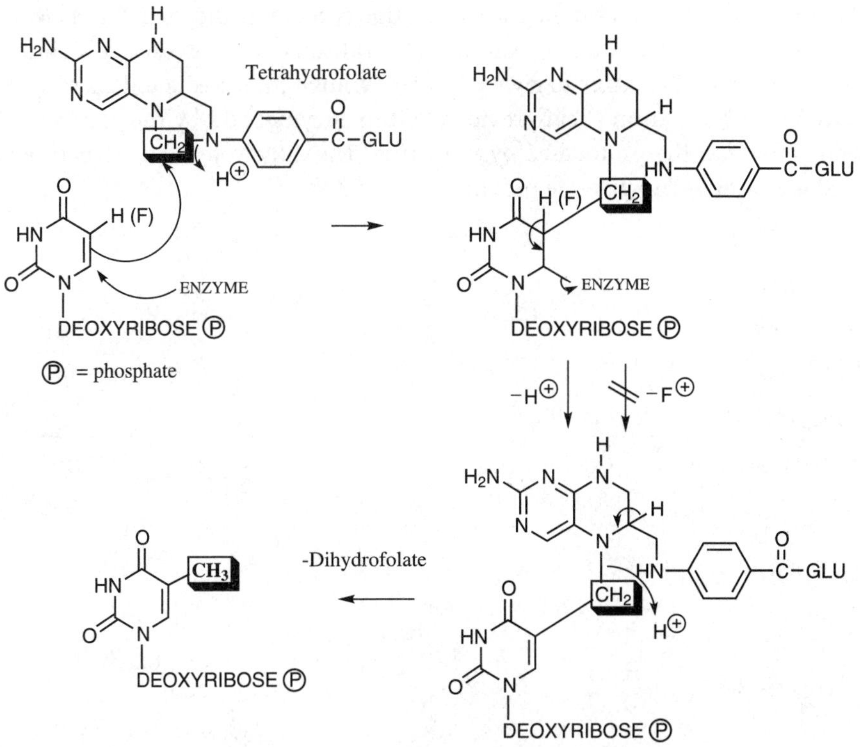

Fig. 4.24 Use of 5-fluorouracil as a transition-state inhibitor.

Transition-state analogues can be viewed as *bona fide* visitors to an enzyme's active site, but which become stubborn squatters once they have arrived. Other apparently harmless visitors can turn into lethal assassins once they have bound to their target enzyme. Once again, it is the enzyme mechanism itself which causes the transformation. One example of this is provided by the irreversible inhibition of the enzyme alanine transaminase by trifluoroalanine (Fig. 4.25).

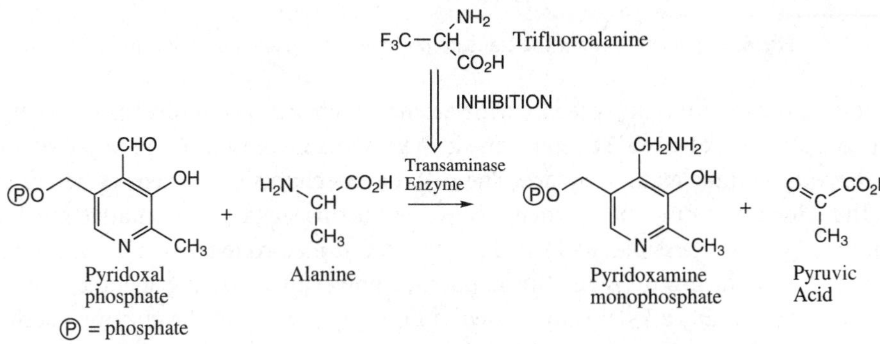

Fig. 4.25 Irreversible inhibition of the enzyme alanine transaminase.

The normal mechanism for the transamination reaction is shown in Fig. 4.26 (R=H) and involves the condensation of alanine and pyridoxal phosphate to give an imine. A proton is lost from the imine to give a dihydropyridine intermediate. This reaction is catalysed by a basic amino acid provided by the enzyme, and by the electron withdrawing effects of the protonated pyridine ring. The dihydropyridine structure now formed is hydrolysed to give the products.

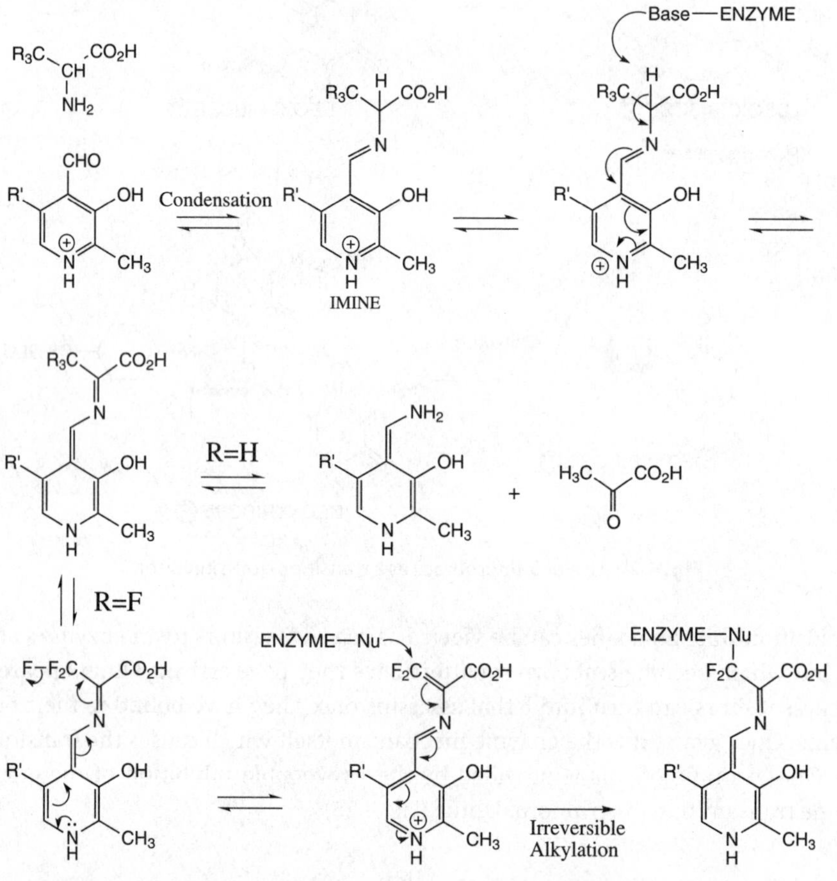

Fig. 4.26 Mechanism for the transamination reaction and its inhibition.

Trifluoroalanine contains three fluorine atoms which are very similar in size to the hydrogen atoms in alanine. The molecule is therefore able to fit into the active site of the enzyme and take alanine's place. The reaction mechanism proceeds as before to give the dihydropyridine intermediate. However, at this stage, an alternative mechanism now becomes possible (R=F). A fluoride atom is electronegative and can therefore act as a leaving group. When this happens, a powerful alkylating agent is formed which can irreversibly alkylate any nucleophilic group present in the enzyme's active site. A covalent bond is now formed and the active site is unable to accept further substrate. As a result, the enzyme is irreversibly inhibited.

Drugs which operate in this way are often called **suicide substrates** since the enzyme is committing suicide by reacting with them. The great advantage of this approach is that the alkylating agent is generated at the site where it is meant to act and is therefore highly selective for the target enzyme. If the alkylating group had not been disguised in this way, the drug would have alkylated the first nucleophilic group it met in the body and would have shown little or no selectivity. (The uses of alkylating agents and the problems associated with them are also discussed in Chapter 7.)

Few useful therapeutic drugs have been designed by this approach so far. Inhibiting the transaminase enzyme has no medicinal application since the enzyme is crucial to mammalian biochemistry and inhibiting it would be toxic to the host. The main use for suicide substrates has been in labelling specific enzymes. The substrates can be labelled with radioactivity and reacted with their target enzyme to locate the enzyme in tissue preparations. However, the suicide substrate approach has potential therapeutic applications against enzymes which are unique to 'foreign invaders' such as bacteria, protozoa, and fungi.

The ability of some drugs to act as suicide substrates at enzymes other than the desired target enzyme can explain their toxic side-effects. For example, the diuretic agent tienilic acid (Fig. 4.27) had to be withdrawn from the market because it was found to act as a suicide substrate at the cytochrome P450 enzymes involved in drug metabolism (see Chapter 10). Unfortunately, the metabolic reaction carried out by these enzymes converted tienilic acid to a thiophene sulphoxide which proved highly electrophilic. This encouraged a Michael reaction leading to alkylation of a thiol group in the enzyme's active site. Loss of water from the thiophene sulphoxide restored the thiophene ring and resulted in tienilic acid being covalently linked to the enzyme, thus inhibiting the enzyme irreversibly.

4.5.4 Cofactors

Many enzymes require additional non-protein substances called cofactors for the reaction to take place (e.g. tetrahydrofolate; Fig 4.24). These cofactors are either metal ions (e.g. zinc) or small organic molecules called coenzymes (e.g. NAD^+, $NADP^+$, pyridoxal phosphate).[2] Most coenzymes are bound by ionic bonds and other non-covalent bonding interactions, but some are bound covalently and are called **prosthetic groups**. Coenzymes are derived from water-soluble vitamins and are really a second substrate undergoing a reaction themselves. For example, lactate dehydrogenase requires the coenzyme nicotinamide adenine dinucleotide (NAD^+) (Fig. 4.28) to catalyse the dehydrogenation of lactic acid to pyruvic acid. NAD^+ is bound to the active site along with lactic acid and acts as the oxidizing agent. During the reaction it is converted to its reduced form (NADH) (Fig. 4.29). Conversely, NADH can bind to the enzyme and act as a reducing agent when the enzyme catalyses the reverse reaction.

---

[2] $NADP^+$ and NADPH are phosporylated analogues of NAD^+ and NADH, respectively, and carry out redox reactions by the same mechanism. NADPH is used almost exclusively for reductive biosynthesis, whereas NADH is used primarily for the generation of ATP.

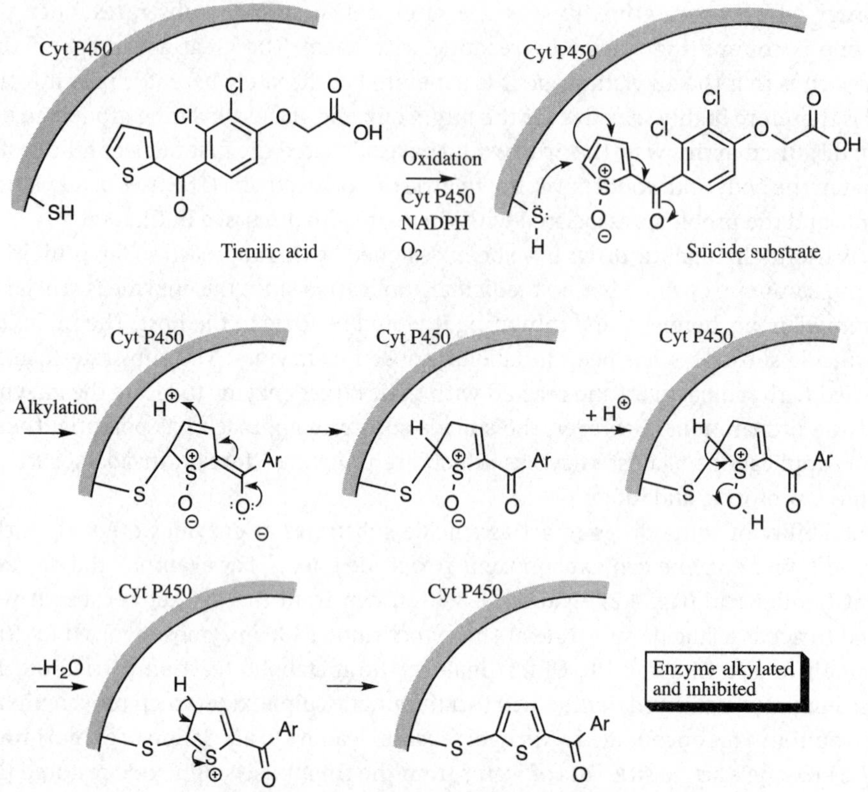

Fig. 4.27 Inhibition of cytochrome P450 by tienilic acid.

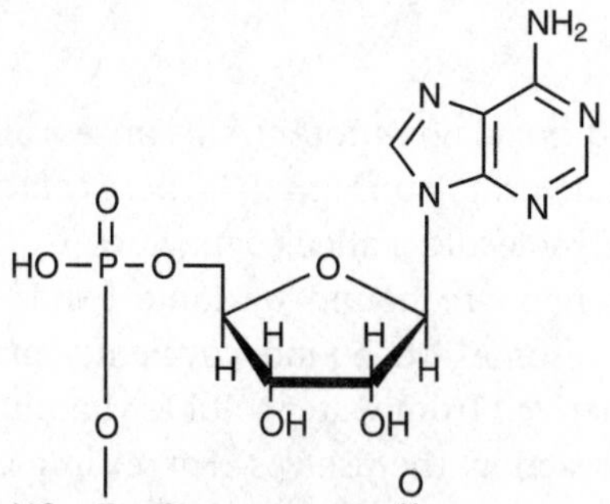

Fig. 4.28 Nicotinamide adenine dinucleotide (NAD$^+$).

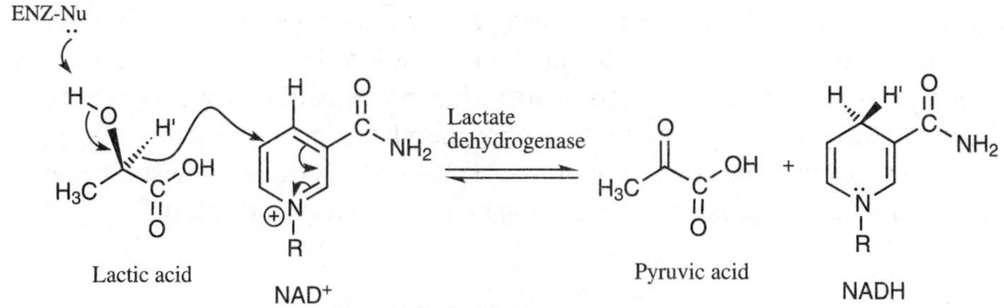

Fig. 4.29 NAD$^+$ as a coenzyme.

Deficiency of coenzymes can arise from a poor diet. This leads to loss of enzyme activity and appearance of disease (e.g. scurvy).

A knowledge of how the coenzyme binds to the enzyme allows the possibility of designing drugs which will fit the region of the active site normally occupied by the coenzyme (see Chapter 13).

4.6 Regulation of enzymes

Virtually all enzymes are controlled by allosteric interactions which can either stimulate or inhibit catalytic activity. Such control reflects the local conditions within the cell. For example, the enzyme phosphorylase *a* catalyses the breakdown of glycogen (a polymer of glucose monomers) to glucose-1-phosphate subunits (Fig. 4.30). It is stimulated by adenosine 5'-monophosphate (AMP) and inhibited by glucose-1-phosphate. Thus, rising levels of the product (glucose-1-phosphate) from this reaction act as a self-regulating 'brake' on the enzyme.

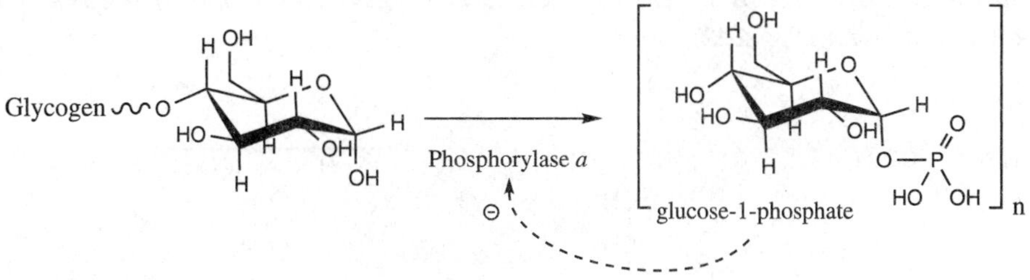

Fig. 4.30 Internal control of phosphorylase *a*.

This is an internal control. However, many enzymes can also be regulated externally. We shall be looking at this in more detail in Chapters 5 and 6 but, in essence, cells receive chemical messages from their environment which trigger a cascade of signals within the cell. These in turn ultimately activate a set of enzymes known as protein kinases. The protein kinases play an important part in controlling enzyme activity within the cell by phosphorylating amino acids such as serine, threonine or tyrosine in target enzymes – a covalent modification. For example, the hormone adrenaline is an external messenger which triggers a signalling sequence resulting in the activation of a protein kinase which phosphorylates an enzyme called phosphorylase b – the inactive form of phosphorylase a (Fig. 4.31). Phosphorylase a is generated and remains active until it is dephosphorylated back to its inactive form.

Fig. 4.31 External control of phosphorylase a.

In this case, phosphorylation of the target enzyme leads to activation. Other enzymes may be deactivated by phosphorylation. For example, glycogen synthase – the enzyme which catalyses the **synthesis** of glycogen from glucose-1-phosphate – is inactivated by phosphorylation and activated by dephosphorylation. The latter is effected by the hormone insulin which triggers a different signalling cascade from that of adrenaline.

The external control of enzymes is usually initiated by external chemical messengers which do not enter the cell. However, there is an exception to this. In recent years it has been discovered that cells can generate the gas nitric oxide by the reaction sequence shown in Fig. 4.32.

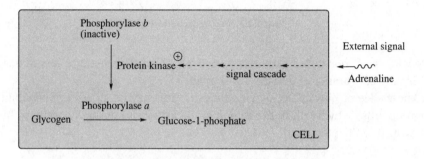

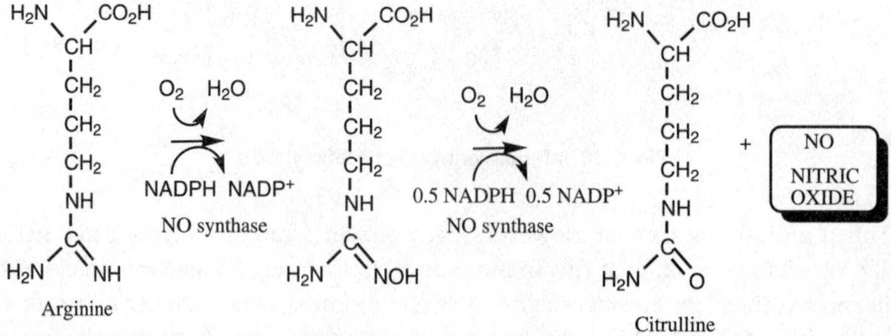

Fig. 4.32 Synthesis of nitric oxide.

Since nitric oxide is a gas it can easily diffuse through cell membranes into target cells. There, it activates enzymes called cyclases to generate cyclic GMP from GTP (Fig. 4.33). Cyclic GMP then acts as a secondary messenger to influence other reactions within the cell (see Chapter 6). By this process, nitric oxide has an influence on a diverse range of physiological processes, including blood pressure, neurotransmission, and immunological defence mechanisms. The well publicized drug viagra (Fig. 4.34) prevents the decomposition of cyclic GMP and so prolongs the signal initiated by nitric oxide.

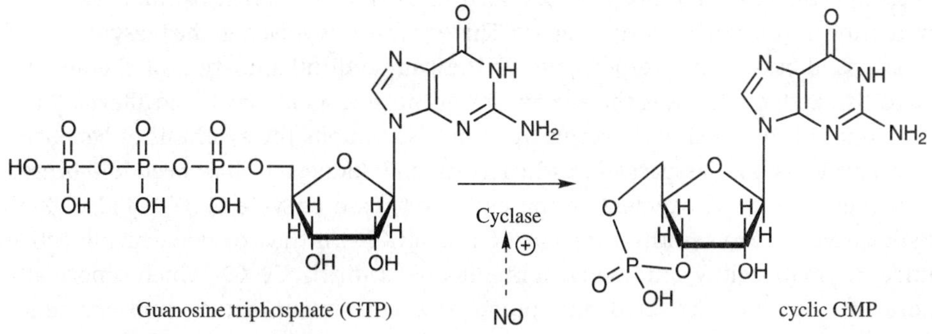

Fig. 4.33 Activation of cyclase enzymes by nitric oxide.

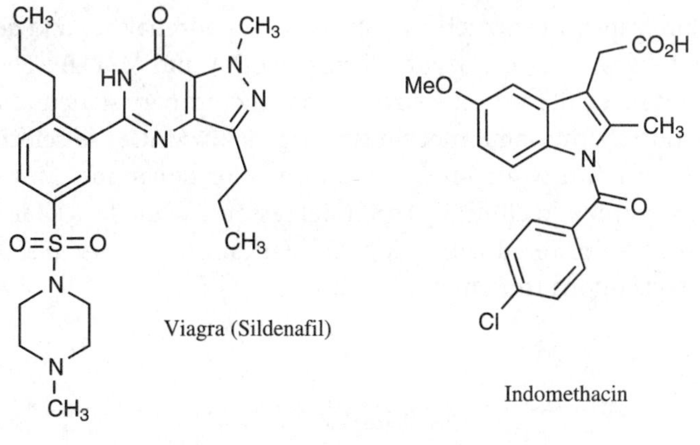

Fig. 4.34 Viagra and indomethacin.

4.7 Isozymes

Enzymes having a quaternary structure are made up of different subunits and the combination of these subunits can differ in different tissues. Such variations are called isozymes. Isozymes catalyse the same reaction but differ in their properties. For example, there are different isozymes of lactate dehydrogenase. This enzyme is made up of four identical subunits, but the subunits which make up lactate dehydrogenase in muscle are different in amino acid composition from those making up lactate dehydrogenase in the heart. Both enzymes catalyse the conversion of lactic acid to pyruvic acid, but the isozyme in muscle is twice as active. Moreover, the isozyme in the heart is inhibited by excess pyruvic acid while the isozyme in muscle is not.

Identification of isozymes which are selective for certain tissues allows the possibility of designing enzyme inhibitors which are tissue-selective.

For example, the non-steroidal anti-inflammatory drug indomethacin (Fig. 4.34) is used to treat inflammatory diseases such as rheumatoid arthritis, and works by inhibiting the enzyme cyclo-oxygenase. This enzyme is involved in the biosynthesis of prostaglandins – agents responsible for the pain and inflammation of rheumatoid arthritis – and so inhibiting the enzyme lowers prostaglandin levels and alleviates the symptoms of the disease. However, the drug also inhibits the synthesis of beneficial prostaglandins in the gastrointestinal tract and kidney. It has been discovered that there are two isozymes for the cyclo-oxygenase enzyme (COX-1 and COX-2). Both isozymes carry out the same reactions, but COX-1 is the isozyme which is active under normal healthy conditions. In rheumatoid arthritis, COX-2, which is normally dormant, becomes activated and produces excess inflammatory prostaglandins. Therefore, drugs are being developed to be selective for the COX-2 isozyme, so that only the production of inflammatory prostaglandins is reduced.

Monoamine oxidase (MAO) is one of the enzymes responsible for the metabolism of important neurotransmitters such as dopamine, noradrenaline, and serotonin (see Chapter 5) and exists in two isozymic forms (MAO-A and MAO-B). These isozymes differ in substrate specificity, tissue distribution, and primary structure but carry out the same reaction by the same mechanism (Fig. 4.35). MAO-A is selective for noradrenaline and serotonin while MAO-B is selective for dopamine. MAO-A inhibitors such as clorgyline are used clinically as antidepressants while MAO-B inhibitors such as selegiline are administered with L-dopa for Parkinson's disease (Fig. 4.36). MAO-B inhibition protects L-dopa from metabolism.

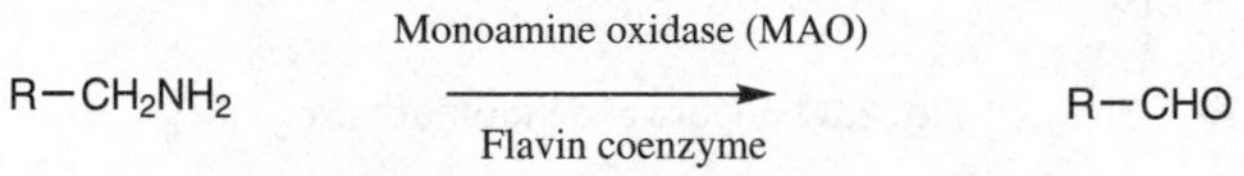

Fig. 4.35 Reaction catalysed by monoamine oxidase.

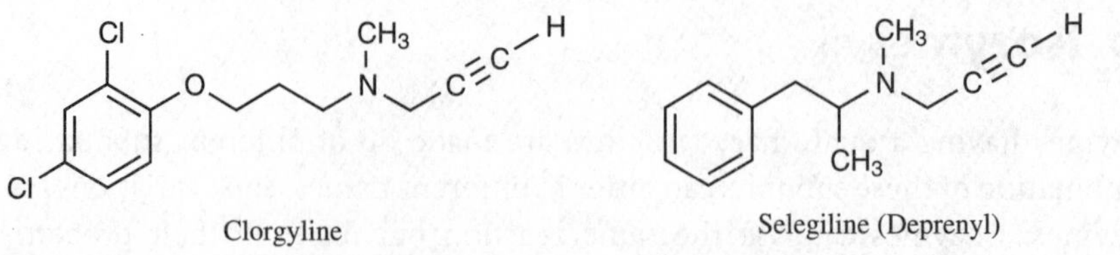

Clorgyline   Selegiline (Deprenyl)

Fig. 4.36 Clorgyline and selegiline.

4.8 Medicinal uses of enzyme inhibitors

4.8.1 Enzyme inhibitors against microorganisms

Inhibitors of enzymes have been extremely successful in the war against infection. If an enzyme is crucial to a microorganism, then switching it off will clearly kill the cell or prevent it from growing. Ideally, the enzyme chosen should be one which is not present in our own bodies, and fortunately such enzymes exist because of the significant biochemical differences between bacterial cells and our own. Nature, of course, is well ahead in this game. For example, many fungal strains produce metabolites (e.g. penicillin) which are toxic to bacteria but not to themselves. This gives the fungi an advantage over their microbiological competitors when competing for nutrients.

Although it is preferable to target enzymes which are unique to the foreign invader, it is still possible to target enzymes which are present in both bacterial and mammalian cells, as long as there are significant differences between the corresponding enzymes. Such differences are perfectly feasible. Although the enzymes in both species may have evolved from a common ancestral protein, they have evolved and mutated separately over several million years. Identifying these differences allows the medicinal chemist to design drugs which can take advantage of these differences, and selectively act against the bacterial enzyme.

Chapter 14 covers antibacterial agents such as the sulphonamides, penicillins, and cephalosporins, all of which act by inhibiting enzymes.

4.8.2 Enzyme inhibitors against viruses

Enzyme inhibitors are also extremely important in the battle against viral infections, (e.g. the herpes virus and HIV). Successful antiviral drugs include acyclovir for herpes and zidovudine for HIV (Fig. 4.37) (see also Section 7.5).

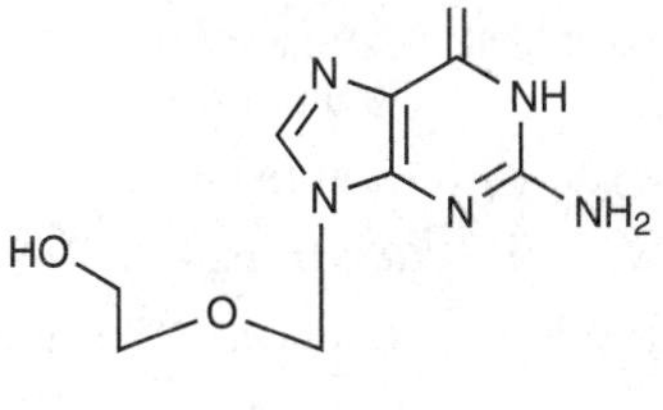

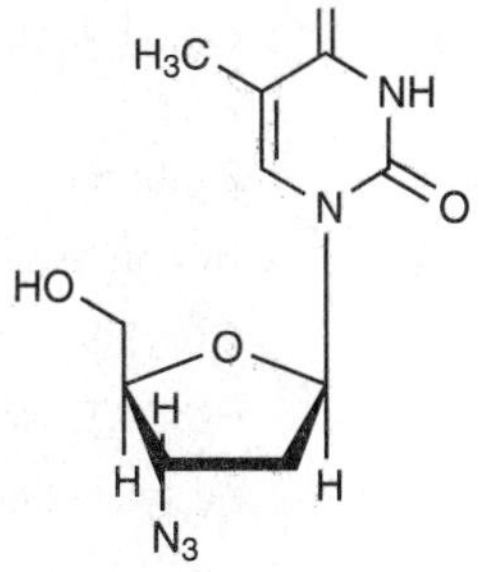

Fig. 4.37 Acyclovir and zidovudine.

4.8.3 Enzyme inhibitors against the body's own enzymes

Drugs which act against the body's own enzymes are also important in medicine and some examples are given in Table 4.1. Chapter 15 also considers agents known such as anticholinesterases which have uses in a variety of diseases including Alzheimer's disease.

The search continues for new enzyme inhibitors, especially inhibitors which are selective for a specific isozyme or which act against recently discovered enzymes. Some current research projects include investigations into inhibitors of the COX-2 isozyme (see Section 4.7), matrix metalloproteinases (anti-arthritic drugs), aromatases (anticancer agents), and caspases. The caspases are enzymes which are implicated in the processes leading to cell death. Inhibitors of caspases may have potential in the treatment of stroke victims.

Table 4.1 Drugs and their target enzymes

Drug	Target enzyme	Field of therapy
Aspirin	Cyclo-oxygenase	Anti-inflammatory
Captopril and enalapril	Angiotensin converting enzyme (ACE)	Antihypertension
Simvastatin	HMG-CoA reductase	Lowering of cholesterol levels
Desipramine	Monoamine oxidase	Antidepression
Clorgyline	Monoamine oxidase A	Antidepression
Selegiline	Monoamine oxidase B	Treatment of Parkinson's disease
Methotrexate	Dihydrofolate reductase	Anticancer
5-Fluorouracil	Thymidylate synthase	Anticancer
Viagra	Phosphodiesterase enzyme (PDE5)	Treatment of male erectile dysfunction
Allopurinol	Xanthine oxidase	Treatment of gout
U75875	HIV protease	AIDS therapy
Ro41–0960	Catechol-O-methyltransferase	Treatment of Parkinson's disease
Omeprazole	H^+/K^+ ATPase proton pump	Ulcer therapy
Organophosphates	Acetylcholinesterase	Treatment of myasthenia gravis, glaucoma and Alzheimer's disease
Acetazolamide	Carbonic anhydrase	Diuretic
Zileutin	5-Lipoxygenase	Anti-asthmatic

4.9 Enzyme kinetics

Before leaving this chapter, it is worth mentioning a little bit about enzyme kinetics since such studies are extremely useful in determining the properties of an enzyme inhibitor (e.g. whether an inhibitor is competitive or non-competitive).

4.9.1 The Michaelis–Menten equation

The Michaelis–Menten equation holds for enzymes (E) which combine with their substrate (S) to form an enzyme substrate (ES). The enzyme substrate (ES) can then either dissociate back to E and S, or go on to form a product (P). It is assumed that formation of the product is irreversible.

$$E + S \underset{k_2}{\overset{k_1}{\rightleftharpoons}} ES \xrightarrow{k_3} E + P$$

(k_1–k_3 are rate constants).

For enzymes such as these, a plot of the rate of enzyme reaction versus substrate concentration [S] would look like that shown in Fig. 4.38.

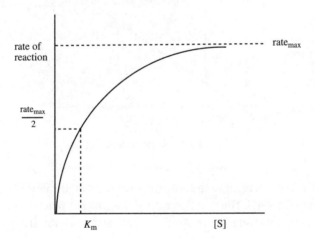

Fig. 4.38 Reaction rate versus substrate concentration.

This plot would show that at low substrate concentrations the rate of reaction increases almost proportionally to the substrate concentration, whereas at high substrate concentration, the rate becomes almost constant and approaches a maximum rate (rate$_{max}$), which is independent of substrate concentration. This reflects a situation where there is more substrate present than active sites available, and so increasing the amount of substrate will have little effect.

The Michaelis–Menten equation relates the rate of reaction versus the substrate concentration for plots of this sort.

$$\text{Rate} = \text{Rate}_{max} \frac{[S]}{[S] + K_M}$$

The derivation of this equation is not covered here but can be found in most biochemistry textbooks. The constant K_M is known as the Michaelis constant and is equal to the substrate concentration at which the reaction rate is half of its maximum value[3].

4.9.2 Lineweaver–Burk plots

The Michaelis–Menten equation can be modified by taking its reciprocal such that a straight line plot is obtained (the Lineweaver–Burk plot) (Fig. 4.39).

$$\frac{1}{\text{Rate}} = \frac{1}{\text{Rate}_{max}} + \frac{K_M}{\text{Rate}_{max} [S]}$$

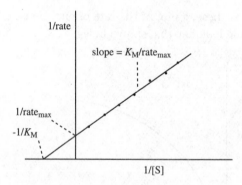

Fig. 4.39 Lineweaver Burk plot.

Plots such as these can be used to determine whether an inhibitor is competitive or non-competitive (Fig. 4.40). The reaction rate of the enzyme reaction is measured with respect to varying substrate concentration with and without the presence of the inhibitor.

In the case of competitive inhibition, the lines cross the y axis at the same point (i.e. the maximum rate of the enzyme is unaffected), but the slopes are different. The fact that the maximum rate is unaffected reflects the fact that the inhibitor and substrate are competing for the same active site and that increasing the substrate

---

[3] The value of K_M can be taken as a measure of the binding strength of the ES complex if it is known that the dissociation rate k_2 is much greater than the rate of formation of product (k_3).

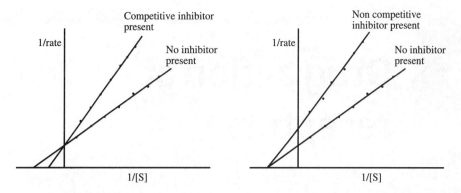

Fig. 4.40 Distinguishing a competitive inhibitor from a non-competitive inhibitor.

concentration sufficiently will overcome the inhibition. The increase in the slope which results from adding an inhibitor is a measure of how strongly the inhibitor binds to the enzyme.

In the case of a non-competitive inhibitor, the lines have the same intercept point on the x axis (i.e. K_M is unaffected), but have different slopes and different intercepts on the y axis. Therefore, the maximum rate for the enzyme has been reduced.

Lineweaver–Burk plots are extremely useful in determining the nature of inhibition, but they have their limitations and are not applicable to enzymes which are under allosteric control.

5 Drug action at receptors

5.1 The receptor role

Enzymes are one major target for drugs. Receptors are another. Drugs which interact with receptors are amongst the most important in medicine and provide treatment for ailments such as pain, depression, Parkinson's disease, psychosis, heart failure, asthma, and many other problems.

What are these receptors and what do they do?

Cells are all individual, but in a complex organism such as a human, they have to 'get along' with their neighbours. There has to be some sort of communication system. After all, it would be pointless if individual heart cells were to contract at different times. The heart would then be a wobbly jelly and totally useless in its function as a pump. Communication is essential to ensure that all heart muscle cells contract at the same time. The same holds true for all the body's organs and functions. Communication is essential if the body is to operate in a coordinated and controlled fashion.

Control and communication come primarily from the brain and spinal column (the central nervous system – CNS) which receives and sends messages via a vast network of nerves (Fig. 5.1). The detailed mechanism by which nerves transmit messages along their length need not concern us here (see Appendix 2). It is sufficient for our purposes to think of the message as being an electrical 'pulse' which travels down the nerve cell towards the target, whether that be a muscle cell or another nerve. If that was all there was to it, it would be difficult to imagine how drugs could affect this communication system. However, there is one important feature of this system which is crucial to our understanding of drug action. The nerves do not connect directly to their target cells. They stop just short of the cell surface. The distance is minute, about 100 Å, but it is a space which the electrical impulse is unable to 'jump'.

Therefore, there has to be a way of carrying the message across the gap between the nerve ending and the cell. The problem is solved by the release of a chemical messenger (neurotransmitter) from the nerve cell (Fig. 5.2). Once released, this

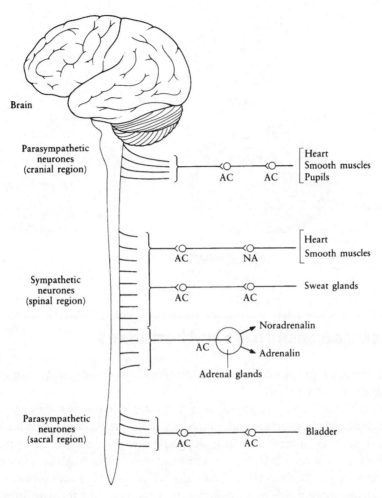

Brain

Parasympathetic
neurones
(cranial region)

AC   AC

Heart
Smooth muscles
Pupils

Sympathetic
neurones
(spinal region)

AC   NA

Heart
Smooth muscles

AC   AC

Sweat glands

AC

Noradrenalin
Adrenalin

Adrenal glands

Parasympathetic
neurones
(sacral region)

AC   AC

Bladder

Fig. 5.1 The central nervous system. (AC=acetylcholine; NA = Noradrenaline.) (Taken from
J. Mann, *Murder, magic, and medicine*, Oxford University Press (1992), with permission.)

chemical messenger can diffuse across the gap to the target cell, where it can bind and
interact with a specific protein (receptor) embedded in the cell membrane. This
process of binding leads to a series or cascade of secondary effects which result either
in a flow of ions across the cell membrane or in the switching on (or off) of enzymes
inside the target cell. A biological response then results, such as the contraction of a
muscle cell or the activation of fatty acid metabolism in a fat cell.

We shall consider these secondary effects and how they result in a biological action
at a later stage (see Chapter 6), but for the moment the important thing to note is that
the communication system depends crucially on a chemical messenger. Since a chemical
process is involved, it should be possible for other chemicals (drugs) to interfere or
to take part in the process.

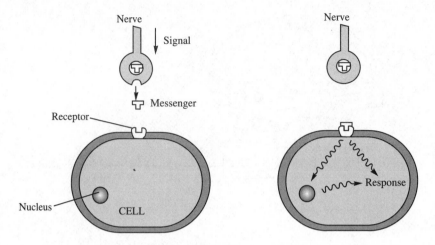

Fig. 5.2 Neurotransmitter action.

5.2 Neurotransmitters and hormones

Let us now look a bit closer at neurotransmitters and receptors, and consider first the messengers. What are they?

There is a large variety of messengers that interact with receptors and that vary quite significantly in structure and complexity. Some neurotransmitters are simple in structure, such as monoamines (e.g. acetylcholine, noradrenaline, dopamine, and serotonin) or amino acids (e.g. γ-aminobutanoic acid (GABA), glutamic acid, and glycine) (Fig. 5.3). Even the calcium ion can act as a chemical messenger. Other chemical messengers are more complex in structure, and include lipids such as prostaglandins, purines such as adenosine or ATP (see Chapter 7), neuropeptides such as endorphins and enkephalins (see Chapter 17), peptide hormones such as angiotensin or bradykinin, and even enzymes such as thrombin (see Chapter 6).

In general, a nerve releases mainly one type of neurotransmitter[1] and the receptor which awaits it on the target cell will be specific for that messenger. However, this does not mean that the target cell has only one type of receptor protein. Each target

---

[1] In the past, it has been assumed that only one type of neurotransmitter is released from any one type of nerve cell. This is now known not to be true. Certainly, as far as the amine neurotransmitters (i.e. acetylcholine, noradrenaline, glycine, serotonin, GABA, and dopamine) are concerned, it is generally true that only one of these messengers is released by any one nerve cell.

However, there is now a growing list of peptide co-transmitter substances which appear to be released from nerve cells along with the above neurotransmitters. For example, somatostatin, cholecystokinin, vasointestinal peptide, substance P, and neurotensin have all been identified as co-transmitters of acetylcholine in a variety of situations.

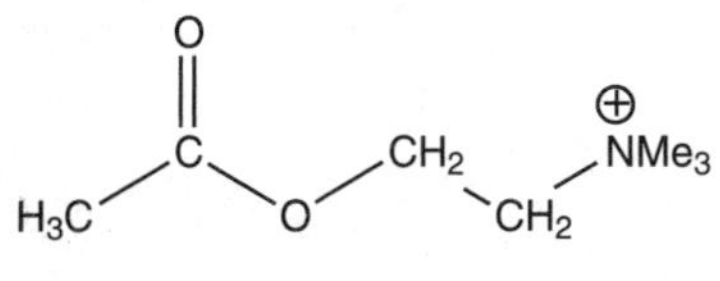

Acetylcholine

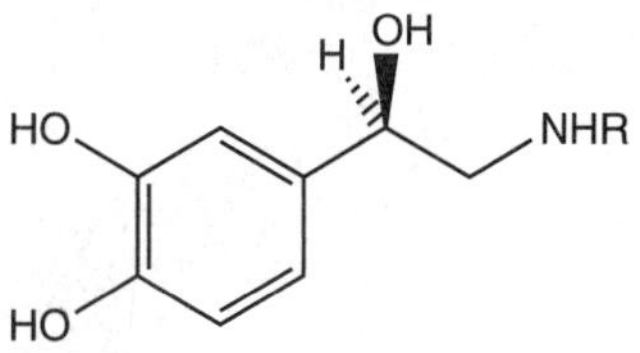

R=H Noradrenaline
R=Me Adrenaline

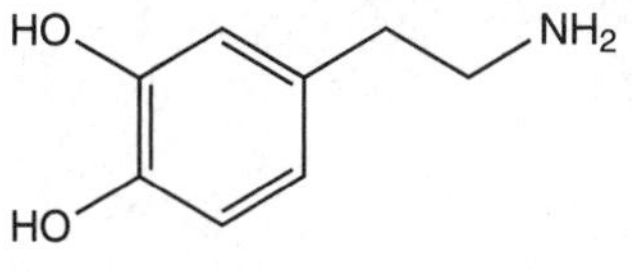

Dopamine

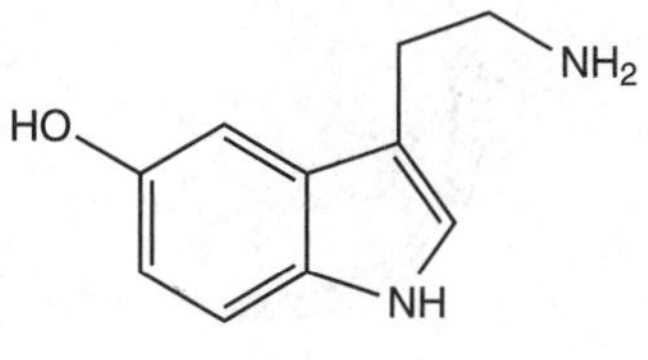

Serotonin

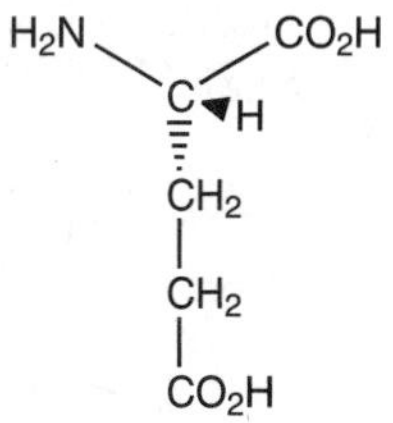

Glutamic Acid

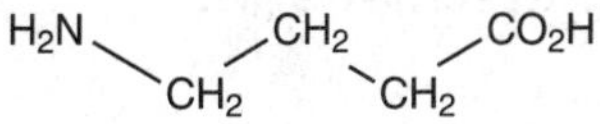

γ-Aminobutanoic acid

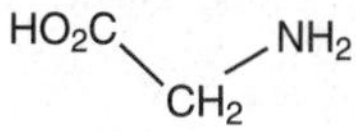

Glycine

Fig. 5.3 Examples of neurotransmitters.

cell has a large number of nerves communicating with it and they do not all use the same neurotransmitter (Fig. 5.4). Therefore, the target cell will have other types of receptors specific for those other neurotransmitters. It may also have receptors waiting to receive messages from chemical messengers which have longer distances to travel. These are the hormones released into the circulatory system by various glands in the body. The best known example of a hormone is adrenaline. When danger or exercise is anticipated, the adrenal medulla gland releases adrenaline into the bloodstream where it is carried round the body, preparing it for vigorous exercise.

Hormones and neurotransmitters can be distinguished by the route they travel and by the way they are released, but their action when they reach the target cell is the same. They both interact with a receptor and a message is received. The cell responds to that message, adjusts its internal chemistry accordingly and a biological response results.

Communication is clearly essential for the normal working of the human body, and if the communication should become faulty then it could lead to such ailments as depression, heart problems, schizophrenia, muscle fatigue, and many other problems. What sort of things *could* go wrong?

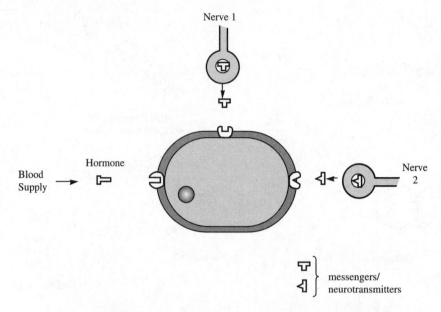

Fig. 5.4 Target cell containing receptors specific to each type of messenger.

One problem would be if too many messengers were released. The target cell could start to 'overheat'. Alternatively, if too few messengers were sent out, the cell could become 'sluggish'. It is at this point that drugs can play a role by either acting as replacement messengers (if there is a lack of the body's own messengers), or by blocking the receptors for the natural messengers (if there are too many host messengers). Drugs of the former type are known as **agonists**. Those of the latter type are known as **antagonists**.

What determines whether a drug is an agonist or an antagonist, and is it possible to predict whether a new drug will act as one or the other?

To answer this question, we have to move down to the molecular level and consider what happens when a small molecule such as a drug or a neurotransmitter interacts with a receptor protein. Let us first look at what happens when one of the body's own messengers (neurotransmitters or hormones) interacts with its receptor.

5.3 Receptors

A receptor is a protein molecule embedded within the cell membrane with part of its structure facing the outside of the cell. The protein surface is a complicated shape containing hollows, ravines, and ridges, and somewhere amidst this complicated geography there is an area which has the correct shape to accept the incoming messenger. This area is known as the binding site and is analogous to the active site of an enzyme. When the chemical messenger fits into this site, it 'switches on' the receptor molecule and a message is received (Fig. 5.5).

However, there is an important difference between enzymes and receptors in that the chemical messenger does not undergo a chemical reaction. It fits into the binding site of the receptor protein, passes on its message and then leaves unchanged. If no reaction takes place, what *has* happened? How does the chemical messenger tell the receptor its message and how is this message conveyed to the cell?

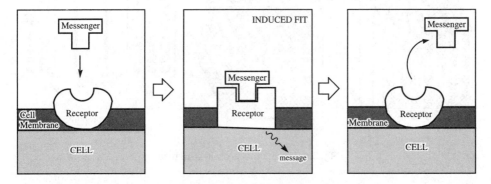

Fig. 5.5 Binding of a messenger to a receptor.

5.4 How does the message get received?

It all has to do with shape. Put simply, the messenger binds to the receptor and induces it to change shape. This change subsequently affects other components of the cell membrane and leads to a biological effect.

There are two main components involved:

1. ion channels
2. membrane-bound enzymes.

5.4.1 Ion channels and their control

Some neurotransmitters operate by controlling ion channels. What are these ion channels and why are they necessary? Let us look again at the structure of the cell membrane.

As described in Chapter 2, the membrane is made up of a bilayer of fatty molecules, and so the middle of the cell membrane is 'fatty' and hydrophobic. Such a barrier makes it difficult for polar molecules or ions to move in or out of the cell. Yet it is important that these species should cross. For example, the movement of sodium and potassium ions across the membrane is crucial to the function of nerves (see Appendix 2). It seems an intractable problem, but once again the ubiquitous proteins come up with the answer by forming ion channels.

Ion channels are protein complexes (Fig. 5.6) which traverse the cell membrane and consist of several protein subunits. The centre of the complex is hollow and is lined with polar amino acids to give a hydrophilic pore.

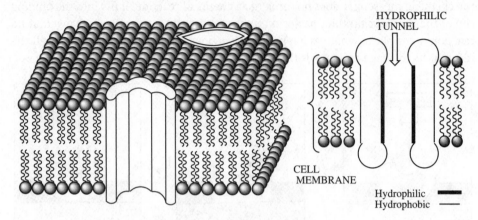

Fig. 5.6 Ion channel protein structure.

Ions can now cross the fatty barrier of the cell membrane by moving through these hydrophilic channels or tunnels. But there has to be some control. In other words, there has to be a 'lock-gate' which can be opened or closed as required. It makes sense that this lock-gate should be controlled by a receptor protein sensitive to an external chemical messenger and this is exactly what happens with many ion channels[2]. In fact, the receptor protein is an integral part of the ion channel complex and is one of the constituent protein subunits. In the resting state, the ion channel is closed (i.e. the lock-gate is shut). However, when a chemical messenger binds to the external binding site of the receptor protein, the receptor protein changes shape. This in turn causes the protein complex to change shape, opening up the lock-gate and allowing ions to pass through the ion channel (Fig. 5.7).

It is worth emphasizing one important point at this stage. If the messenger is to make the receptor protein change shape, then the binding site cannot be a 'negative image' of the messenger molecule. This must be true, otherwise how could a change of shape result? Therefore, when the binding site receives its messenger it is 'moulded' into the correct shape for the ideal fit. (This theory of an induced fit has already been described in Chapter 4 to explain the interaction between enzymes and their substrates.)

The operation of an ion channel explains why the relatively small number of neurotransmitter molecules released by a nerve is able to have such a significant biological effect on the target cell. By opening a few ion channels, several thousand ions are mobilized for each neurotransmitter molecule involved.

---

[2] Ion channels controlled by chemical messengers are called ligand-gated ion channels. Ion channels in excitable cells (e.g. nerves) are controlled by the membrane potential of the cell and are called voltage-gated ion channels (see Appendix 2).

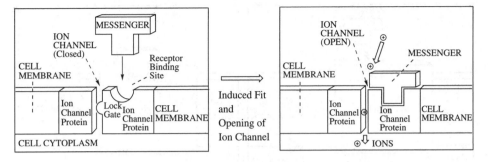

Fig. 5.7 Lock-gate mechanism for opening ion channels.

5.4.2 Membrane-bound enzymes – activation

The second way in which a receptor can pass on a message from a chemical messenger to the cell is if the receptor also doubles up as an enzyme (Fig. 5.8). In this situation, the receptor protein is embedded within the cell membrane, with part of its structure exposed on the outer surface of the cell and part of its structure exposed on the inner surface of the membrane. The outer surface contains the binding site for the chemical messenger while the inner surface has an active site which is closed in the resting state. When a chemical messenger binds to the receptor, it causes a change in shape which opens up the active site and leads to a chemical reaction within the cell. As long as the messenger is bound, the active site remains open and so a single neurotransmitter molecule can amplify its signal by binding long enough for several reactions to take place.

Some receptors control the activity of enzymes in an indirect manner, through the use of a protein messenger called a G-protein (Fig. 5.9). Once again, the receptor is embedded within the membrane, with the binding site for the chemical messenger

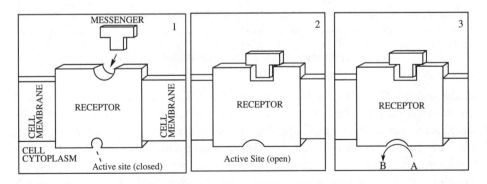

Fig. 5.8 Enzyme activation.

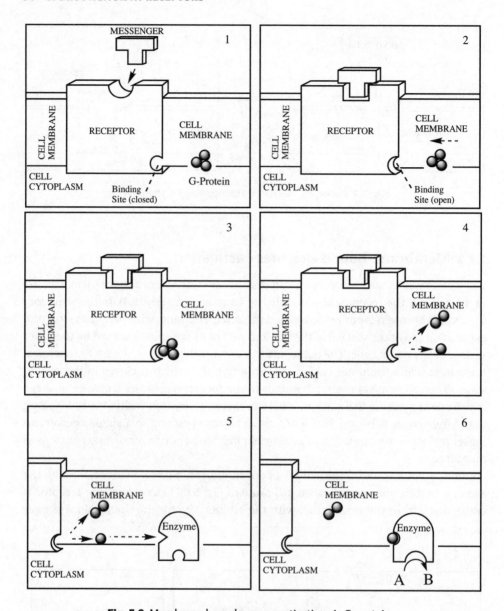

Fig. 5.9 Membrane-bound enzyme activation via G-proteins.

on the outer surface. On the inner surface, there is another binding site which is normally closed (1). When the chemical messenger binds to its binding site, the receptor protein changes shape, opening up the binding site on the inner surface (2). This new binding site is recognized by the G-protein which then binds to it (3). The G-protein is made up of three protein subunits but once it binds to the receptor, the complex is destabilized and fragments (4). One of the subunits then travels through the membrane to a membrane-bound enzyme and binds to an allosteric binding site

(5). This causes the enzyme to alter shape, resulting in the opening of an active site and the start of a new reaction within the cell (6).

There are several different G-proteins which are recognized by different types of receptor. Some of these G-proteins can have an inhibitory effect on a membrane-bound enzyme rather than a stimulatory effect. Nevertheless, the mechanism is much the same – the only difference being that the binding of the G-protein subunit to the enzyme closes the active site.

At first sight, G-proteins may be considered to be unnecessary middle men in the activation or deactivation of enzymes. However, they do have distinct advantages. For a start, each receptor will fragment several G-proteins for as long as the messenger molecule is bound to it. This results in an amplification of the signal initiated by the messenger. This is on top of any amplification resulting from activation of the enzyme. Secondly, the existence of G-protein messengers allows different receptors to control the same membrane-bound enzyme. In other words, a receptor which interacts with one type of chemical messenger can activate a specific enzyme by involving one type of G-protein, while another receptor which interacts with a different chemical messenger could deactivate the same enzyme by involving a different type of G-protein. In this way, the same enzyme is under dual control. This is more efficient than having the enzyme specifically linked to each different type of receptor.

Regardless of the specific mechanism involved in the activation or deactivation of an enzyme, the overall result is the same. A change in receptor shape (or tertiary structure) leads eventually to the activation (or deactivation) of enzymes. Since enzymes can catalyse the reaction of a large number of molecules, there is an amplification of the signal.

To conclude, the mechanisms by which neurotransmitters pass on their message involve changes in molecular shape rather than chemical reactions. These changes in shape ultimately lead to some sort of chemical reaction involving enzymes. This in turn results in an amplification of the original message such that a relatively small number of neurotransmitter molecules can lead to a biological result. This topic is covered more fully in Chapter 6.

We come now to a question which has so far been avoided.

5.5 How does a receptor change shape?

We have seen already that it is the messenger molecule which induces the receptor to change shape, but how does it do that? It is not simply, as one might think, a moulding process whereby the receptor wraps itself around the shape of the messenger molecule. The answer lies rather in specific binding interactions between messenger and receptor. These are the same interactions that have already been described in Chapter 4 for enzyme–substrate binding, i.e. ionic bonding, hydrogen bonding, van der Waals interactions, dipole–dipole interactions, and induced dipole interactions. The messenger and receptor proteins both take up conformations or shapes to maximize these bonding forces. As with enzyme–substrate binding, there is a fine balance involved in receptor–messenger binding. The bonding forces must be large enough to change the shape of the receptor in the first place, but not so strong that the messenger is unable to leave again. Most neurotransmitters bind quickly to their receptors, then 'shake themselves loose again' as soon as the message has been received.

As an example of the various binding forces involved, let us consider a hypothetical neurotransmitter and receptor as shown in Fig. 5.10. The neurotransmitter has an aromatic ring which can take part in van der Waals interactions, an alcoholic group which can take part in hydrogen bonding interactions, and a charged nitrogen centre which can take part in ionic or electrostatic interactions.

The hypothetical receptor protein has three complimentary binding groups and is part of an ion channel. When the neurotransmitter is absent, the receptor is shaped such that it seals the ion channel (Fig. 5.11).

Since the binding site has complementary binding groups for the groups described above, the drug can fit into the binding site and bind strongly. This is all very well, but now that it has docked, how does it make the receptor change shape? As before, we have to surmise that the fit is not quite exact or else there would be no reason for the receptor to change shape. In this example, we can envisage our messenger molecule fitting into the binding site and binding well to two of the three possible binding groups. The third binding group, however (the ionic one), is not quite in the right position (Fig. 5.12). It is close enough to have a weak interaction, but not close enough for the optimum interaction. The receptor protein is therefore forced to alter shape to obtain the best binding interaction. The carboxylate group is pulled closer to the positively charged nitrogen on the messenger molecule and, as a result, the lock-gate is opened and will remain open until the messenger molecule detaches from the binding site, allowing the receptor to return to its original shape.

In reality, the conformational changes involved in the opening of an ion channel are quite complex and the 'lock-gate' is at some distance from the receptor binding site. Nevertheless, the same principles hold. Binding of the chemical messenger induces a

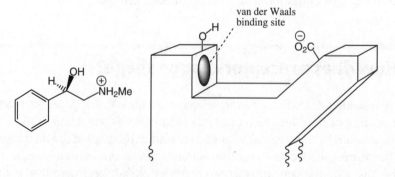

Fig. 5.10 A hypothetical neurotransmitter and receptor.

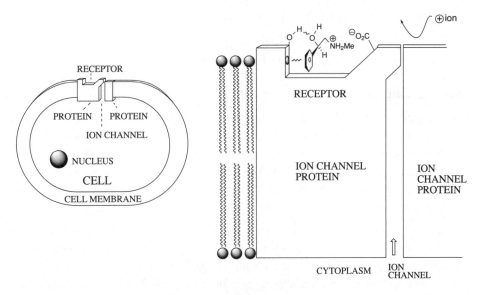

Fig. 5.11 Receptor protein positioned in the cell membrane.

conformational change which has a whole series of knock-on effects, resulting in the final opening of the ion channel. We shall look at this again in Chapter 6.

5.6 The design of agonists

We are now at the stage of understanding how drugs might be designed such that they mimic natural neurotransmitters. Assuming that we know what binding groups are present in the receptor site and where they are, drug molecules can be designed to interact with the receptor. Let us look at this more closely and consider the following requirements in turn:

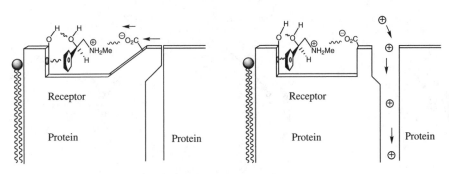

Fig. 5.12 Opening of the 'lock-gate'.

1. The drug must have the correct binding groups.

2. The drug must have these binding groups correctly positioned.

3. The drug must be the right size for the binding site.

5.6.1 Binding groups

If we consider our hypothetical receptor and its natural neurotransmitter, then we might reasonably predict which of a series of molecules would interact with the receptor and which would not. For example, consider the structures in Fig. 5.13. They all look different, but they all contain the necessary binding groups to interact with the receptor. Therefore, they may well be potential agonists or alternatives for the natural neurotransmitter.

However, the structures in Fig. 5.14 lack one or more of the required binding groups and might therefore be expected to have poor activity. We would expect them to drift into the receptor site and then drift back out again, binding only weakly, if at all.

Of course, we are making an assumption here; that all three binding groups are essential. It might be argued that a compound such as structure II in Fig. 5.14 might be effective even though it lacks a suitable hydrogen bonding group. Why, for example, could it not bind initially by van der Waals forces alone and then alter the shape of the receptor protein via ionic bonding?

In fact, this seems unlikely when we consider that neurotransmitters appear to bind, pass on their message and then leave the binding site very quickly. In order to do that, there must be a fine balance in the binding interactions between receptor and neurotransmitter. They must be strong enough to bind the neurotransmitter effectively such that the receptor changes shape. However, the binding interactions cannot be too strong, or else the neurotransmitter would not be able to leave and the receptor would not be able to return to its original shape. Therefore, it is reasonable to assume that a neurotransmitter needs all of its binding interactions to be effective. The lack of even one of these interactions would lead to a significant loss in activity.

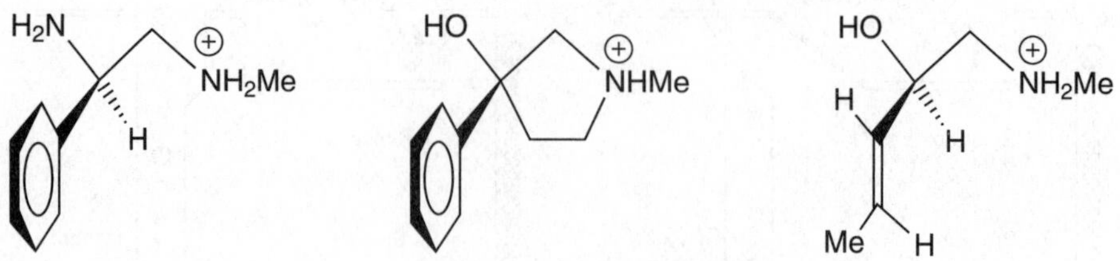

Fig. 5.13 Possible agonists.

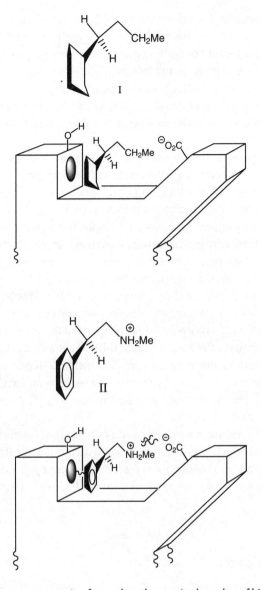

Fig. 5.14 Structures possessing fewer than the required number of binding groups.

5.6.2 Position of binding groups

The molecule may have the correct binding groups, but if they are in the wrong positions, they will not all be able to form bonds at the same time. As a result, bonding would be weak and the molecule would very quickly drift away. Result – no activity.

A molecule such as the one shown in Fig. 5.15 obviously has its binding groups in the wrong position, but there are more subtle examples of molecules which do not have the correct arrangement of binding groups. For example, the mirror image of our hypothetical neurotransmitter would not fit (Fig. 5.16). The structure has the same formula and the same constitutional structure as our original structure. It will have the same physical properties and undergo the same chemical reactions, but it is not the same shape. It is a non-superimposable mirror image and it cannot fit the receptor binding site (Fig. 5.17).

Compounds which have non-superimposable mirror images are termed chiral or asymmetric. There are only two detectable differences between the two mirror images (or enantiomers) of a chiral compound. They rotate plane-polarized light in opposite directions and they interact differently with other chiral systems such as enzymes. This has very important consequences for the pharmaceutical industry.

Pharmaceutical agents are usually synthesized from simple starting materials using simple achiral (symmetrical) chemical reagents. These reagents are incapable of distinguishing between the two mirror images of a chiral compound. As a result, most chiral drugs are synthesized as a mixture of both mirror images (a racemate). However, we have seen from our own simple example that only one of these enantiomers will interact with a receptor. What happens to the other enantiomer?

At best, it floats about in the body doing nothing. At worst, it interacts with a totally different receptor and results in an undesired side-effect. Herein lies the explanation for the thalidomide tragedy. One of the enantiomers was an excellent sedative. The other reacted elsewhere in the body as a poison and was teratogenic (i.e. it induced abnormalities in human embryos).

Even if the 'wrong' enantiomer does not do any harm, it seems to be a great waste of time, money and effort to synthesize drugs which are only 50% efficient. That is why one of the biggest areas of chemical research in recent years has been in the field of asymmetric synthesis – the selective synthesis of a single enantiomer of a chiral compound.

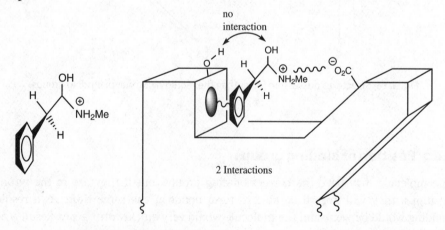

Fig. 5.15 Molecule with binding groups in incorrect positions.

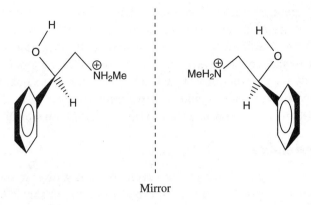

Fig. 5.16 Mirror image of hypothetical neurotransmitter.

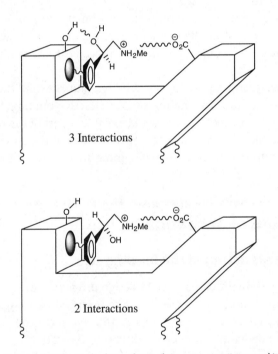

Fig. 5.17 Interactions between the hypothetical neurotransmitter and its mirror image with the receptor site.

Of course, nature has been at it for millions of years. Since nature has chosen to work predominantly with the 'left-handed' enantiomer of amino acids[3], enzymes (made up of left-handed amino acids) are also present as single enantiomers and therefore catalyse enantiospecific reactions – reactions which give only one enantiomer.

The importance of having binding groups in the correct position has led medicinal chemists to design drugs based on what is considered to be the important **pharmacophore** of the messenger molecule. In this approach, it is assumed that the correct positioning of the binding groups is what decides whether the drug will act as a messenger or not, and that the rest of the molecule serves only to hold the groups in these positions. Therefore, the activity of apparently disparate structures at a receptor can be explained if they all contain the correct binding groups at the correct positions. Totally novel structures or molecular frameworks to hold these binding groups in the correct positions could then be designed, leading to a new series of drugs. There is, however, a limiting factor to this which will now be discussed.

5.6.3 Size and shape

It is possible for a compound to have the correct binding groups in the correct positions and yet fail to interact effectively if it is the wrong size or shape. As an example, let us consider the structure shown in Fig. 5.18 as a possible candidate for our hypothetical receptor system.

The structure has a *meta*-methyl group on the aromatic ring and a long alkyl chain attached to the nitrogen atom. By considering size factors alone, we could conclude that both these features would prevent this molecule from binding effectively to the receptor.

The *meta*-methyl group would act as a buffer and prevent the structure from 'sinking' deep enough into the binding site for effective binding. Furthermore, the long alkyl chain on the nitrogen atom would make that part of the molecule too long for the space available to it. A thorough understanding of the space available in the binding site is therefore necessary when designing analogues to fit it.

5.7 The design of antagonists

5.7.1 Antagonists acting at the binding site

We have seen how it might be possible to design drugs to mimic the natural neurotransmitters (agonists) and how these would be useful in treating a shortage of the natural neurotransmitter. However, suppose that we have too much neurotransmitter operating in the body. How could a drug counteract that?

There are several strategies, but in theory we could design a drug (an antagonist) which would be the right shape to bind to the receptor site, but which would either fail to change the shape of the receptor protein or would distort it too much. Consider the following scenario.

---

[3] Naturally occurring amino acids exist mainly as the one enantiomer, termed the L-enantiomer. This terminology is historical and defines the absolute configuration of the asymmetric carbon present at the 'head group' of the amino acid. The current terminology for asymmetric centres is to define them as *R* or *S* according to a set of rules known as the Cahn–Ingold–Prelog rules. Naturally occurring amino acids exist as the (*S*)-configuration, but the older terminology still dominates in the case of amino acids. Experimentally, the L-amino acids were found to rotate plane-polarized light anticlockwise or to the left. Hence the expression left-handed amino acids.

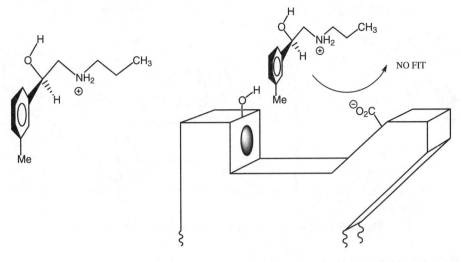

Fig. 5.18 Structure with a *meta* methyl group.

The compound shown in Fig. 5.19 fits the binding site perfectly and, as a result, does not cause any change of shape. Therefore, there is no biological effect and the binding site is blocked to the natural neurotransmitter. In such a situation, the antagonist has to compete with the agonist for the receptor, but usually the antagonist will get the better of this contest since it often binds more strongly.

To sum up, if we know the shape and make-up of receptor binding sites, then we should be able to design drugs to act as agonists or antagonists. Unfortunately, it is not as straightforward as it sounds. Finding the receptor and determining the layout of its binding site is no easy task. In reality, the theoretical shapes of many receptor sites have been worked out by synthesizing a large number of compounds and considering those molecules which fit and those which do not – a bit like a 3D jigsaw.

However, the advent of computer-based molecular modelling and the availability of X-ray crystallographic data now allow a more accurate representation of proteins and their binding sites (see Chapter 13) and have heralded a new phase of drug development.

5.7.2 Antagonists acting outwith the binding site

Even if the 'layout' of a binding site is known, it may not help in the design of antagonists. There are many examples of antagonists which bear no apparent structural similarity to the native neurotransmitter and could not possibly fit the geometrical requirements of the binding site. Such antagonists frequently contain one or more aromatic rings, suggesting van der Waals interactions are important in their binding, yet there may not be any corresponding area in the binding site. How then do such antagonists work? There are two possible explanations.

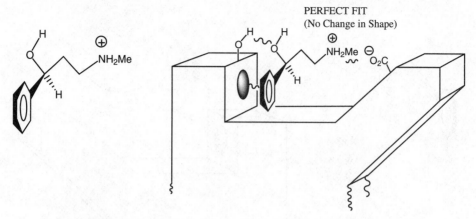

Fig. 5.19 Compound acting as an antagonist at the binding site.

Allosteric antagonists

The antagonist may bind to a totally different part of the receptor. The process of binding could alter the shape of the receptor protein such that the neurotransmitter binding site is distorted and is unable to recognize the natural neurotransmitter (Fig. 5.20). Therefore, binding between neurotransmitter and receptor would be prevented and the message lost. This form of antagonism is non-competitive since the antagonist is not competing with the neurotransmitter for the same binding site (compare allosteric inhibitors of enzymes: see Chapter 4).

Antagonism by the 'umbrella' effect

It must be remembered that the receptor protein is bristling with amino acid residues, all of which are capable of interacting with a visiting molecule. Therefore, it is unrealistic to think of the neurotransmitter binding site as an isolated 'landing pad', surrounded by a bland, featureless 'no go zone'. There will almost certainly be areas close to the binding site which are capable of binding through van der Waals, ionic, or hydrogen bonding interactions.

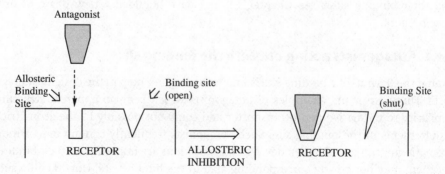

Fig. 5.20 Allosteric antagonists.

These areas may not be used by the natural neurotransmitter, but they can be used by other molecules. If these molecules bind to such areas and happen to lie over, or partially lie over, the neurotransmitter binding site, then they will act as antagonists and prevent the neurotransmitter reaching its binding site (Fig. 5.21). This form of antagonism has also been dubbed the 'umbrella' effect and is a form of competitive antagonism since the normal binding site is directly affected.

Many antagonists are capable of binding to both the normal binding site and the neighbouring sites. Such antagonists will clearly bind far more strongly than agonists. Because of this stronger binding, antagonists have been useful in the isolation and identification of specific receptors present in tissues. A further tactic in this respect is to incorporate a highly reactive chemical group – usually an electrophilic group – into the structure of a powerful antagonist. The electrophilic group will then react with any convenient nucleophilic group on the receptor surface and alkylate it to form a strong covalent bond. The antagonist will then be irreversibly tied to the receptor and can act as a molecular label. One example is tritium-labelled propylbenzilylcholine mustard – used to label the muscarinic acetylcholine receptor (Fig. 5.22) (see also Chapter 15).

5.8 Partial agonists

Frequently a drug is discovered which cannot be defined as a pure antagonist or a pure agonist. The compound acts as an agonist and produces a biological effect, but that effect is not as great as would be obtained with a full agonist. Therefore the compound is called a **partial agonist**. There are several possible explanations for this.

Clearly, a partial agonist must bind to the receptor to have an agonist effect. However, it may be binding in such a way that the conformational change induced is not ideal and the subsequent effects of receptor activation are decreased. In

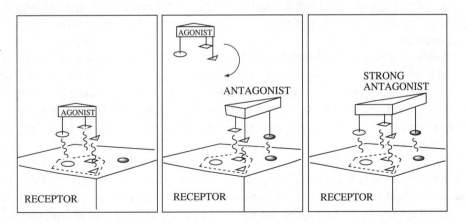

Fig. 5.21 Antagonism by the 'umbrella effect'.

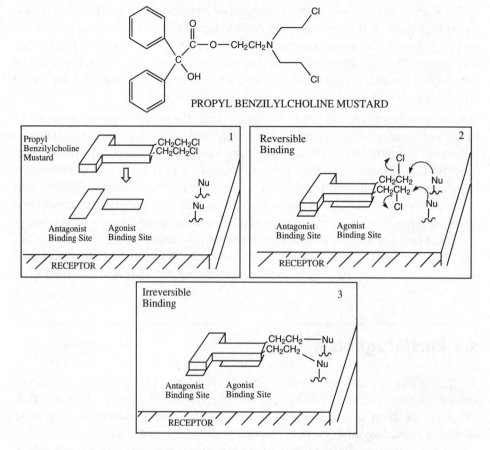

PROPYL BENZILYLCHOLINE MUSTARD

Fig. 5.22 Antagonist used as a molecular label.

our hypothetical situation (Fig. 5.23), we could imagine a partial agonist being a molecule which is almost a perfect fit for the binding site, such that binding results in only a very slight distortion of the receptor. This would then only partly open the ion channel.

Another explanation for partial agonism is that the molecule in question might be capable of binding to a receptor in two different ways by using different binding regions in the binding site. One method of binding would activate the receptor (agonist effect), while the other would not (antagonist effect). The balance of agonism versus antagonism would then depend on the relative proportions of molecules binding by either method (see Section 5.9).

Thirdly, the partial agonist in question may be capable of distinguishing between different types of receptor. In Chapter 6, we shall see that receptors which bind the same chemical messenger are not all the same. There are various types and subtypes. A partial agonist might distinguish between these, acting as an agonist at one subtype, but as an antagonist at another subtype.

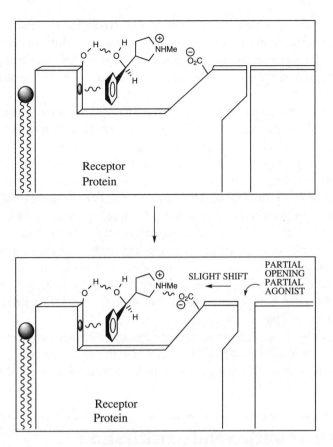

Fig. 5.23 Partial agonism.

Examples of partial agonists in the opiate and antihistamine fields are discussed in Chapters 17 and 18, respectively.

5.9 Inverse agonists

Many antagonists which bind to a receptor binding site are in fact more properly defined as inverse agonists. An inverse agonist has the same effect as an antagonist in that it binds to a receptor, fails to activate it and prevents the normal chemical messenger from binding. However, there is more to an inverse agonist than that. Some receptors (e.g. the GABA (γ-aminobutyric acid) and dihydropyridine receptors) are found to have an inherent activity even in the absence of the chemical messenger. An inverse agonist is capable of preventing this activity as well.

The discovery that some receptors have an inherent activity has important implications in receptor theory. It suggests that these receptors do not have a 'fixed' inactive conformation, but are continually changing shape such that there is an equilibrium between the active conformation and different inactive conformations. In that equilibrium, most of the receptor population is in an inactive conformation but a small proportion of the receptors are in the active conformation. The action of agonists and antagonists is then explained by how that equilibrium is affected by binding preferences (Fig. 5.24).

If an agonist is introduced (B), it binds preferentially to the active conformation and stabilizes it, shifting the equilibrium to the active conformation and leading to an increase in the biological activity associated with the receptor.

In contrast, it is proposed that an antagonist binds equally well to all receptor conformations (both active and inactive) (C). There is no change in biological activity since the equilibrium is unaffected. The introduction of an agonist has no effect either since all the receptor binding sites are already occupied by the antagonist. Antagonists such as these will have some structural similarity to the natural agonist.

An inverse agonist is proposed to have a binding preference for an inactive conformation. This stabilizes the inactive conformation and shifts the equilibrium away from the active conformation, leading to a drop in inherent biological activity (D). An inverse agonist need have no structural similarity to an agonist since it could be binding in a different part of the receptor.

A partial agonist has a slight preference for the active conformation over any of the inactive conformations. The equilibrium is shifted to the active conformation but not to the same extent as with a full agonist and so the increase in biological activity is less (E).

5.10 Desensitization and sensitization

Some drugs bind relatively strongly to a receptor, switch it on, but then subsequently block the receptor after a certain time period. Thus, they are acting as agonists, then antagonists. The mechanism of how this takes place is not clear, but it is believed that prolonged binding of the agonist to the receptor results in phosphorylation of hydroxyl or phenolic groups in the receptor. This causes the receptor to alter shape to an inactive conformation, despite the binding site being occupied (Fig. 5.25). This altered tertiary structure is then maintained as long as the binding site is occupied by the agonist. When the drug eventually leaves, the receptor is dephosphorylated and returns to its original resting shape.

On even longer exposure to a drug, the receptor–drug complex may be removed completely from the cell membrane by a process called endocytosis. Here, the relevant portion of the membrane is 'nipped out', absorbed into the cell and metabolized.

Prolonged activation of a receptor can also result in the cell reducing its synthesis of the receptor protein.

Consequently, it is generally true that the best agonists bind swiftly to the receptor, pass on their message and then leave quickly. Antagonists, in contrast, tend to be slow to add and slow to leave.

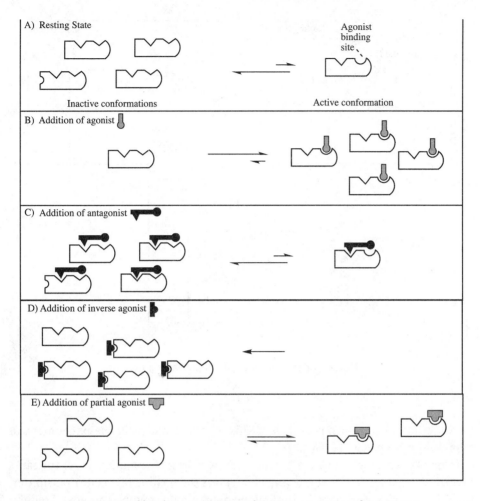

A) Resting State

Agonist binding site

Inactive conformations

Active conformation

B) Addition of agonist

C) Addition of antagonist

D) Addition of inverse agonist

E) Addition of partial agonist

Fig. 5.24 Equilibria between active and inactive receptor conformations.

Prolonged exposure of a target receptor to an antagonist may lead to the opposite of desensitization (i.e. sensitization). This is where the cell synthesizes more receptors to compensate for the receptors which are blocked. This is known to happen when some β-blockers are given over long periods (see Chapter 16).

5.11 Tolerance and dependence

As mentioned above, 'starving' a target receptor of its neurotransmitter may induce that cell to synthesize more receptors. By doing so, the cell gains a greater sensitivity

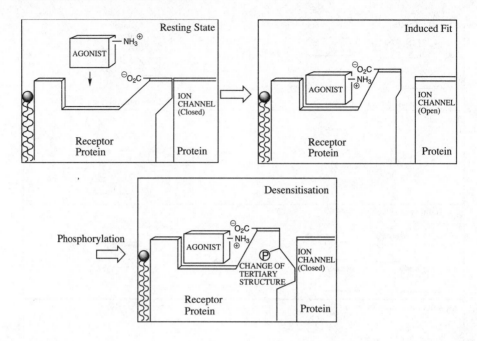

Fig. 5.25 Desensitization, ℗ = phosphate group

for what little neurotransmitter is left. This process can explain the phenomena of tolerance and dependence (Fig. 5.26).

Tolerance is a situation where higher levels of a drug are required to get the same biological response. If a drug is acting to suppress the binding of a neurotransmitter, then the cell may respond by increasing the number of receptors. This would require increasing the dose to regain the same level of antagonism.

If the drug is suddenly stopped, then all the receptors suddenly become available. There is now an excess of receptors which would make the cell supersensitive to normal levels of neurotransmitter. This would be equivalent to receiving a drug over-dose. The resulting biological effects would explain the distressing withdrawal symptoms which follow from stopping certain drugs. These withdrawal symptoms would continue until the number of receptors returned to their original level. During this period, the patient may be tempted to take the drug again in order to 'return to normal' and will have then gained a dependence on the drug. The problem of tolerance is also discussed in Chapter 6.

It is only in recent years that medicinal chemists have begun to understand receptors and drug–receptor interactions and this increased understanding has revolutionized the approach to medicinal chemistry.

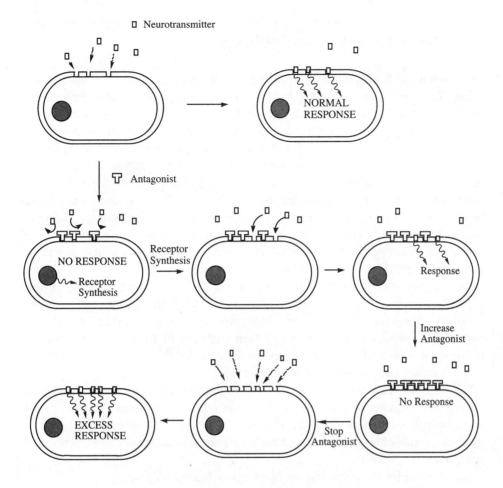

Fig. 5.26 Process of increasing cell sensitivity.

5.12 Cytoplasmic receptors

Not all receptors are located in the cell membrane. Some receptors are within the cell and their chemical messengers are required to pass through the cell membrane to reach them. Such messengers must be fatty in character to do this. The steroid hormones are examples of chemical messengers which have to cross the cell membrane to reach their target receptor (e.g. the oestrogen receptor) (See also Section 6.8).

5.13 Receptor types and subtypes

Receptors are identified by the specific neurotransmitter or hormone which activates them. Thus, the receptor activated by dopamine is called the dopaminergic receptor. The receptor activated by acetylcholine is called the cholinergic receptor, and the receptor activated by adrenaline or noradrenaline is called the adrenergic receptor.

However, not all receptors activated by the same chemical messenger are exactly the same throughout the body. For example, the adrenergic receptors in the lungs are slightly different from the adrenergic receptors in the heart. These differences arise from slight variations in amino acid composition and, if the variations are in the binding site, allow medicinal chemists to design drugs which can distinguish between them. For example, adrenergic drugs can be designed to be lung- or heart-specific. In general, there are various **types** of a particular receptor and various **subtypes** of these. They are normally identified by numbers or letters. However, for historical reasons the main types of cholinergic receptor are named after natural products (see Chapter 15). Some examples of receptor types and subtypes are given in Table 5.1.

The identification of many of these subtypes is relatively recent and the current emphasis in medicinal chemistry is to design drugs which are as specific as possible for receptor types and subtypes so that the drugs are as tissue-specific as possible in order to reduce the possibility of side-effects.

For example, there are five types of dopaminergic receptor. All clinically effective antipsychotic agents (e.g. clozapine, olanzapine and risperidone) antagonize the dopaminergic receptors D2 and D3. However, blockade of D2 receptors may lead to some of the side-effects observed and a selective D3 antagonist may have better properties as an antipsychotic.

Other examples of current research projects include the following:

1. Muscarinic (M_2) agonists for the treatment of heart irregularities.

2. Adrenergic (β_3) agonists for the treatment of obesity.

3. NMDA[4] antagonists for the treatment of stroke.

4. Cannabinoid (CB1) antagonists for treatment of memory loss.

---

[4] NMDA = N-Methyl-D-aspartate

Table 5.1 Some examples of receptor types and subtypes.

Receptor	Type	Subtype	Examples of agonist therapies	Examples of antagonist therapies
Cholinergic	Nicotinic (N) Muscarinic (M)	Nicotinic (four subtypes) M_1–M_5	Stimulation of GI-tract motility (M_1) Glaucoma (M)	Neuromuscular blockers and muscle relaxant (N) Peptic ulcers (M_1) Motion sickness (M)
Adrenergic (Adreno-receptors)	Alpha (α_1, α_2) Beta (β)	α_{1A}, α_{1B}, α_{1D} α_{2A}– α_{2C} (β_1, β_2, β_3)	Anti-asthmatics (β_2)	β blockers (β_1)
Dopamine		D1, D2, D3, D4, D5	Parkinson's disease	Anti-depressant (D2, D3)
Histamine		H_1–H_3	Vasodilatation (limited use)	Treatment of allergies, anti-emetics, sedation (H_1) Antiulcer (H_2)
Opioid and opioid-like		μ, κ, δ, ORL1	Analgesics (κ)	Antidote to morphine overdose
5-hydroxy-tryptamine (serotonin)	5-HT_1–5-HT_7	5-HT_{1A}, 5-HT_{1B}, 5-HT_{1D-1F} 5-HT_{2A-2C} 5-HT_{5A} 5-HT_{5B}	Antimigraine (5-HT_{1D}) Stimulation of GI tract motility (5-HT_4)	Anti-emetics (5-HT_3) Ketanserin (5-HT_2)
Oestrogen			Contraception	Breast cancer

6 Receptor structure and signal transduction

In Chapter 5, we discussed how chemical messengers such as neurotransmitters and hormones (ligands) interact with receptor proteins to change the tertiary structure of these proteins. This change of structure can result in the opening or closing of ion channels, or the activation or deactivation of enzymes, both of which lead to a change in chemistry within the cell. In Chapter 5, this was represented as a fairly simple process. In reality, the interaction of a receptor with its chemical messenger is the first step in a complex chain of events which involves several secondary messengers, proteins and enzymes. In this chapter, we shall look more closely at receptor structure and the events which occur after receptor–ligand binding. These events are referred to as **signal transduction**.

6.1 Receptor families

There are a large number of different membrane-bound receptors, but they can be grouped into three 'superfamilies', based on their structure and function.

1. Ion channel receptors (2-TM, 3-TM, and 4-TM receptors).[1]

2. G-protein-coupled receptors (7-TM receptors).

3. Kinase-linked receptors (1-TM receptors).

The structure and signalling pathways of these superfamilies differ as described in the following sections. They also differ in their speed of response. The signalling process involving ion channels is a rapid response which is measured in a matter of milliseconds. These are the receptors involved in the rapid synaptic transmission

---

[1] TM stands for transmembrane. Thus, 2-TM indicates that the proteins involved in these receptors contain two segments which traverse the cell membrane.

between nerves. The signals arising from G-protein-coupled receptors result in a response measured in seconds while the kinase-linked receptors lead to responses in a matter of minutes. There is a fourth superfamily of intracellular receptors which are not membrane-bound and which lead to responses in a matter of hours or days. These are the receptors in the cytoplasm which interact with steroid hormones.

6.2 Receptors that control ion channels (ligand-gated ion channel receptors)

Receptors which control ion channels are part of a five-protein ion channel structure (Fig. 6.1) and are called ligand-gated ion channel receptors. They are glycoproteins which traverse the cell membrane and the fact that the receptor is part of the ion channel structure means that binding of a chemical messenger leads to the rapid response which is crucial to the speed and efficiency of nerve transmission. When the receptor binds a ligand, it changes shape and this has a knock-on effect on the protein complex which causes the ion channel to open – a process called **gating**.

Ion channels can select between different ions. There are cationic ion channels for Na^+, K^+, and Ca^{++} ions. When these channels are open, they are generally excitatory and lead to depolarization of the cell (see Appendix 2). Some of the receptors involved in the control of cationic ion channels include nicotinic receptors (see Chapter 15), glutamate receptors, $5HT_3$ receptors and ATP receptors. There are also anionic ion channels for the chloride ion and these ion channels are controlled by the $GABA_A$ and glycine receptors. These ion channels generally have an inhibitory effect when they are open.

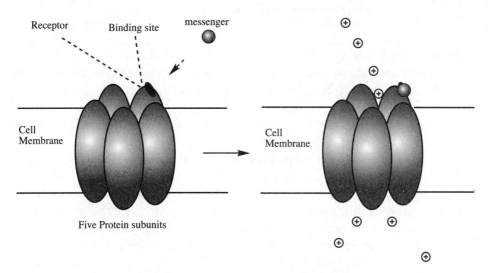

Fig. 6.1 Opening of an ion channel – gating.

The ion selectivity of different ion channels is dependent on the amino acids lining the ion channel and, interestingly enough, the mutation of one amino acid in this area is sufficient to change a cationic selective ion channel to an anionic selective channel.

There are at least three families of receptors involved in the control of ion channels. These are classified by the number of transmembrane domains (4-TM, 3-TM, 2-TM) which they contain.

6.2.1 Structure and function of 4-TM ion channel receptors

The 4-TM ion channel receptors include the nicotinic acetylcholine (nACh) receptor, the serotonin ($5HT_3$) receptor, the glycine receptor, and the $GABA_A$ receptor. The ion channel controlled by the nicotinic acetylcholine receptor is made up of four different types of subunits ($2 \times \alpha$, β, γ, δ) while the ion channel controlled by the glycine receptor is made up of two different types ($3 \times \alpha$, $2 \times \beta$) (Fig. 6.2).

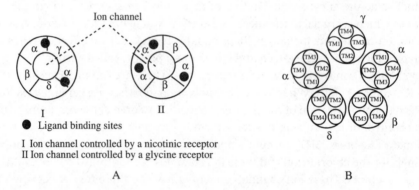

Fig. 6.2 A) Pentameric structure of ion channels (transverse view).
B) Transverse view including TM regions.

The receptor protein in the ion channel controlled by glycine is the α subunit. There are three such subunits present, all capable of interacting with glycine. However, the situation is slightly more complex in the ion channel controlled by acetylcholine. Most of the acetylcholine binding site is on the α subunit but there is some involvement from neighbouring subunits. Therefore, in this case the ion channel complex as a whole might be viewed as the receptor.

Each protein subunit in the ion channel complex has a lengthy N-terminal extracellular chain which (in the case of the α subunit) contains the ligand-binding site. There are also four hydrophobic sections which cross the membrane (transmembrane domains) and which distinguish these ion channels (4-TM). Since the protein chain crosses the membrane four times, there is one extracellular and two intracellular loops connecting these regions as well as an extracellular C-terminal chain. The transmembrane regions are numbered 1–4 from the N-terminal end and at least one of these regions is an α helix (Fig. 6.3).

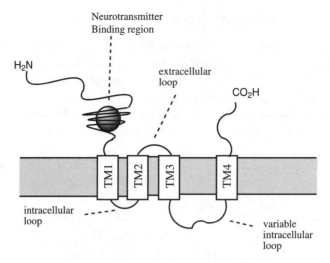

Fig 6.3 Structure of 4-TM receptor subunit.

The binding of a neurotransmitter to its binding site causes a conformational change in the receptor which eventually opens up the central pore and allows ions to flow. This conformational change is quite complex involving several knock-on effects from the initial binding process. This must be so since the binding site is quite far from the 'lock-gate'.

Studies have shown that the 'lock-gate' is made up of five 'kinked' α helices where one helix is contributed from each of the five protein subunits. In the closed state the kinks are pointing towards each other. The conformational change induced by ligand binding causes each of these helices to rotate such that the kink points the other way, thus opening up the pore (Fig. 6.4).

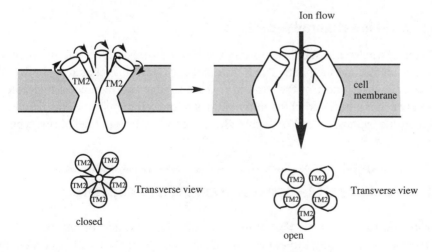

Fig. 6.4 Opening of the 'lock-gate' in the nACh receptor controlled ion channel.

6.2.2 3-TM ion channel receptors

The ion channels controlled by these receptors also consist of five protein subunits. However, in this case each protein subunit has three transmembrane segments which are labelled TM1, TM3, and TM4. This peculiar labelling is due to the presence of a hydrophobic segment which is thought to be embedded in the intracellular side of the membrane and is labelled TM2 (Fig. 6.5). The ligand binding site seems to involve both the *N*-terminal chain and the extracellular loop (regions A and B).

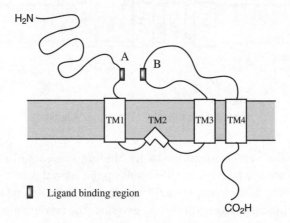

Fig. 6.5 Structure of 3-TM ion channel receptor subunits.

The calcium ion channels controlled in the central nervous system by L-glutamate are examples of this type of ion channel and are important in several processes, including learning and memory.

6.2.3 2-TM ion channel receptors

2-TM ion channels consist of five protein subunits where each of the subunits contains two transmembrane segments. The *N*-terminal and *C*-terminal chains are both inside the cell. However, most of the protein is extracellular and includes a hydrophobic region which is embedded in the outer surface of the cell membrane (Fig. 6.6).

Adenosine triphosphate (ATP) is thought to control an ion channel of this type.

6.3 Structure of G-protein-coupled receptors

G-protein-coupled receptors do not directly affect ion channels or enzymes. Instead they activate a signalling protein called a G-protein which then initiates a signalling cascade involving a variety of enzymes.

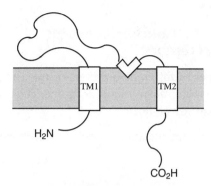

Fig. 6.6 Structure of 2-TM ion channel receptor subunits.

6.3.1 Structure of G-protein-coupled receptors

The G-protein-coupled receptors (or 7-TM receptors) are proteins which are embedded in the cell membrane, and which have regions exposed both to the outside and the inside of the cell (Fig. 6.7). The protein chain winds back and forth through the cell membrane seven times – hence the term 7-TM (transmembrane). Each of the seven transmembrane sections is hydrophobic and helical in shape, and it is usual to assign these helices with Roman numerals (I, II etc) starting from the *N*-terminal end of the protein. The winding of the protein back and forth through the membrane results in three extracellular and three intracellular loops. These loops are fairly constant in length except for the intracellular loop connecting helices V and VI which varies depending on the specific receptor. The *N*-terminal chain is to the exterior of the cell and is variable in length depending on the receptor while the *C*-terminal chain is to the interior.

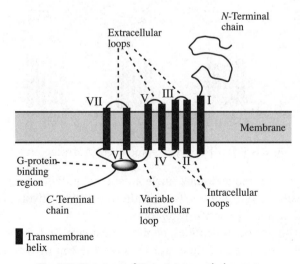

Fig. 6.7 Structure of G-protein-coupled receptors.

6.3.2 Neurotransmitters and hormones for G-protein-coupled receptors

There are a large number of different G-protein-coupled receptors, each of which interact with a different neurotransmitter or hormone. The chemical messengers can be grouped by their structure as follows:

- Monoamines (e.g. dopamine, histamine, serotonin, acetylcholine, noradrenaline)
- Nucleotides
- Lipids
- Neuropeptides, peptide hormones, and protein hormones
- Glycoprotein hormones
- Glutamate
- Calcium ions.

   In general, these receptors mediate the action of hormones and slow acting neuro-transmitters. They include the muscarinic receptor (see Chapter 15) and adrenergic receptors (see Chapter 16).

6.3.3 Ligand binding

Despite the large variety of G-protein receptors, their overall structure is remarkably similar, and it was originally thought that there might be a common binding site into which the different chemical messengers (or ligands) would fit in different ways. However, it is now known that the different structural ligand groups fit their specific receptors in different areas, as shown in Fig. 6.8.

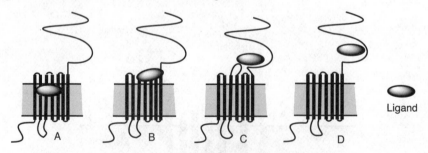

Fig. 6.8 Ligand binding to G-protein-coupled receptors.

   A) The binding site for monoamines, nucleotides and lipids is a deep binding pocket between the transmembrane helices. The specific interactions involved depend on the nature of the ligand and the specific receptor.

   B) The binding site for neuropeptides, peptide hormones, and chemokines is closer to the surface of the cell membrane and involves the top of the transmembrane helices, some of the extracellular loops and the N-terminal chain.

C) The binding site for glycoprotein hormones and some peptide hormones is mostly made up by the region of the *N*-terminal chain closest to the surface of the cell membrane, along with some regions of the extracellular loops.

D) The binding site for glutamate and calcium ions is made up entirely by the *N*-terminal chain above the surface of the cell membrane.

A curious case is the thrombin receptor, which has a 'hidden' intrinsic ligand already present in the *N*-terminal chain. This is revealed when thrombin (an enzyme) binds and cleaves off part of the *N*-terminal chain. This reveals a pentapeptide section of the remaining *N*-terminal chain which is then able to bind to the binding site in the extracellular loops of the receptor, resulting in receptor activation (Fig. 6.9).

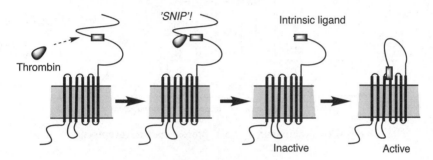

Fig. 6.9 Action of thrombin.

Another receptor in this superfamily which can be viewed as having an inbuilt agonist is the visual receptor rhodopsin – the most studied of the 7-TM receptors. This receptor contains *trans* retinal in a deep pocket. The receptor is activated when a photon of light isomerizes retinal from the *trans* isomer to the *cis* isomer.

6.3.4 The rhodopsin-like family of G-protein-coupled receptors

The G-protein-coupled receptors include the receptors for some of the best known chemical messengers in medicinal chemistry (e.g. glutamic acid, GABA, noradrenaline, dopamine, acetylcholine, serotonin, prostaglandins, adenosine, endogenous opiates, angiotensin, bradykinin, and thrombin). Considering the structural variety of the chemical messengers involved, it is remarkable that the overall structures of the G-protein-coupled receptors are so similar. Nevertheless, despite the similar overall structure, the amino acid sequences of the receptors vary quite significantly. This implies that these receptors have evolved over many years from an ancient common ancestral protein. Comparing the amino acid sequences of the receptors allows us to construct an evolutionary tree and to group the receptors of this superfamily into various families. The most important of these as far as medicinal chemistry is concerned is the rhodopsin-like family – so called because the first receptor of this family to be studied in detail was the rhodopsin receptor itself, a receptor involved in the visual process.

This family includes some of the most well studied receptors in medicinal chemistry (e.g. the cholinergic, dopaminergic, and adrenergic receptors) . A study of the evolutionary tree of rhodopsin-like receptors throws up some interesting observations (Fig. 6.10).

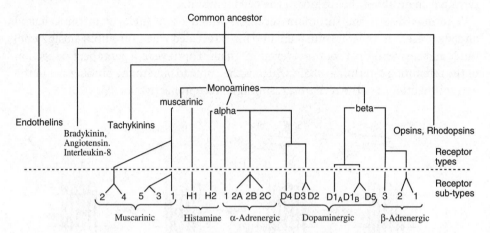

Fig. 6.10 Evolutionary tree of G-protein-coupled receptors.

First, the evolutionary tree illustrates the similarity between different kinds of receptors based on their relative positions on the tree. Thus, the muscarinic, α-adrenergic, β-adrenergic, histamine, and dopamine receptors have evolved from a common branch of the evolutionary tree and have greater similarity to each other than to any receptors arising from an earlier evolutionary branch (e.g. the angiotensin receptor). Such receptor similarity may prove a problem in medicinal chemistry. Although the receptors are distinguished by different neurotransmitters or hormones in the body, a drug may not manage to make this distinction. Therefore, it is important to ensure that any new drug which is aimed at one kind of receptor (e.g. the dopamine receptor) does not interact with a similar kind of receptor (e.g. the muscarinic receptor).

Receptors have further evolved to give receptor **types** and **subtypes** which recognize the same chemical messenger, but which have structural differences between each other. For example, there are two types of adrenergic receptor (α and β), each of which can be split into various subtypes (α_1, α_{2A}, α_{2B}, α_{2C}, β_1, β_2, β_3). There are two types of cholinergic receptor – nicotinic (an ion channel receptor) and muscarinic (a 7-TM receptor). Five subtypes of the muscarinic cholinergic receptor have been identified.

The existence of receptor subtypes allows the possibility of designing drugs against one receptor subtype over another. This is important since one specific receptor subtype may be prevalent in one part of the body (e.g. the gut) while a different receptor subtype is prevalent in another part (e.g. the heart). Therefore, a drug which is designed to interact specifically with the receptor subtype in the gut is less likely to have side-effects on the heart. Even if the different receptor subtypes are present in the same part of the body, it is still important to make drugs as specific as possible since different receptor subtypes frequently activate different signalling systems leading to different biological results.

A closer study of the evolutionary tree reveals some curious facts about the origins of receptor subtypes. As one might expect, various receptor subtypes have diverged from a common evolutionary branch (e.g. the dopamine subtypes D2, D3, D4). This is known as **divergent evolution** and there should be close structural similarity between these subtypes. However, receptor subtypes are also found in separate branches of the tree. For example, the dopamine receptor subtypes (D1$_A$, D1$_B$, and D5) have developed from a different evolutionary branch. In other words, the ability of a receptor to bind dopamine has developed in different evolutionary branches – an example of **convergent evolution**.

Consequently, there may sometimes be greater similarities between receptors which bind different ligands but which have evolved from the same branch of the tree, than there are between the various subtypes of receptors which bind the same ligand. For example, the histamine H1 receptor resembles a muscarinic receptor more closely than it does the histamine H2 receptor. This again has important consequences in drug design since there is an increased possibility that a drug aimed at a muscarinic receptor may also interact with a histamine H1 receptor and lead to unwanted side effects.

6.4 Signal transduction pathways for G-protein-coupled receptors

The sequence of events leading from the combination of receptor and ligand (the chemical messenger) to the final activation of a target enzyme is quite lengthy and so we shall look at each stage of the process in turn.

6.4.1 Interaction of the 7-TM receptor–ligand complex with G-proteins

The first stage in the process is the binding of the ligand to the receptor, followed by the binding of a G-protein to the receptor–ligand complex (Fig. 6.11). G-proteins are membrane-bound proteins situated at the inner surface of the cell membrane and are made up of three protein subunits (α, β, and γ). The α-subunit has a binding pocket which can bind guanyl nucleotides (hence the name G-protein) and which binds GDP when the G-protein is in the resting state. There are several types of G-protein (e.g. G_s, G_i/G_o, $G_{q/11}$) and several subtypes of these. Specific G-proteins are recognized by specific receptors. For example, G_s is recognized by the β-adrenoreceptor but not the α-adrenoreceptor. However, in all cases, the G-protein acts as a molecular 'relay runner', carrying the message received by the receptor to the next target in the signalling pathway.

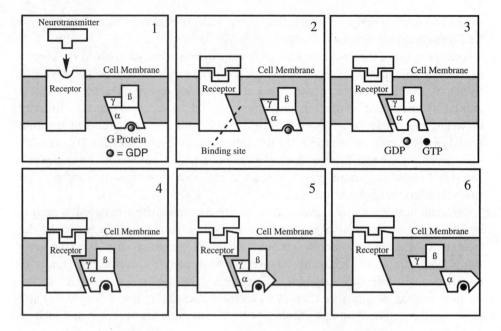

Fig. 6.11 Interaction of 7-TM receptors with G-proteins.

We shall now look at what happens in detail.

First of all, the receptor binds its neurotransmitter or hormone (frame 1). As a result. the receptor changes shape and exposes a new binding site on its inner surface (frame 2).

The newly exposed binding site now recognizes and binds a specific G-protein. (Note that the cell membrane structure is a fluid structure and so it is possible for different proteins to 'float' through it.) The binding process between the receptor and the G-protein causes the latter to change shape, which in turn changes the shape of the guanyl nucleotide binding site. This weakens the intermolecular bonding forces holding GDP and so GDP is released (frame 3).

However, the binding pocket does not stay empty for long because it is now the right shape to bind GTP (guanosine triphosphate). Therefore, GTP replaces GDP (frame 4). Binding of GTP results in another conformational change in the G-protein (frame 5) which weakens the links between the protein subunits such that the α-subunit (with its GTP attached) splits off from the β- and γ-subunits (frame 6). Both the α-subunit and the β γ-dimer then depart the receptor.

The receptor–ligand complex is able to activate several G-proteins in this way before the ligand departs, switching off the receptor. Therefore, this leads to an amplification of the signal.

Both the α-subunit and the β γ-dimer are now ready to enter the second stage of the signalling mechanism. We shall first consider what happens to the α-subunit.

6.4.2 Signal transduction pathways involving the α-subunit

The first stage of signal transduction (i.e. the splitting of a G-protein) is common to all of the 7-TM receptors. However, subsequent stages depend on what type of G-protein is involved and which specific α-subunit is formed (Fig. 6.12). Different α-subunits (there are at least 20 of them) have different targets and different effects, e.g.

α_s stimulates adenylate cyclase

α_i inhibits adenylate cyclase and may also activate K$^+$ channels

α_o activates receptors that inhibit neuronal Ca^{2+} ion channels

α_q activates phospholipase C.

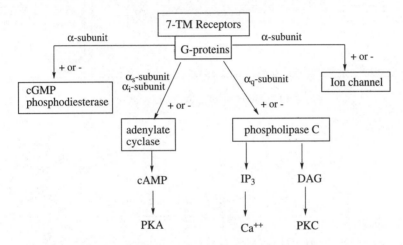

Fig. 6.12 Signalling pathways arising from the splitting of G-proteins.

We do not have the space to look at all these pathways in detail. Instead, we shall look at two of the pathways – the activation of adenylate cyclase and the activation of phospholipase C.

6.5 Signal transduction involving G-protein-coupled receptors and cyclic AMP

6.5.1 Activation of adenylate cyclase by the α_s-subunit

The α_s-subunit binds to a membrane-bound enzyme called adenylate cyclase (or adenylyl cyclase) and 'switches' it on (Fig. 6.13). This enzyme now catalyses the synthesis of a molecule called cyclic AMP (Fig. 6.14). Cyclic AMP is an example of a **secondary messenger** which moves into the cell's cytoplasm and carries the signal from the cell membrane into the cell itself. The enzyme will continue to be active as long as the α_s-subunit is present and this results in the synthesis of several hundred cyclic AMP molecules, representing another substantial amplification of the signal. However, the α_s-subunit has intrinsic GTPase activity (i.e. it can catalyse the hydrolysis of its bound GTP to GDP) and so it deactivates itself after a certain period by returning to the resting state. The α_s-subunit then departs the enzyme and recombines with the $\beta\,\gamma$-dimer to reform the G_s-protein, while the enzyme returns to its inactive conformation.

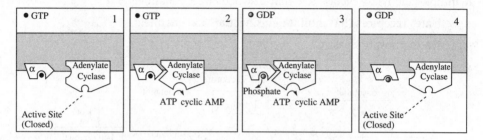

Fig. 6.13 Interaction of α_s with adenylate cyclase.

Fig. 6.14 Synthesis of cyclic AMP.

6.5.2 Activation of protein kinase A

Cyclic AMP (cAMP) now proceeds to activate an enzyme called protein kinase A (PKA) (Fig. 6.15).

Protein kinase A belongs to a group of enzymes called the **serine-threonine kinases** which catalyse the phosphorylation of proteins and enzymes at their serine and threonine residues (Fig. 6.16). Protein kinase A is inactive in the resting state and consists of four protein subunits, two of which are regulatory and two of which are catalytic. The activation of protein kinase A occurs as shown in Fig. 6.17.

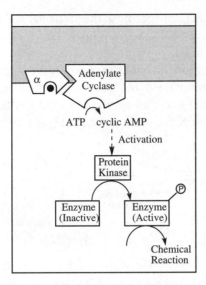

Fig. 6.15 Activation of protein kinase A, Ⓟ Phosphate.

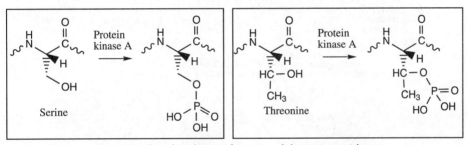

Fig. 6.16 Phosphorylation of serine and threonine residues.

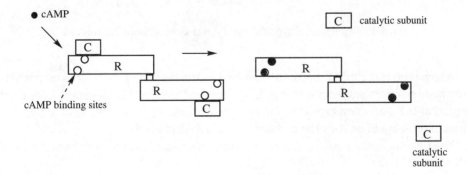

Fig. 6.17 Activation of protein kinase A.

The regulatory subunits have two binding sites for cyclic AMP and, when this binds, the catalytic subunits are split from the protein complex and become active. Once the protein kinase catalytic subunits become active, they phosphorylate and activate further enzymes with functions specific to the particular cell or organ in question. For example, a protein kinase would activate lipase enzymes in fat cells which would catalyse the breakdown of fat.

There may be several more enzymes involved in the signalling pathway between the activation of protein kinase A and the activation (or deactivation) of the target enzyme. For example, the enzymes involved in glycogen breakdown and glycogen synthesis in a liver cell are regulated as shown in Fig. 6.18.

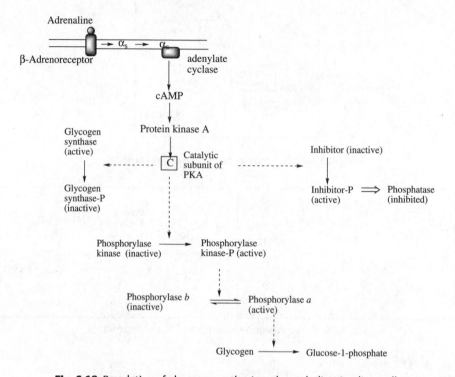

Fig. 6.18 Regulation of glycogen synthesis and metabolism in a liver cell.

Adrenaline is the initial hormone involved in the regulation and is released when the body requires immediate energy in the form of glucose. The hormone initiates a signal at the β-adrenoreceptor leading to the synthesis of cyclic AMP and the activation of protein kinase A by the mechanism already discussed.

The catalytic subunit of protein kinase A now phosphorylates three enzymes within the cell as follows:

• An enzyme called phosphorylase kinase is phosphorylated and is activated as a result. This enzyme then catalyses the phosphorylation of an inactive enzyme called phosphorylase b which is converted to its active form – phosphorylase a. Phosphorylase a now catalyses the breakdown of glycogen by splitting off glucose-1-phosphate units.

- Glycogen synthase is phosphorylated to an inactive form, thus preventing the synthesis of glycogen.
- A molecule called phosphorylase inhibitor is phosphorylated. Once phosphorylated, it acts as an inhibitor for the phosphatase enzyme responsible for the conversion of phosphorylase a back to phosphorylase b. Therefore, the lifetime of phosphorylase a is prolonged.

The overall result of these different phosphorylations is a coordinated inhibition of glycogen synthesis and enhancement of glycogen metabolism to generate glucose in muscle cells. Note that the effect of adrenaline on other types of cell may be quite different. For example, adrenaline can activate the β-adrenoreceptors in fat cells, leading to the activation of protein kinases as before. This time, however, phosphorylation activates lipase enzymes. These then catalyse the breakdown of fat to act as another source of glucose.

6.5.3 The G_i-protein

We have seen how the enzyme adenylate cyclase is activated by the α_s-subunit of the G_s-protein. Adenylate cyclase can also be inhibited by a different G-protein – the G_i-protein. The G_i-protein interacts with different receptors from those which interact with the G_s-protein, but the mechanism leading to inhibition is the same as that leading to activation. The only difference is that the α_i-subunit released binds to adenylate cyclase to inhibit the enzyme rather than to activate it.

Receptors which bind G_i-proteins include the muscarinic M_2-receptor of cardiac muscle, α_2- adrenoreceptors in smooth muscle, and opioid receptors in the central nervous system.

The existence of G_i and G_s-proteins means that the generation of the secondary messenger cyclic AMP is under the dual control of 'brake and accelerator' and this explains the process by which two different neurotransmitters can have opposing effects at a target cell. A neurotransmitter which stimulates the production of cyclic AMP forms a receptor–ligand complex which activates a G_s-protein, whereas a neurotransmitter which inhibits the production of cyclic AMP forms a receptor–ligand complex which activates a G_i-protein.

Since there are various different types of receptor for a particular neurotransmitter, it is actually possible for that neurotransmitter to activate cyclic AMP in one type of cell but to inhibit it in another. For example, noradrenaline interacts with its β-adrenoreceptor to activate adenylate cyclase since the β-adrenoreceptor binds the G_s-protein. On the other hand, noradrenaline interacts with its α_2-adrenoreceptor to inhibit adenylate cyclase since this receptor binds the G_i-protein. This example illustrates the point that it is the receptor, rather than the neurotransmitter or hormone, which determines which G-protein is activated.

It is also worth pointing out that enzymes such as adenylate cyclase and the kinases are never fully active or inactive. At any one time a certain proportion of the enzymes are active and the role of the G_s and G_i-proteins is either to increase or decrease that proportion. In other words, the control is graded rather than an all or nothing effect.

6.5.4 General points about the signalling cascade involving cyclic AMP

The signalling cascade involving the G_s-protein, cyclic AMP and protein kinase A appears very complex and you might wonder whether a simpler signalling process would be more efficient. There are several points worth noting about the process as it stands.

First, the action of the G-protein and the generation of a secondary messenger explains how a message delivered to the outside of the cell surface can be transmitted to enzymes within the cell – enzymes which have no direct association with the cell membrane or the receptor. Such a signalling process avoids the difficulties involved in a messenger molecule (which is commonly hydrophilic) having to cross a hydrophobic cell membrane.

Second, the process involves a 'molecular relay runner' (the G-protein) and several different enzymes in the signalling cascade. At each of these stages, the action of one protein or enzyme results in the activation of a much larger number of enzymes. Therefore, the effect of one neurotransmitter interacting with one receptor molecule results in a final effect several factors larger than one might expect. For example, each molecule of adrenaline is thought to generate 100 molecules of cyclic AMP and each cyclic AMP molecule starts off an amplification effect of its own within the cell.

Third, there is an advantage in having the receptor, the G-protein, and adenylate cyclase as separate entities. The G-protein can bind to several different types of receptor–ligand complexes. This means that different neurotransmitters and hormones interacting with different receptors can switch on the same G-protein, leading to activation of adenylate cyclase. Therefore, there is an economy of organization involved in the cellular signalling chemistry since the adenylate cyclase signalling pathway can be used in many different cells, and yet respond to different signals. Moreover, different cellular effects will result depending on the type of cell involved (i.e. cells in different tissues will have different receptor types and subtypes, and the signalling system will switch on different target enzymes). For example, glucagon activates G_s-linked receptors in the liver leading to gluconeogenesis in the liver, adrenaline activates G_s-linked β_2-adrenoreceptors leading to lipolysis in fat cells, while vasopressin interacts with G_s-linked vasopressin (V_2) receptors in the kidney to affect Na/water reabsorption. Adrenaline acts on $G_{i/o}$-linked α_2-receptors leading to contraction of smooth muscle, and acetylcholine acts on $G_{i/o}$-linked M_2-receptors leading to relaxation of heart muscle. All these effects are mediated by the cyclic AMP signalling pathway.

Last, the dual control of 'brake and accelerator' provided by the G_s and G_i-proteins allows fine control of the adenylate cyclase activity.

6.5.5 The role of the β γ dimer

If you've managed to follow the complexity of the G-protein signalling pathway so far, well done. Unfortunately there's more! You may remember that when the

G-protein binds to a receptor–ligand complex, it breaks up to form the α-subunit and a β γ dimer. So far we have said nothing about the dimer! Until recently, the β γ dimer was viewed merely as an anchor for the α-subunit to ensure that it remained in the cell membrane. However, it has now been found that the β γ dimers from both the G_i and the G_s-proteins can themselves activate or inhibit adenylate cyclase. There are actually six different types or isozymes of adenylate cyclase, and activation or inhibition depends on the isozyme involved. Moreover, adenylate cyclase is not the only enzyme which can be controlled by the β γ dimer. The β γ dimer is more promiscuous than the α-subunits as far as its targets are concerned and can have a variety of different effects. This sounds like a recipe for anarchy. However, there is some advantage in the dimer having a signalling role since it adds an extra subtlety to the signalling process. For example, it is found that higher concentrations of the dimer are required to result in any effect compared to the α-subunit. Therefore, regulation by the dimers becomes more important when a greater number of receptors are activated.

By now it should be becoming clear that the activation of a cellular process is more complicated than the interaction of one type of neurotransmitter with one type of receptor. In reality, the cell is receiving a whole myriad of signals from different chemical messengers via various receptors and receptor–ligand interactions. The final signal depends on the number and type of G-proteins activated at any one time, as well as the various signal transduction pathways which these proteins initiate.

6.5.6 Phosphorylation

As we have seen above, phosphorylation is a key reaction in the activation or deactivation of enzymes. Phosphorylation occurs on the phenolic group of tyrosine residues when catalysed by tyrosine kinases, and on the alcohol groups of serine and threonine residues when catalysed by serine-threonine kinases. These functional groups are all capable of participating in hydrogen bonding, but if a bulky phosphate group is added to the OH group, hydrogen bonding is disrupted. Furthermore, the phosphate group is usually ionized at physiological pH and so phosphorylation introduces two negatively charged oxygens. These charged groups can now form strong ionic bonds with any positively charged residues in the protein causing the enzyme to change its tertiary structure. This change in shape results in the exposure or closure of the active site (Fig. 6.19).

Phosphorylation by kinase enzymes also accounts for the desensitization of G-protein-linked receptors. Phosphorylation of serine and threonine residues occurs on the intracellular *C*-terminal chain after prolonged ligand binding. Since the *C*-terminal chain is involved in G-protein binding, phosphorylation changes the conformation of the protein in that region and prevents the G-protein from binding. Thus the receptor–ligand complex is no longer able to activate the G-protein.

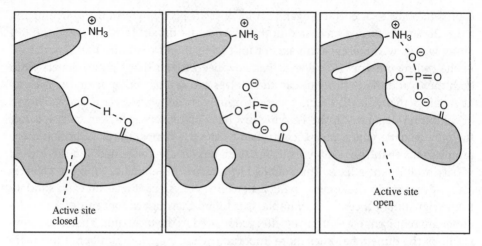

Fig. 6.19 Conformational changes induced by phosphorylation.

6.5.7 Drugs acting on the cyclic AMP signal transduction process

Some drugs and toxins are known to interact with the cyclic AMP signal transduction process. For example, the bacterial toxin responsible for cholera causes persistent activation of adenylate cyclase. This leads to excessive secretion of fluid from epithelial cells in the gut leading to diarrhoea. Theophylline and caffeine (Fig. 6.20) are thought to inhibit the phosphodiesterases responsible for metabolizing cyclic AMP, thus prolonging the action of cyclic AMP.

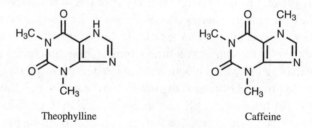

Theophylline   Caffeine

Fig. 6.20 Phosphodiesterase inhibitors.

6.6 Signal transduction involving G-protein-coupled receptors and phospholipase C

6.6.1 Activation by phospholipase C

Certain receptors bind Gs or Gi-proteins and initiate a signalling pathway involving adenylate cyclase (see Section 6.5). Other 7-TM receptors bind a different G-protein called a G_q-protein which intiates a different signalling pathway. This pathway involves the activation or deactivation of a membrane-bound enzyme called phospholipase C. The first part of the signalling mechanism is the interaction of the G-protein with a receptor–ligand complex as described previously in Fig. 6.11. This time however, the G-protein is a G_q-protein rather than a G_s or G_i-protein and so an α_q-subunit is released. Depending on the nature of the released α_q-subunit, a membrane-bound enzyme called phospholipase C is activated or deactivated (Fig. 6.21). If activated, phospholipase C catalyses the hydrolysis of phosphatidylinositol diphosphate (an integral part of the cell membrane structure) to generate the two secondary messengers inositol triphosphate (IP$_3$) and diacylglycerol (DG) (Fig. 6.22).

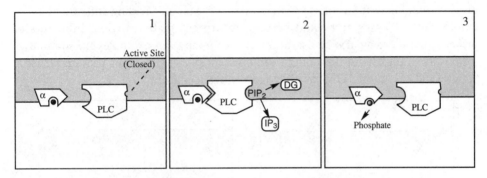

Fig. 6.21 Activation of phospholipase C (PLC).

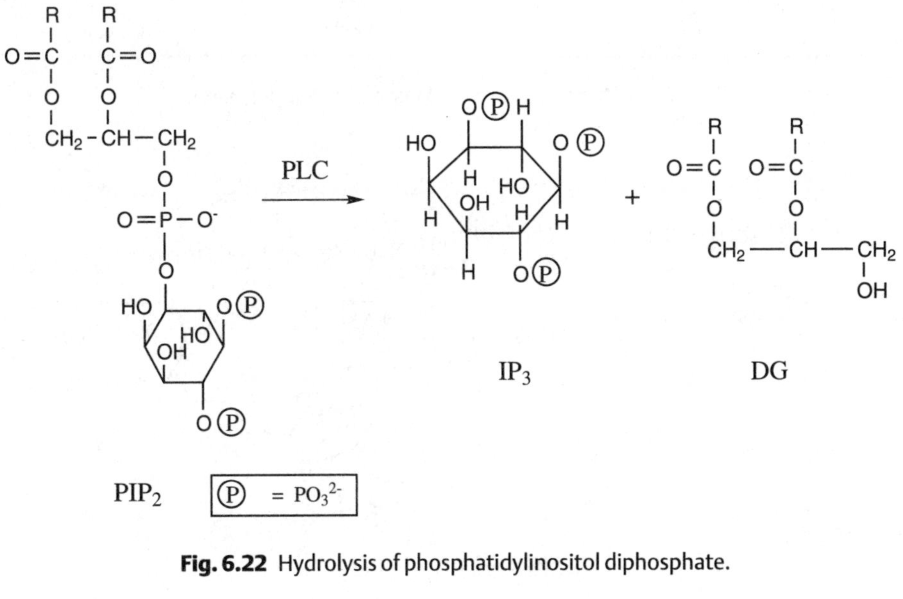

Fig. 6.22 Hydrolysis of phosphatidylinositol diphosphate.

6.6.2 Action of the secondary messenger – diacylglycerol

Diacylglycerol is a hydrophobic molecule and remains in the cell membrane once it is formed (Fig. 6.23). There, it activates an enzyme called protein kinase C, which moves from the cytoplasm to the cell membrane and then catalyses the phosphorylation of serine and threonine residues of enzymes within the cell. Once phosphorylated, these enzymes are activated and catalyse specific reactions within the cell. These induce effects such as tumour propagation, inflammatory responses, contraction or relaxation of smooth muscle, the increase or decrease of neurotransmitter release, the increase or decrease of neuronal excitability, and receptor desensitizations. Designing drugs to inhibit protein kinase C may offer a means of tackling inflammatory or autoimmune diseases and could be used to prevent graft rejection. Inhibitors of protein kinase C may also be useful in cancer therapy. For example, bryostatin (Fig. 6.24) is a complex natural product from a sea moss which inhibits protein kinase C activity and has entered clinical trials as an anticancer agent.

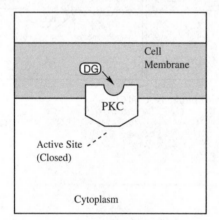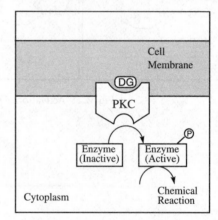

Fig. 6.23 Activation of protein kinase C by diacylglycerol.

Fig. 6.24 Bryostatin.

Since so many extracellular and intracellular signals converge on this enzyme, inhibiting this one target may be better than blocking a range of separate receptors.

6.6.3 Action of the secondary messenger – inositol triphosphate

Inositol triphosphate is a hydrophilic molecule and moves into the cytoplasm (Fig. 6.25). This messenger works by mobilizing calcium from calcium stores in the endoplasmic reticulum. It does so by binding to a receptor and opening up a calcium ion channel. Once the ion channel is open, calcium ions flood the cell and activate calcium-dependent protein kinases which in turn phosphorylate and activate cell-specific enzymes. The released calcium ions can also bind to a calcium-binding protein called calmodulin, which then activates calmodulin-dependent protein kinases which also phosphorylate and activate cellular enzymes. Calcium also has effects on contractile proteins and ion channels, but it is not possible to cover these effects in detail in this text. Suffice it to say that the release of calcium is crucial to a large variety of cellular functions, including smooth muscle and cardiac muscle contraction, secretion from exocrine glands, transmitter release from nerves, and hormone release.

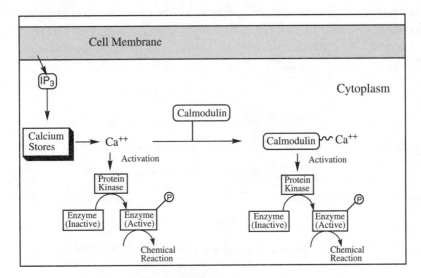

Fig. 6.25 Action of inositol triphosphate.

6.6.4 Resynthesis of phosphatidylinositol diphosphate

Once the inositol triphosphate and diacylglycerol have completed their tasks, they are recombined to form phosphatidylinositol diphosphate. Oddly enough, they cannot be linked directly and both molecules have to undergo several metabolic steps before resynthesis can occur. It is thought that lithium salts control the symptoms of manic depressive illness by interfering with this involved synthesis.

6.7 Kinase-linked (1-TM) receptors

Kinase-linked receptors are a superfamily of receptors which activate enzymes directly and do not require a G-protein to act as a 'relay runner'. They are activated by a large number of polypeptide hormones, growth factors, and cytokines. Loss of function of these receptors can lead to developmental defects or hormone resistance. Overexpression can result in malignant growth disorders.

6.7.1 Structure of kinase-linked receptors

An important family of receptors belonging to this superfamily is the tyrosine kinase receptors. The basic structure of all the tyrosine kinase receptors consists of a single extracellular region (the N-terminal chain) that includes the binding site for the chemical messenger, a single hydrophobic region that traverses the membrane as an α helix of seven turns (just sufficient to traverse the membrane), and the C-terminal chain on the inside of the cell membrane (Fig. 6.26). The C-terminal region contains a catalytic binding site and so these receptors act as a receptor and an enzyme in the one molecule.

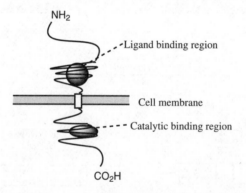

Fig. 6.26 Structure of 1-TM receptors.

These receptors include the receptor for insulin, and receptors for various cytokines and growth factors.

6.7.2 Signalling mechanism for the tyrosine kinase receptor family

As stated above, the tyrosine kinase receptors consist of membrane-bound proteins which have a dual role as receptors and enzymes. In the resting state, the receptor has no catalytic activity within the cell since the active site is hidden. However, when a ligand binds to the receptor, the receptor changes shape, revealing the active site on the C-terminal chain. An enzymic reaction can now take place within the cell. This reaction is yet another phosphorylation which specifically phosphorylates tyrosine residues and so the receptor is classed as a tyrosine kinase receptor. A specific example of a tyrosine kinase receptor is the receptor for a hormone called epidermal growth factor. Epidermal growth factor is a **bivalent** ligand which can bind to two receptors at the same time. This results in receptor dimerization as well as activation of enzymic activity. This dimerization process is important because the active site on each half of the receptor dimer catalyses the phosphorylation of the accessible tyrosine residues on the other half (Fig. 6.27). If dimerization did not occur, no phosphorylation would take place. Note that these phosphorylations occur on the intracellular portion of the receptor protein chain.

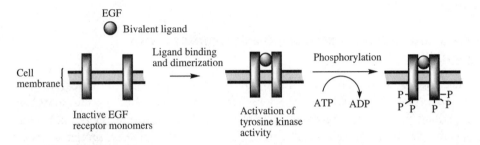

Fig. 6.27 Signal mechanism for the epidermal growth factor receptor.

Dimerization and autophosphorylation is a common theme for receptors in this family. However, some of the receptors in this family already exist as dimers or tetramers and only require binding of the ligand. For example, the insulin receptor is a heterotetrameric complex (Fig. 6.28).

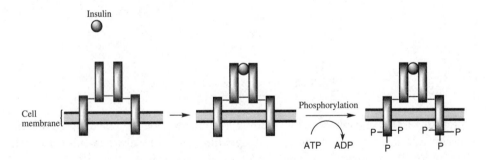

Fig. 6.28 Ligand binding for the insulin receptor.

Some 1-TM kinase receptors bind ligands and dimerize in a similar fashion but do not have inherent catalytic activity in their C-terminal chain. However, once they have dimerized, they can bind and activate a tyrosine kinase enzyme from the cytoplasm. The growth hormone receptor is an example of this type of receptor (a tyrosine kinase-linked receptor) (Fig. 6.29).

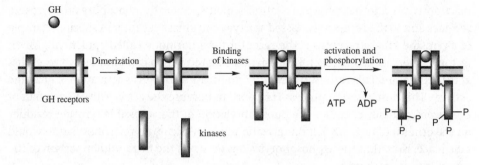

Fig. 6.29 The growth hormone receptor.

6.7.3 Interaction of protein kinase receptors with signalling proteins

Once the kinase receptor (or its associated enzyme) has been phosphorylated, the phosphotyrosine groups and the regions around them act as binding sites for various signalling proteins or enzymes. Each phosphorylated tyrosine region can bind a specific signalling protein or enzyme. Some of these signalling proteins/enzymes become phosphorylated themselves once they are bound and act as further binding sites for yet more signalling proteins (Fig. 6.30).

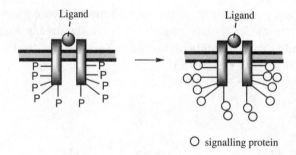

Fig. 6.30 Binding of signalling proteins.

Not all of the phosphotyrosine binding regions can be occupied by signalling proteins at one time and so the type of signalling which results depends on which signalling proteins *do* manage to bind to all the kinase receptors available. There is no room in an introductory text to consider what each and every signalling protein does, but most are the starting point for phosphorylation (kinase) cascades along the same principles as the cascades initiated by G-proteins (Fig. 6.31). Some growth factors activate a specific subtype of phospholipase C (PLCγ), which catalyses phospholipid breakdown, leading to the generation of inositol triphosphate and subsequent calcium release by the same mechanism as described in Section 6.6. Other signalling proteins are chemical 'adaptors' which serve to transfer a signal from the receptor to a wide variety of other proteins, including many involved in cell division and differ-entiation. For example, the principal action of growth hormones is to stimulate transcription of particular genes through a kinase signalling cascade (Fig. 6.32). As an example, a signalling protein called Grb2 binds to a specific phosphorylated site of the receptor–ligand complex and becomes phosphorylated itself. A membrane protein called Ras (with a bound molecule of GDP) interacts with the receptor–ligand–signal protein complex and functions in a similar way to a G-protein (i.e. GDP is lost and GTP is gained). Ras is now activated and activates a serine-threonine kinase called Raf which initiates a serine-threonine kinase cascade which finishes with the activation of MAP kinase.[1] This phosphorylates and activates proteins called transcription factors which enter the nucleus and inititate gene expression, resulting in various responses including cell division. Many cancers can arise from malfunctions along this signalling cascade if the kinases become permanently activated despite the absence of the initial receptor signal. Consequently, targetting this signalling pathway may lead to new drugs for the treatment of cancer.

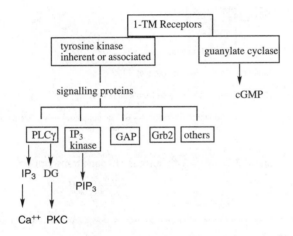

Fig. 6.31 Signalling pathways from 1-TM receptors.

6.7.4 Activation of guanylate cyclase by 1-TM receptors

Some 1-TM receptors have the ability to catalyse the formation of cyclic GMP from GTP. Therefore, they are both receptor and enzyme (guanylate cyclase). The mem-brane-bound receptor/enzyme spans the membrane and has a single transmembrane segment. It has a receptor-binding site on the cell exterior and a guanylate cyclase active site on the cell interior. Its ligands are α atrial natriuretic peptide and brain natriuretic peptide.

Cyclic GMP appears to open sodium ion channels in the kidney, promoting the excretion of sodium.

---

[1] MAP = Mitogen Activated Protein

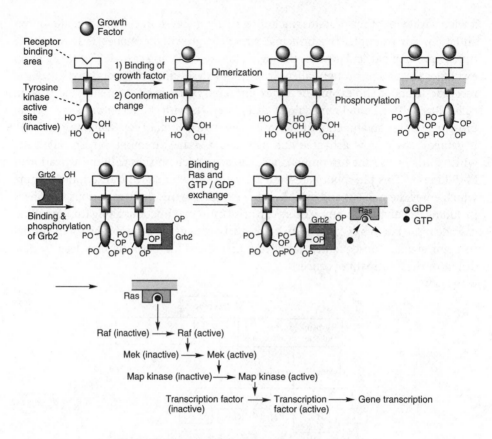

Fig. 6.32 From growth hormone to gene transcription.

6.8 Intracellular receptors

As discussed above, there are three superfamilies of membrane-bound receptors. There is also a fourth superfamily of receptors. However, the receptors in this super-family are not membrane-bound and are present in the cytoplasm or the nucleus of the cell. These receptors are particularly important in regulating gene transcription, and their chemical messengers are required to pass through the cell membrane to reach them.

The receptors for steroid hormones are examples of cytoplasmic receptors. Steroids are hydrophobic molecules which can diffuse through the cell membrane to reach the steroid receptors. The steroid receptor is a single protein containing a binding site for the steroid at the *C*-terminal end of the protein and a binding region for DNA near the centre (Fig. 6.33). The DNA binding region contains two loops of about 15 residues each (called the 'zinc fingers') knotted together by cysteine residues surrounding a zinc atom. When the receptor and steroid combine, the receptor changes shape, resulting in dimerization of the steroid–receptor complex. The dimer then travels across the nuclear membrane into the nucleus whereupon it binds to an acceptor site on the cell's DNA. (The zinc fingers are so called because they are believed to wrap round the DNA helix.) This binding then switches on transcription and the synthesis of mRNA. mRNA then acts as a code for new enzymes or receptors.

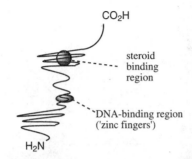

Fig. 6.33 Structure of steroid receptors.

The oestrogen receptor is one example of an intracellular steroid receptor and is the target for the anticancer agent tamoxifen (Nolvadex) (Fig. 6.34), used widely to treat breast cancer.

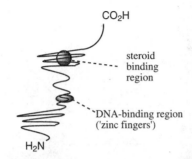

Fig. 6.34 Tamoxifen.

The oestrogen receptor is situated in the cytoplasm and is associated with another protein (Fig. 6.35). When the female hormone oestrogen enters the cell (stage 1), it binds to the receptor–protein complex (stage 2), resulting in dissociation of the protein from the receptor (stage 3). The receptor–ligand complex then enters the nucleus (stage 4) and binds to specific DNA sequences (stage 5), thus activating the transcription of certain genes and the production of mRNA (stage 6). The mRNA molecules produced are then translated to form various functional and structural proteins (stage 7).

Tamoxifen binds to the oestrogen receptor and acts as an antagonist to prevent the production of these proteins. It is remarkably effective in treating some cases of hormone-dependent breast cancer.

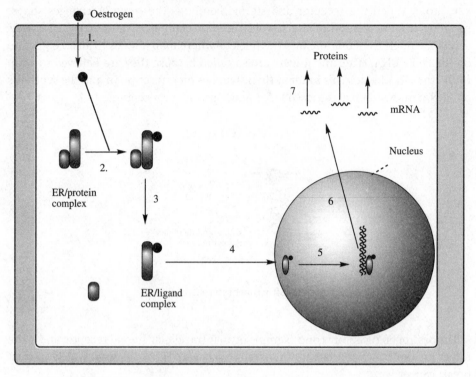

Fig. 6.35 Function of the oestrogen receptor. ER = oestrogen receptor.

7 Nucleic acids

Although the majority of drugs act on protein structures, there are several examples of important drugs which act directly on nucleic acids to disrupt replication, transcription, and translation.

There are two types of nucleic acid: DNA (deoxyribonucleic acid) and RNA (ribonucleic acid). We shall first consider the structure of DNA and the drugs which act on it.

7.1 Structure of DNA

As with proteins, DNA has a primary, secondary, and tertiary structure.

7.1.1 The primary structure of DNA

The primary structure of DNA is the way in which the DNA building blocks are linked together. Whereas proteins have over 20 building blocks from which to choose, DNA has only four – the nucleosides deoxyadenosine, deoxyguanosine, deoxycytidine, and deoxythymidine (Fig. 7.1).

Each nucleoside is constructed from two components – a deoxyribose sugar and a base. The sugar is the same in all four nucleosides and only the base is different. The four possible bases are two bicyclic purines (adenine and guanine), and two smaller pyrimidine structures (cytosine and thymine) (Fig. 7.2).

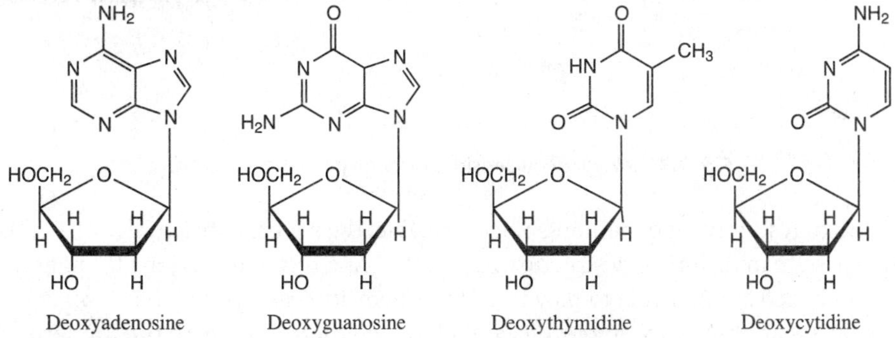

Deoxyadenosine   Deoxyguanosine   Deoxythymidine   Deoxycytidine

Fig. 7.1 The building blocks of DNA – nucleosides.

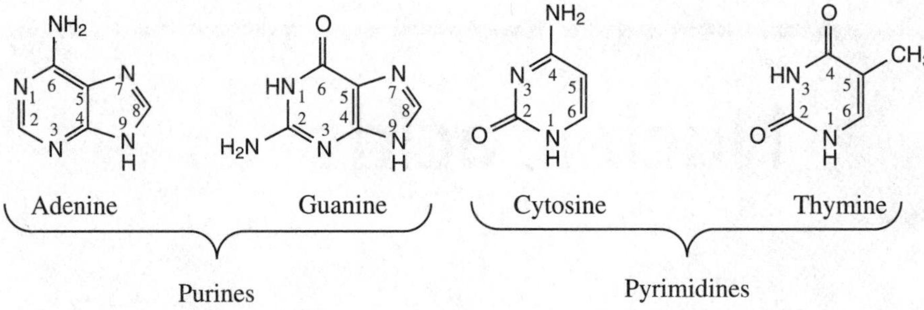

Fig. 7.2 The four bases of nucleosides.

The nucleoside building blocks are joined together through phosphate groups which link the 5´-hydroxyl group of one nucleoside unit to the 3´-hydroxyl group of the next (Fig. 7.3).

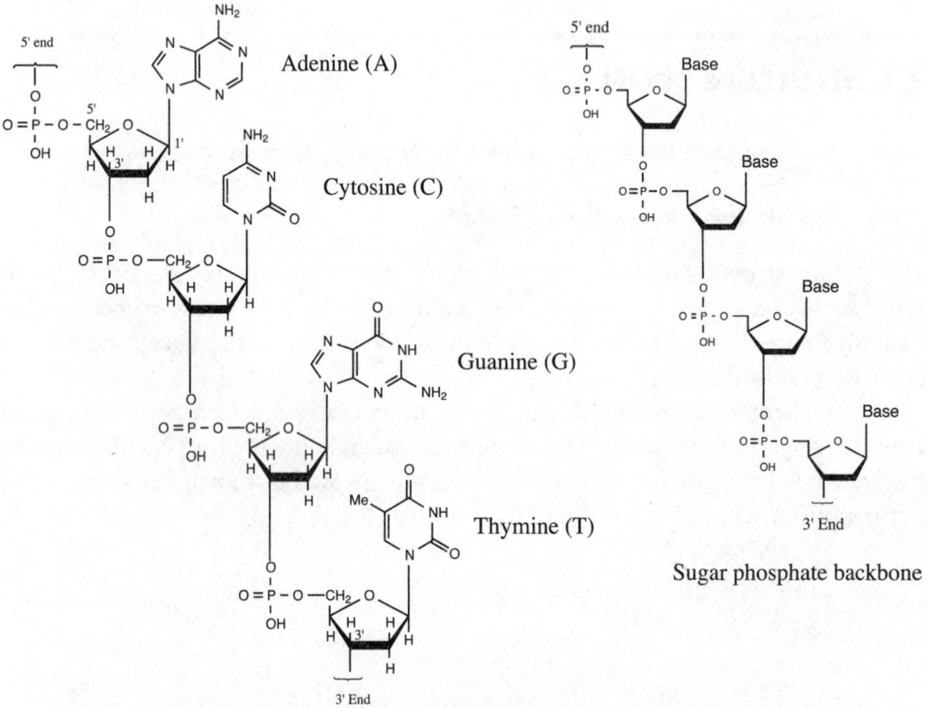

Fig. 7.3 Linkage of nucleosides through phosphate groups.

With only four types of building block available, the primary structure of DNA is far less varied than the primary structure of proteins. As a result, it was long thought that DNA only had a minor role to play in cell biochemistry, since it was hard to see how such an apparently simple molecule could have anything to do with the mysteries of the genetic code. The solution to this mystery lies in the secondary structure of DNA.

7.1.2 The secondary structure of DNA

Watson and Crick solved the secondary structure of DNA by building a model that fitted all the known experimental results. The structure consists of two DNA chains arranged together in a double helix of constant diameter (Fig. 7.4). The double helix can be seen to have a major groove and a minor groove which are of some importance to the action of several antibacterial agents (see later).

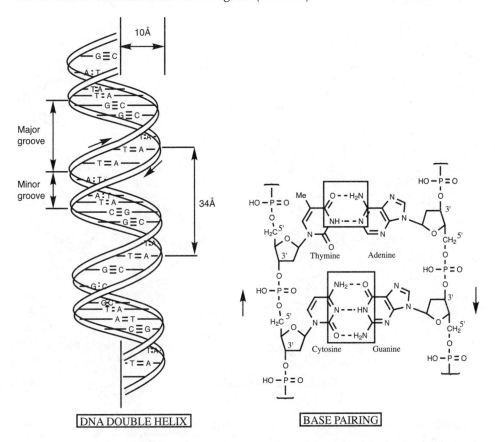

Fig. 7.4 The secondary structure of DNA.

The structure relies crucially on the pairing up of nucleic acid bases between the two chains. Adenine pairs only with thymine via two hydrogen bonds, whereas guanine pairs only with cytosine via three hydrogen bonds. Thus, a bicyclic purine base is always linked with a smaller monocyclic pyrimidine base to allow the constant diameter of the double helix. The double helix is further stabilized by the fact that the base pairs are stacked one on top of each other, allowing hydrophobic interactions between the heterocyclic rings. The polar sugar–phosphate backbone is placed to the outside of the structure and therefore can form favourable polar interactions with water.

The fact that adenine always binds to thymine, and cytosine always binds to guanine means that the chains are complementary to each other. In other words, one chain can be visualized as a negative image of its partner. It is now possible to see how replication (the copying of the genetic information) is feasible. If the double helix unravels, then a new chain can be constructed on each of the original chains (Fig. 7.5). In other words, each of the original chains can act as a template for the construction of a new and identical double helix.

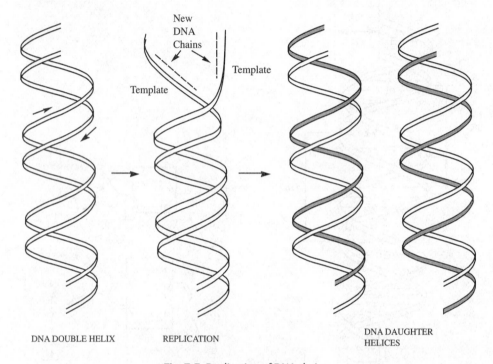

New
DNA
Chains

Template

Template

DNA DOUBLE HELIX   REPLICATION

DNA DAUGHTER
HELICES

Fig. 7.5 Replication of DNA chains.

It is less obvious how DNA can code for proteins. How can four nucleotides code for over twenty amino acids?

The answer lies in the triplet code. In other words, an amino acid is not coded by one nucleotide but by a set of three. There are 64 ways in which four nucleotides can be arranged in sets of three – more than enough for the task required.

7.1.3 The tertiary structure of DNA

The tertiary structure of DNA is often neglected or ignored, but it is important to the action of the quinolone group of antibacterial agents (see Chapter 14). The double helix is able to coil into a 3D shape and this is known as supercoiling. During replication, the double strand of DNA must unravel but, because of the tertiary supercoiling this leads to a high level of strain which has to be relieved by the temporary cutting, then repair, of the DNA chain. These procedures are enzyme-catalysed and these are the enzymes which are inhibited by the quinolone antibacterial agents.

7.2 Drugs acting on DNA

In general, we can classify the drugs which act on DNA into three groups:

1. intercalating cytostatic agents
2. alkylating agents
3. chain 'cutters'.

7.2.1 Intercalating agents

Intercalating drugs are compounds which are capable of slipping between the layers of nucleic acid base pairs and disrupting the shape of the double helix. This disruption prevents replication and transcription. One example is the antibacterial agent proflavine (Fig. 7.6).

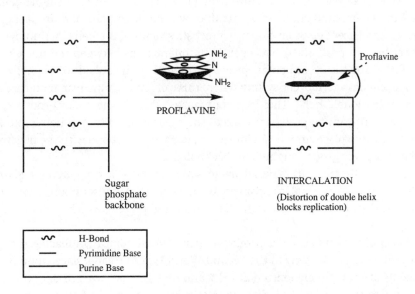

Fig. 7.6 Action of intercalating drugs.

Drugs which work in this way must be flat in order to fit between the base pairs, and must therefore be aromatic or heteroaromatic in nature. Some drugs prefer to approach the helix via the major groove, whereas others prefer access via the minor groove. Several antibiotics, such as the antitumour agents actinomycin D and adriamycin (Fig. 7.7), operate by intercalating DNA.

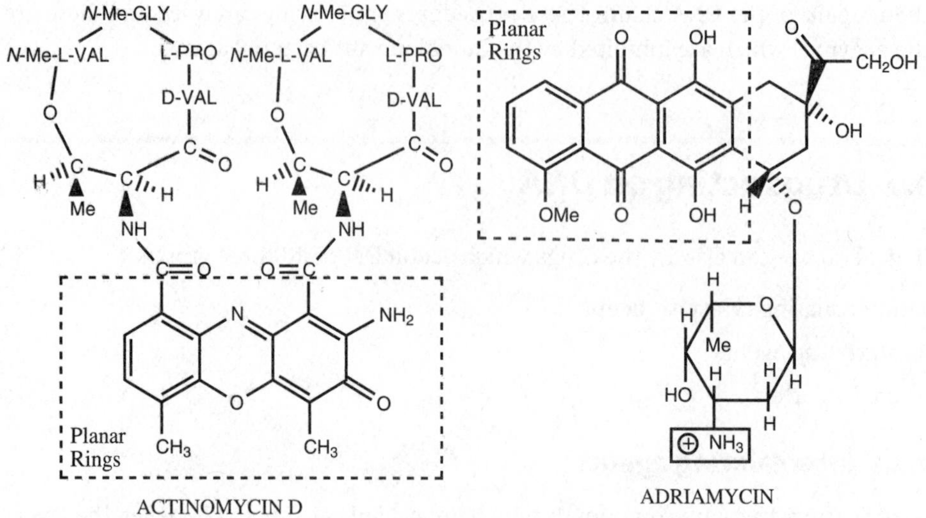

Fig. 7.7 Antibiotics which operate by intercalating DNA.

Actinomycin D contains two cyclic pentapeptides, but the important feature is the flat, tricyclic, heteroaromatic structure which is able to slide into the double helix. It appears to favour interactions with guanine–cytosine base pairs and, in particular, between two adjacent guanine bases on alternate strands of the helix. Actinomycin D is further held in position by hydrogen bond interactions between the nucleic acid bases of DNA and the cyclic pentapeptides positioned on the outside of the helix.

Adriamycin has a tetracyclic system where three of the rings are planar and are able to fit into the double helix. The drug approaches DNA via the major groove of the double helix. The amino group attached to the sugar is important in helping to lock the antibiotic into place since it can ionize and form an ionic bond with the negatively charged phosphate groups on the DNA backbone.

The highly effective antimalarial agent chloroquine – a drug developed from quinine – can attack the malarial parasite by blocking DNA transcription as part of its action. Once again a flat heteroaromatic structure is present which can intercalate DNA (Fig. 7.8).

Aminoacridine agents such as proflavine (Fig. 7.9) are topical antibacterial agents which were used particularly in the Second World War to treat surface wounds. The best agents are completely ionized at pH 7 and they interact with DNA in the same way as adriamycin. The flat tricyclic ring intercalates between the DNA base pairs and interacts by van der Waals forces, while the amino cations form ionic bonds with the negatively charged phosphate groups on the sugar–phosphate backbone.

7.2.2 Alkylating agents

Alkylating agents are highly electrophilic compounds which will react with nucleophiles to form strong covalent bonds. There are several nucleophilic groups in DNA

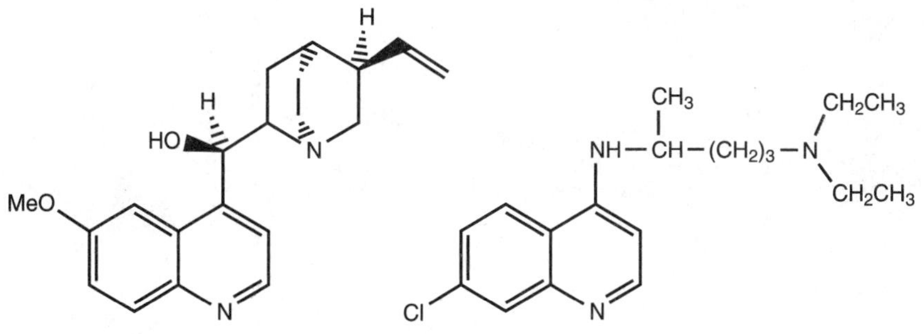

Fig. 7.8 Intercalating antimalarial drugs.

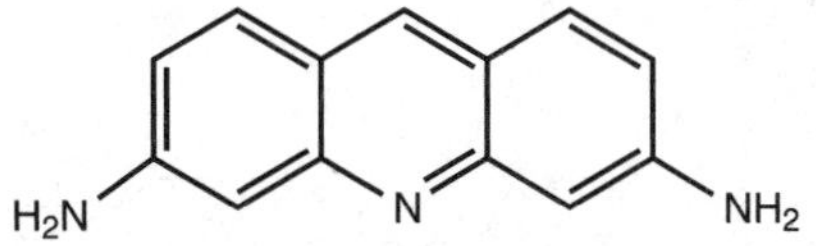

Fig. 7.9 Proflavine.

and in particular the 7-nitrogen of guanine. Drugs with two such alkylating groups could therefore react with a guanine on each chain and cross-link the strands such that they cannot unravel during replication or transcription.

Alternatively, the drug could link two guanine groups on the same chain such that the drug is attached like a limpet to the side of the DNA helix. Such an attachment would mask that portion of DNA and block access to the necessary enzymes required for DNA function.

Miscoding due to alkylated guanine units is also possible. The guanine base usually exists as the keto tautomer and base pairs with cytosine. Once alkylated, however, guanine prefers the enol tautomer and is more likely to base pair with thymine. Such miscoding ultimately leads to an alteration in the amino acid sequence of proteins, which in turn leads to disruption of protein structure and function.

Since alkylating agents are very reactive, they will react with any good nucleophile and so are not very selective in their action. They will alkylate proteins and other macromolecules as well as DNA. Nevertheless, alkylating drugs have been useful in the treatment of cancer. Tumour cells often divide more rapidly than normal cells and so disruption of DNA function will affect these cells more drastically than normal cells.

The nitrogen mustard compound mechlorethamine (Fig 7.10) was the first alkylating agent to be used (1942). The nitrogen atom is able to displace a chloride ion intramolecularly to form the highly electrophilic aziridine ion. Alkylation of DNA can then take place. Since the process can be repeated, cross-linking between chains will occur.

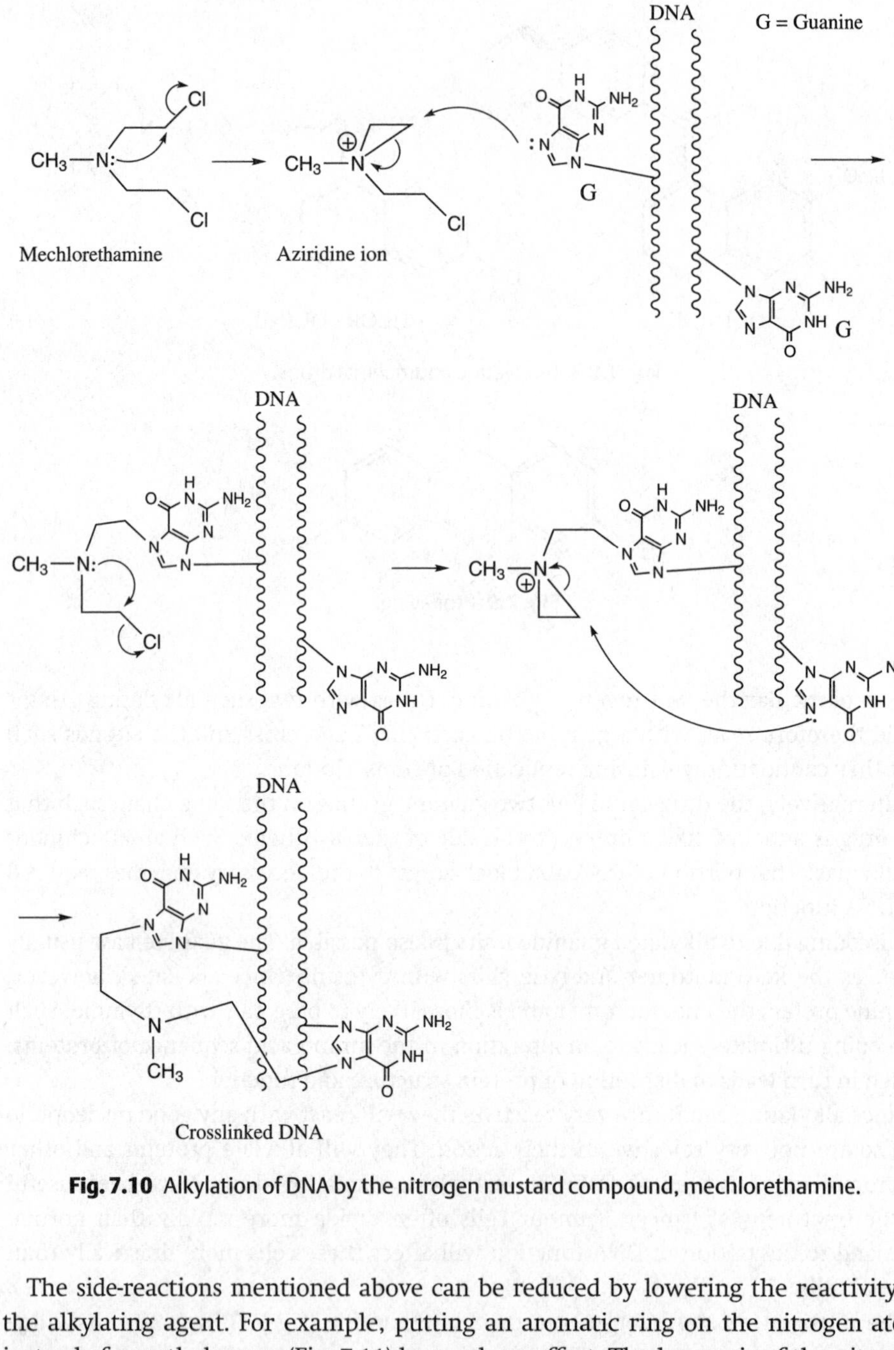

Fig. 7.10 Alkylation of DNA by the nitrogen mustard compound, mechlorethamine.

The side-reactions mentioned above can be reduced by lowering the reactivity of the alkylating agent. For example, putting an aromatic ring on the nitrogen atom instead of a methyl group (Fig. 7.11) has such an effect. The lone pair of the nitrogen is 'pulled into' the ring and is less available to displace the chloride ion. As a result, the intermediate aziridine ion is less easily formed and only strong nucleophiles such as guanine will now react with it.

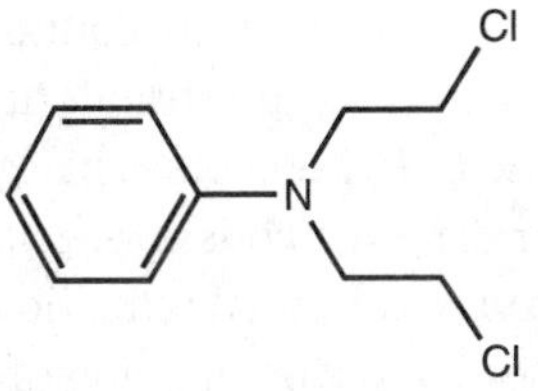

Fig. 7.11 Method of reducing reactivity of the alkylating agent.

Another approach which has been used to direct these alkylating agents more specifically to DNA has been to attach a nucleic acid building block onto the molecule. For example, uracil mustard (Fig. 7.12) contains one of the nucleic acid bases. This drug has been used successfully in the treatment of chronic lymphatic leukaemia and has a certain selectivity for tumour cells over normal cells. This is because tumour cells generally divide faster than normal cells. As a result, nucleic acid synthesis is faster and tumour cells are 'hungrier' for the nucleic acid building blocks. The tumour cells therefore take more than their share of the building blocks and of any cytotoxic drug which mimics the building blocks. Unfortunately, this approach has not so far succeeded in achieving the high levels of selectivity desired for effective eradication of tumour cells.

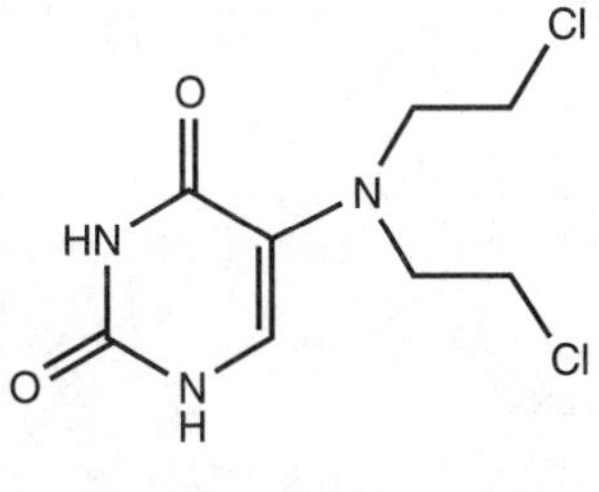

Fig. 7.12 Uracil mustard.

Cisplatin (Fig. 7.13) is a very useful antitumour agent for the treatment of testicular and ovarian tumours. Its discovery was fortuitous in the extreme, arising from research carried out to investigate the effects of an electric current on bacterial growth. During these experiments, it was discovered that bacterial cell division was inhibited. Further research led to the discovery that an electrolysis product from the platinum electrodes was responsible for the inhibition and the agent was eventually identified as *cis* diammonia dichloroplatinum (II), known as cisplatin.

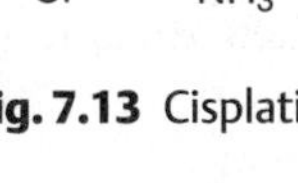

Fig. 7.13 Cisplatin.

Cisplatin binds strongly to DNA in regions containing several guanidine units, binding in such a way as to form links within strands (intrastrand binding) rather than between them. Unwinding of the DNA helix takes place and transcription is inhibited.

Mitomycin C is an anticancer drug which is converted to an alkylating agent in the body. This is initiated by an enzyme catalysed reduction of the quinone ring system to a hydroquinone. Loss of methanol and ring opening of the three-membered aziridine ring then takes place to generate the alkylating agent. Guanine residues on different DNA strands are then alkylated, leading to inhibition of DNA replication and cell division (Fig. 7.14).

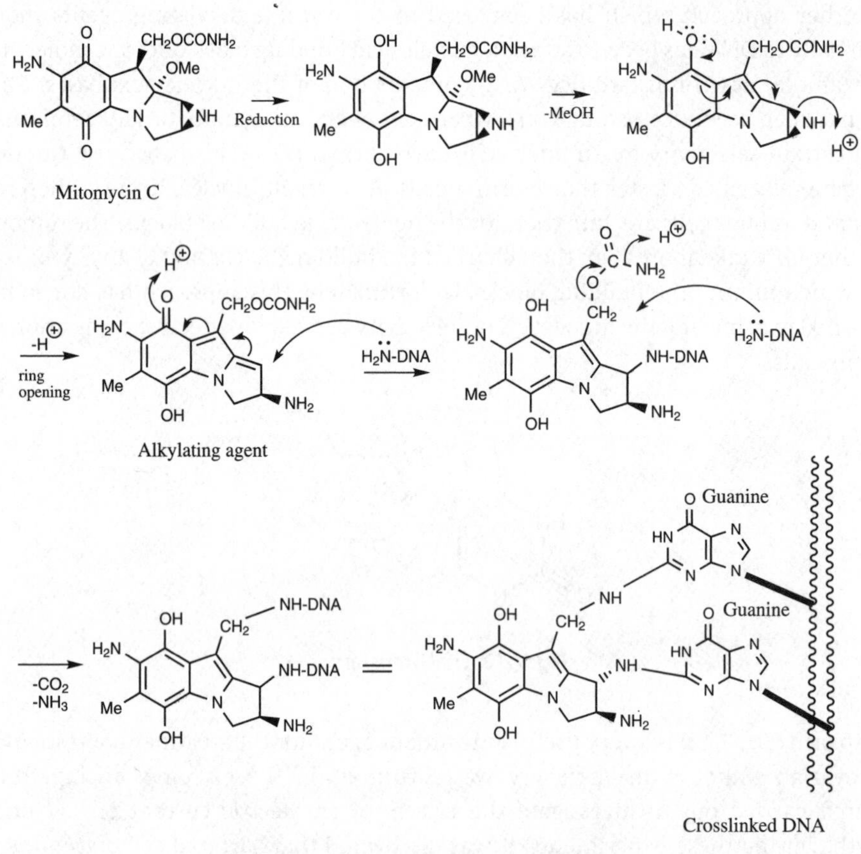

Fig. 7.14 DNA cross-linking by mitomycin C.

7.2.3 Drugs acting by chain 'cutting'

Bleomycin (Fig. 7.15) is a large glycoprotein which appears to be able to cut the strands of DNA and then prevent the enzyme DNA ligase from repairing the damage. It seems to act by abstracting hydrogen atoms from DNA. The resultant radicals react with oxygen to form peroxy species which then fragment. The drug is useful against certain types of skin cancer.

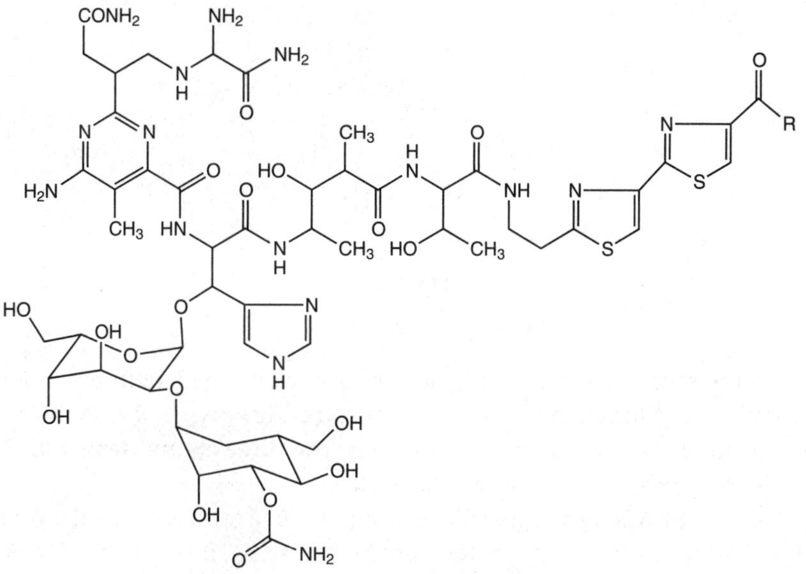

BLEOMYCIN A$_2$ R = NHCH$_2$CH$_2$CH$_2$SMe$_2$
BLEOMYCIN B$_2$ R = NHCH$_2$CH$_2$CH$_2$CH$_2$NHC(NH$_2$)=NH

Fig. 7.15 Bleomycin.

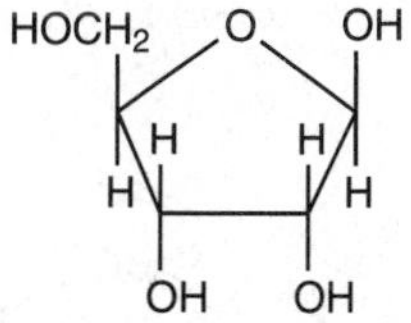

Fig. 7.16 Ribose

7.3 Ribonucleic acid

The primary structure of RNA is the same as DNA, with two exceptions. Ribose (Fig. 7.16) is the sugar component rather than deoxyribose, while uracil (Fig. 7.17) replaces thymine as one of the bases.

Base pairing between nucleic acid bases can occur in RNA, with adenine pairing to uracil and cytosine pairing to guanine. However, the pairing is between bases within the same chain, and it does not occur for the whole length of the molecule (e.g. Fig. 7.19). Therefore, RNA is not a double helix, but it does have regions of helical secondary structure.

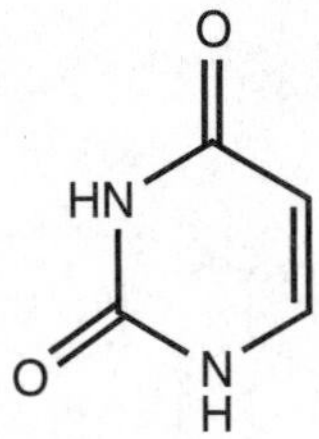

Fig. 7.17 Uracil.

Since the secondary structure is not uniform along the length of the RNA chain, more variety is allowed in RNA tertiary structure. Three types of RNA molecules have been identified with different cell functions. The three are messenger RNA (mRNA), transfer RNA (tRNA), and ribosomal RNA (rRNA).

Messenger RNA is responsible for relaying the code for one particular protein from the DNA genetic bank to the protein 'production site'. The segment of DNA required is copied by a process called transcription. The DNA double helix unravels and the stretch which is exposed acts as a template on which the mRNA can be built (Fig. 7.18). Once complete, the mRNA leaves to seek out rRNA, while the DNA reforms the double helix.

Ribosomal RNA can be looked upon as the production site for protein synthesis. It binds to one end of the mRNA molecule, then travels along it to the other end, reading the code and constructing the protein molecule one amino acid at a time as it moves

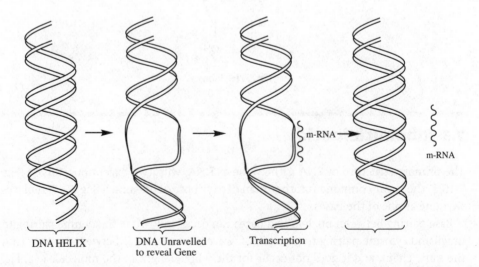

DNA HELIX   DNA Unravelled   Transcription
                 to reveal Gene

Fig. 7.18 Formation of mRNA.

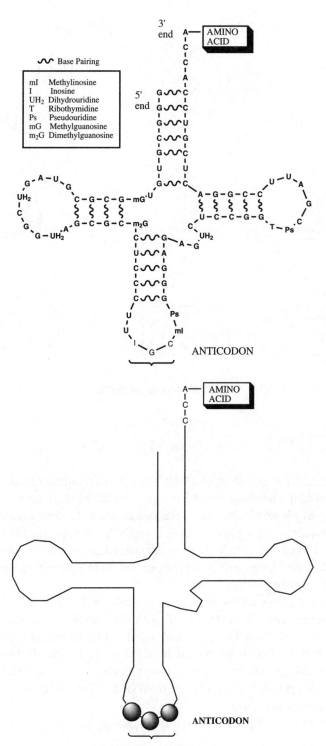

Fig. 7.19 Yeast alanine tRNA.

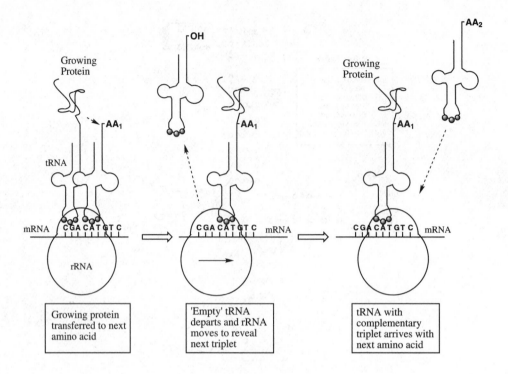

Fig. 7.20 Protein synthesis.

along (see Fig. 7.20). There are two segments to the rRNA, known as the 50S and 30S subunits.

Transfer RNA is the crucial adaptor unit which links the triplet code on mRNA to a specific amino acid. Therefore, there has to be a different tRNA for each amino acid. All the tRNAs are clover-leaf in shape with two different binding regions at opposite ends of the molecule (Fig. 7.19). One binding region is for a specific amino acid where the amino acid is covalently linked to the terminal adenosyl residue. The other is a set of three nucleic acid bases (anticodon) which will base pair with a complementary triplet on the mRNA molecule.

As rRNA travels along mRNA, it reveals the triplet codes on mRNA (Fig. 7.20). As a triplet code is revealed (e.g. CAT), a tRNA with the complementary GTA triplet will bind to it and bring the specific amino acid coded by that triplet. The growing peptide chain will then be grafted on to that amino acid (Fig. 7.21). The rRNA will shift along the chain to reveal the next triplet and so the process continues until the whole strand is read. The new protein is then released from rRNA, which is now available to start the process (translation) again.

The overall process of protein biosynthesis can be summarized as shown in Fig. 7.22.

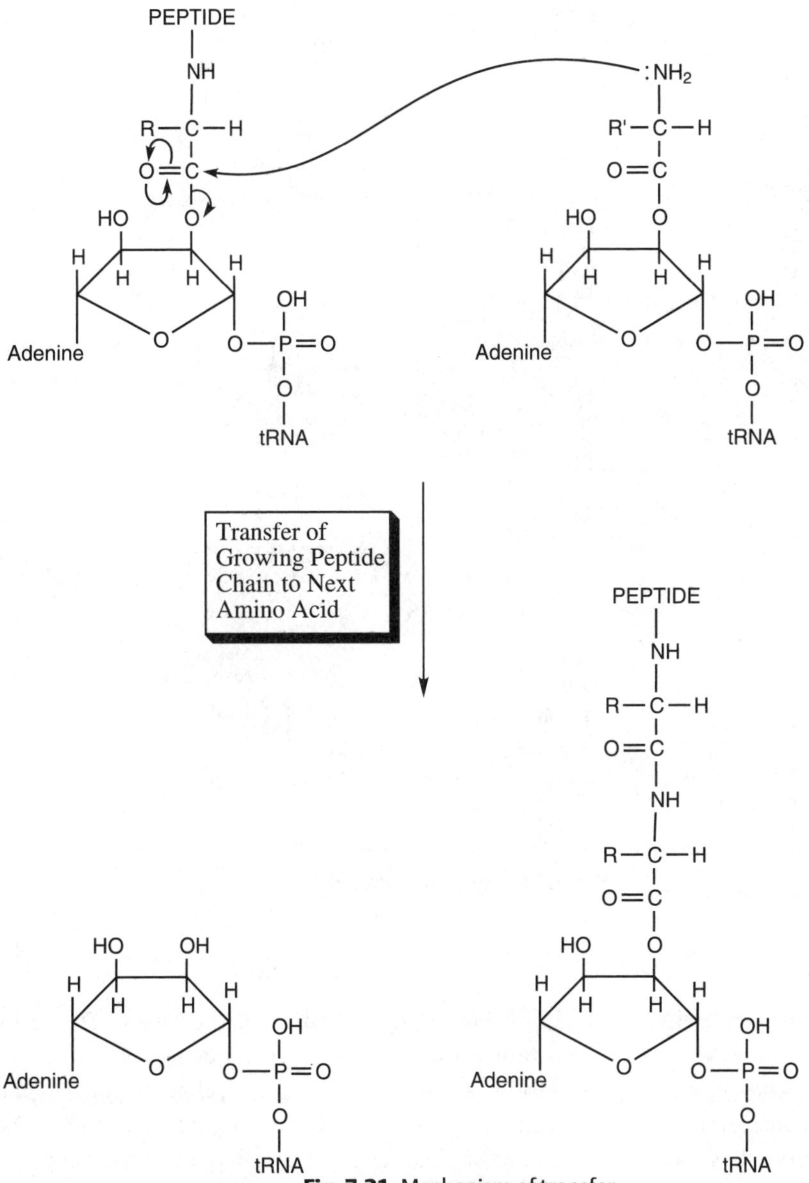

Fig. 7.21 Mechanism of transfer.

7.4 Drugs acting on RNA

Several antibiotic agents are capable of acting on RNA molecules and interfering with transcription and translation. These are discussed in Chapter 14.

A great deal of research has also been carried out into the possibility of using oligonucleotides to block the coded message carried by mRNA. This is an approach

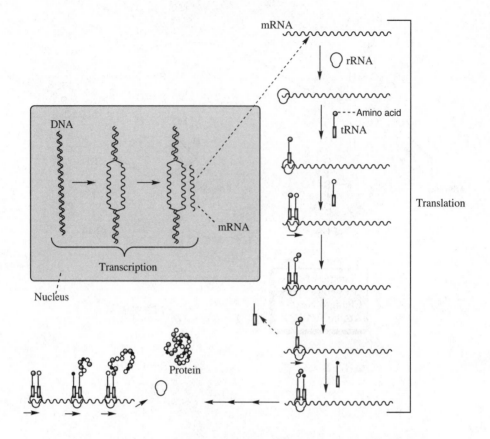

mRNA

rRNA

Amino acid

tRNA

Translation

DNA

mRNA

Transcription

Nucleus

Protein

Fig. 7.22 Protein biosynthesis.

known as **antisense therapy** and has interesting potential for the future. The rationale is as follows (Fig. 7.23). Assuming that the primary sequence of an mRNA
molecule is known, an oligonucleotide can be synthesized such that its nucleic acid
bases are complementary to a specific stretch of the mRNA molecule. Since the
oligonucleotide has a complementary base sequence, it is called an antisense oligonucleotide. When mixed with mRNA, the antisense oligonucleotide recognizes its
complementary section in the mRNA, interacts with it and forms a duplex structure
such that the bases pair up by hydrogen bonding. This section can now hinder the
enzymes involved in the translation process and block protein synthesis.

There are several advantages to this approach. First of all, it can be highly specific.
Statistically, an oligonucleotide of 17 nucleotides should be specific for a single mRNA
molecule and result in the inhibition of a single protein. Secondly, since one mRNA
leads to several copies of the same protein, inhibiting mRNA should be more efficient
than inhibiting the resulting protein. Both these factors should result in fewer side-

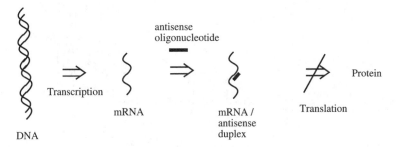

Fig. 7.23 Antisense therapy.

effects than conventional protein inhibition.

However, there are several difficulties involved in designing suitable antisense drugs. mRNA is a large molecule with a secondary and tertiary structure. Care has to be taken to choose a section which is 'exposed'. There are also problems related to the pharmacokinetics of nucleotides with respect to their stability and their ability to enter cells (see Chapter 10).

Nevertheless, antisense oligonucleotides provide the possibility of promising antiviral and anticancer agents for the future since they should be capable of preventing the biosynthesis of 'rogue' proteins and have fewer side-effects than currently used drugs. The first antisense oligonucleotide to be approved for clinical trials in 1992 was for the treatment of genital warts caused by the papilloma virus.

7.5 Drugs related to nucleic acids and nucleic acid building blocks

There are several important drugs which are related structurally to nucleic acids or to the building blocks for nucleic acids. Antisense oligonucleotides (see Section 7.4) and uracil mustard (see Section 7.2.2) are two such examples. Another important group of drugs which falls into this category are the various nucleoside-like structures used as antiviral agents. The vast majority of these are not active themselves but are phosphorylated by cellular enzymes to form an active nucleotide. For example, AZT (zidovudine or Retrovir) (Fig. 7.24) was the first drug to be approved for use in the treatment of AIDS. It is an analogue of thymidine and is phosphorylated to form a triphosphate. This compound inhibits a viral enzyme called reverse transcriptase which is required for the synthesis of viral DNA. Furthermore, the triphosphate is attached to the growing DNA chain. Since the sugar unit has an azide substituent instead of the required hydroxyl group, the nucleic acid chain cannot be extended any further.

Other antiviral agents such as acyclovir (Zovirax) and famciclovir (Famvir) (Fig. 7.25) have an incomplete sugar ring which serves as the chain terminator. Like AZT, these structures are phosphorylated to an active triphosphate form. Both structures are used in the treatment of Herpes simplex infections and shingles.

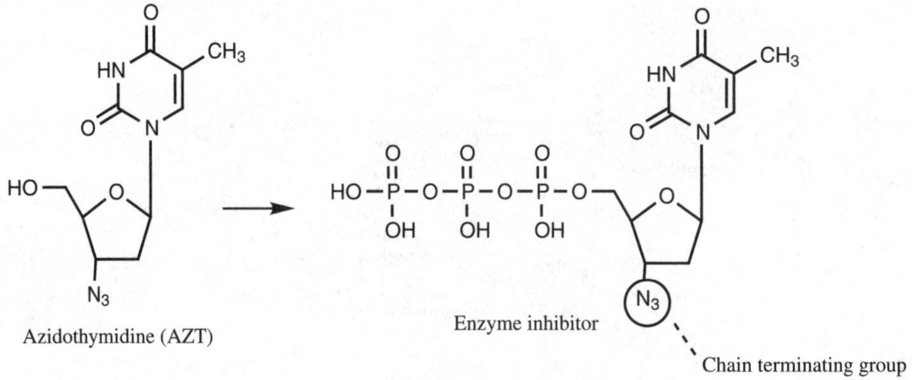

Fig. 7.24 AZT.

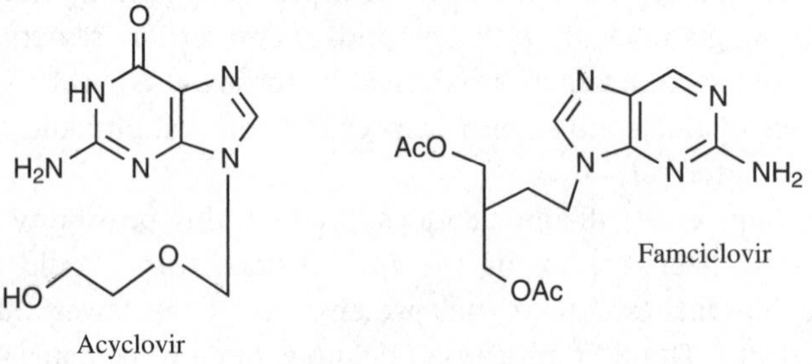

Fig. 7.25 Antiviral agents.

7.6 Molecular biology and genetic engineering

Over the last few years, rapid advances in molecular biology and genetic engineering have had important repercussions for pharmacology and medicinal chemistry. A detailed account of this field is not possible in an introductory text of this sort. Suffice it to say that it is possible to clone specific genes and to include these genes into the DNA of fast growing cells such that the proteins encoded by these genes are expressed in the modified cell. Since the cells are fast growing, this leads to a significant quantity of the desired protein, which permits its isolation, purification, and structural determination. Before these techniques became available, it was extremely difficult to isolate and purify many proteins from their parent cells because of the small quantities present. Even if one was successful, the low yields inherent in the process made analysis of the protein's structure and mechanism of action very difficult. Advances in molecular biology and recombinant DNA techniques have changed all that.

The following are some of the applications of genetic engineering to the medical field.

'Harvesting' important hormones and growth factors

The genes for important hormones or growth factors, such as insulin and human growth factor, have been included in fast growing unicellular organisms. This allows the 'harvesting' of these proteins in sufficient quantity that they can be marketed and administered to patients who are deficient in these hormones.

Identification of new protein drug targets

Nowadays, it is relatively easy to isolate and identify a range of signalling proteins, enzymes, and receptors by cloning techniques. This has led to the identification of a growing number of isozymes and receptor subtypes which offer potential drug targets for the future. The human genome project involves the mapping of human DNA and has led to the discovery of new proteins previously unsuspected. These too may offer potential drug targets.

Study of the molecular mechanism of target proteins.

Genetic engineering allows the controlled mutation of proteins such that specific amino acids are altered. This allows researchers to identify which amino acids are important to enzyme activity or to receptor binding. This in turn leads to a greater understanding of how enzymes and receptors operate at the molecular level.

Somatic gene therapy

Somatic gene therapy involves the use of a carrier virus to smuggle a healthy gene into cells in the body where the corresponding gene is defective. Once the virus has infected the cell, the healthy gene is inserted into the host DNA where it undergoes transcription and translation. This approach has great therapeutic potential for cancers, AIDS, and genetic abnormalities such as cystic fibrosis. However, the approach is still confined to research labs and there is still a long way to go before it is used clinically. There are several problems still to be tackled such as how to target the viruses specifically to the defective cells, how to insert the gene into DNA in a controlled manner, how to regulate gene expression once it is in the DNA, and how to avoid immune responses to the carrier virus.

7.7 Summary

In this chapter we have concentrated on drugs which are structurally related to the nucleic acid building blocks or which act on transcription, translation, and replication by acting directly on DNA and RNA. There are other drugs (e.g. nalidixic acid) which affect these processes but, since these drugs work by inhibiting enzymes rather than by a direct interaction with DNA or RNA, they have not been mentioned here.

8 Drug discovery and drug development

8.1 Introduction

8.1.1 Drug discovery and development – the past

Before the twentieth century medicines consisted mainly of herbs and potions, and it was not until the mid-nineteenth century that the first serious efforts were made to isolate and purify the active principles of these remedies (i.e. the pure chemicals responsible for the medical properties). The success of these efforts led to the birth of many of the pharmaceutical companies we know today. Since then, many naturally occurring drugs have been obtained and their structures determined (e.g. morphine from opium, cocaine from coca leaves, and quinine from the bark of the cinchona tree).

These natural products sparked off a major synthetic effort in which chemists made literally thousands of analogues in an attempt to improve on what nature had provided. Much of this work was carried out on a trial and error basis, but the results obtained revealed several general principles behind drug design. Many of these principles are described in Chapters 9 and 10.

An overall pattern for drug discovery and drug development also evolved, but there was still a large element of trial and error involved in the process. The mechanism by which a drug worked at the molecular level was rarely understood and drug research very much focused on what is known as the **lead compound** – the active principle isolated from the plant.

8.1.2 Drug discovery and development – the present

In recent years, medicinal chemistry has undergone a revolutionary change. Rapid advances in the biological sciences have resulted in a much better understanding of how the body functions at the cellular and molecular levels. As a result, most research projects in the pharmaceutical industry now begin by identifying a suitable target in the body and designing a drug to interact with that target. An understanding of the structure and mechanism of the target is crucial to this approach and so research is now target-orientated.

Generally, we can identify the following stages in drug discovery and drug development:

- Choose a disease!
- Choose a drug target
- Identify a bioassay
- Find a 'lead compound'
- Isolate and purify the lead compound if necessary
- Determine the structure of the lead compound
- Identify structure–activity relationships (SARs)
- Identify the pharmacophore
- Improve target interactions
- Improve pharmacokinetic properties
- Patent the drug
- Study drug metabolism
- Test for toxicity
- Design a manufacturing process
- Carry out clinical trials
- Market the drug
- Make money!

Many of these stages run concurrently and are dependent on each other. For example, drug metabolism studies, toxicity testing and the development of a large scale synthesis are usually carried out in parallel. Even so, the discovery and development of a new drug can take 10 years or more, involve the synthesis of over 10 000 compounds and cost in the region of $360 million. We shall now look at each stage in turn.

8.2 Choosing a disease

How does a pharmaceutical company decide which disease to target when designing a new drug? Clearly, it would make sense to concentrate on diseases where there is a need for new and/or improved drugs. However, pharmaceutical companies have to consider economic factors as well as medical ones. A huge investment has to be made towards the research and development of a new drug. Therefore, companies must ensure that they get a good financial return for their investment. As a result, research projects tend to be biased towards 'first world' diseases since this is the market best able to afford new drugs. A great deal of research is carried out on ailments such as migraine, depression, ulcers, obesity, flu, cancer, and cardiovascular disease. Less is carried out on the tropical diseases of the third world – diseases which can reduce life expectancies to the thirties or forties. Only when such diseases start to make an impact on western society do the pharmaceutical companies sit up and take notice. For example, there has been a noticeable increase in antimalarial research because of an increase in tourism to more exotic countries and the spread of malaria into the southern states of the USA.

Choosing which disease to tackle is usually a matter for a company's market state-gists. The science becomes important at the next stage.

8.3 Choosing a drug target

8.3.1 Drug targets

Once a particular area of medical need has been determined, the next stage is to iden-tify a suitable drug target (i.e. receptor, enzyme, or nucleic acid). An understanding of which enzymes or receptors are involved in a particular disease state is clearly impor-tant. This allows the medicinal research team to identify whether agonists or antagonists should be designed for a particular receptor, or whether inhibitors should be designed for a particular enzyme. For example, agonists of serotonin receptors are useful for the treatment of migraine, while antagonists of dopamine receptors are useful as antidepressants (see Chapter 5). Sometimes it is not known for certain whether a particular target will be suitable or not. For example, the classic tricyclic antidepressants are known to inhibit the uptake of the neurotransmitter nora-drenaline from nerve synapses by inhibiting the carrier protein responsible for noradrenaline's uptake (see Chapter 16). However, these drugs were also observed to inhibit uptake of a separate neurotransmitter called serotonin, and the possibility arose that inhibiting serotonin uptake might also be beneficial. A search for selective serotonin uptake inhibitors was initiated which led to the discovery of the best-selling antidepressant drug fluoxetine (Prozac), but when this project was initiated it was not known for certain whether serotonin uptake inhibitors would be effective or not.

8.3.2 Discovering drug targets

In the past, the existence of a drug target could only be established if a drug or a poison produced a biological effect, demonstrating the existence of a molecular target. Therefore, the discovery of drug targets depended on finding the drug first. Many early drugs were natural products from plants. However, natural products from a plant are not synthesized to interact with a receptor or enzyme in the human body, so such interactions are due to coincidence rather than design. Therefore, detecting drug targets in this way was very much a hit and miss affair. Later, the body's own chem-ical messengers began to be discovered and these pointed the finger at particular targets. However, relatively few of the body's messengers were identified, either because they were present in such small quantities or because they were too short-lived to be isolated. Indeed, many chemical messengers still remain undiscovered today. This too meant that many of the body's potential drug targets remained hidden.

Advances in molecular genetics have changed all this. The various genome projects which are mapping the DNA of humans and microorganisms are revealing an ever increasing number of new receptors and enzymes which are potential drug targets for the future. Since these targets have managed to stay hidden for so long, their natural chemical messengers are also unknown and, for the first time, medicinal chemistry is faced with new targets, but with no lead compounds to interact with them. The challenge is now to find a chemical which will interact with these targets to find out what their function is and whether they will be suitable as drug targets. This has been one of the main driving forces behind the rapidly expanding area of combinatorial synthesis (see Chapter 12).

As an analogy, one could imagine that the medicinal chemist of the past played the role of an Indiana Jones-like character who has discovered the key and map to some long lost temple in the jungle (Fig. 8.1). The map leads him to the very door which the key (the lead compound) will open, and he is able to enter to win his prize. However, since the temple is overgrown, he has no idea how many other doors (or targets) might be present, so he would have to go away to see if he could find another key and another map.

To continue the analogy, the present day medicinal chemist plays the role of a rather more methodical Indiana Jones who calls in 'the heavy squad' to clear the jungle from the temple, thus revealing all the possible doors. Once revealed, he is faced with a large number of hitherto unsuspected doors. However, he has no keys to fit them. To solve the problem, he can do one or more of the following:

- search for more keys as he did in the past
- get a lockmaker to make a huge number of random keys in the hope that some will fit (the equivalent of combinatorial synthesis)
- have his locksmith study the lock itself to see if a key could be designed to fit (the equivalent of computer-aided design).

An example of recently discovered novel targets are the caspases. These are a family of enzymes which catalyse the hydrolysis of important cellular proteins and which have been found to play a role in inflammation and cell death. As far as the latter role is concerned, it is important to realize that cell death is a normal occurrence in the body and that cells are regularly recycled. Therefore, caspases should not necessarily be seen as 'bad' or 'undesirable' enzymes. Without them, cells could well be more prone to unregulated growth, resulting in diseases such as cancer.

The caspases catalyse the hydrolysis of particular target proteins, such as those involved in DNA repair and the regulation of cell cycles. By understanding how these enzymes operate, there is the possibility of producing new therapies for a variety of

Fig. 8.1 Cartoon.

diseases. For example, agents which promote the activity of caspases and lead to more rapid cell death might be useful in the treatment of diseases such as cancer, autoimmune disease, and viral infections. Alternatively, agents which inhibit caspases and reduce the prevalence of cell death could provide novel treatments for trauma, neurodegenerative disease, and cell death arising from strokes. It is already known that the caspases recognize protein substrates with aspartate linkages and that the mechanism of hydrolysis involves the amino acids cysteine (acting as a nucleophile) and histidine (acting as an acid–base catalyst) in the caspase active site. The mechanism is similar to that used by acetylcholinesterase (see Section 15.16).

8.3.3 Target specificity and selectivity between species

Target specificity and selectivity is a crucial factor in modern medicinal chemistry research. The more selective a drug is for its target, the less chance that it will interact with different targets and the less chance that it will have undesired side-effects.

In the field of antimicrobial agents, the best targets to choose are those which are unique to the microbe and which are not present in man. For example, penicillin targets an enzyme involved in bacterial cell wall biosynthesis. Since mammalian cells do not have a cell wall, this enzyme is absent in human cells and penicillin has few side-effects (see Chapter 14).

Other cellular features which are unique to microorganisms could also be targeted. For example, the microorganisms which cause sleeping sickness in Africa are propelled by means of a tail-like structure called a flagellum. This feature is not present in mammalian cells and so designing drugs which would bind to the proteins making up the flagellum and prevent it from working could be potentially useful in treating sleeping sickness.

Having said all that, it is still possible to design drugs against targets which are present both in humans and microbes, as long as the drugs show selectivity against the microbial target. Fortunately, there are significant differences between such targets to allow such selectivity. An enzyme which catalyses a reaction in a bacterial cell differs significantly from the equivalent enzyme in a human cell. The enzymes may have been derived from an ancient common ancestor, but several million years of evolution have resulted in significant structural differences. For example, the antifungal agent fluconazole (Fig. 8.2) inhibits a fungal demethylase enzyme involved in steroid biosynthesis. This enzyme is also present in man, but the structural differences between the two enzymes are so significant that the antifungal agent is highly selective for the fungal enzyme.

8.3.4 Target specificity and selectivity within the body

Target selectivity is also important for drugs acting on our own receptors or enzymes. Enzyme inhibitors should only inhibit the target enzyme and not some other enzyme. Receptor agonists/antagonists should ideally interact with a specific kind of receptor (e.g. the adrenergic receptor) rather than a variety of different receptors. However,

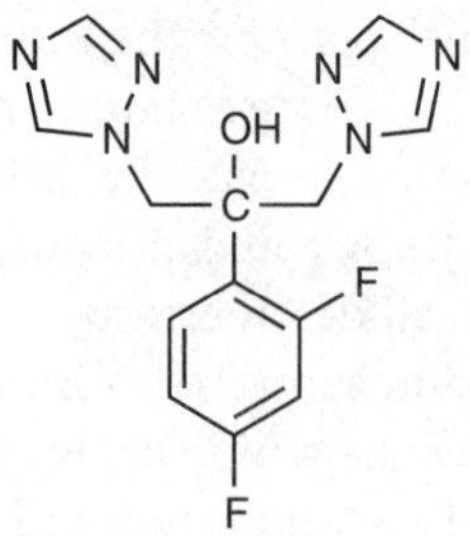

Fig. 8.2 Fluconazole (UK-49858).

nowadays, medicinal chemists aim for even higher standards of target selectivity. Ideally, enzyme inhibitors should show selectivity between the various isozymes of that enzyme. Receptor agonists and antagonists should not only show selectivity for a particular receptor (e.g. an adrenergic receptor) or even a particular receptor type (e.g. the β-adrenergic receptor), but also for a particular receptor subtype (e.g. the β_2-adrenergic receptor).

A current area of research is to find antipsychotic agents with fewer side-effects. Traditional antipsychotic agents act as antagonists of dopamine receptors. However, it has been found that there are five dopamine receptor subtypes and that traditional antipyschotic agents antagonize two of these (D_3 and D_2). There is good evidence that the D_2-receptor is responsible for the undesirable Parkinsonian type side-effects of current drugs, and so research is now under way to find a selective D_3-antagonist.

8.3.5 Targeting drugs to specific organs and tissues

Targeting drugs against specific receptor subtypes often allows drugs to be targeted against specific organs or against specific areas of the brain. This is because the various receptor subtypes are not uniformly distributed around the body, but are often concentrated in particular tissues. For example, the β-adrenergic receptors in the heart are predominantly β_1 while those in the lungs are β_2. This makes it feasible to design drugs which will work on the lungs with a minimal side-effect on the heart, and *vice versa*.

Attaining subtype selectivity is particularly important for drugs which are intended to mimic neurotransmitters. Neurotransmitters are released close to their target receptors and, once they have passed on their message, are quickly deactivated and do not have the opportunity to 'switch on' more distant receptors. Therefore, only those receptors which are fed by 'live' nerves are switched on.

In many diseases, there is a 'transmission fault' in a particular tissue or in a particular region of the brain. For example, in Parkinson's disease dopamine transmission is deficient in certain regions of the brain. However, dopamine transmission elsewhere functions normally. A drug could be given to mimic dopamine in the brain. However, such a drug acts like a hormone rather than as a neurotransmitter since it has to travel round the body to reach its target. This means that the drug could potentially 'switch on' all the dopamine receptors around the body and not just the ones which are suffering the dopamine deficit. Such drugs would have a large number of side-effects and so it is important to make the drug as selective as possible for the particular type or subtype of dopamine receptor affected in the brain. This would target the drug more effectively to the affected area and reduce side-effects elsewhere in the body.

Many research projects set out to discover new drugs with a defined profile of activity against a range of specific targets. For example, a research team may set out to find a drug which has agonist activity for one receptor subtype and antagonist activity at another. A further requirement may be that the drug does not inhibit metabolic enzymes (see Chapter 10).

8.3.6 Pitfalls

A word of caution! It is possible to identify whether a particular enzyme or receptor plays a role in a particular ailment. However, the body is a highly complex system. For any given function, there are usually several messengers, receptors, and enzymes involved in the process. For example, there is no one simple cause for hypertension (high blood pressure). This is illustrated by the variety of receptors and enzymes which can be targeted in its treatment (e.g. β_1-adrenoreceptors, calcium ion channels, angiotensin-converting enzyme (ACE), potassium ion channels, etc.).

As a result, more than one target may need to be addressed for a particular ailment. For example, most of the current therapies for asthma involve a combination of a bronchodilator (β_2-agonist) and an anti-inflammatory agent (e.g. a corticosteroid).

Sometimes it is found that a particular target is not so important to a disease as was first thought. For example, the dopamine D_2-receptor was thought to be involved in causing nausea. Therefore, the D_2-receptor antagonist metoclopramide (Fig. 8.3) was developed as an antiemetic agent. However, further developments showed that more potent D_2-antagonists were less effective, implying that a different receptor might be more important in producing nausea. Since metoclopramide also antagonizes the 5-hydroxytryptamine (5-HT$_3$) receptor, antagonists for this receptor were studied, leading to the development of the antiemetic drugs granisetron and ondansetron (Fig. 8.3).

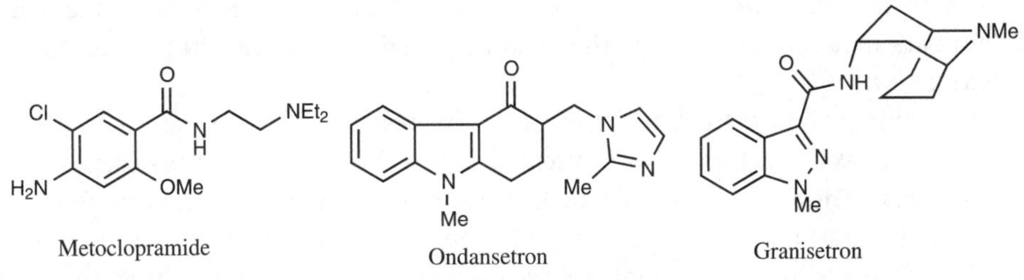

Fig. 8.3 Antiemetic agents.

Sometimes, drugs designed against a specific target become less effective over time. Since cells have a highly complex system of signalling mechanisms, it is possible that the blockade of one part of that system could be bypassed. This could be compared to blocking the main road into town to try and prevent town centre congestion. To begin with, the policy might work but in a day or two commuters would discover alternative routes and town centre congestion would become as bad as ever (Fig. 8.4).

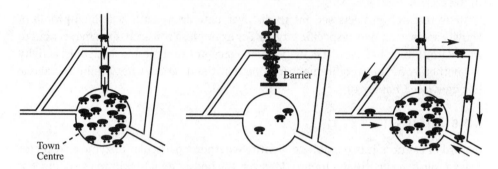

Fig.8.4 Avoiding the jam.

8.4 Identifying a bioassay

8.4.1 Choice of bioassay

Choosing the right bioassay or test system is crucial to the success of a drug research programme. The test should be simple, quick, and relevant since a large number of compounds usually need to be analysed. Human testing is not possible at such an early stage and so the test has to be done *in vitro* (i.e. on isolated cells, tissues, enzymes, or receptors) or *in vivo* (on animals). In general, *in vitro* tests are preferred over *in vivo* tests.

8.4.2 *In vivo* tests

In vivo tests on animals often involve inducing a clinical condition in an animal to produce observable symptoms. The animal is then treated to see whether the drug alleviates the problem by eliminating the observable symptoms. For example, the development of non-steroidal anti-inflammatory drugs was carried out by inducing inflammation on test animals and then testing to see whether the drugs relieved the inflammation.

Transgenic animals are often used in *in vivo* testing. These are animals whose genetic code has been altered. For example, it is possible to replace mouse genes with human genes. The mouse produces the human receptor or enzyme and this allows *in vivo* testing against that target. Alternatively, the mouse's genes could be altered such that the mouse became susceptible to a particular disease (e.g. breast cancer). Drugs could then be tested to see how well they prevent that disease.

There are several problems associated with *in vivo* testing. Such testing is slow and also causes animal suffering. There are the many problems of pharmacokinetics (see Chapter 10) and so the results obtained may be misleading and difficult to rationalize. For example, how can one tell whether a negative result is due to the drug failing to bind to its target or not reaching the target in the first place? The test might not even be valid. It is possible that the observed symptoms might be caused by a different physiological mechanism than the one intended. For example, many promising antiulcer drugs which proved effective in animal testing were ineffective in clinical trials.

Finally, different results may be obtained in different animal species. For example, penicillin methyl ester prodrugs (see Chapter 14) are hydrolysed in the mouse or rat to produce active penicillins, but are not hydrolysed in rabbit, dog or man.

8.4.3 *In vitro* tests

In vitro tests do not involve live animals. Instead, specific tissues, cells, or enzymes are used. Enzyme inhibitors can be tested on the pure enzyme in solution. Receptor agonists and antagonists can be tested on isolated tissues or cells which express the target receptor on their surface. Sometimes these tissues can be used to test drugs for physiological effects. For example, bronchodilator activity can be tested by observing how well compounds inhibit contraction of isolated tracheal smooth muscle. Alternatively, the affinity of drugs for receptors (how strongly they bind) can be measured by radioligand studies. For example, a drug could be applied to the tissue and allowed to bind to the target receptor. A radiolabelled ligand for the receptor could then be applied to show how many receptors have not been occupied by the test compound.

Many *in vitro* tests have been designed using molecular genetics where the gene coding for a specific enzyme or receptor is identified, cloned, and expressed in fast dividing cells such as bacterial, yeast, or tumour cells. For example, HIV protease has been cloned and expressed in the bacterium *Escherichia coli.*

8.4.4 Test validity

Sometimes the validity of testing procedures is easy and clearcut. For example, an antibacterial agent can be tested *in vitro* by measuring how effectively it kills bacterial cells. A local anaesthetic can be tested *in vitro* on how well it blocks action potentials in isolated nerve tissue. In other cases, the testing procedure is more difficult. For example, how do you test a new antipsychotic drug? There is no animal model for this condition and so a simple *in vivo* test is not possible. One way round this problem is to propose which receptor or receptors might be involved in a medical condition and to carry out *in vitro* tests against these in the expectation that the drug will have the desired activity when it comes to clinical trials. One problem with this approach is that it is not always clearcut whether a specific receptor or enzyme is as important to the targeted disease as one might think (e.g. metoclopramide; see Section 8.3.6).

8.4.5 High throughput screening

Robotics and the miniaturization of *in vitro* tests on genetically modifed cells has led to a process called high throughput screening (HTS). This involves the automated testing of large numbers of compounds against a large number of targets where, typically, several thousand compounds can be tested in 30–50 biochemical tests at once. It is important that the test should produce an easily measurable effect which can be detected and measured automatically. This effect could be cell growth, an enzyme-catalysed reaction which produces a colour change, or displacement of radioactively labelled ligands from receptors.

Receptor antagonists can be studied using modified cells which contain the target receptor in their cell membrane. Detection is possible by observing how effectively the test compounds inhibit the binding of a radiolabelled ligand. Another approach is to use yeast cells which have been modified such that activation of a target receptor results in the production of an enzyme which, when supplied with a suitable substrate, catalyses the release of a dye, resulting in a colour change.

8.4.6 Screening by nuclear magnetic resonance

Nuclear magnetic resonance (NMR) spectroscopy can be used to detect whether a compound binds to a protein target. In an NMR spectrum, a compound is radiated with a short pulse of energy and its nuclei are promoted to an excited state. The nuclei then slowly relax to the ground state, giving off energy as they do so. This energy can be measured to produce a spectrum. The time taken by different nuclei to relax back to their resting state is called the relaxation time and this varies depending on the size of the molecule. Small molecules such as drugs have long relaxation times, whereas large molecules have short relaxation times. Therefore it is possible to delay the measurement of energy emission such that only small molecules are detected. This is the key to the detection of protein binding.

First, the NMR spectrum of the drug is taken, then the protein is added and the spectrum is rerun, introducing a delay in the measurement such that the protein signals are not detected. If the drug fails to bind to the protein, then its NMR spectrum will still be detected. If the drug binds to the protein, it essentially becomes part of the protein. As a result, its nuclei will now have a shorter relaxation time and no NMR spectrum will be detected.

This screening method can also be applied to a mixture of compounds resulting from a natural extract or from a combinatorial synthesis. If any of the compounds present bind to the protein, its relaxation time is shortened and so signals due to that compound will disappear from the spectrum. This will show that a component of the mixture is active and identify whether it is worthwhile separating the mixture or not.

There are several advantages in using NMR as a detection system:

• It is possible to screen 1000 small molecular weight compounds a day.

• The method can detect weak binding which would be missed by conventional screening methods.

• It can identify small molecules (epitopes) binding to small regions of the binding site.

• It is complimentary to high throughput screening. The latter may give false-positive results, but these can be checked by NMR to ensure that the compounds concerned are binding in the correct binding site (see Section 8.5.10).

• The identification of weakly binding molecules allows the possibility of using them as building blocks for the construction of larger molecules which bind more strongly (see Section 8.5.10).

• Screening can be done on a new protein without needing to know its function.

There are also limitations to screening by NMR, the main one being that it is limited to proteins which can be obtained in quantities greater than 200 mg.

8.5 Finding a lead compound

Once a target and a testing system have been chosen, the next stage is to find a 'lead compound' – a compound which shows the desired pharmaceutical activity. The level of activity may not be very great and there may be undesirable side-effects, but the lead compound provides a start for the drug development process. There are many ways in which a lead compound might be discovered.

• Screening of natural materials
• Medical folklore
• Screening of synthetic 'banks'
• Existing drugs
• Starting from the natural ligand or modulator
• Combinatorial synthesis
• Computer-aided design
• Serendipity and the prepared mind
• Computerized searching of structural databanks
• Designing lead compounds by NMR

8.5.1 Screening of natural products

Natural products have provided biologically active compounds for many years and many of today's medicines are either obtained directly from a natural source or were developed from a lead compound originally obtained from a natural source. Most biologically active natural products are secondary metabolites with quite complex structures. This has the advantage that they are extremely novel lead compounds. Unfortunately, this complexity also makes their synthesis extremely difficult and the compound usually has to be extracted from its natural source – a slow, expensive, and inefficient process. As a result, there is usually an advantage in designing simpler analogues (see Chapter 9).

In general, natural products are particularly good at providing radically new chemical structures which no chemist would dream of synthesizing. For example, the antimalarial drug artemisinin (Fig. 8.5) is one such example, containing as it does an extremely unstable-looking trioxane ring – one of the most unlikely structures to have appeared in recent years.

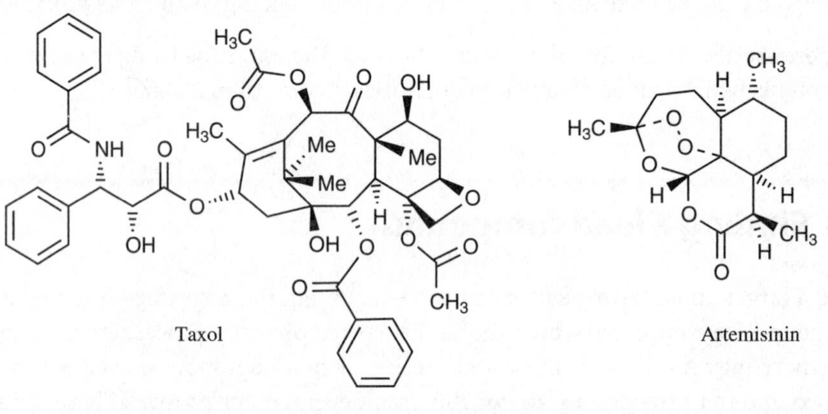

Taxol   Artemisinin

Fig. 8.5 Natural products as drugs.

8.5.1.1 The plant kingdom

Plants and trees have always been a rich source of lead compounds (e.g. morphine, cocaine, digitalis, quinine, tubocurarine, nicotine, muscarine, and many others). Many of these lead compounds are useful drugs in themselves (e.g. morphine and quinine), while others have been the basis for synthetic drugs (e.g. local anaesthetics developed from cocaine). Plants still remain a promising source of new drugs and will continue to be so. Clinically useful drugs which have recently been isolated from plants include the anticancer agent taxol from the yew tree, and the antimalarial agent artemisinin from a Chinese plant (Fig. 8.5).

Plants provide a bank of rich, complex and highly varied structures which are unlikely to be synthesized in laboratories. Furthermore, evolution has already carried out a screening process whereby plants are more likely to survive if they contain potent compounds which deter animals or insects from eating them. Considering the debt medicinal chemistry owes to the natural world, it is sobering to think that very few plants have been fully studied and the vast majority have not been studied at all. The rainforests of the world are particularly rich in plant species which have still to be discovered, let alone studied. Who knows how many exciting new lead compounds await discovery for the fight against cancer, AIDS or any of the other myriad of human afflictions? This is why the destruction of rainforests and other ecosystems is so tragic. Once these ecosystems are destroyed, unique plant species are destroyed with them. Medicine has already lost potentially useful plants for ever. For example, silphion – a plant cultivated near Cyrene in North Africa and famed as a contraceptive agent in ancient Greece – is now extinct. It is almost certain that many more useful plants have become extinct without medicine ever being aware of them.

8.5.1.2 The microbiological world

Microorganisms such as bacteria and fungi have also provided rich pickings for drugs and lead compounds. These organisms produce a large variety of antimicrobial agents which have evolved to give their hosts an advantage over their competitors in the microbiological world. The screening of microorganisms became highly popular following the discovery of penicillin. Soil and water samples were collected from all round the world to study new fungal or bacterial strains, leading to an impressive arsenal of antibacterial agents such as the cephalosporins, tetracyclines, aminoglycosides, rifamycins, and chloramphenicol (see Chapter 14). Although most of the drugs derived from microorganisms are used in antibacterial therapy, some microbial metabolites have provided lead compounds in other fields. For example, asperlicin (Fig. 8.6), isolated from *Aspergillus alliaceus,* proved to be a novel antagonist of a peptide hormone called cholecystokinin which is involved in the control of appetite. Cholecystokinin also acts as a neurotransmitter in the brain and is thought to be involved in panic attacks. Therefore, analogues of asperlicin may have potential in treating anxiety.

  Other examples include the fungal metabolites lovastatin (Fig. 8.6), which was the lead compound for a series of drugs which lower cholesterol levels, and cyclosporin (Fig. 8.6), which is used to suppress the immune response when carrying out transplant operations.

8.5.1.3 The marine world

In recent years, there has been great interest in the possibility of finding lead compounds from marine sources. Coral and sponges have a wealth of biologically potent chemicals with interesting inflammatory, antiviral and anticancer activity. For example, curacin A (Fig. 8.7) is obtained from a marine cyanobacterium and shows potent antitumour activity.

8.5.1.4 Animal sources

Animals can sometimes be a source of new lead compounds. For example, a series of antibiotic peptides were extracted from the skin of the African clawed frog, while a potent analgesic compound called epibatidine (Fig. 8.7) was obtained from the skin extracts of the Ecuadorian poison frog.

8.5.1.5 Venoms and toxins

Venoms and toxins from animals, plants, insects, and microorganisms can often be useful lead compounds. These compounds are extremely potent because they often

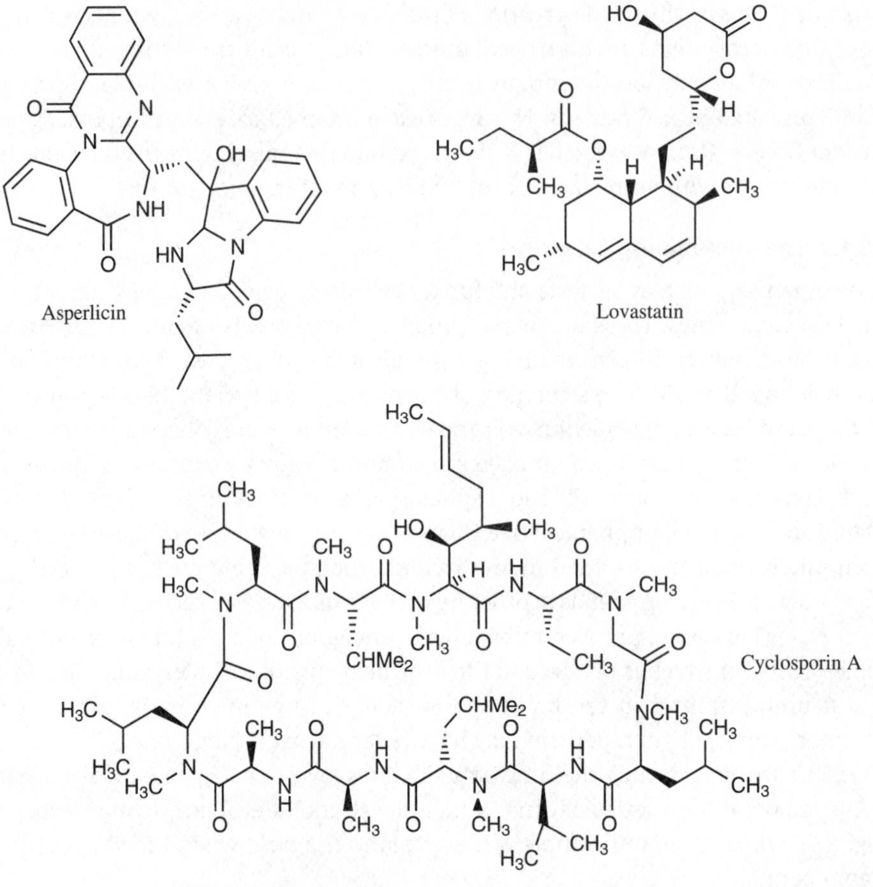

Fig. 8.6 Fungal metabolites of pharmaceutical interest.

have a very specific interaction with a receptor or enzyme in the body. As a result, they have proved important tools in studying receptors, ion channels, and enzymes. Many of these toxins are polypeptides (e.g. α-bungarotoxin from cobras). However, non-peptide toxins are also extremely potent (e.g. tetrodotoxin from puffer fish; see Fig. 8.7).

Venoms and toxins have been used as lead compounds in the development of novel drugs. For example, teprotide is a peptide which was isolated from the venom of the Brazilian viper and was the lead compound for the development of the antihypertensive agents cilazapril and captopril (Fig. 8.8).

The neurotoxins from *Clostridium botulinum* are responsible for serious food poisoning (botulism). However, the toxins have a clinical use. They can be injected into specific muscles (such as those controlling the eyelid) to prevent muscle spasm. These toxins prevent cholinergic transmission (see Chapter 15) and could well prove a lead for the development of novel drugs acting against cholinergic transmission.

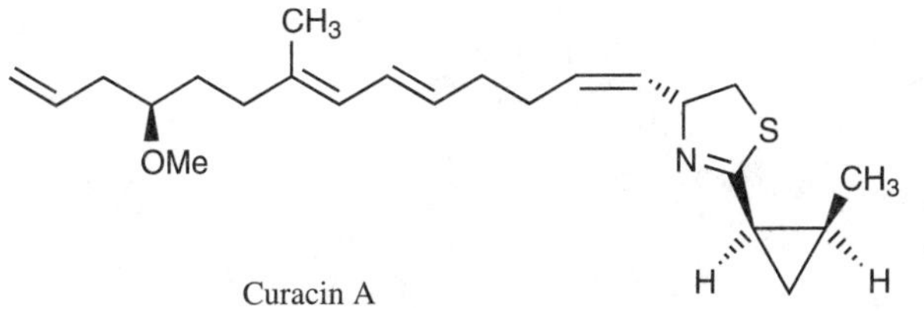

Curacin A

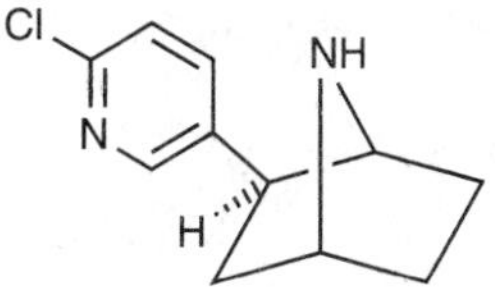

Epibatidine

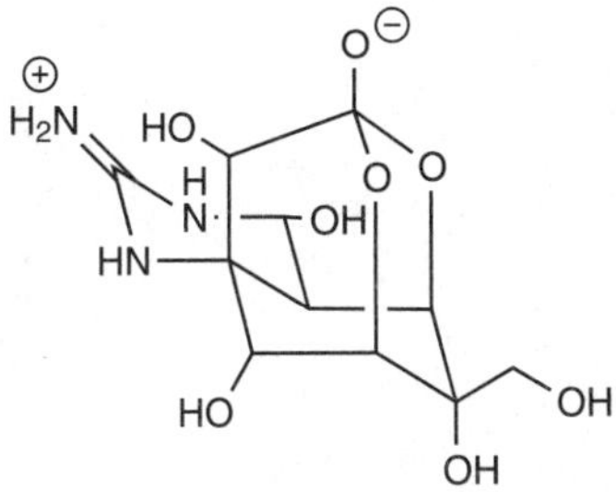

Tetrodotoxin

Fig 8.7 Natural products as drugs.

Glu-Trp-Pro-Arg-Pro-Gln-Ile-Pro-Pro

Teprotide (SQ20881)

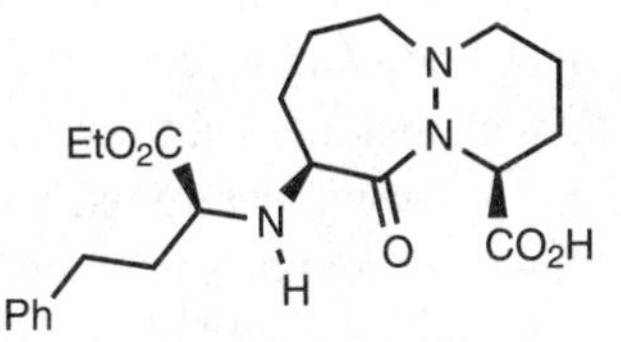

Cilazapril

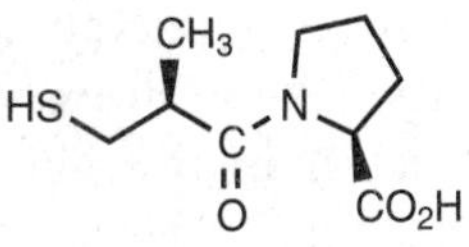

Captopril

Fig. 8.8 Antihypertensive agents.

8.5.2 Medical folklore

It is often worthwhile studying the medical folklore of ancient civilizations throughout the world. In the past, these civilizations depended greatly on local flora and fauna for their survival. They would often experiment with various berries, leaves, and roots to find out what effects they had. As a result, many brews were claimed by the local medicine man or witch doctor to have some medicinal use. More often than not, these concoctions were useless or downright dangerous and if they worked at all it was because the patient willed it to work – a placebo effect. However, some of these concoctions may indeed have a real and beneficial effect, and a study of medical folklore can give clues as to which plants might be worth studying in more detail. Rhubarb root has been used as a purgative for many centuries and in China it was called 'The General' because of its 'galloping charge'! The most significant chemicals in rhubarb root are anthraquinones which were used as the lead compounds in the design of the laxative danthron (Fig. 8.9).

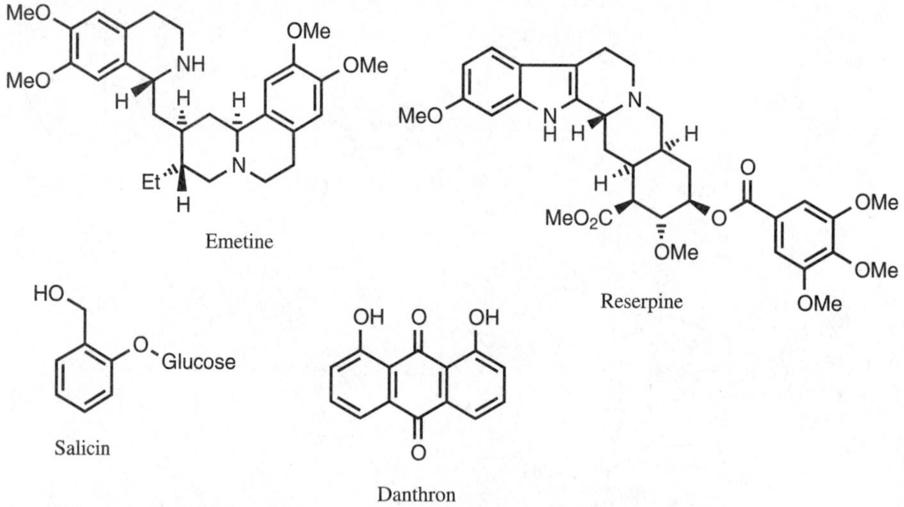

Fig. 8.9 Active compounds resulting from studies of herbs and potions.

The ancient records of Chinese medicine also provided the clue to the novel anti-malarial drug artemisinin mentioned previously. The therapeutic properties of the opium poppy (morphine) were known in Ancient Egypt, as were the therapeutic properties of the Solanaceae plants in ancient Greece (atropine and hyoscine), the snakeroot plant in India (reserpine; Fig. 8.9), the willow tree (salicin; Fig. 8.9) and foxglove (digitalis) in mediaeval England, and the ipecacuanha root (emetine; Fig. 8.9), coca bush (cocaine), and cinchona bark (quinine) to the Aztec and Mayan cultures of South America.

8.5.3 Screening synthetic 'banks'

The thousands of compounds which have been synthesized by the pharmaceutical companies over the years are another source of possible lead compounds. The vast majority of these compounds never made the market place, but they have been stored in compound 'banks' and are still available for testing. Pharmaceutical companies often screen their bank of compounds whenever they study a new target. However, it has to be said that the vast majority of these compounds are merely variations on a theme (e.g. a thousand or so different penicillin structures). This reduces the chances of finding a novel lead compound.

Pharmaceutical companies often try to diversify their range of structures by purchasing novel compounds prepared by research groups elsewhere – a useful source of revenue for hard pressed university departments! These compounds may never have been synthesized with medicinal chemistry in mind and may be intermediates in a purely synthetic research study, but there is always the chance that they may have useful biological activity. The antitubercular drug isoniazid (Fig. 8.10) was discovered in this way. Similarly, a series of quinoline-3-carboxamide intermediates (Fig. 8.10) were found to have antiviral activity.

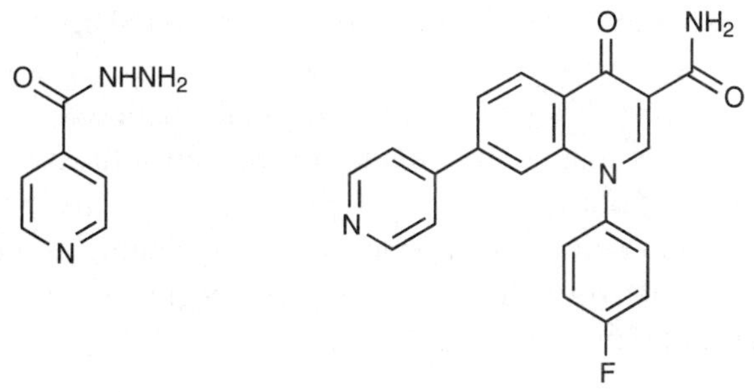

Isoniazid   Quinoline-3-carboxamides

Fig. 8.10 Pharmaceutically active compounds discovered from synthetic intermediates.

8.5.4 Existing drugs

8.5.4.1 'Me too' drugs

Many companies use established drugs from their competitors as lead compounds to design a drug which gives them a foothold in the same market area. The aim is to modify the structure sufficiently such that it avoids patent restrictions, retains activity and ideally has improved therapeutic properties. For example, the antihypertensive drug captopril (Fig. 8.8) was used as a lead compound by various companies to produce their own antihypertensive agents (Fig. 8.11).

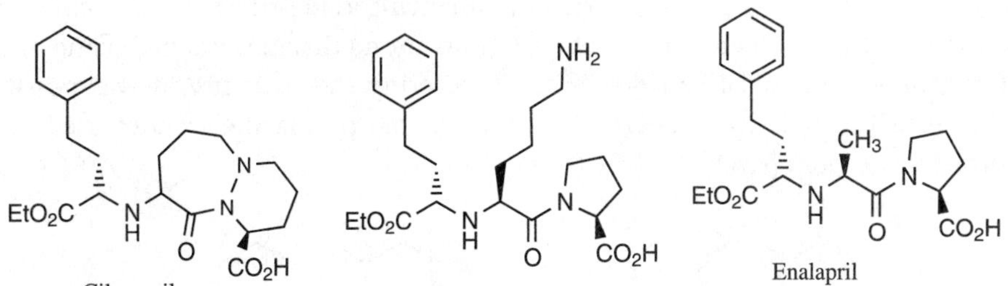

Cilazapril   Lisinopril   Enalapril
(Hoffmann-LaRoche)   (Merck)   (Merck)

Fig. 8.11 'Me too' drugs.

Such drugs are often labelled unfairly as 'me too' drugs, but they can often offer improvements over the original drug. For example, modern penicillins are more selective, more potent and more stable than the original penicillins.

8.5.4.2 Enhancing a side-effect

An existing drug may have a minor property or an undesirable side-effect which might be of use in another area of medicine. As such, the drug could act as a lead compound based on its side-effects. The aim would then be to enhance the desired side-effect and to eliminate the major biological activity.

For example, most sulphonamides are used as antibacterial agents. However, some sulphonamides with antibacterial activity could not be used clinically because they had convulsive side-effects brought on by hypoglycaemia (lowered glucose levels in the blood). Clearly, this is an undesirable side-effect for an antibacterial agent, but the ability to lower blood glucose levels would be useful in the treatment of diabetes. Therefore, structural alterations were made to the sulphonamides concerned to eliminate the antibacterial activity and to enhance the hypoglycaemic activity. This led to the antidiabetic agent tolbutamide (Fig. 8.12).

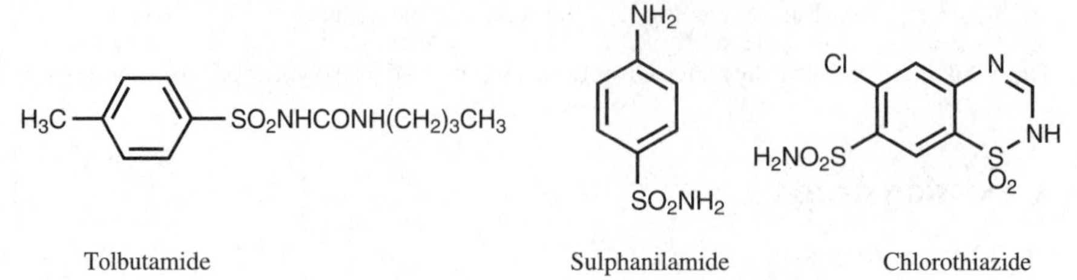

| Tolbutamide | Sulphanilamide | Chlorothiazide |

Fig. 8.12 Drugs developed by enhancing a side-effect.

Similarly, the development of sulphonamide diuretics such as chlorothiazide (Fig. 8.12) arose from the observation that sulphanilamide has a diuretic effect in large doses (due to its action on an enzyme called carbonic anhydrase).

Another example of a drug developed by enhancing a side-effect is that of chlorpromazine (Fig. 8.13), which was developed from promethazine. Promethazine is an antihistamine with sedative side-effects. By altering the structure, it was possible to enhance the sedative properties such that chlorpromazine is now used as a neuroleptic agent in psychiatry (Fig. 8.13).

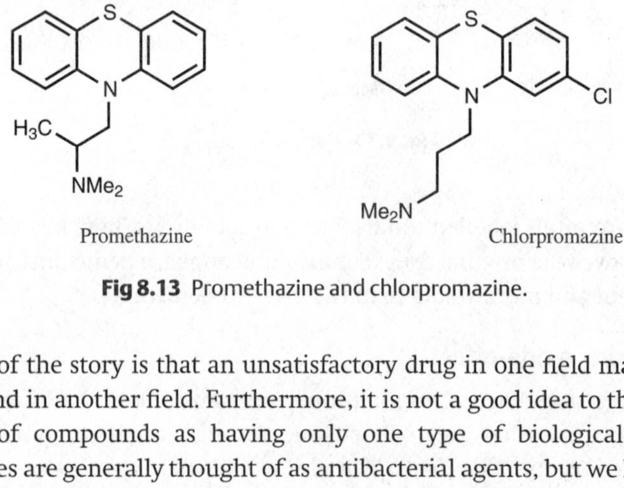

| Promethazine | Chlorpromazine |

Fig 8.13 Promethazine and chlorpromazine.

The moral of the story is that an unsatisfactory drug in one field may provide the lead compound in another field. Furthermore, it is not a good idea to think of a structural group of compounds as having only one type of biological activity. The sulphonamides are generally thought of as antibacterial agents, but we have seen that they can have other properties as well.

8.5.5 Starting from the natural ligand or modulator

8.5.5.1 Natural ligands for receptors

The natural ligand of a target receptor has sometimes been used as the lead compound. The natural neurotransmitters adrenaline and noradrenaline were the starting points for the development of adrenergic β-agonists such as salbutamol, dobutamine, and xamoterol (see Chapter 16), while 5-hydroxytryptamine (5-HT) was the starting point for the development of the 5-HT-agonist sumatriptan (Figs 8.14 and 8.15).

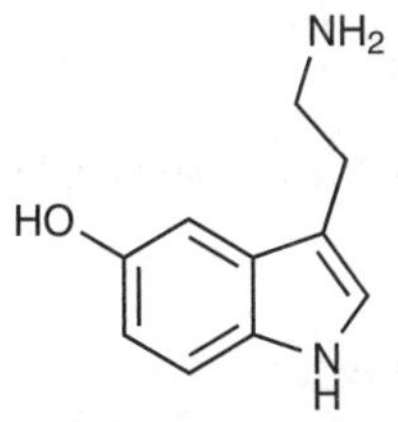

Fig. 8.14 5-Hydroxytryptamine.

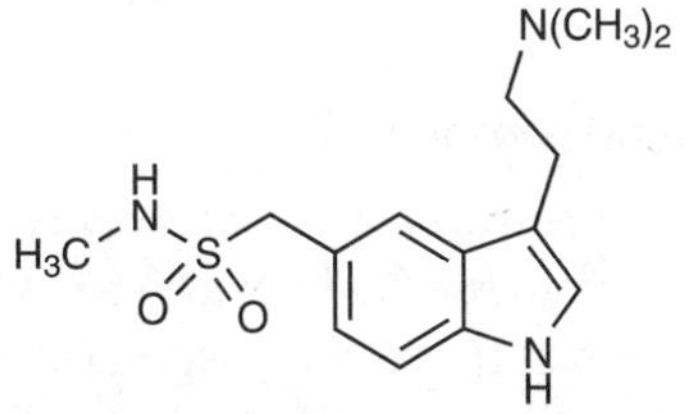

Fig. 8.15 Sumatriptan (5HT$_1$-agonist).

The natural ligand of a receptor can also be used as the lead compound in the design of an antagonist. For example, histamine was used as the original lead compound in the development of the H$_2$-histamine antagonist cimetidine (see Chapter 18). Turning an agonist into an antagonist is frequently achieved by adding extra binding groups to the lead structure. Examples include the development of the adrenergic antagonist pronethalol (see Chapter 16), the H$_2$-antagonist burimamide (see Chapter 18), and the 5-HT$_3$-antagonists ondansetron and granisetron (Fig. 8.3).

Sometimes the natural ligand for a receptor is not known (an orphan receptor) and the search for it can be a major project in itself. However, if the search is successful it opens up a brand new area of drug design. For example, the identification of the opiate receptors for morphine led to a search for endogenous opiates (natural body painkillers) which eventually led to the discovery of endorphins and enkephalins as the natural ligands and their use as lead compounds (see Chapter 17).

The discovery of cannabinoid receptors in the early 1990s led to the discovery of two endogenous cannabinoid messengers – arachidonylethanolamine (anandamide) (Fig. 8.16) and 2-arachidonylglycerol. These have now been used as lead compounds for developing agents which will interact with cannabinoid receptors. Such agents may prove useful in suppressing nausea during chemotherapy or stimulating appetite in AIDS patients.

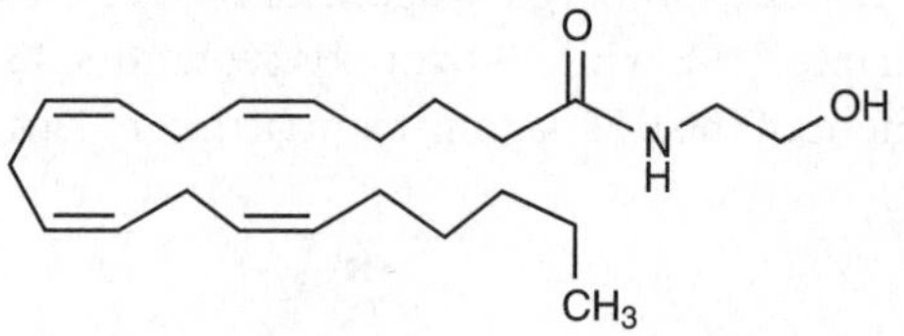

Fig. 8.16 Anandamide.

8.5.5.2 Natural substrates for enzymes

The natural substrate for an enzyme can be used as the lead compound in the design of an enzyme inhibitor. For example, enkephalins have been used as lead compounds for the design of enkephalinase inhibitors. Enkephalinases are enzymes which metabolize enkephalins and their inhibition should prolong the activity of enkephalins (see Chapter 17).

8.5.5.3 Enzyme products as lead compounds

It should be remembered that enzymes catalyse a reaction in both directions and so the product(s) of an enzyme-catalysed reaction can also be used as lead compounds for inhibitors. For example, the design of the carboxypeptidase inhibitor L-benzylsuccinic acid (Fig. 8.17) was based on the products arising from the carboxypeptidase-catalysed hydrolysis of peptides (see Chapter 9).

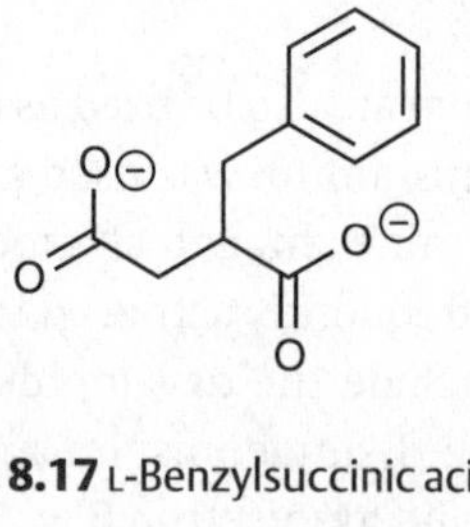

Fig. 8.17 L-Benzylsuccinic acid.

8.5.5.4 Natural modulators as lead compounds

Many receptors and enzymes are under allosteric control (see Chapters 4 and 5). The natural chemicals which exert this control (modulators) could also act as lead compounds.

In some cases, a modulator for an enzyme or receptor is suspected but has not yet been found. For example, it is known that the benzodiazepines modulate the receptor for γ-aminobutyric acid (GABA) by binding to an allosteric binding site not used by GABA itself. The benzodiazepines are synthetic compounds not produced in the body. The fact that they bind to an allosteric binding site suggests the possibility of a natural chemical messenger which *does* bind to this site. So far, this has not been found but, if it is, it will serve as a lead compound for further novel drugs which might have the same activity as the benzodiazepines.

8.5.6 Combinatorial synthesis

The growing number of potentially new drug targets arising from genomic projects has meant that there is an urgent need to find new lead compounds to interact with them. Unfortunately, the traditional sources of lead compounds have not managed to keep pace and, in the last decade or so, the pharmaceutical companies have invested greatly in a process called combinatorial synthesis to tackle this problem. Combinatorial synthesis is an automated solid phase procedure aimed at producing as many different structures as possible in as short a time span as possible. The reactions are carried out on a very small scale and are often designed to produce mixtures of compounds. In a sense, combinatorial synthesis aims to mimic what plants do, i.e. produce a pool of chemicals, one of which may prove to be a useful lead compound. Combinatorial science has developed so swiftly that it is almost a branch of chemistry in itself and a separate chapter is devoted to it (see Chapter 12).

8.5.7 Computer-aided design

A detailed knowledge of a target binding site significantly aids in the design of novel lead compounds intended to bind with that target. In cases where enzymes or receptors can be crystallized, it is possible to determine the structure of the protein and its binding site by X-ray crystallography. Molecular modelling software programmes can then be used to study the binding site, and to design molecules which will fit and bind.

In some cases, the enzyme or receptor cannot be crystallized and so X-ray crystallography cannot be carried out. However, if the structure of an analogous protein has been determined, this can be used as the basis for a computer-generated model of the protein under study. This is covered in more detail in Chapter 13.

8.5.8 Serendipity and the prepared mind

Frequently, lead compounds are found as a result of serendipity (i.e. chance). However, it still needs someone with an inquisitive nature or a prepared mind to recognize the significance of chance discoveries and to take advantage of these events. The discovery of cisplatin (see Chapter 7) and penicillin (see Chapter 14) are two such examples, but there are many more.

In World War Two, an American ship which was carrying mustard gas exploded. It was observed that many of the survivors who had inhaled the gas had lost their natural defences against microbes. Further study showed that their white blood cells had been destroyed. It is perhaps hard to see how a drug which wipes out white blood cells and weakens the immune system could be useful. However, there is one disease where such drugs are useful – leukaemia. Leukaemia is a form of cancer which results in the excess proliferation of white blood cells. A drug which kills these cells would therefore be useful in fighting the disease. As a result, a series of mustard-like drugs were developed based on the structure of the original mustard gas (see also Chapter 7).

Another example involved the explosives industry, where it was quite common for workers to suffer severe headaches. These headaches resulted from dilatation of blood vessels in the brain, caused by handling TNT. Once again, it is hard to see how drugs which give you headaches could be useful. However, the crucial point here was the dilatation of the blood vessels. Dilating blood vessels in the brain may not be particularly beneficial, but dilating the blood vessels in the heart certainly could be for some patients. As a result, drugs were developed which could be taken to dilate coronary blood vessels and to alleviate the pain of angina.

A third example involves workers in the rubber industry who found that they often acquired a disgust for alcohol! This was caused by an antioxidant used in the rubber manufacturing process which was finding its way into workers' bodies and preventing the normal oxidation of alcohol in the liver. As a result, there was a build up of acetaldehyde which was so unpleasant that workers preferred not to drink. The antioxidant became the lead compound for the development of disulfiram (antabuse), used for the treatment of chronic alcoholism (Fig. 8.18).

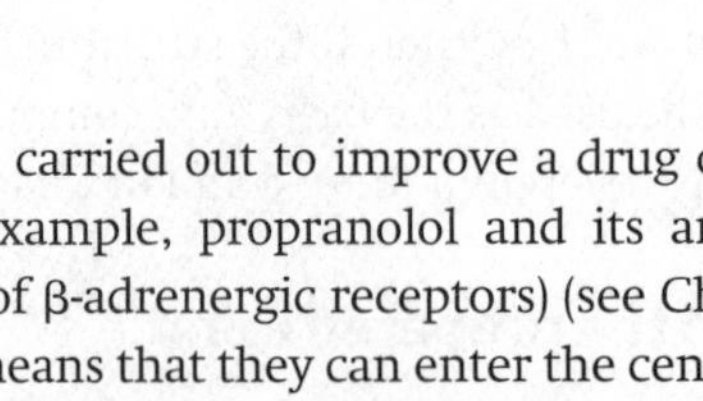

Fig. 8.18 Disulfiram (Antabuse).

Sometimes, the research carried out to improve a drug can have unexpected and beneficial spin offs. For example, propranolol and its analogues are effective β-blocking drugs (inhibitors of β-adrenergic receptors) (see Chapter 16). However, they are also lipophilic, which means that they can enter the central nervous system (CNS) and cause side-effects. In an attempt to cut down entry into the CNS, it was decided to add a hydrophilic amide group to the molecule and so inhibit passage through the blood–brain barrier. One of the compounds made was practolol (Fig. 8.19). As expected, this compound had less CNS side-effects, but, more importantly, it was found to be a selective antagonist for the β-receptors of the heart over β-receptors in

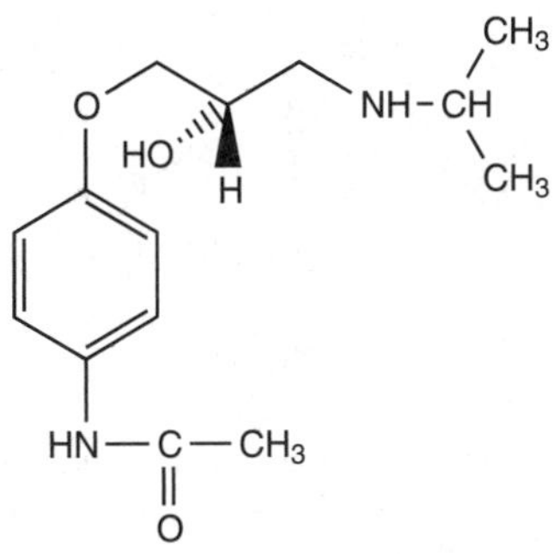

Fig. 8.19 Practolol.

other organs – a result that was highly desirable, but not the one that was being looked for at the time.

Frequently, new lead compounds have arisen from research projects in a totally different field of medicinal chemistry. This emphasizes the importance of keeping an open mind, especially when testing for biological activity. For example, we have already described the development of the antidiabetic drug tolbutamide (Fig. 8.12), based on the observation that some antibacterial sulphonamides could lower blood glucose levels.

There are many other well known examples, of which the following are a selection.

The anti-impotence drug Viagra was discovered by chance from a project aimed at developing a new heart drug (Fig. 8.20).

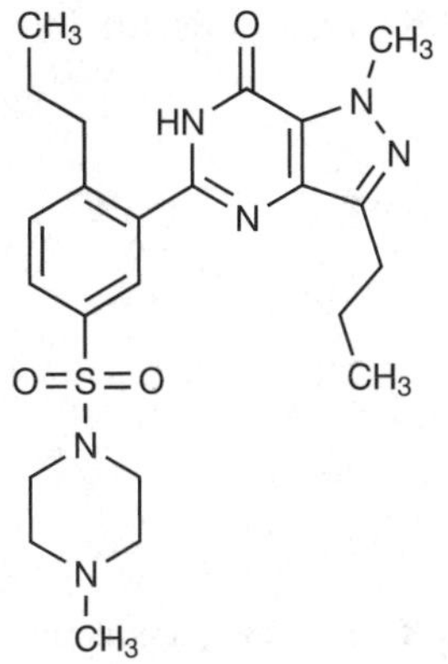

Fig. 8.20 Viagra (Sildenafil)

Clonidine (Fig. 8.21) was originally designed to be a nasal vasoconstrictor to be used in nasal drops and shaving soaps. However, clinical trials revealed a marked fall in blood pressure and it became an important antihypertensive instead.

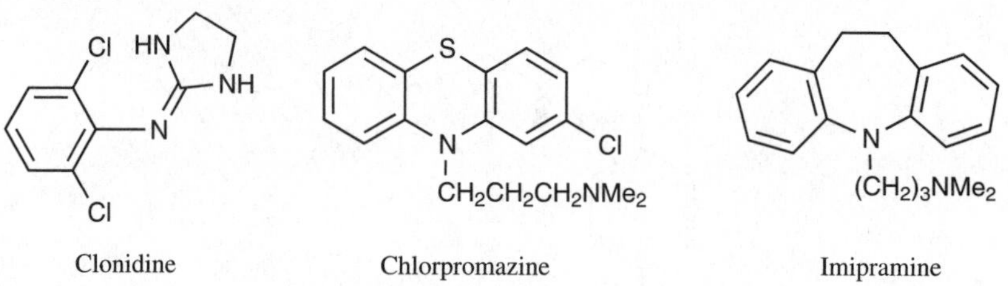

Fig. 8.21 Drugs used in fields other than the one originally intended.

Isoniazid (Fig. 8.10) was originally developed as an antituberculosis agent but patients were found to be very cheerful and this led to the drug being used as the lead compound for a series of drugs known as the monoamine oxidase inhibitors for the treatment of depression.

Chlorpromazine (Fig. 8.21) was synthesized as an antihistamine for possible use in preventing surgical shock and was found to make patients relaxed and unconcerned. This led to the drug being tested on manic depressants where it was found to have tranquillizing effects, resulting in it becoming the first of the neuroleptic drugs (major tranquillizers) used for schizophrenia.

Imipramine (Fig. 8.21) was synthesized as an analogue of chlorpromazine, initally for use as an antipsychotic. However, it was found to be effective in alleviating depression and this led to the development of a series of compounds classified as the tricyclic antidepressants.

Aminoglutethimide was prepared as a potential anti-epileptic drug, but is now used as an anticancer agent (Fig. 8.22). The drug stops the production of oestrogens by inhibiting the enzyme aromatase and is useful in treating oestrogen-dependent tumours.

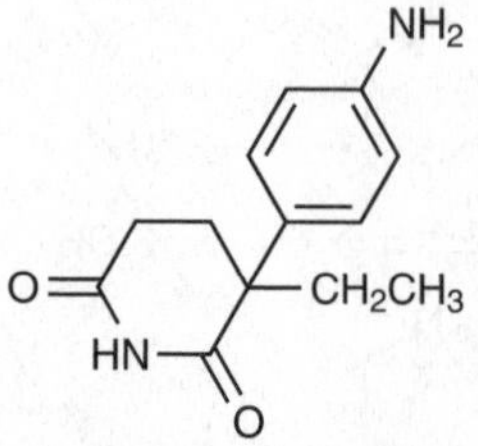

Fig. 8.22 Aminoglutethimide.

Cyclosporin A (Fig. 8.6) suppresses the immune system and is used during organ and bone marrow transplants, such that the body is less likely to reject the donor organs. The compound was isolated from a soil sample as part of a study aimed at finding new antibiotics. Fortunately, the compounds were screened for other activities apart from antibacterial activity. This illustrates the importance of screening new compounds for a variety of biological activities.

In a similar vein, the alkaloids vincristine and vinblastine (Fig. 8.23) were discovered by chance when searching for compounds which could lower blood sugar levels. Vincristine (Oncovin) is used in the treatment of Hodgkin's disease.

Fig. 8.23 Vincristine (R=CHO) and vinblastine (R=CH₃).

8.5.9 Computerized searching of structural databanks

New lead compounds can be found by carrying out computerized searches of structural databanks. To carry out such a search, it is necessary to know the pharmacophore of the drug (see below). This type of database searching is also known as **database mining** and is covered in more detail in Chapter 13.

8.5.10 Designing lead compounds by NMR

So far we have described methods by which a lead compound can be discovered from a natural or synthetic source, but in all these methods we rely on the active compound already being present. Unfortunately, there is no guarantee that such a lead compound will be present. Recently, NMR spectroscopy has been used to design a lead compound rather than to discover one. In essence, the method sets out to find small molecules (epitopes) which will bind to specific but different regions of a protein's binding site. These molecules will have no activity in themselves since they only bind to one part of the binding site but, if a larger molecule is designed which links these molecules together, then a lead compound may be created which binds to the whole of the binding site and may have activity (Fig. 8.24).

Lead discovery by NMR can be applied to proteins of known structure which are labelled with ¹⁵N such that each amide bond in the protein has an identifiable peak. To make identification of each peak easier, a 2D NMR spectrum is run matching ¹⁵N with ¹H.

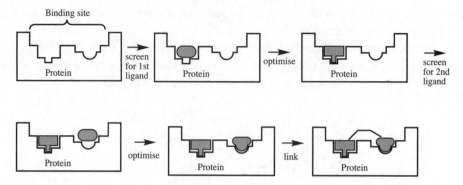

Fig. 8.24 Epitope mapping.

Next, a range of low molecular weight compounds is screened to see whether any of them bind to a specific region of the binding site. Binding can be detected by observing a shift in any of the amide signals which will not only reveal that binding is taking place, but will also show which part of the binding site is occupied. Once a compound (or ligand) has been found which binds to one region of the binding site, the process can be repeated to find another ligand which will bind to a different region of the binding site. This is usually done in the presence of the first ligand to ensure that the second ligand does in fact bind to a distinct region.

Once two ligands (or epitopes) have been identified, the structure of each can be optimized to find the best ligand for each of the binding regions, then a molecule can be designed where the two ligands are linked together.

There are several advantages to this approach. Since the individual ligands are optimized for each region of the binding site before synthesizing the overall structure, a lot of synthetic effort is spared. It is much easier to synthesize a series of small molecular weight compounds to optimize the interaction with specific parts of the binding site than it is to synthesize a range of larger molecules to fit the overall binding site. A high level of diversity is also possible since various combinations of fragments could be used.

The method has been proven by the design of high affinity ligands for FK506 binding protein – a protein involved in the suppression of the immune response. Two optimized epitopes (A) and (B) were discovered which were linked by a propyl link to give structure C (Fig. 8.25) which had higher affinity than either of the individual epitopes.

Lead generation by NMR uses a building block approach but this is different from combinatorial synthesis. In the latter, a large number of linked molecules are made from various building blocks, requiring a large synthetic effort. In the former, the building blocks are optimized first and a small number of linked molecules are then prepared.

Lead generation by NMR is limited to small protein targets (molecular weight < 40 000) which can be obtained in quantities greater than 200 mg, and which can be labelled with [15]N.

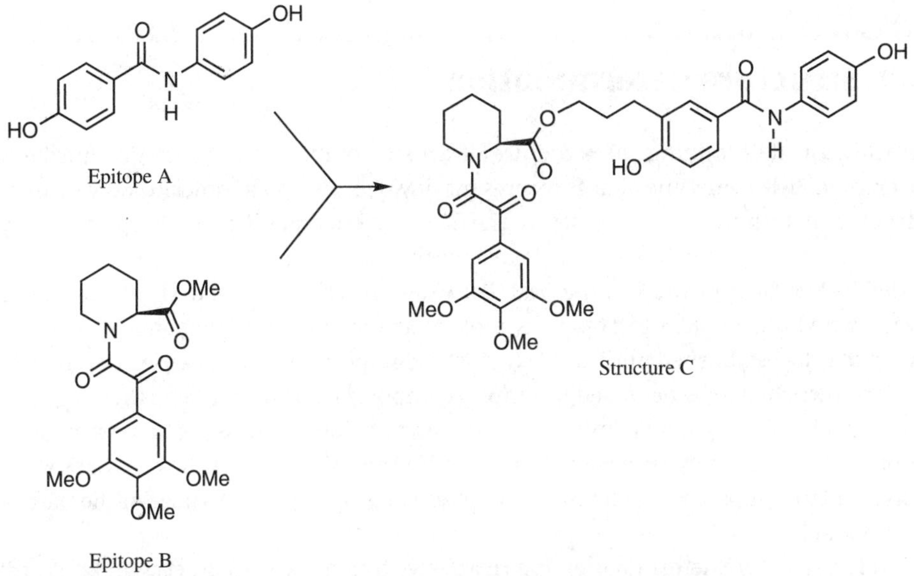

Fig. 8.25 Design of a ligand for FK506.

8.6 Isolation and purification

Isolation and purification of the lead compound (or 'active principle') is necessary if it is present in a mixture of other compounds, whether the mixture be from a natural source or from a combinatorial synthesis (see Chapter 12). The ease with which the active principle can be isolated and purified depends very much on the structure, stability, and quantity of the compound. For example, penicillin proved a difficult compound to isolate and purify because of its instability. Although Fleming recognized the antibiotic qualities of the compound and its remarkable non-toxic nature to man, he disregarded it as a clinically useful drug since he was unable to purify it. He could isolate it in solution, but whenever he tried to remove the solvent, the drug was destroyed. Now that we know the structure of penicillin (see Chapter 14), its instability under the purification procedures of the day is understandable and it was not until the development of a new procedure called freeze-drying that a successful isolation of penicillin was achieved.

Other advances in isolation techniques have occurred since those days and in particular in the field of chromatography. There are now a variety of chromatographic techniques available to help in the isolation and purification of a natural product.

8.7 Structure determination

In the past, determining the structure of a new compound was a major hurdle to overcome. It is sometimes hard for present day chemists to appreciate how difficult structure determinations were before the days of NMR and infrared (IR) spectroscopy. A novel structure which may now take a week's work to determine would have provided 2–3 decades of work in the past. For example, the microanalysis of cholesterol was carried out in 1888 to obtain its molecular formula, but its chemical structure was not fully established until an X-ray crystallographic study was carried out in 1932.

Structures had to be degraded to simpler compounds, which were further degraded to recognizable fragments. From these scraps of evidence, possible structures were proposed, but the only sure way of proving the theory was to synthesize these structures and to compare their chemical and physical properties with those of the natural compound.

Today, structure determination is a relatively straightforward process and it is only when the natural product is obtained in minute quantities that a full synthesis is required to establish its structure.

The most useful analytical techniques are X-ray crystallography and NMR spectroscopy. The former technique comes closest to giving a 'snapshot' of the molecule, but requires a suitable crystal of the sample. The latter technique is used more commonly since it can be carried out on any sample, whether it be a solid, oil or liquid. There are a large variety of different NMR experiments which can be used to establish the structure of quite complex molecules. These include various 2D NMR experiments which involve a comparison of signals from different types of nuclei in the molecule (e.g. carbon and hydrogen). Such experiments can identify how many bonds there are between different atoms and thus allow the chemist to build up a picture of the molecule atom by atom, and bond by bond.

In cases where there is not enough sample for an NMR analysis, mass spectroscopy can be helpful. The fragmentation pattern can give useful clues about the structure, but it does not prove the structure. A full synthesis is still required as final proof.

8.8 Structure–activity relationships

Once the structure of a biologically active compound is known, the medicinal chemist is ready to move on to study the structure–activity relationships of the compound. The aim of such a study is to discover which parts of the molecule are important to biological activity and which are not. The chemist makes a selected number of compounds, which vary slightly from the original molecule, and studies what effect this has on the biological activity.

One could imagine the drug as a chemical knight entering the depths of a forest (the body) to make battle with an unseen dragon (the body's affliction) (Fig. 8.26). The knight (Sir Drugalot) is armed with a large variety of weapons and armour, but since his battle with the dragon goes unseen, it is impossible to tell which weapon he uses or whether his armour is essential to his survival. We only know of his success if he returns unscathed with the dragon slain. If the knight declines to reveal how he slew the dragon, then the only way to find out how he did it would be to remove some of his weapons and armour and to send him in against other dragons to see if he can still succeed.

As far as a drug is concerned, the weapons and armour are the various chemical functional groups present in the structure which can bind to the receptor or enzyme, or which assist and protect it on its journey through the body. We have to be able to recognize these functional groups and to determine which ones are important.

Let us imagine that we have isolated a natural product with the structure shown in Fig. 8.27. We shall name it glipine. There are a variety of groups present in the structure and the diagram shows the potential binding interactions which are possible with a receptor.

It is unlikely that all of these interactions take place, so we have to identify those which do. By synthesizing compounds (such as the examples shown in Fig. 8.28) where one particular group of the molecule is removed or altered, it is possible to find out which groups are essential and which are not. This involves testing all the analogues for biological activity and comparing them with the original compound. If an analogue shows a significantly lowered activity, then the group which has been modified must have been important. If the activity remains very similar, then the group was not important.

The ease with which this task is carried out depends on how easily we can carry out the necessary chemical transformations to remove or alter the relevant group. For example, the importance of an amine group is relatively easy to establish, whereas establishing the importance of an aromatic ring might be more difficult. Let us consider what analogues of common functional groups could be synthesized to establish whether they are involved in binding or not.

8.8.1 The binding role of hydroxyl groups

Hydroxyl groups are commonly involved in hydrogen bonding. Converting such a group to a methyl ether or an ester is straightforward (Fig. 8.29) and will usually destroy or weaken such a bond.

There are several possible explanations for this. The obvious explanation is that the proton of the hydroxyl group is involved in the hydrogen bond to the receptor and if it is removed the hydrogen bond is lost (Fig. 8.30). However, suppose it is the oxygen atom which is hydrogen bonding to a suitable amino acid residue? The oxygen is still present in the ether or the ester analogue, so could we really expect there to be any effect on hydrogen bonding? Well, yes we could. The hydrogen bonding may not be completely destroyed, but we could reasonably expect it to be weakened, especially in the case of an ester.

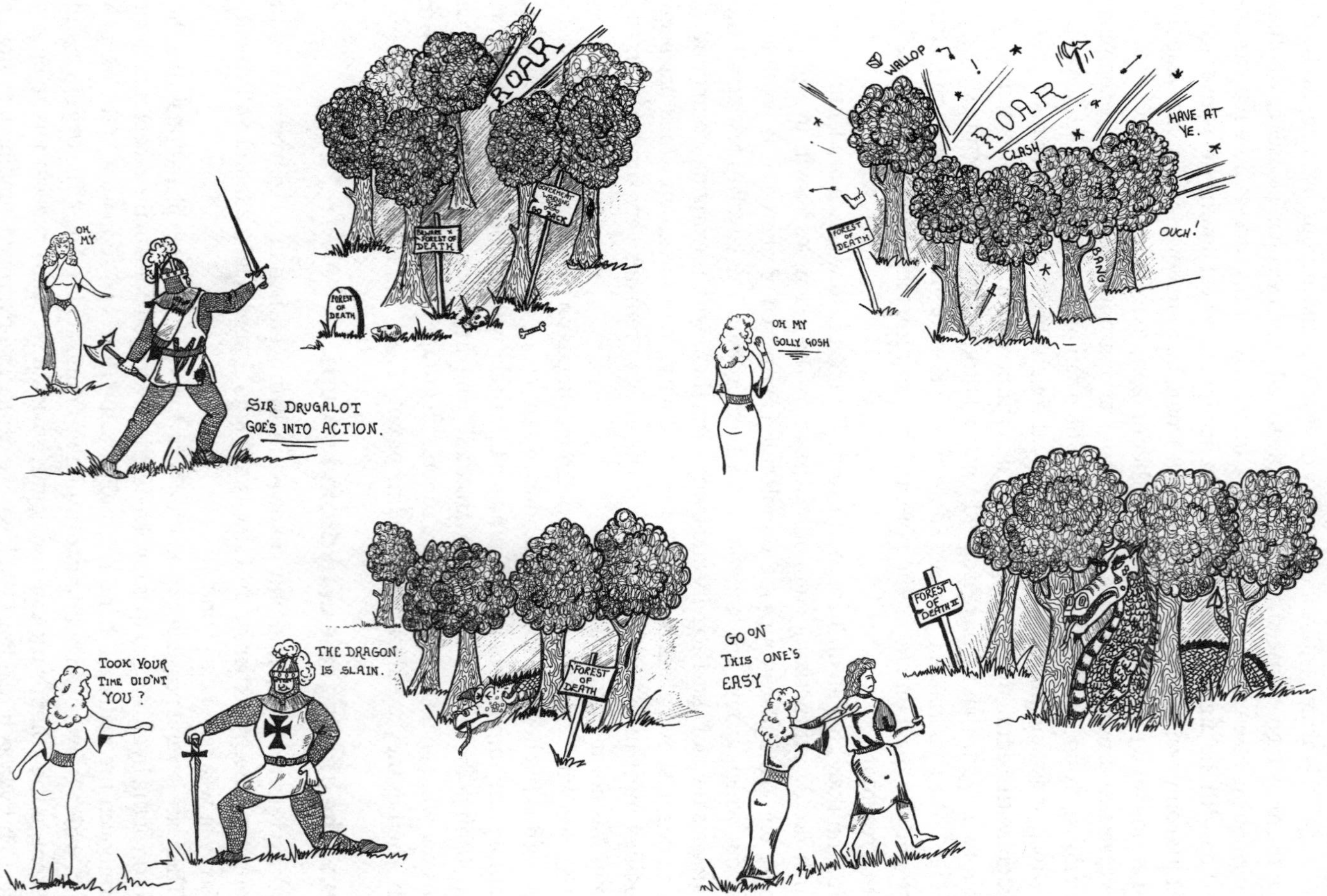
OH MY
ROAR
FOREST OF DEATH
SIR DRUGALOT GOES INTO ACTION.
WALLOP
ROAR
CLASH
HAVE AT YE.
FOREST OF DEATH
BANG
OUCH!
OH MY GOLLY GOSH
TOOK YOUR TIME DIDN'T YOU?
THE DRAGON IS SLAIN.
FOREST OF DEATH
GO ON THIS ONE'S EASY
FOREST OF DEATH II

Fig. 8.26

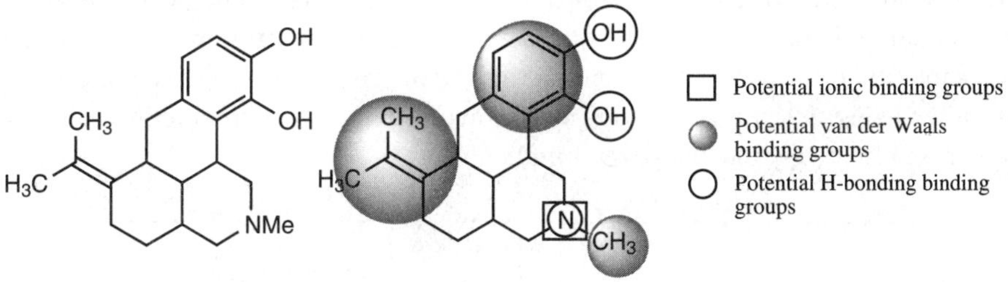

Fig. 8.27 Glipine.

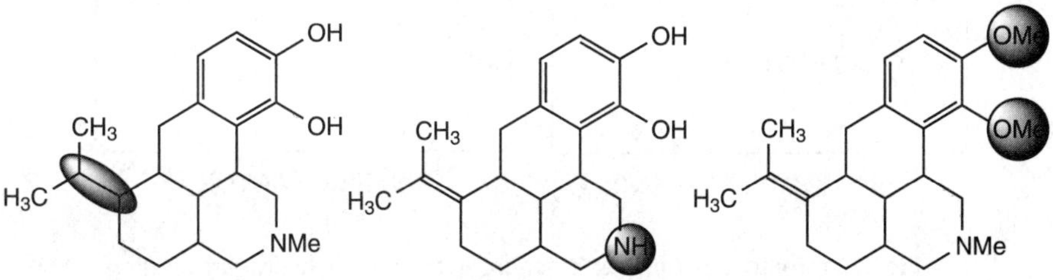

Fig. 8.28 Modifications of glipine.

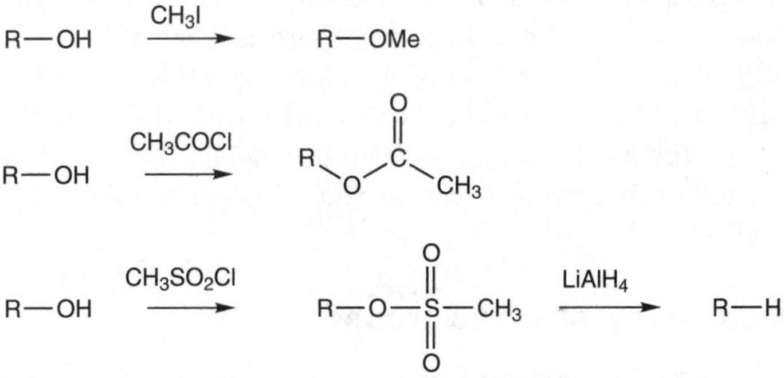

Fig. 8.29 Conversions of hydroxyl groups.

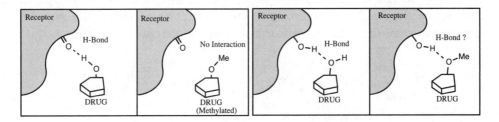

Fig. 8.30 Possible hydrogen bond interactions.

The reason is straightforward. When we consider the electronic properties of an ester compared to an alcohol, then we observe an important difference. The carboxyl group can 'pull' electrons from the neighbouring oxygen to give the resonance structure shown in Fig. 8.31. Since the lone pair is involved in such an interaction, it cannot take part so effectively in a hydrogen bond.

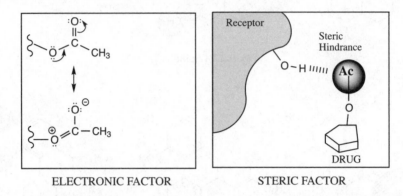

ELECTRONIC FACTOR   STERIC FACTOR

Fig. 8.31 Factors by which an ester group can disrupt hydrogen bonding.

Steric factors also count against the hydrogen bond. The extra bulk of the acyl group will hinder the close approach which was previously attainable. This steric hindrance also explains how a methyl ether could disrupt hydrogen bonding.

If there is still some doubt over whether a hydroxyl group is involved in hydrogen bonding, it could be removed altogether by reacting it with methanesulphonyl chloride followed by lithium aluminium hydride (Fig. 8.29). This would replace the hydroxyl with a proton, but any other group which is prone to reduction would have to be protected first.

8.8.2 The binding role of amino groups

Amines may be involved in hydrogen bonding or ionic bonding, but the latter is more common. The same strategy used for hydroxyl groups works here too. Converting the amine to an amide will prevent the nitrogen's lone pair taking part in hydrogen bonding or taking up a proton to form an ion. Tertiary amines have to be dealkylated first, before the amide can be made. Dealkylation is normally carried out with cyanogen bromide or a chloroformate such as vinyloxycarbonyl chloride (Fig. 8.32).

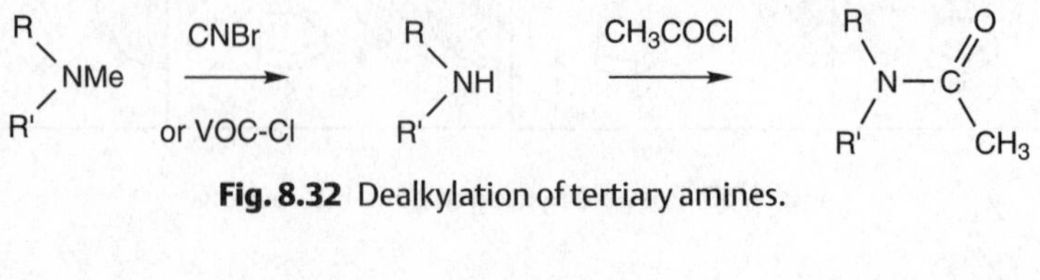

Fig. 8.32 Dealkylation of tertiary amines.

8.8.3 The binding role of aromatic rings

Aromatic rings are commonly involved in van der Waals interactions with flat hydrophobic regions of the binding site. If the ring is hydrogenated to a cyclohexane ring, the structure is no longer flat and interacts far less efficiently with the binding site (Fig. 8.33).

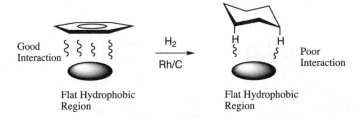

Fig. 8.33 Reduction in the binding efficiency of aromatic rings by hydrogenation.

However, carrying out the reduction may well cause problems elsewhere in the structure, since aromatic rings are difficult to reduce and need forcing conditions.

Replacing the ring with a bulky alkyl group could also reduce van der Waals bonding because of bad steric interactions, but obtaining such compounds could involve a major synthetic effort.

8.8.4 The binding role of double bonds

Unlike aromatic rings, double bonds are easy to reduce and this has a significant effect on the shape of that part of the molecule, since the planar double bond is converted into a bulky alkyl group. If the original alkene was involved in van der Waals bonding with a flat surface on the receptor, reduction should weaken that interaction, since the bulkier product is less able to approach the receptor surface (Fig. 8.34).

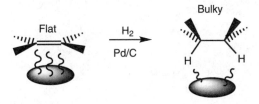

Fig. 8.34 Hydrogenation of alkenes.

8.8.5 The binding role of ketones

A ketone is commonly found in many drugs. This group could interact with its protein target through hydrogen bonding or through dipole–dipole interactions. It is relatively easy to reduce a ketone to an alcohol and this significantly changes the geometry of the functional group from the planar ketone to the tetrahedral alcohol. Such an alteration in geometry may well weaken any previous hydrogen bonding interactions and will certainly weaken any dipole–dipole interactions (Fig. 8.35).

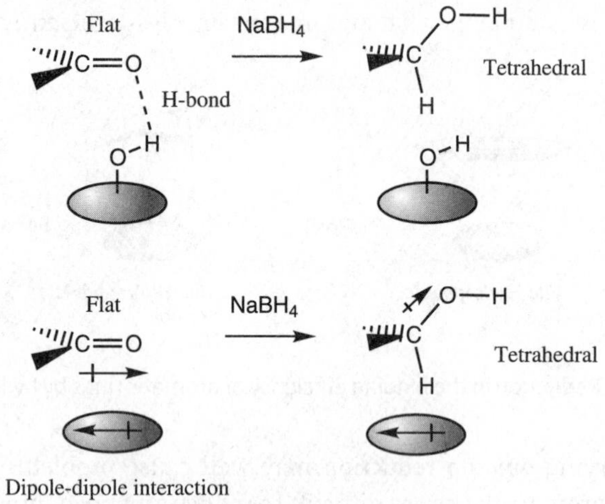

Fig. 8.35 Reduction of ketones.

8.8.6 The binding role of amides

Amides are likely to interact with protein targets through hydrogen bonding. The easiest modification to carry out on this group is hydrolysis. However, since this splits the drug molecule in two, any loss in activity may be due to the loss of other important binding groups. Alternatively, reduction of the amide results in an amine. This will disrupt any hydrogen bonding involving the carbonyl oxygen (Fig. 8.36).

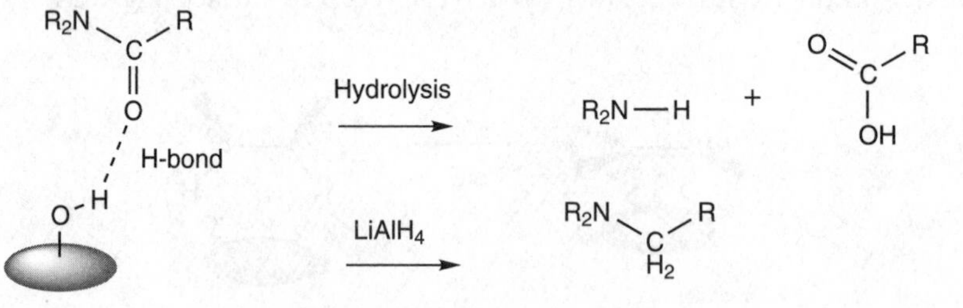

Fig. 8.36 Reactions of amides.

8.8.7 Isosteres

Isosteres are atoms or groups of atoms which have the same valency (or number of outer shell electrons). For example, SH, NH_2, and CH_3 are isosteres of OH, while S, NH, and CH_2 are isosteres of O. Isosteres can be used to determine whether a particular group is an important binding group or not by altering the character of the molecule in as controlled a way as possible. Replacing O with CH_2, for example, will make little difference to the size of the analogue, but will have a marked effect on its polarity, electronic distribution, and bonding. Replacing OH with the larger SH may not have such an influence on the electronic character, but steric factors become more significant.

Isosteric groups could be used to determine whether a particular group is involved in hydrogen bonding. For example, replacing OH with CH_3 would completely destroy hydrogen bonding, whereas replacing OH with NH_2 would not.

The β-blocker propranolol has an ether linkage (Fig. 8.37). Replacement of the OCH_2 segment with the isosteres CH=CH, SCH_2, or CH_2CH_2 eliminates activity, whereas replacement with $NHCH_2$ retains activity (though reduced). These results show that the ether oxygen is important to the activity of the drug and suggests that it is involved in hydrogen bonding with the receptor.

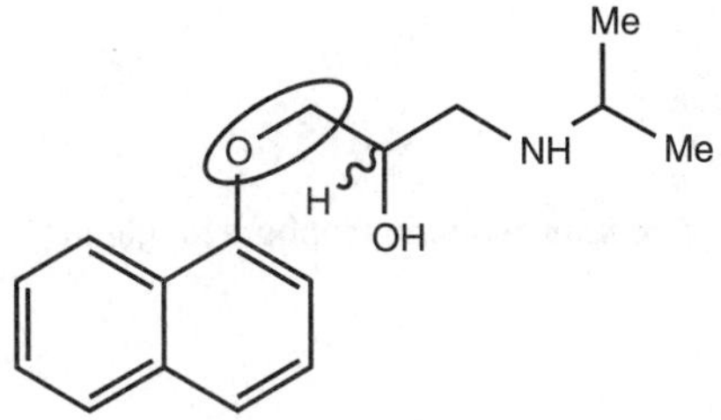

Fig. 8.37 Propranolol.

8.8.8 Testing procedures

When carrying out structure–activity relationships, biological testing should involve *in vitro* tests (e.g. inhibition studies on isolated enzymes or binding studies on membrane-bound receptors). The results then show conclusively which binding groups are important in drug–target interactions. If *in vivo* testing is carried out, the results are less clearcut since loss of activity may be due to the inability of the drug to reach its target rather than reduced drug–target interactions. On the other hand, *in vivo* testing may reveal important groups which are important in protecting or assisting the drug in its passage through the body, which would not be revealed by *in vitro* testing.

8.9 Identification of a pharmacophore

Once it is established which groups are important for a drug's activity, it is possible to move on to the next stage – the identification of the pharmacophore. The pharmacophore summarizes the important functional groups which are required for activity and their relative positions in space with respect to each other. For example, if we discover that the important binding groups for our hypothetical drug glipine are the two phenol groups, the aromatic ring and the nitrogen atom, then the pharmacophore is as shown in Fig. 8.38, where the nitrogen atom is 5.063 Å from the centre of the phenolic ring and lies at an angle of 18° from the plane of the ring. We return to the concept of pharmacophores in Chapters 9 and 13.

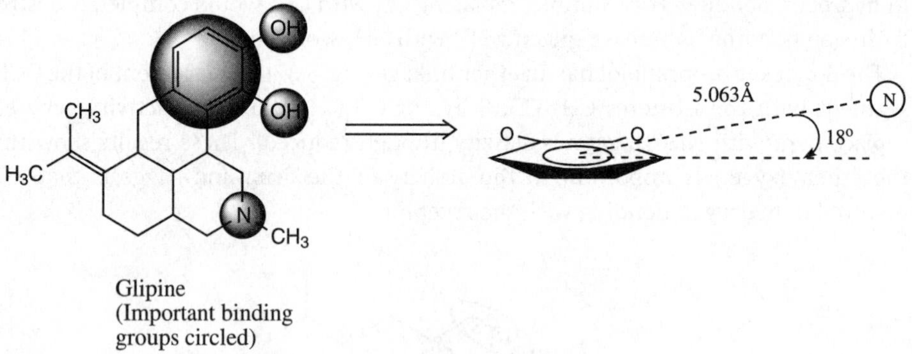

Glipine
(Important binding
groups circled)

Fig. 8.38 A pharmacophore for glipine.

8.10 Target-orientated drug design

Once the important binding groups and pharmacophore have been identified, it is possible to synthesize analogues of the lead compound which contain the same pharmacophore. But why is this stage necessary? If a natural compound such as our hypothetical glipine has useful biological activity, why bother making synthetic analogues? The answer is that very few drugs are ideal. Many have serious side-effects and there is an advantage in finding analogues which lack them. In general, the medicinal chemist is developing drugs with four objectives in mind:

- to increase activity
- to reduce side-effects
- to provide easy and efficient administration to the patient.
- ease of synthesis

Target-orientated drug design aims to modify the lead compound such that it interacts more effectively and selectively with its molecular target in the body. Stronger drug–target interactions should increase the activity of the drug while an increase in target selectivity will lower side-effects. Since this is such a substantial part of medicinal chemistry, the topic is covered more fully in Chapter 9.

8.11 Pharmacokinetic drug design

Target-orientated drug design sets out to optimize the binding interactions of a drug with its target. However, the compound with the best binding interactions is not necessarily the best drug to use in medicine, and some of the most active drugs discovered *in vitro* show no activity at all *in vivo*. This is because a clinically useful drug has to travel through the body to reach its target. There are many barriers and hurdles which can prevent a drug reaching its target and, as far as the drug is concerned, it is a long, strenuous, and dangerous journey.

Pharmacokinetic drug design concentrates on designing drugs to overcome these barriers. This, too, is a large area of medicinal chemistry and is covered in more detail in Chapter 10. In general, the aim here is to design drugs which will survive long enough to have the desired effect, but which will be excreted once they have had that effect.

8.12 Drug metabolism

When drugs enter the body, they are attacked by a whole range of metabolic enzymes, mostly in the liver, whose role is to degrade or modify foreign structures such that they can be excreted. As a result most drugs undergo some form of metabolic reaction, resulting in modified structures known as metabolites.

8.12.1 Phase I and phase II metabolism

Drugs are foreign substances[1] as far as the body is concerned, and it has its own method of ridding itself of such chemical invaders. If the drug is polar, it will be quickly excreted by the kidneys. However, non-polar drugs are not easily excreted and the purpose of drug metabolism is to convert such compounds into more polar molecules which **can** be easily excreted.

Non-specific enzymes (particularly cytochrome P450 enzymes in the liver) are able to add polar functional groups to a wide variety of drugs. Once the polar functional group has been added, the overall drug is more polar and water-soluble, and is therefore more likely to be excreted when it passes through the kidneys.

An alternative set of enzymatic reactions can reveal 'masked' polar functional groups which might be present in a drug. For example, there are enzymes which can demethylate a methyl ether to reveal a more polar hydroxyl group. Once again, the more polar product (metabolite) is excreted more efficiently.

---

[1] Substances which are foreign to the particular biological system under study are known as xenobiotics. The word is derived from the Greek words 'xenos', meaning foreign and 'bios', meaning life.

These reactions are classed as phase I reactions and generally involve oxidation (the P450 enzymes), reduction, and hydrolysis (Fig. 8.39). Most of these reactions occur in the liver but some (such as the hydrolytic reactions of esters and amides) occur in the gut wall, plasma, and the lung. The structures most prone to oxidation are N-methyl groups, aromatic rings, the terminal positions of alkyl chains, and the least hindered positions of alicyclic rings. Nitro and carbonyl groups are prone to reduction by reductases, whilst amides and esters are prone to hydrolysis by esterases.

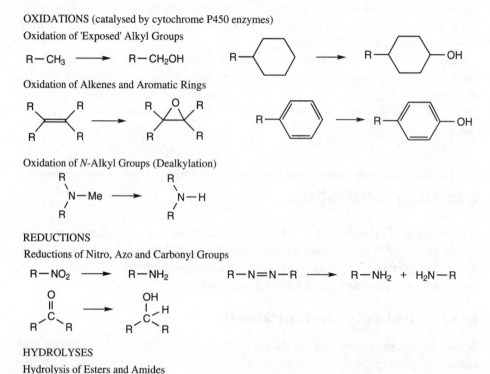

Fig. 8.39 Drug metabolism: phase I reactions.

There is also a series of metabolic reactions classed as phase II reactions; these also occur mainly in the liver (Fig. 8.40). Most of these reactions are conjugation reactions whereby a polar molecule is attached to a suitable polar 'handle' which is already present on the drug or has been placed there by a phase I reaction. The resulting conjugate has increased polarity, thus increasing its excretion rate in urine or bile even further.

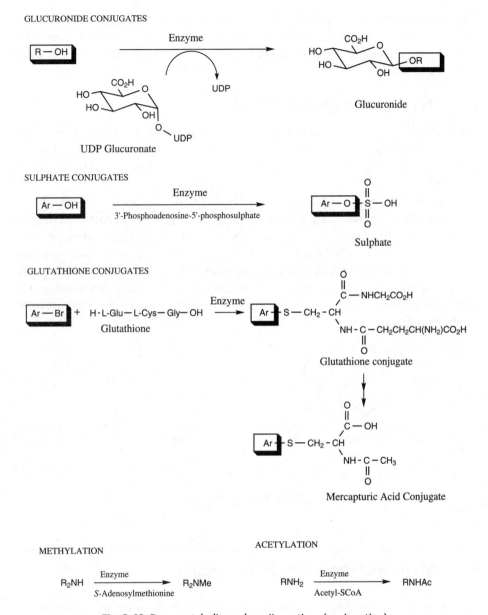

Fig. 8.40 Drug metabolism: phase II reactions (conjugation).

Phenols, alcohols, and amines form *O*- or *N*-glucuronides by reaction with UDP-glucose such that the highly polar glucose molecule is attached to the drug. Phenols, epoxides, and halides can react with the tripeptide glutathione to give mercapturic acids, while some steroids react with sulphates.

However, not all phase II reactions result in increased polarity. Methylation and acetylation are important phase II reactions which **decrease** the polarity of the drug.

The functional groups susceptible to methylation are phenols, amines, and thiols, while primary amines are susceptible to acetylation. The enzyme cofactors involved in contributing the methyl group or acetyl group are *S*-adenosyl methionine and acetyl SCoA respectively.

8.12.2 Testing for drug metabolites

Drugs should be tested on animals and humans to see what metabolites are formed. This is a safety issue, since it is important to ensure that no toxic metabolites are formed. Ideally, any metabolites which are formed should be inactive and quickly excreted.

Drug metabolism studies can sometimes be useful in drug design. On several occasions it has been found that a drug which is active *in vivo* is inactive *in vitro*. This is often a sign that the structure is not really active at all, but is being converted to the active drug by metabolism. The story of oxamniquine (see Chapter 9) illustrates this. Another example was the discovery that the antihypertensive compound (A) (Fig. 8.41) was less active *in vitro* than it was *in vivo*, implying that it was being converted into an active metabolite. Further studies led to the discovery of cromakalim which proved to be superior as an antihypertensive agent.

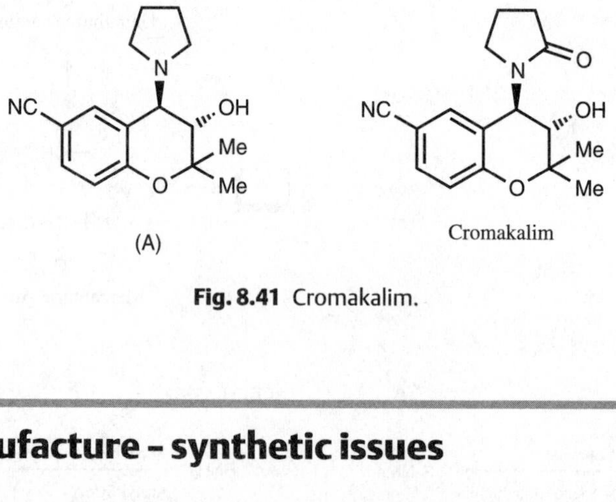

(A)   Cromakalim

Fig. 8.41 Cromakalim.

8.13 Manufacture – synthetic issues

Sometimes the most active drug is not taken to the clinic. There are several reasons for this, but one may be the relative ease and cost of synthesizing it. If a choice must be made between two drugs where one is slightly less active than the other but is easier to synthesize, then the less active structure may well be chosen for clinical trials and further development.

The industrial synthesis of a drug should be efficient and economic. However, such priorities are less important during the drug design phase. Here, the priority is to synthesize compounds as quickly as possible and so less consideration is given to such factors as yield, cost of reagents, safety of reagents, number of steps, experimental procedures, etc. Once it comes to manufacturing the drug on a large scale, economic and safety issues become far more important and it may be necessary to alter completely the original synthesis used to obtain the compound.

There is a particular problem with chiral drugs. The easiest and cheapest method of synthesis is to make the racemate. However, if a chiral drug is marketed as a racemate then both enantiomers have to be tested, which doubles the number of tests to be carried out. This is because different enantiomers can have different activities. For example, compound UH-301 (Fig. 8.42) is inactive as a racemate, whereas its enantiomers have opposing agonist and antagonist activity at the serotonin receptor (5-HT$_{1A}$).

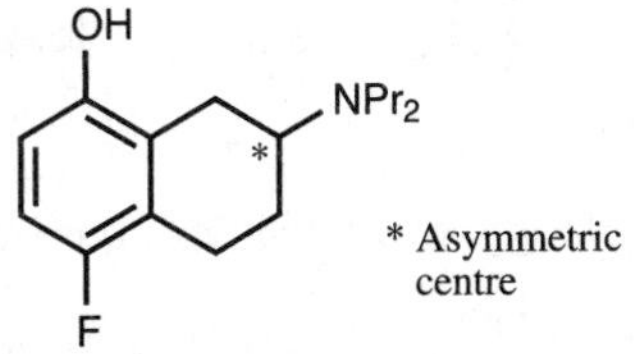

Fig. 8.42 UH-301.

Since the use of racemic drugs is discouraged, the manufacturing process either has to separate the enantiomers of the racemic drug, or carry out an asymmetric synthesis. Both options inevitably add to the cost of the synthesis. On occasions, the drug might return to the design phase to see whether it can be modified to reduce the number of asymmetric centres. For example, the cholesterol lowering agent mevinolin has eight asymmetric centres, but a second generation of cholesterol lowering agents have been developed which contain far fewer (e.g. HR780; Fig. 8.43).

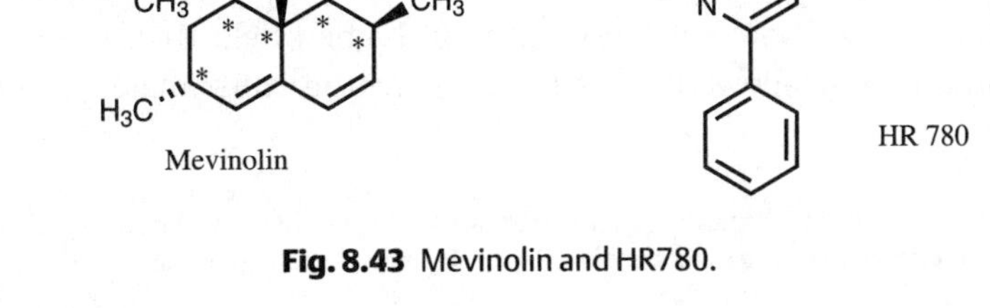

Fig. 8.43 Mevinolin and HR780.

Various tactics can be used to remove asymmetric carbon centres. For example, replacing the carbon centre with nitrogen has been effective in many cases (Fig. 8.44)[2]. An illustration of this can be seen in the design of thymidylate synthase inhibitors, described in Chapter 13. Another tactic is to introduce added symmetry where once there was none. For example, the muscarinic agonist (II) was developed from (I) to remove asymmetry and has the same activity (Fig. 8.45).

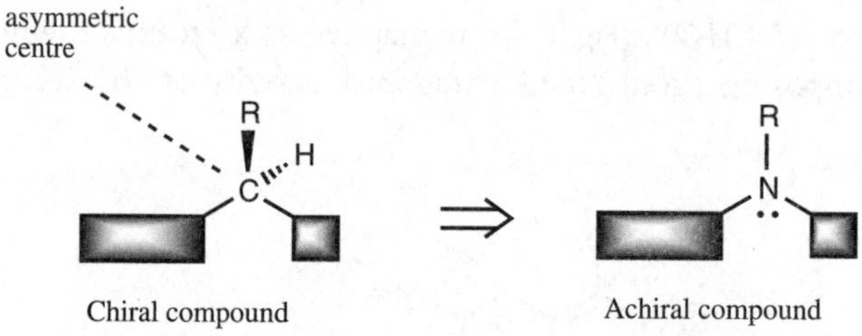

Fig. 8.44 Replacing an asymmetric carbon with nitrogen.

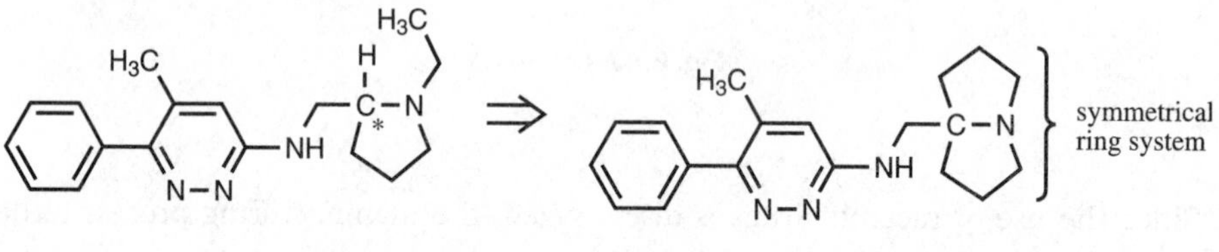

Fig. 8.45 Introducing symmetry.

Not all drugs can be synthesized, however. Many natural products have quite complex structures which are too difficult and expensive to synthesize on an industrial scale. These include drugs such as penicillin, morphine, and taxol. Such compounds can only be harvested from their natural source – a process which can be tedious, time-consuming, and expensive, as well as being wasteful on the natural resource. For example, four mature yew trees have to be cut down to obtain sufficient taxol to treat one patient! Furthermore, the number of structural analogues which can be obtained from harvesting is severely limited.

Semi-synthetic procedures can sometimes get round these problems. This often involves harvesting a biosynthetic intermediate from the natural source, rather than the final compound itself. The intermediate could then be converted to the final product by conventional synthesis. This approach can have two advantages. First, the intermediate may be more easily extracted in higher yield than the final product itself. Second, it may allow the possibility of synthesizing analogues of the final product.

---

[2] It should be noted that the introduction of an amine in this way may well have significant effects on the pharmacokinetics of the drug in terms of logP, basicity, polarity, etc (see Chapter 10).

The semisynthetic penicillins are an illustration of this approach (see Chapter 14).

Another more recent example is that of taxol. It is manufactured by extracting 10-deacetylbaccatin III from the needles of the yew tree, then carrying out a four-stage synthesis (Fig. 8.46).

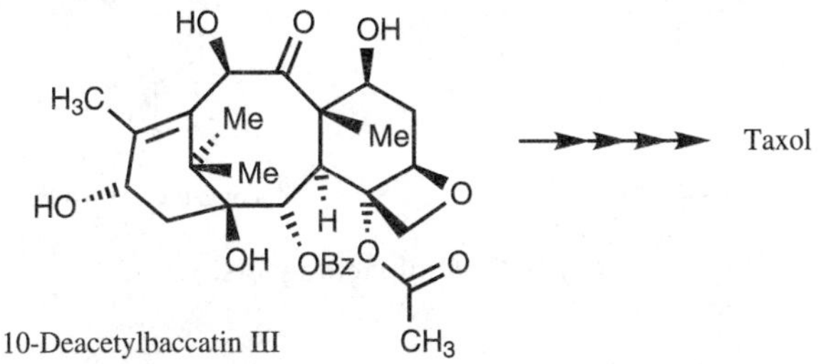

Fig. 8.46 Semi-synthetic synthesis of taxol.

8.14 Toxicity testing

Before the drug moves on to clinical trials it is tested for toxicity. Safety assessment often starts with *in vitro* and *in vivo* testing on genetically engineered cell cultures or transgenic mice to examine any effects on cell reproduction and to identify potential carcinogens.

This is usually followed by administration to lab animals to check for toxicity over a period of months. The animals are then killed and the tissues analysed by pathologists for any sign of cell damage or cancer.

The toxicity of a drug used to be measured by its LD_{50} value (the lethal dose required to kill 50% of a group of animals). However, this is a poor measure of a drug's toxicity since it fails to pick up non-lethal or long-term toxic effects. Therefore, toxicity testing involves a large variety of different *in vitro* and *in vivo* tests which are designed to reveal different types of toxicity. This is not foolproof, however, and a new and unexpected toxic effect may appear during clinical trials which will require the development of a new test. For example, when thalidomide was developed, nobody appreciated that drugs could cause foetal deformities and so there was no test for this.

Many promising drugs will fail toxicity testing – a fustrating experience indeed for the drug design teams. For example, the antifungal agent UK47 265 was an extremely promising antifungal agent, but *in vivo* tests on mice, dogs, and rats showed that it had liver toxicity and was potentially teratogenic. The design team had to synthesize more analogues and finally discovered the clinically useful drug fluconazole (Fig. 8.47).

It should also be borne in mind that it is rare for a drug to be 100% pure. There are bound to be minor impurities present arising from the synthetic route used, and these may well have an influence on the toxicity of the drug. The toxicity results of a drug prepared by one synthetic route may not be the same for the same drug synthesized

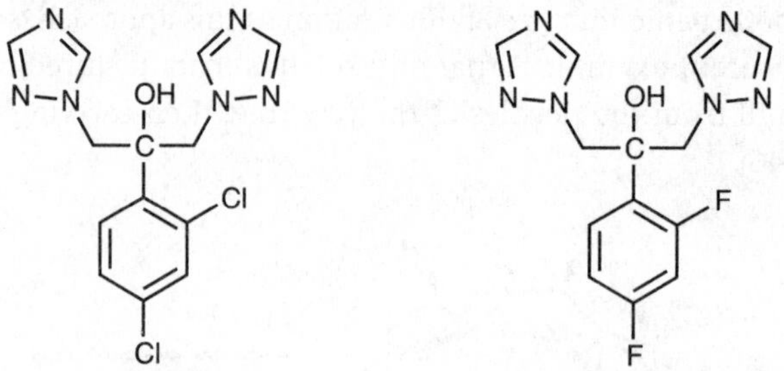

UK-47265   Fluconazole (UK-49858)

Fig. 8.47 Antifungal agents.

by a different route, and so it is important to establish the manufacturing synthesis as quickly as possible.

Finally, it is unlikely that the thorny problem of animal testing will disappear for a long time. There are so many variables involved in a drug's interaction with the body that it is impossible to anticipate them all. One has also to take into account that the drug will be metabolized to other compounds, all with their own range of biological properties.

It appears impossible, therefore, to predict whether a potential drug will be safe by *in vitro* tests alone. Therein lies the importance of animal experiments. Only animal tests can test for the unexpected. Unless we are prepared to volunteer ourselves as guinea pigs, animal experiments will remain an essential feature of drug development for many years to come.

8.15 Clinical trials

Clinical trials are the province of the clinician rather than the scientist. However, this does not mean that the research team can wash its hands of the candidate drug and concentrate on other things. Many promising drug candidates fail this final hurdle and, if this happens, further analogues may need to be prepared before a clinically acceptable drug is achieved. Clinical trials involve testing the drug on volunteers and patients. Therefore, the procedures involved must be ethical and beyond reproach. These trials can take many years to carry out, involve hundreds to thousands of patients, and be extremely expensive. There are four phases of clinical trial.

8.15.1 Phase I studies

In phase I studies, healthy volunteers take the drug to test whether the drug has the effect claimed. Tests are also carried out to test the drug's potency, pharmacokinetics, and side-effects. In some situations (e.g. anticancer drugs), patients may be used in phase I studies.

8.15.2 Phase II studies

In phase II studies, the drug is tested on a small group of patients to see if it has any effect and to find out what dose levels should be used.

8.15.3 Phase III studies

In phase III studies, the drug is tested on a much larger sample of patients and compared with other available treatments. Aternatively, they may be compared with a placebo (i.e. a preparation which has no effect at all). Comparative studies of this sort must be carried out without bias and this is achieved by random selection of patients – those who will receive the new drug and those who will receive the alternative treatment or placebo. Nevertheless, there is always the possibility of a mismatch between the two groups with respect to factors such as age, race, sex, and disease severity, and so the greater the number in the trial the better. The trials themselves are carried out by a 'double blind' technique where neither the patient nor the investigator knows which treatment is being applied. In the past, it has been found that investigators can unwittingly 'give the game away' if they know which patient is getting the actual drug.

Phase III studies establish whether the drug is really effective or whether any beneficial effects are psychological. They also allow further 'tweaking' of dose levels to achieve the optimum dose. Last, any side-effects not previously detected may be picked up with a larger sample of patients.

If the drug succeeds in passing phase III, it can be licensed and marketed. In certain circumstances where the drug shows a clear beneficial effect early on, the phase III trials may be terminated earlier than planned.

8.15.4 Phase IV studies

The drug is now placed on the market and can be prescribed. However, the drug is still monitored for its effectiveness and for any unexpected side-effects. In a sense, this phase should be a never ending process since rare and unexpected side-effects may crop up many years after the introduction of the drug. In the UK, the medicines committee runs a voluntary yellow card scheme where doctors and pharmacists report suspected adverse reactions to drugs. This system revealed serious side-effects for a number of drugs after they had been put on the market. For example, the β-blocker practolol (Fig. 8.19) had to be withdrawn after several years of use since some patients suffered blindness and even death. The toxic effects were unpredictable and are still not understood, and so it has not been possible to develop a test for this effect.

The diuretic agent tienilic acid (Fig. 8.48) had to be withdrawn from the market because it damaged liver cells in one out of every 10 000 patients, while the anti-inflammatory agent phenylbutazone (Fig. 8.48) can cause a rare but fatal side-effect in 22 patients out of every million treated with the drug! Such a rare toxic effect would clearly not be detected during phase III trials.

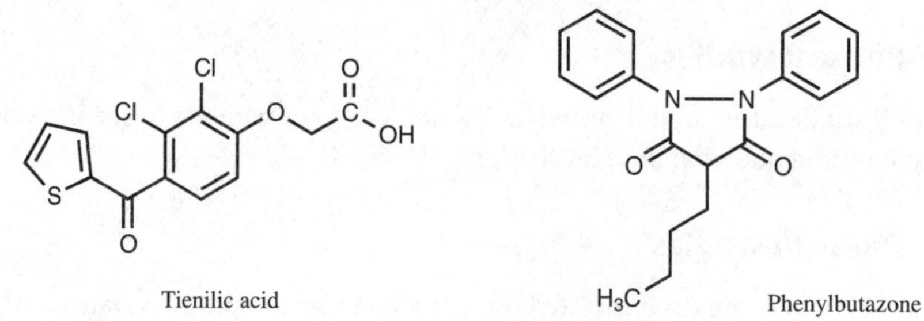

Fig. 8.48 Tienilic acid and phenylbutazone.

8.15.5 Ethical issues

In phases I–III of clinical trials, the permission of the patient is mandatory. However, ethical problems can still arise. For example, unconscious patients and mentally ill patients cannot give consent, but might benefit from the improved therapy. Should one include them or not? The ethical problem of including children in clinical trials is also a thorny issue, and so most clinical trials exclude them. However, this means that most licensed drugs have been licensed for adults. When it comes to prescribing for children, clinicians are left with the problem of deciding which dose levels to use and simple arithmetic mistakes can have tragic consequences.

8.16 Patents

Having spent enormous amounts of time and money on research and development, a pharmaceutical company quite rightly wants to recoup its costs and reap the benefit of all its hard work. To do so, it needs to have the exclusive rights to sell and manufacture its products for a reasonable period of time, and at a price which will not only recoup its costs, but will generate sufficient profits for further research and development. Without these rights, any competitor could step in and synthesize the same products at a much lower price.

Patents allow companies to gain these rights. A patent will grant a pharmaceutical company the exclusive right to the use and profits of a novel pharmaceutical for a limited term. To gain a patent, the company has first to submit or 'file' the patent, revealing what the new pharmaceutical is, for what use it is intended, and how it can be synthesized. This is no straightforward task. Each country has its own patents, and so the company has first to decide in which countries it is going to market its new drug and then file the relevant patents. Patent law is also very precise and varies from country to country. Therefore, submitting a patent is best left to specialists in the field, i.e. patent attorneys and lawyers.

Once a patent has been filed, the patent authorities will then decide whether the claims within are novel and whether they satisfy the necessary requirements for that patent body. One golden rule which is universal to patents worldwide is the requirement that the information supplied has not previously been revealed, either in print or by word of mouth. As a result, pharmaceutical companies will only reveal their work at conferences or in scientific journals after the structures involved have been safely patented.

It is important that a patent is filed as soon as possible. Such is the competition between the pharmaceutical companies that it is highly likely that a novel agent discovered by one company may be discovered by a rival company only weeks or months later. This means that patents are filed as soon as a novel agent or series of agents is found to have significant activity. Usually, the patent is filed before the research team has had the chance to start all the extensive tests which need to be carried out on novel drugs (e.g. toxicity testing, drug metabolism studies). It may not even have synthesized all the possible structures which it is intending to make. Therefore, the team is in no position to identify which specific compound in a series of interesting looking compounds is likely to be the best drug candidate. As a result, most patents are designed to cover a series of compounds belonging to a particular structural class, rather than one specific structure. Even if a specific structure has been identified as the best drug candidate, it is best to write the patent to cover a series of analogues. This prevents a rival company making a close analogue of the specified structure and selling it in competition. All the structures to be protected by the patent should be specified in the patent, but only a representative few need to be described in detail.

Patents in most countries run for 20 years after the date of filing. This sounds a reasonable time span, but it has to be remembered that the protection period starts from the time of filing, not from when the drug comes onto the market. A significant period of patent protection is lost because of the time required for drug development, drug testing, clinical trials, and regulatory approval. This often involves a period of 6–10 years. In some cases, this period may threaten to be even longer, in which case the company may decide to abandon the project since the duration of patent protection would be deemed too short to make sufficient profits. This illustrates the point that not all patents lead to a commercially successful product.

Patents can be taken out to cover specific products, the medicinal use of the products, the synthesis of the products, or all three of these aspects. It is best to cover all three. Taking out a patent which only covers the synthesis of a novel product offers poor patent protection. A rival company could quite feasibly develop a different synthesis to the same structure and then sell it quite legally.

8.17 Herbal medicine

Earlier we described how useful drugs and lead compounds can be isolated from natural sources, so where does this place herbal medicine? Are there any advantages or disadvantages in using herbal medicines instead of the drugs developed from their active principles? There are no simple answers to this. Herbal medicines contain a large variety of different compounds, several of which may have biological activity, so there is a significant risk of side-effects and even toxicity. The active principle is also present only in small quantity, and so the herbal medicine may be expected to be less active than the pure compound.

However, there are also advantages to herbal medicine. If the herbal extract contains the active principle in small quantities, there is an inbuilt safety limit to the dose levels received. Different compounds within the extract may also have roles to play in the medicinal properties of the plant and enhance the effect of the active principle. Alternatively, some plant extracts have a wide variety of different active principles which act together to produce a beneficial effect. The Aloe plant (the 'wand of heaven') is an example of this. It is a cactus-like plant found in the deserts of Africa and Arizona and has long been revered for its curative properties. Aloe preparations are still used today to treat burns, irritable bowel syndrome, rheumatoid arthritis, and asthma. They are also useful in treating chronic leg ulcers, itching, eczema, psoriasis, and acne, thus avoiding the undesirable side-effects of long-term steroid use. The preparation contains analgesic, anti-inflammatory, antimicrobial, and many other agents which all contribute to the overall effect, and trying to isolate each active principle would detract from this.

9 Drug design and drug–target interactions

9.1 Drug design

In Chapter 8, we discussed the overall procedure involved in the discovery and development of new drugs. A crucial stage in this process is the design of new drugs in which the aim is to improve the activity and properties of the lead compound. In this chapter, we shall describe the drug design strategies used to improve the binding interactions between a drug and its target. Such improvements should increase activity and may also reduce side-effects if the improved interactions lead to increased selectivity between different targets.

There are various strategies which can be used to improve the interactions between a drug and its target:

- variation of substituents
- extension of the structure
- chain extensions/contractions
- ring expansions/contractions
- ring variations
- ring fusions
- isosteres
- simplification of the structure
- rigidification of the structure.

9.2 Variation of substituents

Varying easily accessible substituents is a common method of fine tuning the binding interactions of a drug. The following are routine variations which can be carried out.

9.2.1 Alkyl substituents

Certain alkyl subsitutents can be varied more easily than others. For example, the alkyl substituents of ethers, amines, esters, and amides are easily varied as shown in Fig. 9.1. In these cases, the alkyl substituent already present can be removed and replaced with another substituent. Alkyl substituents which are part of the carbon skeleton of the molecule are not easily removed and it is usually necessary to carry out a full synthesis to vary them.

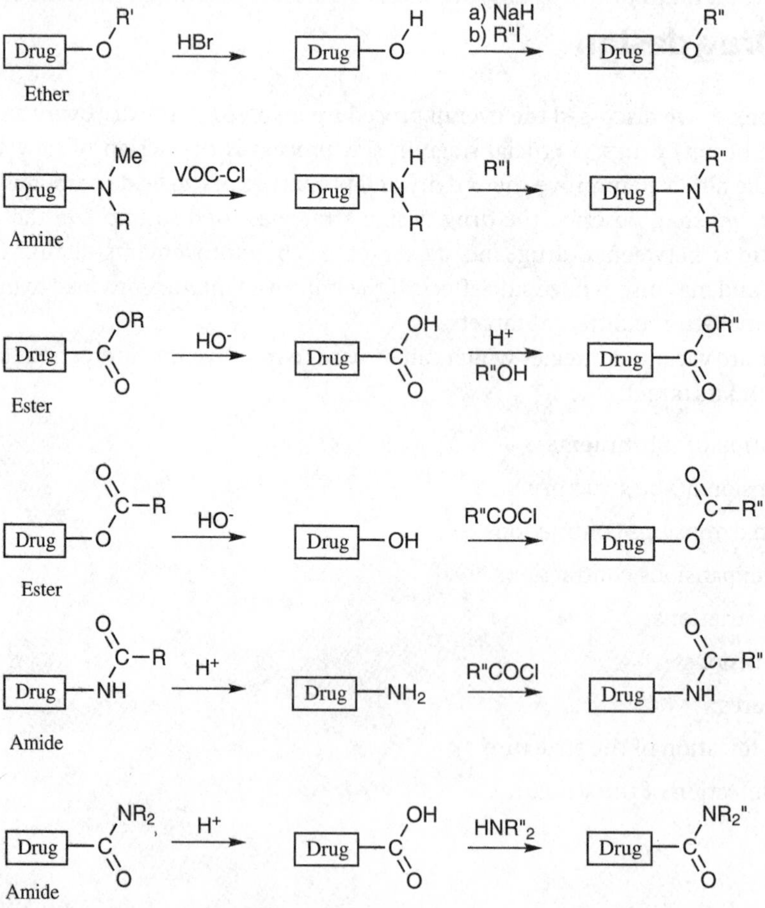

Fig. 9.1 Alkyl modifications.

Alkyl substituents such as methyl, ethyl, propyl, butyl, isopropyl, isobutyl or *tert*-butyl are often used to investigate the effect of chain length and bulk on binding. If these groups are interacting with a hydrophobic pocket present in the target receptor, then varying the alkyl group allows one to probe how deep and wide the pocket might be. Increasing the length and/or bulk of the alkyl chain to take advantage of such a pocket will then increase the binding interaction (Fig. 9.2).

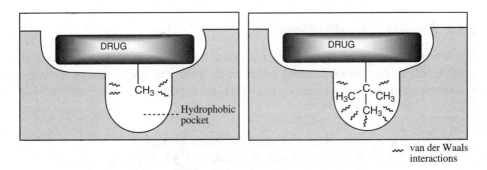

Fig. 9.2 Variation of alkyl chain to fill a hydrophobic pocket.

Larger alkyl groups increase the bulk of the compound as a whole and this may also confer selectivity on the drug. For example, in the case of a compound which interacts with two different receptors, a bulkier alkyl substituent may prevent the drug from binding to one of those receptors and so cut down side-effects (Fig. 9.3).

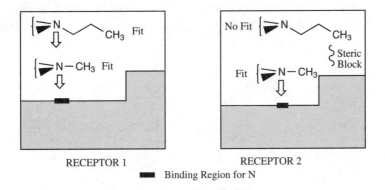

Fig. 9.3 Use of a larger alkyl group to confer selectivity on a drug.

For example, isoprenaline is an analogue of adrenaline where a methyl group has been replaced by an isopropyl group, resulting in selectivity for adrenergic β-receptors over adrenergic α-receptors (Fig. 9.4).

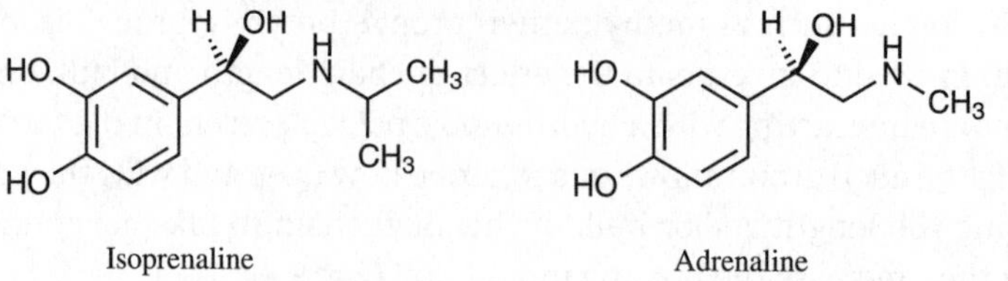

Isoprenaline   Adrenaline

Fig. 9.4 Introducing selectivity for β-adrenoreceptors over α-adrenoreceptors.

9.2.2 Aromatic substitutions

If a drug contains an aromatic ring, it is relatively easy to vary the position of the substituents to find better binding interactions. This may give increased activity if they are not already in the ideal positions for bonding (Fig. 9.5).

Fig. 9.5 Aromatic substitutions.

For example, the anti-arrhythmic activity of a series of benzopyrans was found to be best when the sulphonamide substituent was at position 7 of the aromatic ring (Fig. 9.6).

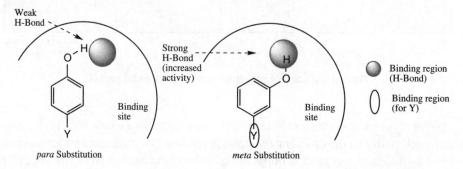

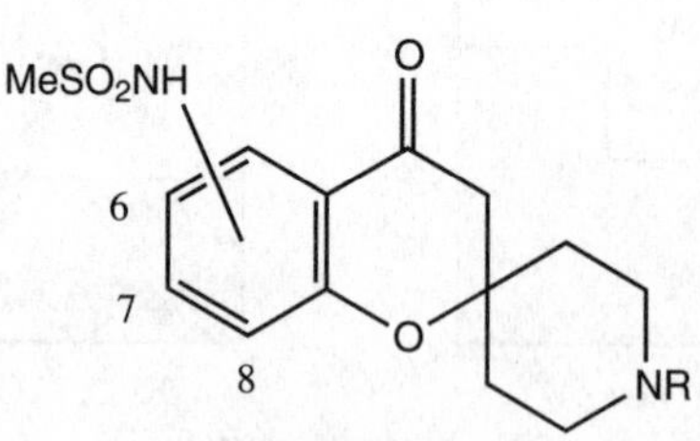

Fig. 9.6 Benzopyrans.

Changing the position of one substituent may have an important effect on another substituent. For example, an electron withdrawing nitro group will affect the basicity of an aromatic amine more significantly if it is in the *para* position rather than the *meta* position (Fig. 9.7). At the *para* position, the nitro group will make the amine a

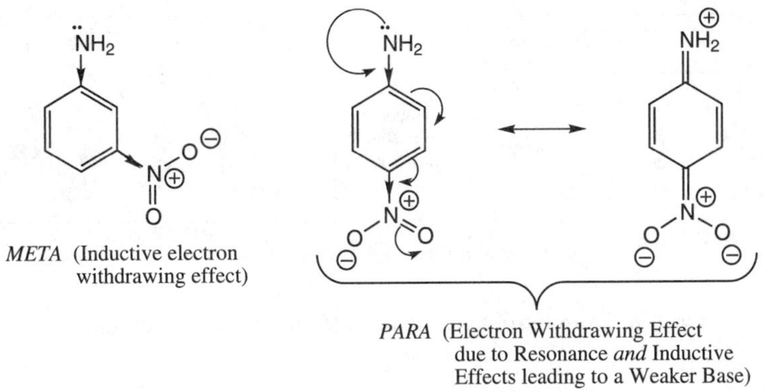

Fig. 9.7 Electronic effects of aromatic substitutions.

weaker base and less liable to protonate. This would decrease the amine's ability to interact with ionic binding groups in the binding site, and lower activity.

If the substitution pattern is ideal, then we can try varying the substituents themselves. Substituents having different steric, hydrophobic, and electronic properties are usually tried to see if these factors have any effect on activity. It may be that activity is improved by having a more electron-withdrawing substituent, in which case a chloro substituent might be tried in place of a methyl substituent.

The chemistry involved in these procedures is usually straightforward and so these analogues are made as a matter of course whenever a novel drug structure is developed. Furthermore, the variation of aromatic or aliphatic substituents is open to quantitative structure–activity studies (QSARs), as described in Chapter 11.

9.3 Extension of the structure

The strategy of extension involves the addition of another functional group to the lead compound to probe for extra binding interactions with the target. Lead compounds are capable of fitting the binding site and have the necessary functional groups to interact with some of the important binding groups which are present. However, it is possible that they do not interact with all the binding groups that are available. For example, a lead compound may bind to three binding regions in the binding site but fail to use a fourth (Fig. 9.8). Therefore, why not add extra functional groups to probe for that fourth binding interaction?

Extension tactics are often used to find extra hydrophobic regions in a binding site by adding various alkyl or arylalkyl groups. Such groups could be added to functional groups such as alcohols, phenols, amines, and carboxylic acids should they be present in the drug. Other functional groups could be added to probe for extra hydrogen bonding interactions and/or ionic interactions.

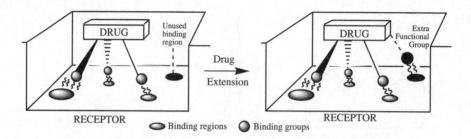

Fig. 9.8 Extension of a drug to provide a fourth binding group.

Extension tactics were used in the development of antihypertensive agents which inhibit an enzyme known as angiotensin converting enzyme (ACE). Adding a phenethyl group to structure (I) resulted in a thousandfold improvement in inhibition (Fig 9.9), demonstrating that the extra aromatic ring was binding to a hydrophobic pocket in the enzyme's active site.

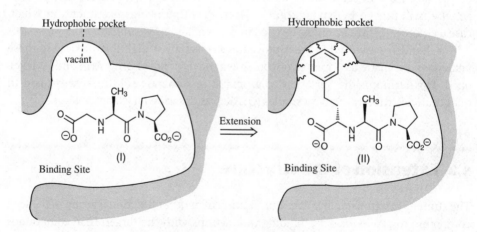

Fig. 9.9 ACE inhibitors.

The extension tactic has also been employed successfully to produce more active analogues of morphine (see Chapter 17).

Frequently, extension strategies are used to convert an agonist into an antagonist. This may happen because an extra binding interaction is found which is not used by the natural agonist or substrate. Since the overall binding interaction is different, antagonist activity results rather than agonist activity.

9.4 Chain extensions/contractions

Some drugs have two important binding groups linked together by a chain, in which case it is possible that the chain length is not ideal for the best interaction. Therefore, shortening or lengthening the chain length is a useful tactic to try (Fig. 9.10).

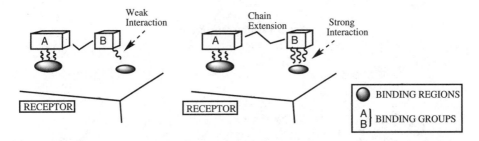

Fig. 9.10 Chain extension.

For example, consider the ACE inhibitor (II) described in Fig. 9.9. The aromatic ring here is connected to the peptide portion of the molecule by an alkyl chain. This chain was varied in length to find the ideal separation for these groups. A chain with two carbons resulted in higher activity than a chain with one or three carbons.

9.5 Ring expansions/contractions

If a drug has a ring, it is generally worth synthesizing analogues where one of these rings is expanded or contracted by one unit. The principle behind this approach is much the same as varying the substitution pattern of an aromatic ring. Expanding or contracting the ring puts the binding groups in different positions relative to each other and may lead to better interactions with the binding site (Fig. 9.11).

For example, during the development of the antihypertensive agent cilazaprilat (another ACE inhibitor), the bicyclic structure I showed promising activity (Fig. 9.12). The important binding groups were the two carboxylate groups and the amide group. By carrying out various ring contractions and expansions, cilazaprilat was identified as the structure having the best interaction with the binding site.

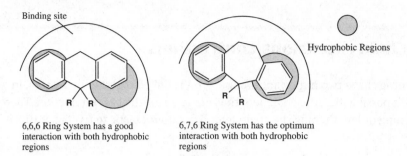

6,6,6 Ring System has a good
interaction with both hydrophobic
regions

6,7,6 Ring System has the optimum
interaction with both hydrophobic
regions

Fig. 9.11 Ring expansion.

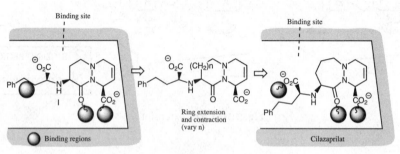

Fig. 9.12 Development of cilazaprilat.

9.6 Ring variations

A popular strategy used for compounds containing an aromatic or heteroaromatic ring is to replace the original ring with a range of other heteroaromatic rings of different ring size and heteroatom positions. For example, several non-steroidal anti-inflammatory agents (NSAIDs) have been reported, all consisting of a central ring with 1,2-diaryl substitution. Different pharmaceutical companies have varied the central ring to produce a range of active compounds (Fig. 9.13).

Admittedly, a lot of these changes are merely ways of avoiding patent restrictions ('me too' drugs), but there can often be significant improvements in activity, increased selectivity, and reduced side-effects. For example, the antifungal agent (I) (Fig. 9.14) acts against a fungal enzyme which is also present in man. Replacing the imidazole ring of structure (I) with a 1,2,4-triazole ring to give UK 46245 resulted in better selectivity against the fungal form of the enzyme.

One advantage of altering an aromatic ring to a heteroaromatic ring is that it introduces the possibility of an extra hydrogen bonding interaction with the binding site should a suitable binding region be available. For example, structure I (Fig. 9.15) was the lead compound for a project looking into novel antiviral agents. Replacing the aromatic ring with a pyridine ring resulted in an additional binding interaction with the target enzyme. Further development led eventually to the antiviral agent nevirapine (Fig. 9.15).

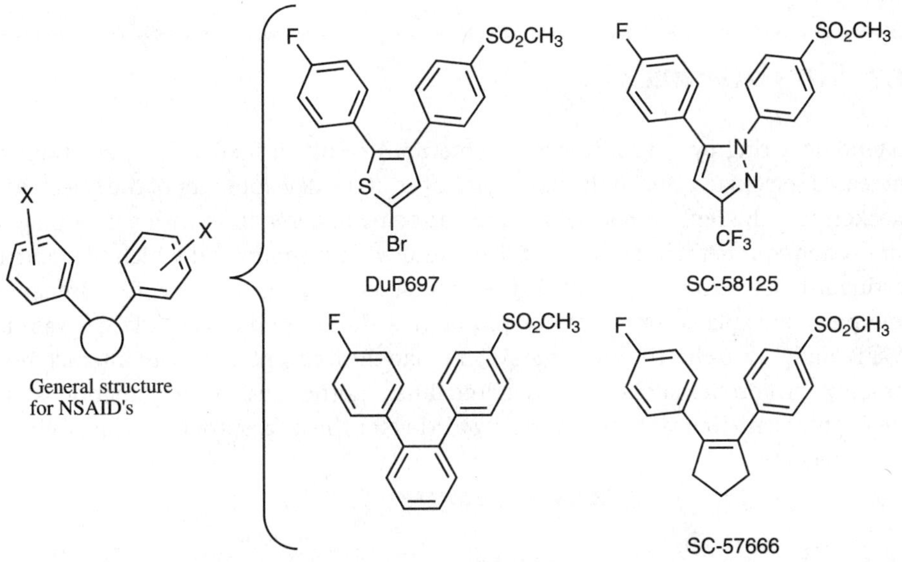

Fig. 9.13 Non-steroidal anti-inflammatory drugs (NSAIDS).

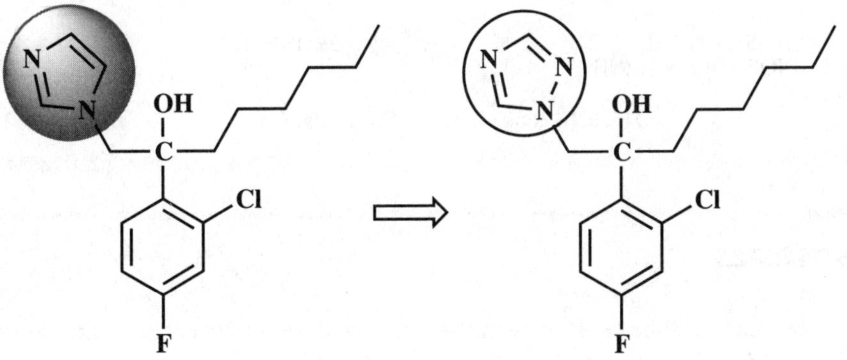

Fig 9.14 Development of UK 46245.

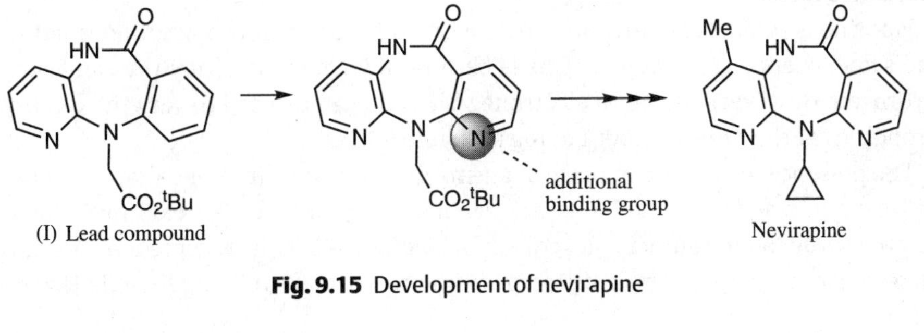

Fig. 9.15 Development of nevirapine

9.7 Ring fusions

Extending a ring by ring fusion can sometimes result in increased interactions or increased selectivity. One of the major advances in the development of the selective β-blockers was the replacement of the aromatic ring in adrenaline with a naphthalene ring system (pronethalol) (Fig. 9.16). This resulted in a compound which was able to distinguish between two very similar receptors, the α- and β-receptors for adrenaline. One possible explanation for this could be that the β-receptor has a larger van der Waals binding area for the aromatic system than the α-receptor and can interact more strongly with pronethalol than with adrenaline. Another possible explanation is that the naphthalene ring system is sterically too big for the α-receptor but is just right for the β-receptor.

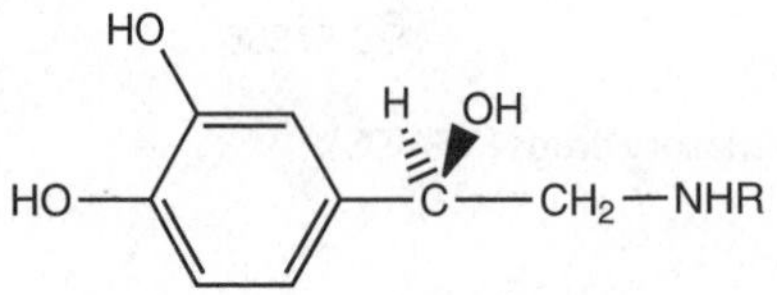

R = Me   ADRENALINE
R = H   NORADRENALINE

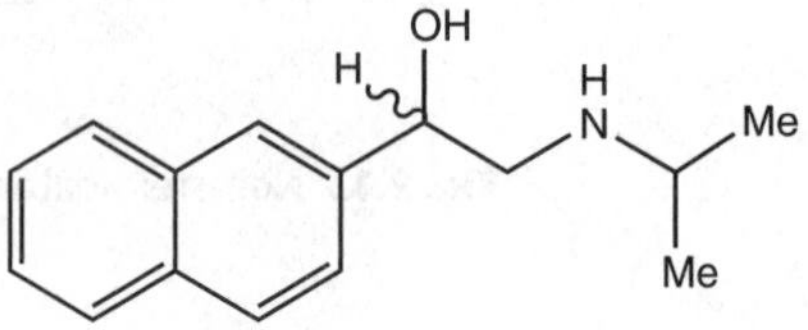

PRONETHALOL

Fig. 9.16 Ring variation of adrenaline.

9.8 Isosteres

Isosteres (see Chapter 8) have often been used in drug design to vary the character of the molecule with respect to such features as size, polarity, electronic distribution, and bonding in such a way that any alteration in drug character can be rationalized. Therefore, some isosteres would be used to determine the importance of size towards activity, whereas a different isostere would be used to determine the importance of electronic factors.

Fluorine is often considered an isostere of hydrogen even though it does not have the same valency as hydrogen. This is because fluorine is virtually the same size as hydrogen. However, it is more electronegative and can be used to vary the electronic properties of the drug without having any steric effect.

The presence of fluorine in place of an enzymically labile hydrogen can also disrupt an enzymic reaction since C–F bonds are not easily broken. For example, the anti-tumour drug 5-fluorouracil, described in Section 4.5.3, is accepted by its target enzyme since it appears little different from the normal substrate (uracil). However, the mechanism of the enzyme-catalysed reaction is totally disrupted since fluorine has replaced a labile hydrogen atom which is lost as a proton during the normal enzyme mechanism. There is no chance of fluorine departing as a positively charged species.

Several non-classical isosteres have become accepted as replacements for particular functional groups and sometimes have a beneficial effect. For example, a pyrrole ring has been used as a replacement for an amide. Carrying out such a replacement on the dopamine antagonist sultopride led to an increase in activity and selectivity towards the dopamine D_3-receptor over the dopamine D_2-receptor (Fig. 9.17). Such agents are promising lead compounds for antipsychotic agents acting on the D_3-receptor without the side-effects associated with the D_2-receptor.

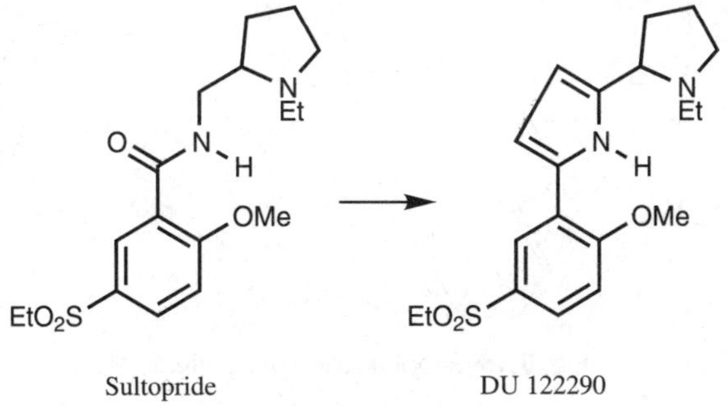

Sultopride   DU 122290

Fig. 9.17 Isosteric change for an amide group.

9.9 Simplification of the structure

Simplification is a strategy which is commonly used on the often complex lead compounds arising from natural sources. Once the essential groups of such a drug have been identified by structure–activity studies (SARs), then it is usually possible to discard the non-essential parts of the structure without losing activity.

Simplification is best carried out in small stages. For example, let us consider our hypothetical natural product glipine (Fig. 9.18). The essential groups have been marked and so we might aim to synthesize simplified compounds in the order shown in Fig. 9.18. These have simpler structures, but still retain the groups which we have identified as being essential.

Simplification was used successfully with the alkaloid cocaine (Fig. 9.19). It was well known that cocaine had local anaesthetic properties and the hope was to develop a local anaesthetic based on a simplified structure of cocaine which could be easily synthesized in the laboratory. Success resulted with the discovery of procaine (or Novocain) in 1909. Simplification tactics have also proved effective in the design of simpler morphine analogues (see Chapter 17).

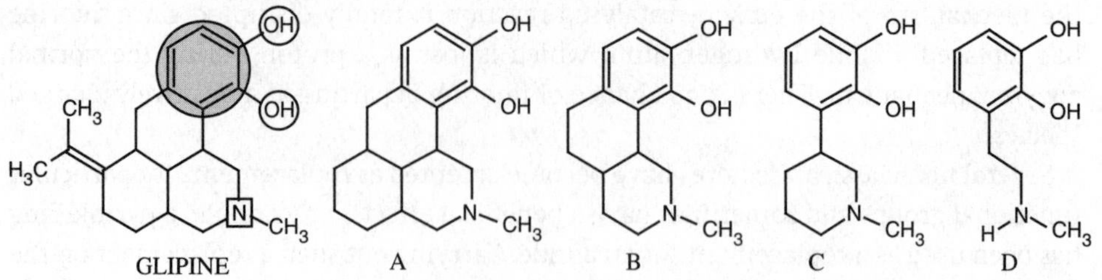

Fig. 9.18 Glipine analogues.

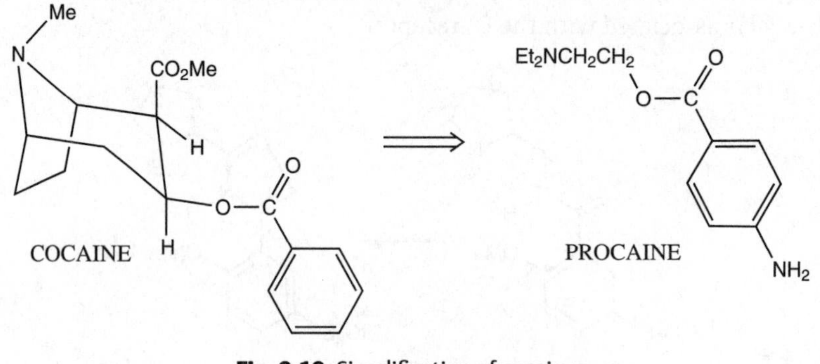

Fig. 9.19 Simplification of cocaine.

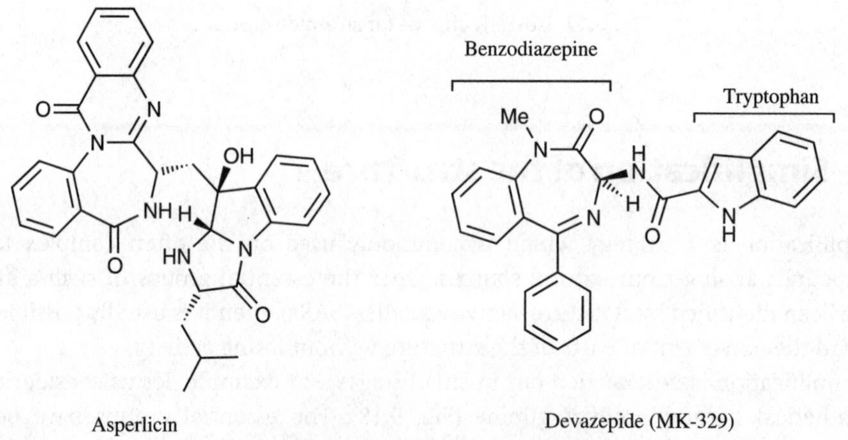

Fig. 9.20 Simplification of asperlicin.

More recently, the microbial metabolite asperlicin was simplified to devazepide, retaining the benzodiazepine and indole skeletons inherent in the structure (Fig. 9.20). Both asperlicin and devazepide act as antagonists of a neuropeptide chemical messenger called cholecystokinin (CCK), which has been implicated in causing panic attacks. Therefore, antagonists may be of use in treating such attacks.

The advantages with simpler structures are that they are much easier, quicker, and cheaper to synthesize in the laboratory. Usually the complex lead compounds obtained from natural sources are impractical to synthesize and have to be extracted from the source material – a slow, tedious, and expensive business.

However, there are possible disadvantages in oversimplifying molecules. Simpler molecules sometimes bind differently to their targets compared to the original lead compound, resulting in different effects. This is why it is best to simplify in small stages to check that the desired activity is retained at each stage. Oversimplification may also result in reduced activity, increased side-effects and reduced selectivity. We shall see why in the next section, on rigidification.

9.10 Rigidification of the structure

Rigidification has been a popular tactic used to increase the activity of a drug or to reduce its side-effects. To understand why, let us consider again our hypothetical neurotransmitter from Chapter 5 (Fig. 9.21). This is quite a simple molecule and is highly flexible. Bond rotation can lead to a large number of conformations or shapes. However, as seen from the receptor–messenger interaction, conformation I is the conformation accepted by the receptor. Other conformations such as II have the ionized amino group too far away from the anionic centre to interact efficiently and so this is an inactive conformation for our model receptor site. However, it is quite possible that a different receptor exists which *is* capable of binding conformation II. If this is the case, then our model neurotransmitter could switch on two different receptors and give two different biological responses.

The body's own neurotransmitters are highly flexible molecules (see Chapter 5), but fortunately the body is quite efficient at releasing them close to their target receptors, then quickly inactivating them so that they do not make the journey to other receptors. However, this is not the case for drugs. They have to be sturdy enough to travel through the body and consequently will interact with all the receptors which are prepared to accept them. The more flexible a drug molecule is, the more likely it will interact with more than one receptor and produce other biological responses (side-effects).

The strategy of rigidification is to 'lock' the drug molecule into a more rigid conformation such that it cannot take up these other shapes or conformations. Consequently, other receptor interactions and side-effects are eliminated. This same strategy should also increase activity since, by locking the drug into the active conformation, the drug is ready to fit its target receptor site and does not need to 'find' the correct conformation. Incorporating the skeleton of a flexible drug into a ring is the usual way of 'locking' a conformation and so, for our model compound, the analogue shown in Fig. 9.22 would be suitably rigid.

Actually, we have already seen an example of rigidification in the development of DU 122290 from sultopride (Fig. 9.17). Here two rotatable bonds in sultopride are locked within a pyrrole structure in DU 122290 and can no longer rotate. This limits

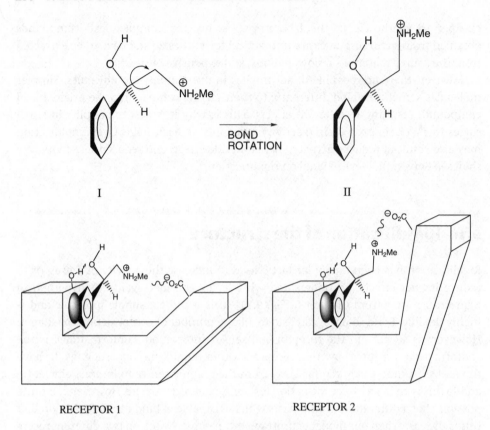

Fig. 9.21 Two conformations of a neurotransmitter which are capable of binding with different receptors.

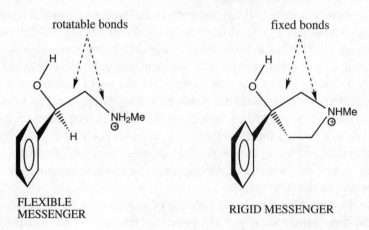

Fig. 9.22 Locked analogue.

the number of possible conformations and, as we have seen, this results in an increase in receptor activity and selectivity.

Similar rigidification tactics have been useful in the development of the antihypertensive agent cilazapril (see pages 338–9) and the development of the sedative etorphine (see Chapter 17).

Locking a rotatable bond into a ring is not the only way a structure can be rigidifed. A flexible side chain can be partially rigidified by incorporating a rigid functional group such as a double bond, alkyne, amide, or aromatic ring.

For example, this tactic was used to rigidify structure (I) in Fig. 9.23. This compound is an inhibitor of platelet aggregation and binds to its target receptor by means of a guanidine functional group and a diazepine ring system. These binding groups are linked together by a highly flexible chain and it was considered desirable to make this chain less flexible. Structures (II) and (III) are examples of active compounds in which the connecting chain has been partially rigidified by the introduction of rigid functional groups.

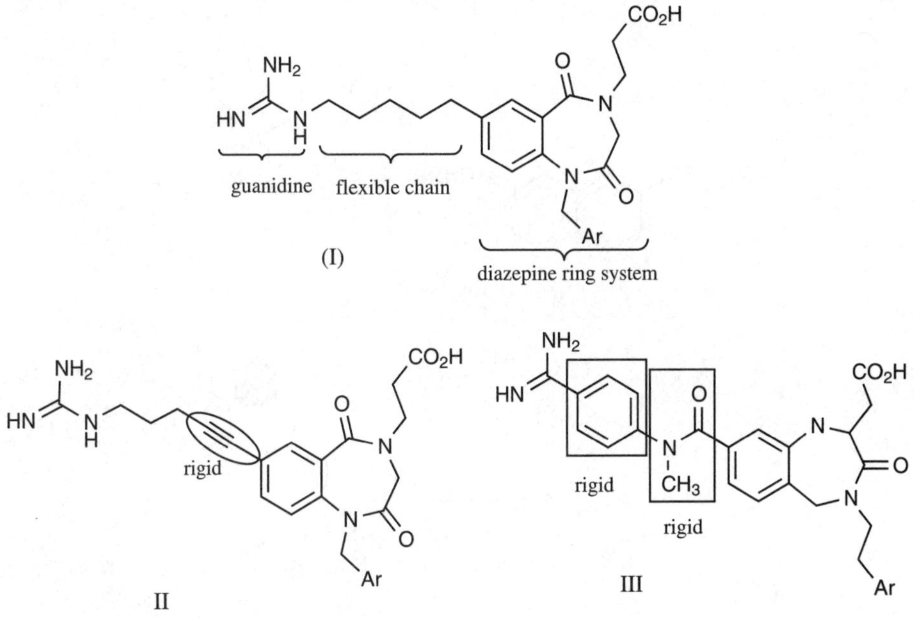

Fig. 9.23 Rigidification of flexible chains.

The advantages of rigidification have already been mentioned, but there are also potential disadvantages. Rigidified structures may be more complicated to synthesize. There is also no guarantee that rigidification will retain the active conformation. It is quite possible that rigidification will lock the compound into inactive conformations.

9.11 Conformation blockers

We have already seen how rigidification tactics can restrict the number of conformations which are possible for a compound. Another tactic which can have the same effect involves conformational blockers. In certain situations, a quite simple substituent can hinder the free rotation of a single bond. For example, introducing a methyl substituent to the dopamine (D_3) antagonist (I) (Fig. 9.24) gives structure (II) and results in a dramatic reduction in affinity. The explanation lies in a bad steric clash between the new methyl group and an *ortho* proton on the neighbouring ring which prevents both rings being in the same plane. Free rotation around the bond between the two rings is no longer possible and so the structure adopts a conformation where the two rings are at an angle to each other. In structure I, free rotation around the connecting bond allows the molecule to adopt a conformation where the aromatic rings are coplanar – the active conformation for the receptor.

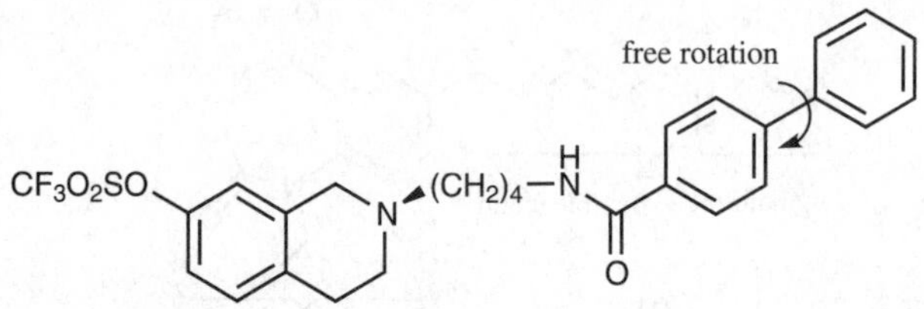

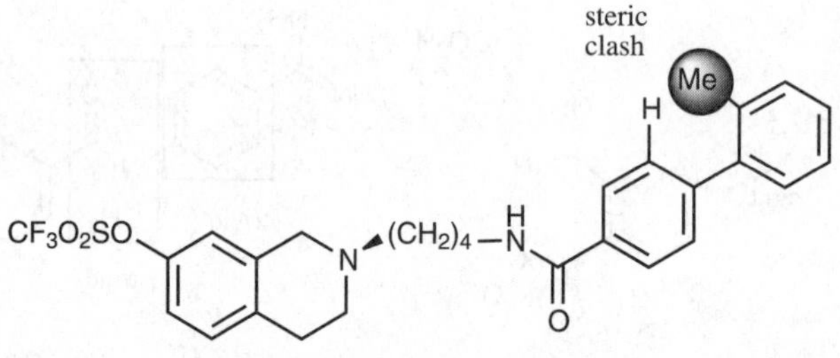

Fig. 9.24 Conformational blocking.

In this case, a conformational blocker 'rejects' the active conformation. An example of a situation in which a conformational blocker favours the active conformation can be seen in the case study described in Chapter 13.

9.12 X-ray crystallographic studies

So far we have discussed the traditional strategies of drug design. These were frequently carried out with no knowledge of the target structure, and the results obtained were useful in providing information about the target protein and its binding site. Clearly, if a drug has an important binding group, there must be a complementary binding group present in the binding site of the receptor or enzyme.

If the target protein can be isolated and crystallized, then an X-ray structure of the protein is of enormous benefit to the whole drug design process, since this can reveal the actual structure of the binding site. Unfortunately, the X-ray structure of the protein itself does not reveal where the binding site is, and so it is better to obtain an X-ray structure of the macromolecule with an inhibitor or antagonist (ligand) bound to the binding site. This structure can then be downloaded onto a computer and the complex studied by molecular modelling to see how the ligand binds. Once the important binding groups in the binding site have been identified, the modelling software can be used to remove the ligand and add potential drugs to see how well they fit. This also allows the identification of regions in the binding site which are not occupied by the drug and guides the medicinal chemist as to what modifications and additions can be made to the lead compound. This topic is covered in more detail in Chapter 13.

Unfortunately, not all enzymes and receptors can be crystallized, especially those which are membrane-bound. However, the structural and mechanistic information obtained from analogous receptors or enzymes can be useful. For example, the enzyme ACE (angiotensin converting enzyme) is a membrane-bound enzyme which has been difficult to isolate and study. It is a member of a group of enzymes called the zinc metalloproteinases and catalyses the hydrolysis of a dipeptide fragment from the end of a decapeptide called angiotensin I to give the octapeptide angiotensin II (Fig. 9.25).

Asp-Arg-Val-Tyr-Ile-His-Pro-Phe-His-Leu $\xrightarrow{\text{ACE}}$ Asp-Arg-Val-Tyr-Ile-His-Pro-Phe + His-Leu

Angiotensin I   Angiotensin II

Fig. 9.25 Reaction catalysed by ACE.

Angiotensin II is an important hormone, causing blood vessels to constrict which results in a rise in blood pressure. Therefore, ACE inhibitors are potential antihypertensive agents since they inhibit the production of angiotensin II. Although the enzyme ACE could not be isolated, the design of ACE inhibitors was helped by studying the structure and mechanism of another zinc metalloproteinase – which had an enzyme called carboxypeptidase. This enzyme splits a terminal amino acid from a peptide chain, as shown in Fig. 9.26, and is inhibited by L-benzylsuccinic acid.

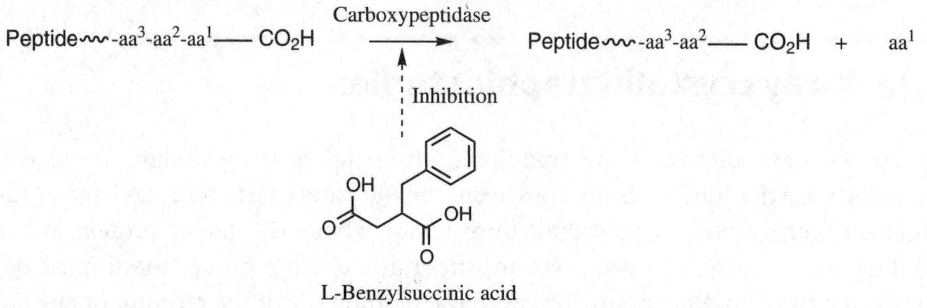

Peptide ∿ -aa³-aa²-aa¹ —— CO_2H $\xrightarrow{\text{Carboxypeptidase}}$ Peptide ∿ -aa³-aa² —— CO_2H + aa¹

Inhibition

L-Benzylsuccinic acid

Fig. 9.26 Hydrolysis by carboxypeptidase.

The active site of carboxypeptidase (Fig. 9.27) contains a charged arginine unit (Arg 145) and a zinc ion which are both crucial in binding the substrate peptide. The peptide binds such that the terminal carboxylic acid is ionically bound to the arginine unit, while the carboxyl group of the terminal peptide bond is bound to the zinc ion. There is also a pocket called the S1′ pocket which can accept the side chain of the terminal amino acid (Phe in the example shown). Hydrolysis of the terminal peptide then takes place.

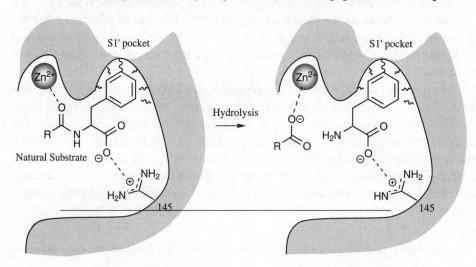

Fig. 9.27 Binding site interactions for carboxypeptidase.

The design of the carboxypeptidase inhibitor L-benzylsuccinic acid was based on the hydrolysis products arising from this enzymic reaction. The benzyl group was included to occupy the S1′ pocket, while the adjacent carboxylate anion was present to form an ionic interaction with Arg145. The second carboxylate was present to act as a ligand to the zinc ion, mimicking the carboxylate ion of the hydrolysis product.

L-benzylsuccinic acid binds to the binding site as shown in Fig. 9.28. However, hydrolysis of L-benzylsuccinic acid is not possible since there is no peptide bond present, and so the enzyme is inhibited for as long as the compound stays attached.

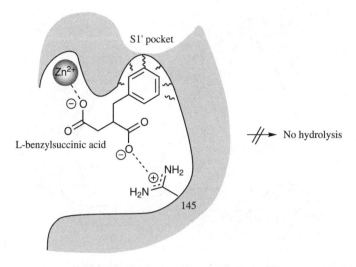

Fig. 9.28 Inhibition by L-benzylsuccinic acid.

An understanding of the above mechanism and inhibition helped in the design of ACE inhibitors. First of all, it was assumed that the active site contained the same zinc ion and arginine group. However, since ACE splits a dipeptide unit from the peptide chain rather than an amino acid, these groups are likely to be further apart and so an analogous inhibitor to benzylsuccinic acid would be a succinyl-subsituted amino acid. Succinyl proline was chosen since proline is present on the terminus of teprotide (a known inhibitor of ACE; Fig. 9.29).

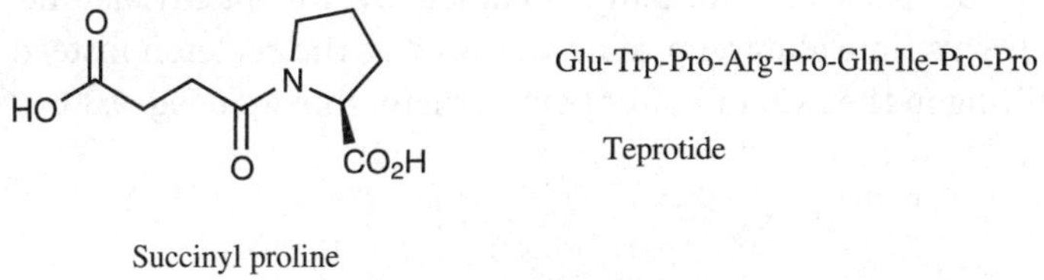

Glu-Trp-Pro-Arg-Pro-Gln-Ile-Pro-Pro

Teprotide

Succinyl proline

Fig. 9.29 ACE inhibitors.

Succinyl proline did indeed inhibit ACE and it was proposed that both carboxylate groups were ionized, one interacting with the arginine group and one with the zinc ion (Fig. 9.30). It was now argued that there must be pockets available to accommodate amino acid side chains (pockets S1 and S1′). The strategy of extension was now employed to find a group which would fit the S1′ pocket and increase the binding affinity. A methyl group fitted the bill and resulted in an increase in activity (Fig. 9.31).

The next step was to see whether there was a better group than the carboxylate ion to interact with zinc and it was discovered that a thiol group led to increased activity. This resulted in captopril, the first non-peptide ACE inhibitor to become commercially available.

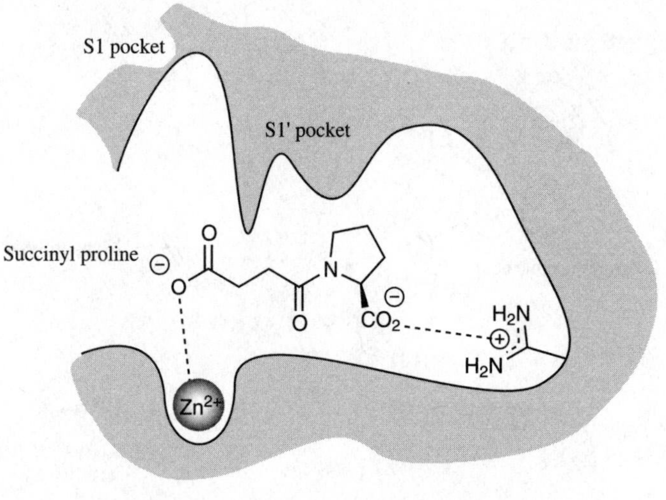

Fig. 9.30 Binding site interaction for ACE.

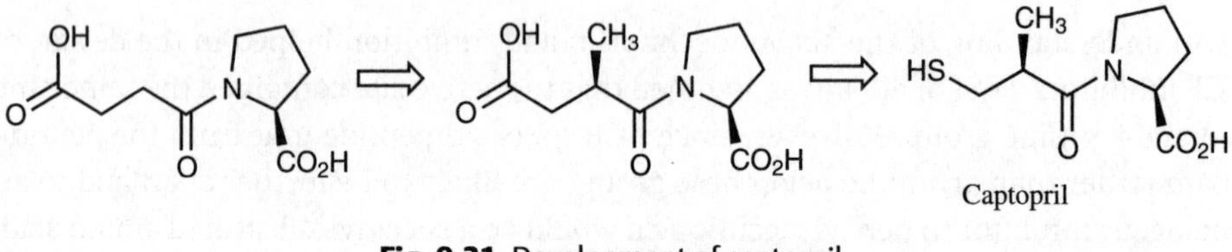

Fig. 9.31 Development of captopril.

The next advance involved extension strategies aimed at finding a group which would fit the S1 pocket – normally occupied by the phenylalanine residue in angiotensin I. This time glutaryl proline was used as the skeleton instead of succinyl proline, resulting in the ACE inhibitor (enalaprilate) shown in Fig. 9.32.

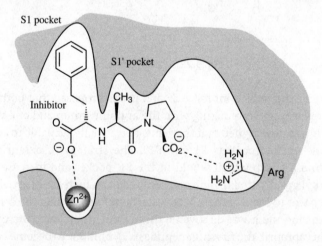

Fig. 9.32 Enalaprilate.

9.13 Molecular modelling studies

Molecular modelling studies have revolutionized medicinal chemistry in recent years and a separate chapter is devoted to the topic (Chapter 13). In brief, molecular modelling can be used for drug discovery, design, and development. We have already mentioned the use of molecular modelling in studying X-ray structures of enzymes and receptors and their binding sites. However, molecular modelling can be useful in drug design even if the structure of the target molecule is unknown. Different compounds interacting with the same target can be compared and the important pharmacophore identified, allowing the design of novel structures containing the same pharmacophore. Compound databanks can be searched for those pharmacophores to identify novel lead compounds.

There are many other uses for computers in medicinal chemistry, some of which are described in Chapter 13. However, a warning! It is important to appreciate that molecular modelling studies usually tackle only one part of a much bigger problem – the design of an effective drug. True, one might design a compound which binds perfectly to a particular enzyme or receptor but, if the compound cannot be synthesized or never reaches the target protein in the body, it is a useless drug.

9.14 Drug design by nuclear magnetic resonance

The use of nuclear magnetic resonance (NMR) in designing lead compounds has already been discussed in Chapter 8. This can also be seen as a method of drug design since the focus is not only on designing a lead compound but in designing a *potent* lead compound. Usually, drug design aims to optimize a lead compound once it has been discovered. In the NMR method, optimization of the component parts (epitopes) is carried out first to maximize binding interactions; then they are linked together to produce the final compound.

9.15 The elements of luck and inspiration

It is true to say that drug design has become more rational, but it has not yet eliminated the role of chance or the need for hard working, mentally alert bench chemists. The vast majority of drugs still on the market were developed by a mixture of rational design, trial and error, hard graft, and pure luck. The drugs which were achieved by purely rational design [e.g. ACE inhibitors (Section 9.12), thymidylate synthase inhibitors (see Chapter 13), cimetidine (see Chapter 18) and pralidoxime (see Chapter 15)] are still in the minority.

Frequently, the development of drugs has been based on 'ringing the changes', watching the literature to see what works on related compounds and what doesn't, then trying out similar alterations to one's own work. It is very much a case of groping in the dark, with the chemist asking whether adding a group at a certain position will have a steric effect, an electronic effect, or a bonding effect. Even when drug design is followed on rational lines, good fortune often has a role to play.

The development of the β-blocker propranolol (Fig. 9.33) was aided by such a slice of good fortune. Chemists at Imperial Chemical Industries (ICI) were trying to improve on a drug called pronethalol (Fig. 9.16), the first β-blocker to reach the market. It was known that the naphthalene ring and the ethanolamine segment were important both to activity and selectivity, so these groups had to be retained. Therefore, a decision was taken to study what would happen if the distance between these two groups was extended (chain extension). Perhaps by doing so, the two groups would interact with their respective binding sites more efficiently.

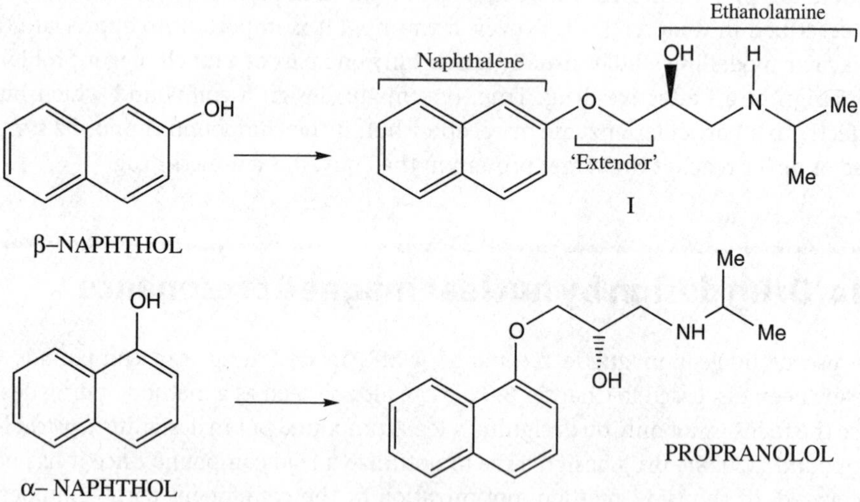

Fig. 9.33 The development of the β-blocker propranolol.

Various segments were to be inserted, one of which was the OCH_2 moiety. The analogue which would have been obtained is shown as structure I in Fig. 9.29. β-Naphthol was the starting material, but was not immediately available, and rather than waste the day, α-naphthol was used instead. The result was propranolol, which has proved to be a successful drug in the treatment of angina for many years. When the original target was eventually made, it showed little improvement over the original compound – pronethalol.

A further interesting point concerning this work is that the propranolol skeleton had been synthesized some years earlier. However, the workers involved had not been searching for β-blocking activity and had not recognized the potential of the compound.

9.16 A case study – oxamniquine

The development of oxamniquine (Fig. 9.34) is a nice example of how traditional strategies were used in the development of a drug. It also demonstrates that strategies can be used in any order and may be used more than once.

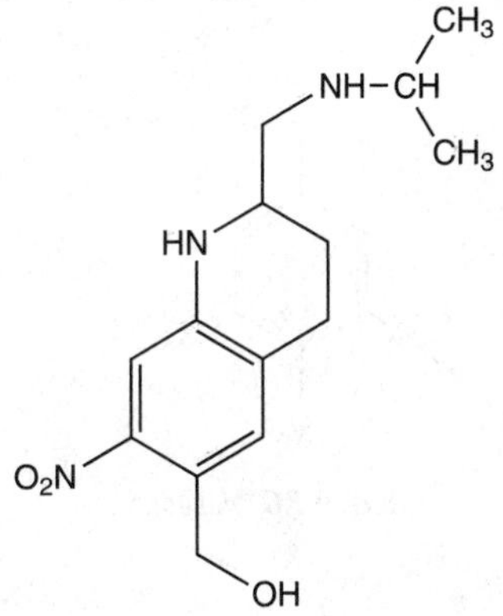

Fig. 9.34 Oxamniquine.

Oxamniquine is an important Third World drug used in the treatment of schistosomiasis (bilharziasis). This disease affects an estimated two hundred million people and is contracted by swimming or wading in infected water. The disease is carried by a snail whose flukes can penetrate human skin and enter the blood supply. There, eggs are produced which become trapped in organs and tissues, and this in turn leads to the symptoms of the disease.

The first stage in the development of oxamniquine was to find a lead compound, and so a study was made of compounds which were active against the parasite. The tricyclic structure lucanthone (Fig. 9.35) was chosen. It was known to be effective against some forms of the disease, but it was also toxic and had to be injected at regular intervals to remain effective. Therefore, the goal was to increase the activity of the drug, broaden its activity, reduce side-effects, and make it orally active.

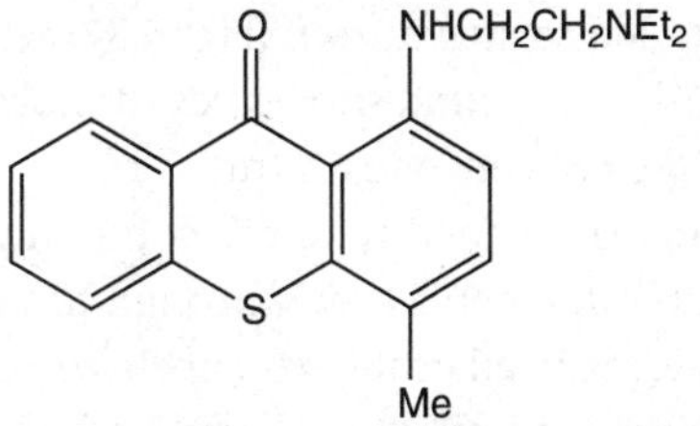

Fig. 9.35 Lucanthone.

Having found a lead compound, it was decided to try *simplifying* the structure to see whether the tricyclic system was really necessary. Several compounds were made, and the most interesting structure was one where the two 'left-hand' rings had been removed. This gave a compound called mirasan (Fig. 9.36) which retained the 'right-hand' aromatic ring containing the methyl and β-aminoethylamino side-chains *para* to each other. *Varying substituents* showed that an electronegative chloro substituent positioned where the sulfur atom had been was beneficial to activity. Mirasan was active against the schistosome parasite in mice, but not in humans.

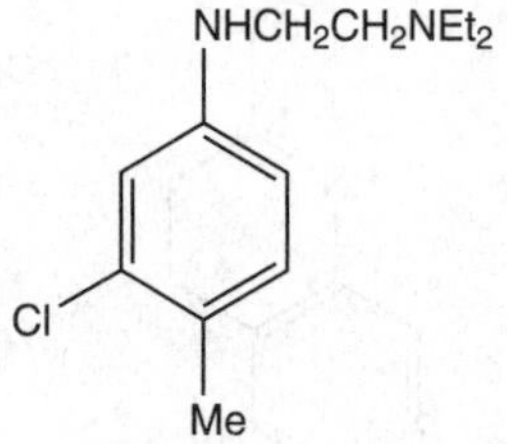

NHCH$_2$CH$_2$NEt$_2$

Cl

Me

Fig. 9.36 Mirasan.

It was now reasoned that the β-aminoethylamino side chain was important to receptor binding and would adopt a particular conformation to bind efficiently. This conformation would only be one of the many conformations which are available to a flexible molecule such as mirasan and so there would only be a limited chance of it being adopted at any one time. Therefore, the decision was taken to try and restrict the number of possible conformations by incorporating the side chain into a ring (*rigidification*). This would cut down the number of available conformations and increase the chance of the molecule having the correct conformation when it approached the receptor. There was the risk, however, that the active conformation itself would be disallowed by this tactic. Therefore, rather than incorporate the whole side chain into a ring structure, compounds were designed initially such that only portions of the chain were included.

The bicyclic structure (I) (Fig. 9.37) contains one of the side chain bonds fixed in a ring to prevent rotation round that bond. It was found that this gave a dramatic improvement in activity. The compound was still not active in man but, unlike mirasan, it was active in monkeys. This gave hope that the chemists were on the right track. Further rigidification led to structure II (Fig. 9.37) where two of the side chain bonds were constrained. This compound showed even more activity in mouse studies and the decision was taken to concentrate on this.

By now, it can be seen that the structure of the compound has been altered significantly from mirasan. In general, when a breakthrough has been achieved and a novel structure has been obtained, it is advisable to check whether past results still hold true. For example, does the chloro group still have to be *ortho* to the methyl group? Can we change the chloro group for something else? Novel structures may fit the

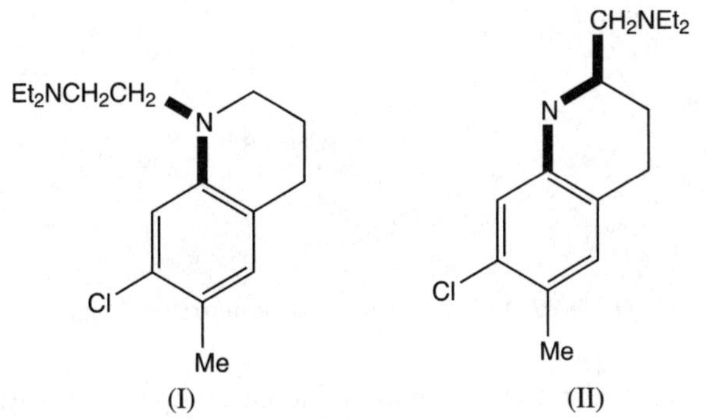

Fig. 9.37 Bicyclic structures I and II.

binding site slightly differently from the lead compound such that the binding groups are no longer in the optimum positions for binding.

Therefore, structure II was modified by *varying substituents and substitution patterns* on the aromatic ring, and by *varying alkyl substituents* on the amino groups. Chains were also *extended* to search for other possible binding sites.

The results and possible conclusions were as follows.

• The substitution pattern on the aromatic ring could not be altered and was essential for activity. Altering the substitution pattern presumably places the essential binding groups out of position with respect to their binding sites.

• Replacing the chloro substituent with more electronegative substituents improved activity, with the nitro group being the best substituent. Therefore, an electron-deficient aromatic ring is beneficial to activity. One possible explanation for this could be the effect of the neighbouring aromatic ring on the basicity of the nitrogen atom. A strongly electron-deficient aromatic ring would 'pull' the cyclic nitrogen's lone pair of electrons into the ring, thus reducing its basicity (Fig. 9.38). This in turn might improve the pK_a of the drug such that it is less easily ionized and is able to pass through cell membranes more easily (see Chapter 10).

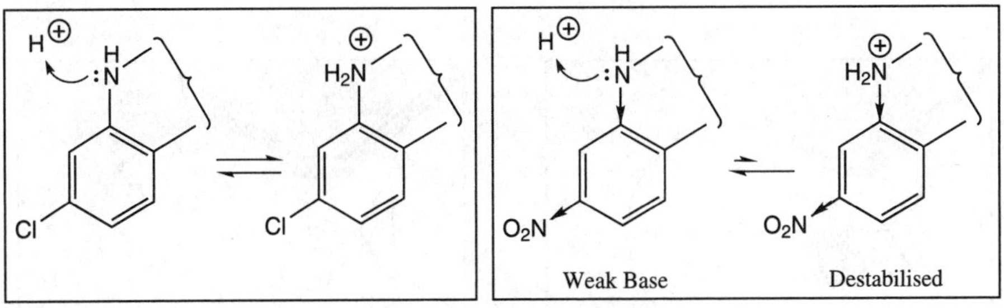

Fig. 9.38 Effect of aromatic substituents on pK_a.

• The best activities were found if the amino group on the side chain was secondary rather than primary or tertiary (Fig. 9.39).

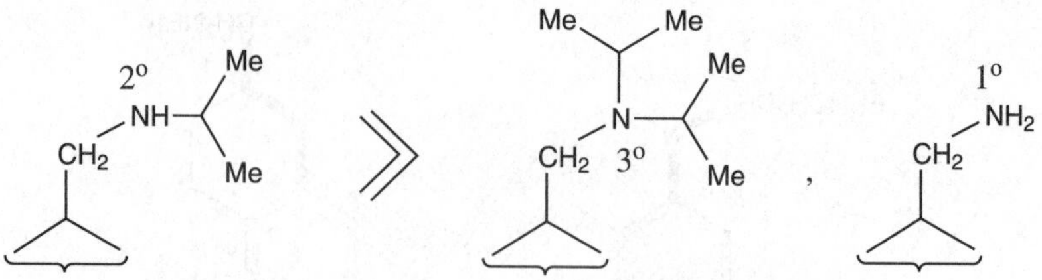

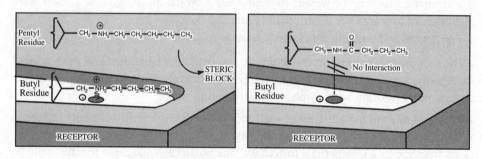

Fig. 9.39 Activity of secondary amino side chain.

• The alkyl group on this nitrogen could be increased up to four carbon units with a corresponding increase in activity. Longer chains led to a reduction in activity. The latter result might imply that large substituents are too bulky and prevent the drug from binding to the binding site. Acyl groups eliminated activity altogether, emphasizing the importance of this nitrogen atom. It is most likely that it is ionized and interacts with the receptor through an ionic bond (Fig. 9.40).

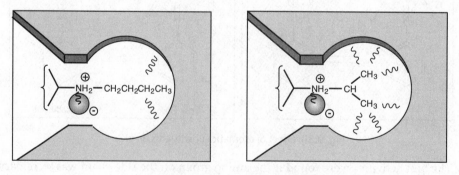

Fig. 9.40 Proposed ionic binding interaction.

• Branching of the alkyl chain increased activity. A possible explanation could be that branching increases van der Waals interactions to a hydrophobic region of the binding site (Fig. 9.41). Alternatively, the lipophilicity of the drug might be increased, allowing easier passage through cell membranes.

Fig. 9.41 Branching of the alkyl chain.

• Putting a methyl group on the side chain eliminated activity (Fig. 9.42). A methyl group is a bulky group compared with a proton and it is possible that it prevents the side chain taking up the correct binding conformation.

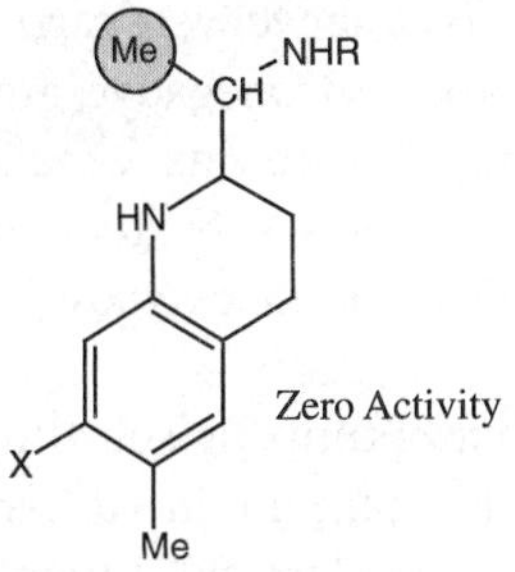

Fig. 9.42 Addition of a methyl group.

• Extending the length of the side chain by an extra methylene group eliminated activity (Fig. 9.43). This tactic was tried in case the binding groups were not far enough apart for optimum binding. This result suggests the opposite.

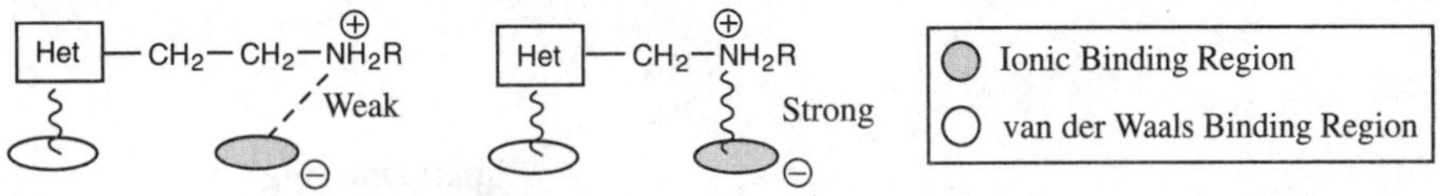

Fig. 9.43 Effect of extension of the side chain.

The optimum structure based on these results was structure III (Fig. 9.44). It has one asymmetric centre and, as one might expect, the activity was much greater in one enantiomer than it was in the other.

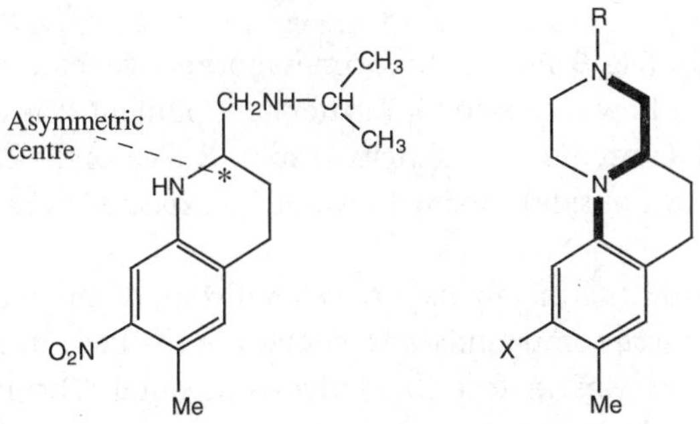

Fig. 9.44 The optimum structure (III) and the tricyclic structure (IV).

The tricyclic structure IV (Fig. 9.44) was also constructed. In this compound, the side chain is fully incorporated into a ring structure, restricting the number of possible conformations drastically (*rigidification*). As mentioned earlier, there was a risk that the active conformation would no longer be allowed, but in this case good activity was still obtained. The same variations as above were carried out to show that a secondary amine was essential and that an electronegative group on the aromatic ring was required. However, some conflicting results were obtained compared with the previous results for structure III. A chloro substituent on the aromatic ring was better than a nitro, and it could be in either of the two possible *ortho* positions relative to the methyl group.

These results demonstrate that optimizing substituents in one structure does not necessarily mean that they will be optimum in a different skeleton.

One possible explanation for the chloro substituent being better than the nitro is that a less electronegative substituent is required to produce the optimum pK_a or basicity for membrane permeability.

Adding a further methyl group on the aromatic ring to give the structure shown in Fig. 9.45 increased activity. It was proposed that the bulky methyl group was interacting with the piperazine ring and causing it to twist out of the plane of the other two rings (*conformational blocker*). The increase in activity which resulted suggests that a better fitting conformation is obtained for the receptor.

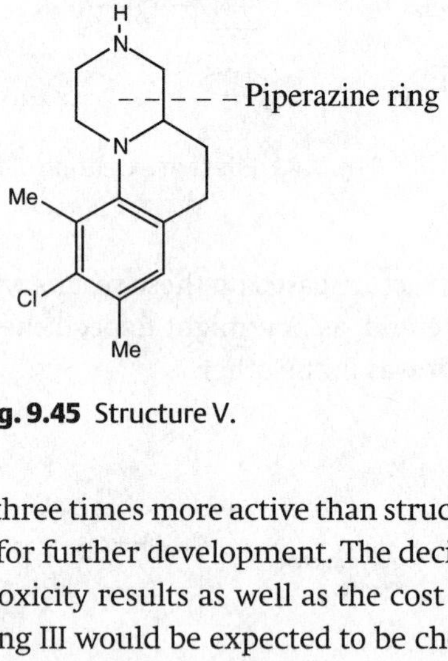

Fig. 9.45 Structure V.

This compound V (Fig. 9.45) was three times more active than structure III (Fig. 9.44). However, structure III was chosen for further development. The decision to choose III over V was based on preliminary toxicity results as well as the cost of producing the compounds. The cost of synthesizing III would be expected to be cheaper since it is a simpler molecule.

Further studies on the metabolism of related compounds then revealed that the methyl group on these compounds was oxidized in the body to a hydroxymethylene group and that this was in fact the active compound. The methyl group on III was replaced with a hydroxymethylene group to give oxamniquine (Fig. 9.34), which was found to be more active than the methyl structure (III). The drug was put on the market in 1975, 11 years after the start of the project.

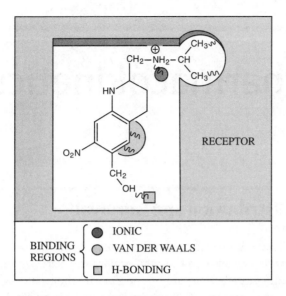

Fig. 9.46 Bonding interactions of oxamniquine.

It is now believed that compound III is totally inactive in itself. This is not as surprising as it may appear, since the metabolic reaction converts a non-polar methyl group to a polar hydroxymethylene group. Presumably the newly gained hydroxyl group forms an important hydrogen bond to the receptor (Fig. 9.46).

10 Pharmacokinetics

10.1 Drug distribution and 'survival'

In Chapter 9, we studied design strategies which could be used to improve the binding interactions of drugs with their molecular targets. However, the compounds with the best binding interactions are not necessarily the best drugs to use in medicine. This is because a drug has to travel through the body to reach its target and in doing so faces many obstacles and hazards which can divert it from its goal. This is particularly true for any drug which is taken orally. First the drug has to be water-soluble if it is to dissolve in the gastrointestinal (GI) tract and the blood supply. However, it also has to be fat-soluble if it is to cross the fatty cell membranes between the GI tract and the blood supply. Second, it has to be stable enough to survive the acids of the stomach, as well as a battery of enzymes which will attack the slightest weakness in its structure. These enzymes include the digestive enzymes in the GI tract, as well as an array of metabolic enzymes in the gut wall, liver, and blood. Even if the drug survives all these attacks, there is still no guarantee that it will reach its target. There are many ways in which a drug can be removed from the blood supply. For example, some drugs are shunted through the bile duct back into the intestines and have to 'try again'. Lipophilic drugs might be taken up by fat tissue, while polar drugs are easily excreted by the kidneys. Anionic drugs can become bound to plasma protein while cationic drugs can become bound to nucleic acids. Drugs which are taken up or bound in these ways are not available to act on their targets; it is the plasma concentration of free unbound drug which is the crucial measure of a drug's availability.

There is an extra barrier facing drugs intended to act in the central nervous system (CNS). The blood vessels in the brain are surrounded by a continuous layer of tightly joined endothelial cells which means that drugs have to pass through these cells to reach the brain. The fatty cell membranes of these cells act as an effective barrier to polar drugs and this is known as the **blood–brain barrier**. Elsewhere in the body, the blood vessels are 'leakier', which allows polar drugs to reach other tissues.

Clearly, a drug's journey to its target is not easy and the science of pharmacokinetics looks at how a drug is absorbed, metabolized, distributed, and eliminated in the body. Only a small percentage of the drug administered may ever reach its target, and the loss of drug involved in its first passage through the GI tract, liver, and blood supply is known as the **'first pass effect'**.

10.2 Pharmacokinetic issues in drug design

A drug's success in reaching its target depends principally on its physical and chemical properties. These include its chemical and metabolic stability, its hydrophilic/hydrophobic character, its ionization, and its size. We shall now consider each of these factors in turn.

10.2.1 Chemical stability

Drugs must be chemically stable and not decompose in aqueous solution. Orally taken drugs must also survive the harsh acid conditions of the stomach. There are several important drugs which have chemically labile functional groups and which are easily hydrolysed in the stomach. For example, penicillins have a chemically labile β-lactam ring, while cholinergic agents have a susceptible ester group. One way round the problem of acid sensitivity is to inject the drug directly into the blood supply. However, it is also possible to design drugs such that the offending functional group is less labile (see Section 10.5).

10.2.2 Metabolic stability

As well as being chemically stable, drugs need to be stable to the various digestive and metabolic enzymes which they will encounter. The greatest challenges here are the metabolic enzymes, mostly to be found in the liver. The study of how these enzymes alter drugs is known as **drug metabolism** and the products obtained are called **drug metabolites**. There are various strategies which can be used to make a drug more resistant to drug metabolism (see Section 10.5), and a knowledge of why drug metabolism takes place, as well as the reactions involved, is important in understanding why these strategies work (see Section 8.12).

Ideally, a drug should be resistant to drug metabolism since the production of metabolites complicates drug therapy. For example, the metabolites formed will usually have different properties from the original drug. In some cases, activity may be lost. In others, the metabolite may prove to be toxic. For example, the metabolites of paracetamol cause liver toxicity, while the carcinogenic properties of some polycyclic hydrocarbons are due to the formation of epoxides.

Another problem arises from the fact that the activity of metabolic enzymes varies from individual to individual. This is especially true of a set of metabolic enzymes known as the cytochrome P450's. These enzymes are haemoproteins (containing haem and iron) which are crucial to the oxidative metabolic reactions carried out in the liver. They catalyse a reaction which splits molecular oxygen, such that one of the oxygen atoms is introduced into the drug substrate and the other ends up as water. There are at least 20 different P450 enzymes, which can be split into three main families called CYP1, CYP2, and CYP3. Within each family there are various subfamilies designated by a letter, and each enzyme within that subfamily is designated a number. For example, CYP3A4 is enzyme 4 in the subfamily A of the main family 3. The profile of P450 enzymes in different patients can vary, resulting in a variation in the extent to which a drug is metabolized.

If this is the case, then the amount of drug which can be safely administered will also vary depending on the level of metabolism. Differences across populations can be quite significant, resulting in different countries having different recommended dose levels for particular drugs.

The situation is further complicated if the activity of the P450 enzymes is affected by other chemicals. For example, certain foods can have an influence on the activity of P450 enzymes. Brussel sprouts and cigarette smoke enhance activity whereas grapefruit juice inhibits activity. This can have a significant effect on the activity of drugs metabolized by P450 enzymes. For example, certain drugs, such as the immunosuppressant drug cyclosporin or the dihydropyridine hypotensive agents, are more efficient when taken with grapefruit juice since their metabolism is reduced. However, serious toxic effects can arise if the antihistamine agent terfenadine is taken with grapefruit juice. Terfenadine is not active itself but is metabolized to form fexofenadine which is the active drug (Fig. 10.1). If metabolism is inhibited by grapefruit juice, then terfenadine persists in the body and can cause serious cardiac toxicity. As a result, fexofenadine itself is now favoured over terfenadine as an antihistamine.

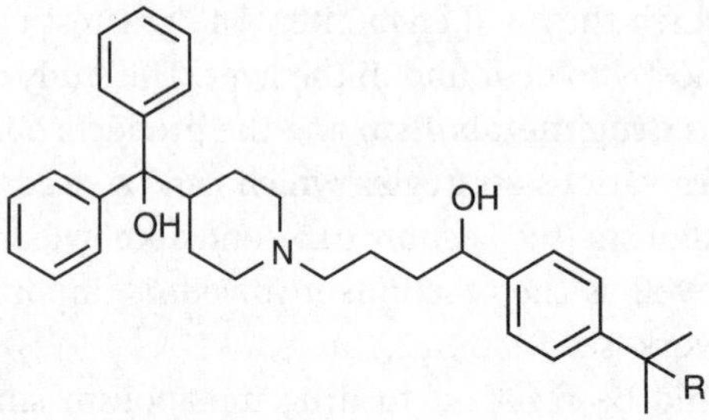

Fig. 10.1 Terfenadine (Seldane) (R=CH$_3$), and fexofenadine (Allegra)   (R=CO$_2$H).

Certain drugs are also capable of inhibiting or promoting P450 enzymes, leading to a phenomenon known as drug–drug interaction where the presence of one drug affects the activity of another. For example, several antibiotics can act as P450 inhibitors and will slow the metabolism of any other drug metabolized by these enzymes. Another example involves drug–drug interactions between the anticoagulant warfarin and the barbiturate phenobarbitone (Fig. 10.2) or the anti-ulcer drug cimetidine (see Chapter 18).

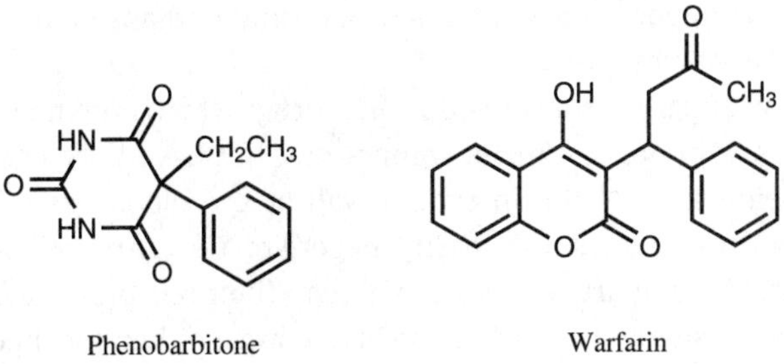

Phenobarbitone   Warfarin

Fig. 10.2 Drug–drug interactions.

Phenobarbitone stimulates P450 enzymes and accelerates the metabolism of warfarin, making it less effective. On the other hand, cimetidine inhibits P450 enzymes, thus slowing the metabolism of warfarin. Such drug–drug interactions affect the plasma levels of warfarin and could cause serious problems if the levels move outwith the normal therapeutic range.

Herbal medicine is not immune from this problem either. St John's wort is a popular remedy used for mild to moderate depression. However, it promotes the activity of cytochrome P450 enzymes and decreases the effectiveness of contraceptives and warfarin.

Because of the problems caused by P450 activation or inhibition, new drugs are usually tested to check whether they have any P450 activity. Furthermore, in many projects, one of the requirements for developing a new drug is that it should have no such activity.

10.2.3 Hydrophilic/hydrophobic balance

A drug must have the correct balance of hydrophilic and hydrophobic properties. Without this balance, drugs suffer several disadvantages. For example, drugs which are too polar are easily excreted by the kidneys and do not easily cross the fatty barriers of cell membranes. On the other hand, drugs which are too lipophilic (fat-loving) show poor solubility in water and are also poorly absorbed from the GI tract since they are likely to coagulate in fatty globules and fail to interact with the gut wall. Even should they reach the bloodstream, they are swiftly removed and stored away in the fatty tissues of the body. This fat solubility can lead to problems. For example, obese patients undergoing surgery require a larger than normal volume of anaesthetic since the gases used are particularly fat-soluble. Unfortunately, once surgery is over and the patient has regained consciousness, the anaesthetics stored in the fat tissues will be released and may render the patient unconscious again. Barbiturates were once seen as potential intravenous anaesthetics which could replace the anaesthetic gases. Unfortunately, they too are fat-soluble and, as a result, it is extremely difficult to estimate a sustained safe dosage. The initial dose can be estimated to allow for the amount of barbiturate taken up by fat cells. However, further doses eventually lead to saturation of the fat depot, and result in a sudden and perhaps fatal increase of barbiturate levels in the blood supply.

We have here an apparent contradiction. The drugs which bind most strongly to the receptor are often very polar, ionized compounds, but have little chance of crossing the fatty cell membranes of the intestinal wall or the blood–brain barrier. On the other hand, the drugs which can easily negotiate the fatty cell membranes get mopped up by fat tissue or are too weak to bind to their receptor sites. Consequently, the best drugs are usually a compromise. They are neither too lipophilic nor too hydrophilic.

10.2.4 Ionization

It is noticeable how many drugs contain an amine functional group. There are very good reasons for this. Amines are often involved in a drug's binding interactions with its target. However, they are also one answer to the problem of balancing the dual requirements of water and fat solubility. Amines are weak bases and, in general, it is found that many of the most effective drugs are amines having a pK_a value in the range 6–8. In other words, they are drugs which are partially ionized at blood pH[1] and can easily equilibrate between their ionized and non-ionized forms. This allows them to cross cell membranes in the non-ionized form, while the presence of the ionized form gives the drug good solubility in water and permits good binding interactions with its receptor (Fig. 10.3).

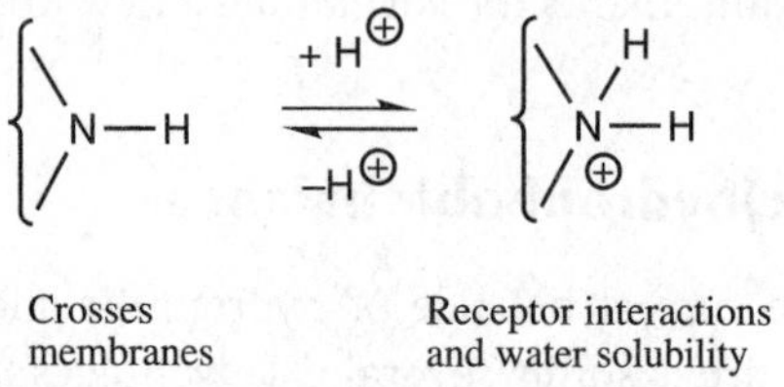

Crosses
membranes

Receptor interactions
and water solubility

Fig. 10.3 Hydrophobic/hydrophilic balance.

A drug which is strongly ionized has difficulty crossing cell membranes and yet there are several important drugs which contain acidic and/or basic groups (e.g. penicillins) or are fully ionized (e.g. cholinergic and anticholinergic drugs). How can such drugs be used effectively?

One method is to inject the drug into the blood supply, thus avoiding the need to cross the gut wall. Another method is to design a prodrug where the ionizable group is temporarily masked (see Section 10.9). A third possibility is to take advantage of the cell's own carrier proteins (see Chapter 3) which are designed to carry the important

---

[1] Weak bases such as amines obey the Henderson–Hasselbalch equation: pH = pKa + log ([RNH$_2$]/[RNH$_3{}^+$]). When the amine is 50% ionized, pH = pKa. Therefore, drugs having a pKa of 6–8 are approximately 50% ionized at blood pH.

polar and ionic molecules (i.e. sugars, amino acids, neurotransmitters and metal ions) which the cell needs for its survival across the cell membrane. If the drug bears some structural resemblance to such a molecule, then it too may be transported. For example, levodopa is transported by the carrier protein for phenylalanine while fluorouracil is transported by carrier proteins for thymine and uracil. Carrier proteins for dipeptides accept the antihypertensive agent lisinopril (Fig. 10.4).

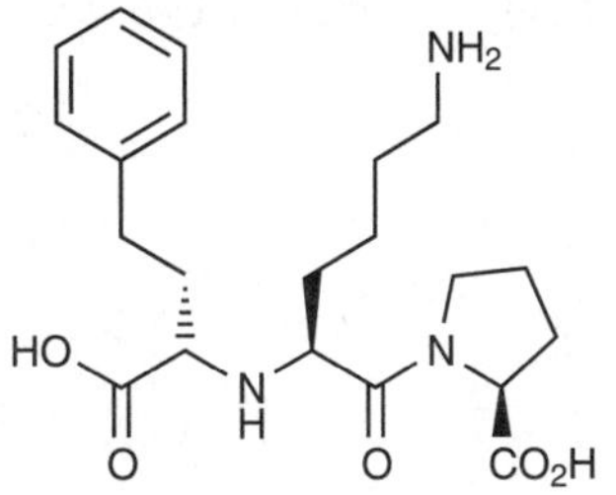

Fig. 10.4 Lisinopril.

10.2.5 Size

In general, most useful drugs have a molecular weight of less than 500. However, size itself is no barrier to absorption. As molecular weight increases, the rate of absorption certainly decreases, but it is still possible for drugs of larger molecular weight to be absorbed. For example, cyclosporin has a molecular weight of about 1200 and is successfully absorbed through cell membranes. A more important limiting factor to absorption is polarity, and it is often the case that larger molecules are poorly absorbed, not because of their size, but because they are more likely to have a larger number of polar functional groups.

Having said that, there are some exceptions to this rule. Some high molecular weight compounds with high polarity can still enter cells and be useful drugs because they can cross cell membranes by a process called **pinocytosis**. This involves the cell membrane budding inwards, then being 'nipped' off such that it forms a vesicle containing the drug. The vesicle may then be broken down to release the drug within the cell, or the vesicle may fuse with the opposite membrane to release the drug on the other side. Insulin crosses the blood–brain barrier by this process (Fig. 10.5).

At the other end of the size spectrum, molecules with a molecular weight less than 200 can enter tissues by 'squeezing' through the gaps between cells rather than by passing through cells. Thus, highly polar molecules can reach tissues without having to cross cell membranes, as long as their molecular weight is less than 200.

10.2.6 Number of hydrogen bonding interactions

The more hydrogen bonding groups present in a molecule, the less likely it is that a molecule will be absorbed. As a rule of thumb, compounds which contain more than five hydrogen bond donating groups or 10 hydrogen bond accepting groups are poorly

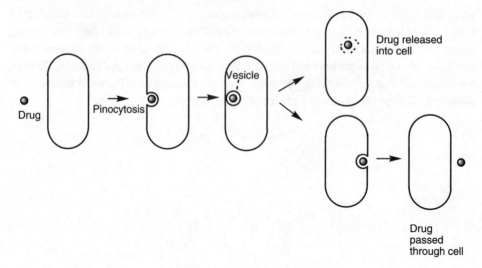

Fig. 10.5 Pinocytosis.

absorbed. There are exceptions. The anticancer agent methotrexate and the antibiotic erythromycin are absorbed better than expected because of active transport by carrier proteins.

10.3 Drug dose levels

Because of the number of pharmacokinetic variables involved, estimating the correct dose levels for a drug can be a difficult problem. There are also other issues to consider. Ideally, the blood levels of any drug should be constant and controlled, but this would require a continuous, intravenous drip which is clearly impractical for most drugs. Therefore, drugs are usually taken at regular time intervals and the doses taken are designed to keep the blood levels of drug within a maximum and minimum level such that they are not too high to be toxic, yet not too low to be ineffective. This works well in most cases, and in general the most successful drugs have a dose size of less than 200 mg and are given once or twice a day. However, there are certain situations where timed doses are not suitable. The treatment of diabetes with insulin is a case in point. Insulin is normally secreted continuously by the pancreas and so the injection of insulin at timed intervals is unnatural and can lead to a whole range of physiological complications.

Other complications include differences of age, sex, and race. Diet, environment, and altitude also have an influence. Body weight is another important factor to be taken into account. Obese people present a particular problem since it can prove very difficult to estimate how much of a drug will be stored in fat tissue and how much will be free drug. The precise time at which drugs are taken may be important since metabolic reaction rates can vary throughout the day.

Drugs can interact with other drugs. For example, some drugs used for diabetes are bound by plasma protein in the blood supply and are therefore not 'free' to react with receptors. However, they can be displaced from the plasma protein by aspirin and this can lead to a drug overdose. A similar phenomenon is observed between anticoagulants and aspirin.

Problems can also occur if a drug which inhibits a metabolic reaction is taken with a drug normally metabolized by that reaction. The latter would then be more slowly metabolized, increasing the risk of an overdose. For example, the antidepressant drug phenelzine (Fig. 10.6) inhibits the metabolism of amines and should not be taken with drugs such as amphetamines or pethidine. Even amine-rich foods can lead to adverse effects, implying that cheese and wine parties are hardly the way to cheer the victim of depression. Other examples have been described above (see Section 10.2.2).

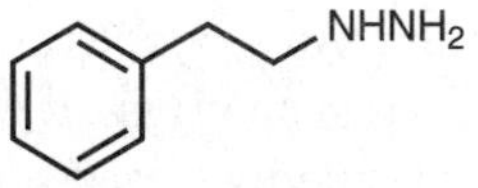

Fig. 10.6 Phenelzine.

When one considers all these complications, it is hardly surprising that individual variation to drugs can differ by as much as a factor of ten.

10.4 Drug design – solubility and membrane permeability

We shall now look at how drug design can be used to tackle some of the pharmacokinetic problems faced by drugs. First of all, we shall consider how the solubility and membrane permeability of a drug can be improved by varying the hydrophilic/hydrophobic balance of the drug. A drug's polarity and ionization are both important in this respect. Drugs which are too polar or strongly ionized do not cross cell membranes easily and are quickly excreted, whereas non-polar drugs are poorly soluble in aqueous solution and get taken up by fat tissue. In general, the polarity and ionization of compounds can be altered by changing easily accessible substituents. Such changes are particularly open to a quantitative approach known as quantitative structure–activity relationships (QSAR), discussed in Chapter 11.

10.4.1 Variation of alkyl or acyl substituents to alter polarity

Molecules which are too polar can be made less polar by masking a polar functional group with an alkyl or acyl group. For example, an alcohol or a phenol can be converted to an ether or ester, a carboxylic acid can be converted to an ester or amide, while primary and secondary amines can be converted to amides or to secondary and tertiary amines. Polarity is decreased not only by masking the polar group but also by the addition of the extra hydrophobic alkyl group, with larger alkyl groups having a greater hydrophobic effect. However, one has to be careful in masking polar groups since these groups are often important in binding the drug to its target, and masking them may prevent binding. If this turns out to be the case, it is often useful to mask the polar group temporarily such that the masking group is removed once the drug is absorbed (see prodrugs). Alternatively, extra alkyl groups could be added to the carbon skeleton of the molecule instead. However, this usually involves a more involved synthesis.

If the molecule is not sufficiently polar, then the opposite strategy can be used; that is, replacing alkyl groups with smaller alkyl groups or removing alkyl groups entirely.

Sometimes there is a benefit in increasing the size of one alkyl group and decreasing the size of another. This is called a **methylene shuffle** and has been used to modify the hydrophobicity of a compound. An example of this is in the design of second generation anti-impotence drugs based on viagra. It was found that adding extra bulk on the right hand side of the molecule increased the drug's selectivity to its target. However, this also made the drug too lipophilic. Therefore a methylene shuffle was carried out such that a propyl and a methyl group were altered to two ethyl groups. This resulted in reduced lipophilicity and better *in vivo* activity. The compound (UK 343664) is now in clinical trials (Fig. 10.7).

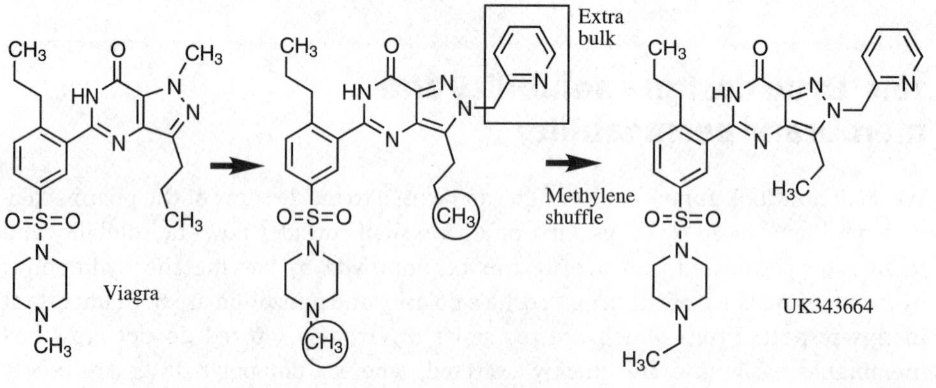

Fig. 10.7 Methylene shuffle.

10.4.2 Varying polar functional groups to alter polarity

A polar functional group could be added to a drug to increase its polarity. For example, the antifungal agent tioconazole is only used for skin infections because it is non-polar and is poorly soluble in blood. Introducing a polar OH group and more polar heterocyclic rings led to the orally active antifungal agent fluconazole with improved solubility and enhanced activity against systemic infection (i.e. in the blood supply) (Fig. 10.8).

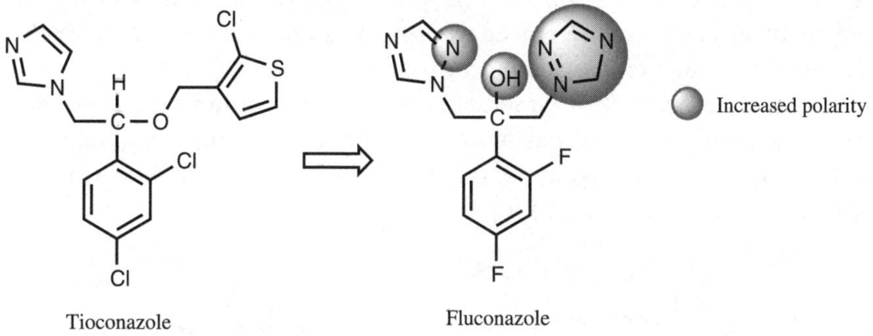

Fig. 10.8 Increasing polarity in antifungal agents.

In contrast, the polarity of an excessively polar drug could be lowered by removing polar functional groups. This strategy has been particularly successful with lead compounds derived from natural sources. However, it is important not to remove functional groups which are important to the drug's binding interactions with its target. In some cases, a drug may have too many essential polar groups which prevent it being used clinically. For example, the antibacterial agent in Fig. 10.9 has good *in vitro* activity but poor *in vivo* activity because of the large number of polar groups. Some of these groups can be removed or masked, but the majority are required for activity. As a result, the drug cannot be used clinically.

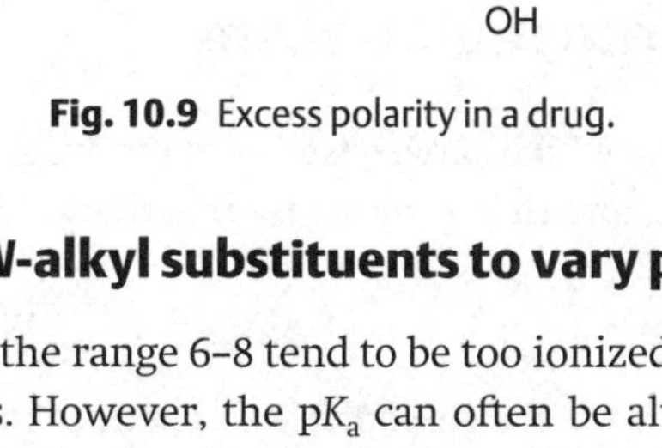

Fig. 10.9 Excess polarity in a drug.

10.4.3 Variation of *N*-alkyl substituents to vary pK_a

Drugs with a pK_a outside the range 6–8 tend to be too ionized and are poorly absorbed through cell membranes. However, the pK_a can often be altered to bring it into the preferred range. For example, the pK_a of an amine can be varied by varying the alkyl substituents. However, it is sometimes difficult to predict how such variations will affect the pK_a. Extra alkyl groups or larger alkyl groups on an amine have an increased electron donating effect which should increase basicity. However, increasing the size or number of alkyl groups also results in an increased steric bulk around the nitrogen atom. This hinders water molecules from solvating the ionized form of the base and prevents stabilization of the ion. This in turn decreases the basicity of the amine. Therefore, there are two different effects acting against each other. Nevertheless, varying alkyl substituents is a useful tactic to try.

A variation of this tactic is to 'wrap up' a basic nitrogen within a ring. For example, the benzamidine structure (I) has antithrombotic activity, but the amidine group present is too basic. Incorporating the group into an isoquinoline ring system (PRO3112) reduces basicity (Fig. 10.10).

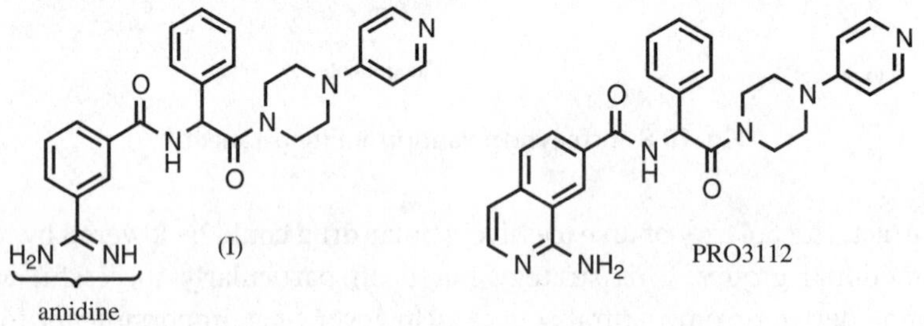

Fig. 10.10 Varying basicity in antithrombotic agents.

10.4.4 Variation of aromatic substituents to vary pK_a

The pK_a of an aromatic amine or carboxylic acid can be varied by adding electron donating or electron withdrawing substituents to the ring. The position of the substituent relative to the amine or carboxylic acid is important if the substituent interacts with the ring through resonance (see Section 11.3.2). An illustration of this can be seen in the development of oxamniquine (see Chapter 9).

10.5 Drug design – making drugs more resistant to hydrolysis and drug metabolism

There are various strategies which can be used to make drugs more resistant to hydrolysis and drug metabolism, and thus prolong their activity.

10.5.1 Steric shields

Some functional groups are more susceptible to chemical and enzymic degradation than others. For example, esters and amides are particularly prone to hydrolysis. A common strategy which is used to protect such groups is to add 'steric shields', designed to hinder the approach of a nucleophile or an enzyme to the susceptible group. These usually involve the addition of a bulky alkyl group close to the functional group. For example, the *tert*-butyl group in the antirheumatic agent D1927 serves as a steric shield and blocks hydrolysis of the terminal peptide bond (Fig. 10.11).

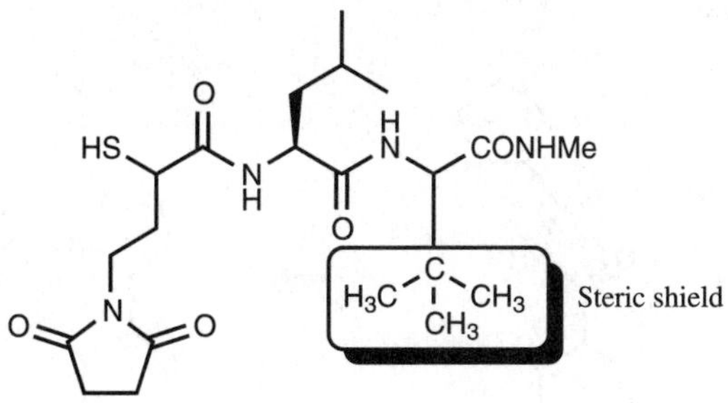

Fig. 10.11 D1927.

10.5.2 Bioisosteres – electronic effects

Another popular tactic used to protect a labile functional group is to stabilize the group electronically using a **bioisostere**. A bioisostere is a chemical group used to replace another chemical group within the drug without affecting the important biological activity. However, other features such as the drug's stability may be improved. Isosteres (see Chapter 8) are frequently used as bioisosteres.

For example, replacing the methyl group of a methyl ester with NH_2 results in a urethane functional group which is more stable than the original ester (Fig. 10.12). The NH_2 group is the same valency and size as the methyl group and therefore has no steric effect. However, it has totally different electronic properties and, as such, can feed electrons into the carboxyl group and stabilize it from hydrolysis. The cholinergic agonist carbachol is stabilized in this way (see Chapter 15).

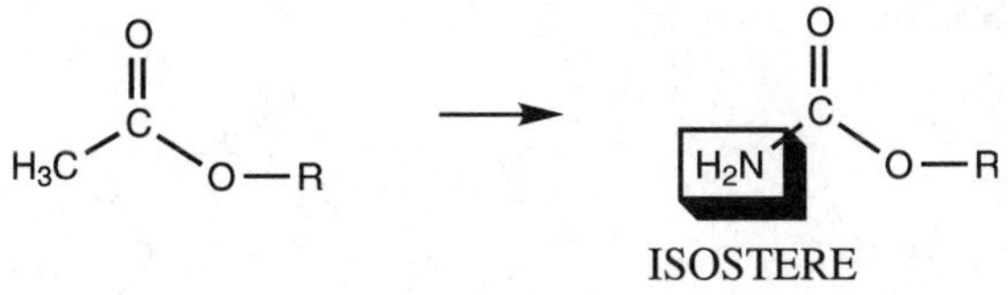

Fig. 10.12 Isosteric replacement of a methyl with an amino group.

Alternatively, a labile ester group could be replaced with an amide group (NH replacing O). Amides are more resistant than esters to hydrolysis, due again to the lone pair of the nitrogen feeding its electrons into the carbonyl group and making it less electrophilic.

However, it is important to realize that bioisosteres are not general and are often specific to a particular field. Replacing an ester with a urethane or an amide may work in one category of drugs but not another. It is also important to realize that bio-isosteres are different from isosteres. It is the retention of important biological activity which determines whether a group is a bioisostere, not the valency. Therefore, non-isosteric groups can be used as bioisosteres. For example, a heterocycle was used as a bioisostere for an amide bond in the development of the dopamine antagonist Du122290 (Fig. 10.13).

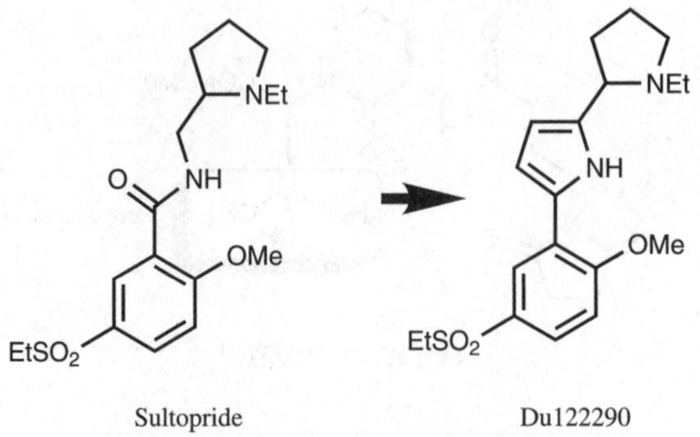

Sultopride   Du122290

Fig. 10.13 Using a bioisostere for an amide group.

10.5.3 Stereoelectronic modifications

Steric hindrance and electronic stabilization have often been used together to stabilize labile groups. For example, procaine is a good local anaesthetic, but it is short lasting because of hydrolysis of the ester group. By changing the ester group to the less reactive amide group, chemical hydrolysis is reduced. Furthermore, the presence of two *ortho* methyl groups on the aromatic ring helps to shield the carbonyl group from attack by nucleophiles or enzymes. This results in the longer acting local anaesthetic lidocaine (Fig. 10.14). Since steric and electronic influences are both involved, the modifications are defined as stereoelectronic. Further successful examples of stereo-electronic modification are demonstrated by oxacillin (see Chapter 14) and betha-nechol (see Chapter 15).

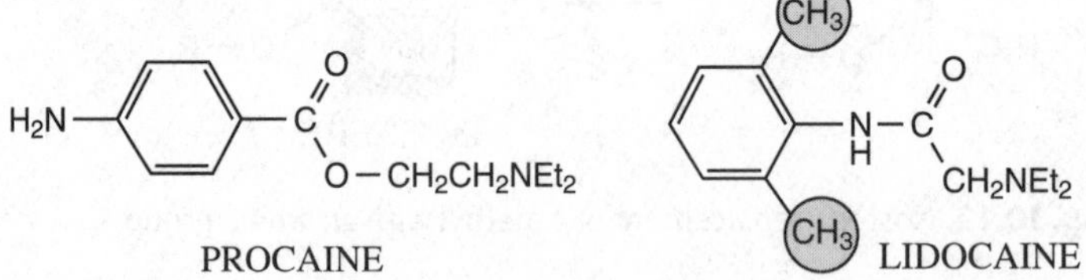

PROCAINE   LIDOCAINE

Fig. 10.14 Stereoelectronic modification.

10.5.4 Metabolic blockers

Some drugs are metabolized by the introduction of polar groups at particular positions in their skeleton. For example, the oral contraceptive megestrol acetate is oxidized at position 6 to give a hydroxyl group at that position. The introduction of a polar hydroxyl group allows the formation of polar conjugates which can be quickly eliminated from the system. By introducing a stable methyl group at position 6 (Fig. 10.15) metabolism is blocked and the activity of the drug is prolonged.

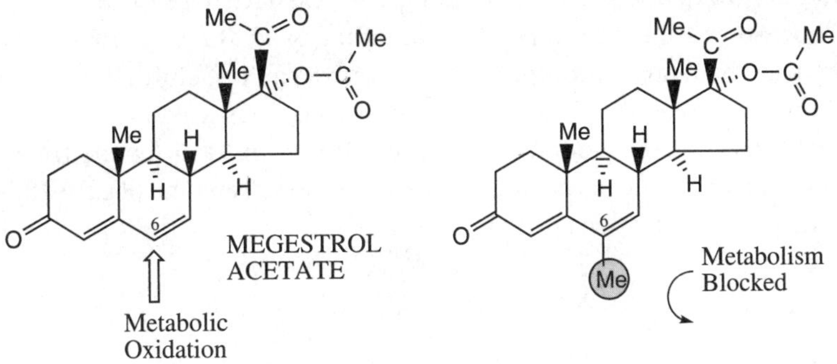

Fig. 10.15 Metabolic blocking.

10.5.5 Removal of susceptible metabolic groups

Certain chemical groups are particularly susceptible to metabolic enzymes. For example, methyl groups on aromatic rings are often oxidized to carboxylic acids (Fig. 10.16). These acids can then be quickly eliminated from the body. Other common metabolic reactions include aliphatic and aromatic C-hydroxylations (Fig. 10.16), N- and S-oxidations, O- and S-dealkylations, and deamination.

Susceptible groups can sometimes be removed or replaced with groups that are stable to oxidation to prolong the lifetime of the drug. For example, the methyl group of the antidiabetic tolbutamide was replaced with a chlorine atom to give chlorpropamide, which is much longer lasting (Fig. 10.17).

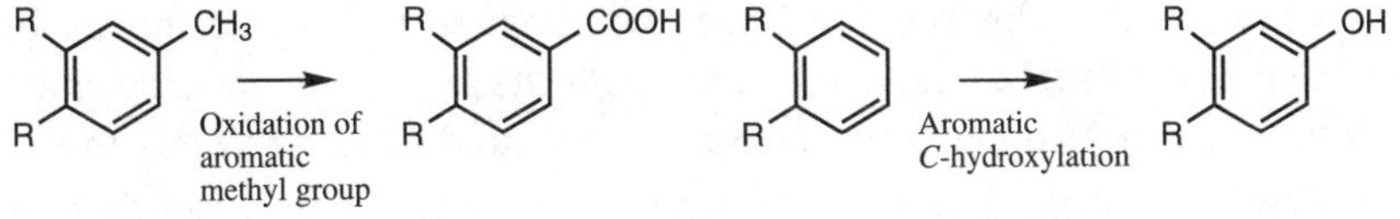

Fig. 10.16 Examples of chemical groups susceptible to metabolic enzymes.

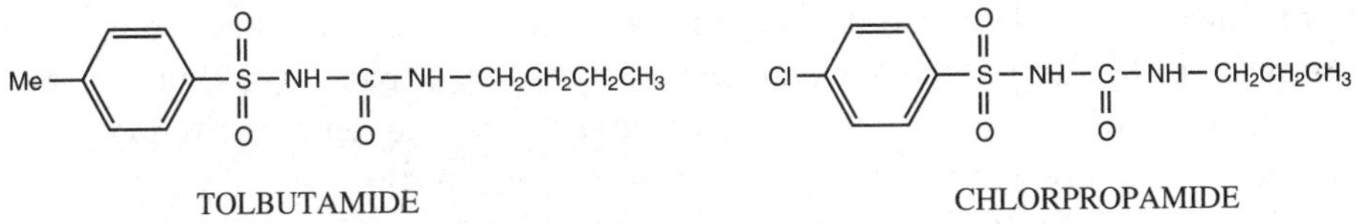

Fig. 10.17 Replacing metabolically labile groups.

10.5.6 Group shifts

Removing or replacing a metabolically vulnerable group is feasible if the group concerned is not involved in important binding interactions with the drug's target. However, if the group *is* important to binding, then we have to use a different strategy.

There are two possible solutions. We can either mask the vulnerable group on a temporary basis by using a prodrug (see later) or we can try 'shifting' the vulnerable group within the molecular skeleton. The latter tactic was used in the development of salbutamol (Fig. 10.18). Salbutamol was introduced in 1969 for the treatment of asthma and is an analogue of the neurotransmitter noradrenaline (Fig. 10.18) – a catechol structure containing two *ortho* phenolic groups.

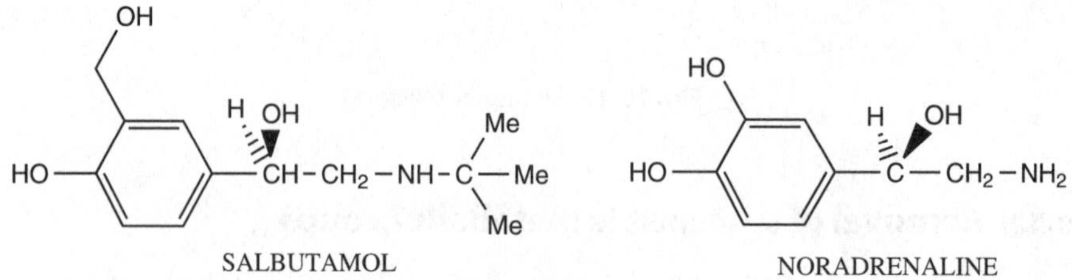

Fig. 10.18 Salbutamol.

One of the problems faced by catechol compounds is metabolic methylation of one of the phenolic groups. Since both phenol groups are involved in hydrogen bonds to the receptor, methylation of one of the phenol groups disrupts the hydrogen bonding and makes the compound inactive. For example, the noradrenaline analogue (I) shown in Fig. 10.19 has useful antiasthmatic activity, but it is of short duration because of its rapid metabolism to the inactive methyl ether (II).

Removing the OH or replacing it with a methyl group may prevent metabolism but will also prevent the important hydrogen bonding. So how could this problem be solved? The answer was to move the vulnerable OH group out from the ring by one carbon unit. This was enough to make the compound unrecognizable to the metabolic enzyme.

Fortunately, the receptor appears to be quite lenient over the position of this hydrogen bonding group and it is interesting to note that a hydroxyethyl group is also acceptable. Beyond that, activity is lost because the OH is 'out of range' or is too large to fit. These results demonstrate that it is better to consider a binding region within the receptor binding site as an available volume rather than imagining it as being fixed at one spot. A drug can then be designed such that the relevant binding group is positioned into any part of that available volume (Fig. 10.20).

Shifting an important binding but metabolically susceptible group worked for salbutamol, but one cannot guarantee that the same tactic will always be successful.

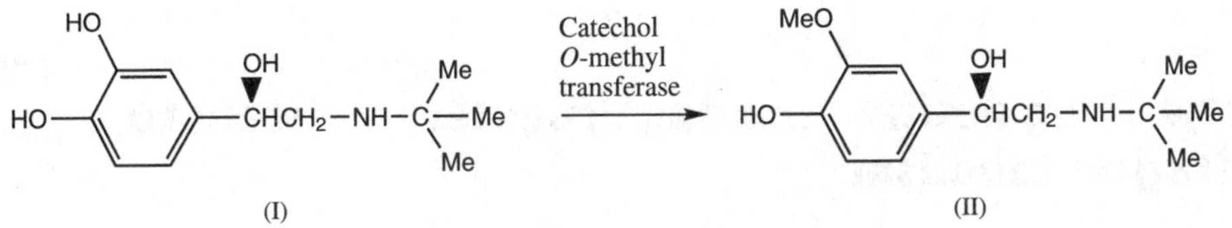

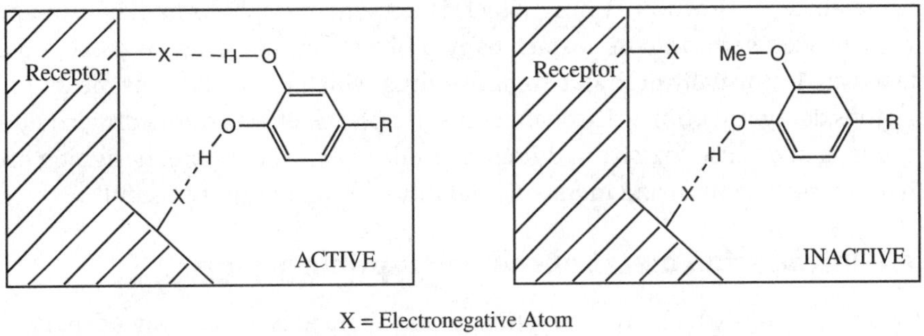

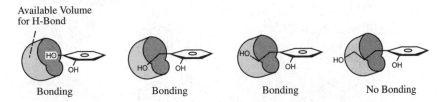

X = Electronegative Atom

Fig. 10.19 Metabolic methylation of a noradrenaline analogue.

Available Volume
for H-Bond

Bonding   Bonding   Bonding   No Bonding

Fig. 10.20 Viewing a binding region as an available volume.

The strategy depends on the important group being shifted in such a manner that it is still recognized by the target receptor or enzyme, but is no longer recognized by metabolic enzymes.

10.5.7 Ring variation

Certain ring systems are often found to be susceptible to metabolism and so varying the ring can improve metabolic stability. For example, the imidazole ring of the antifungal agent ticonazole mentioned previously is susceptible to metabolism, but replacement with a 1,2,4-triazole ring, as in fluconazole, results in improved stability (Fig.10.8).

10.6 Drug design – making drugs less resistant to drug metabolism

So far, we have looked at how the activity of drugs can be prolonged by inhibiting their metabolism. However. a drug which is extremely stable to metabolism and is very slowly excreted can pose just as many problems as one which is susceptible to metabolism. It is usually desirable to have a drug which does what it is meant to do, then stops doing it within a reasonable time. If not, the effects of the drug could last far too long and cause toxicity and lingering side-effects. Therefore, designing drugs with decreased chemical and metabolic stability can sometimes be useful.

10.6.1 Introducing metabolically susceptible groups

Introducing groups which are susceptible to metabolism is a good way of shortening the lifetime of a drug. For example, a methyl group was introduced to the anti-arthritic agent L787257 to shorten its lifetime. The methyl group was metabolically oxidized to a polar alcohol as well as to a carboxylic acid (Fig 10.21).

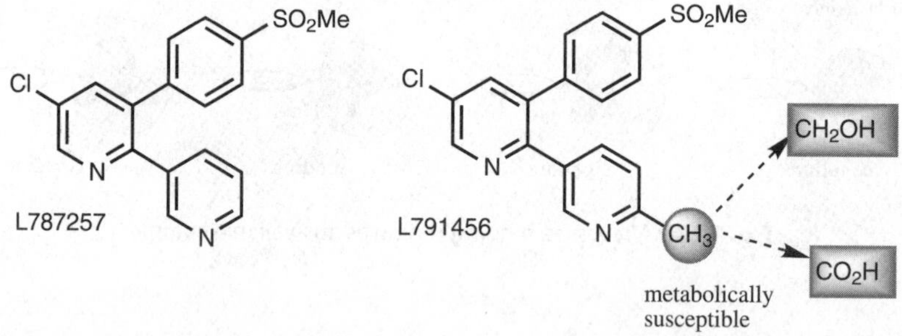

Fig. 10.21 Adding a metabolically labile methyl group.

Another example involves anti-asthmatic drugs. These are usually taken by inhalation to lower the chances of side-effects elsewhere in the body. However, a significant amount is swallowed and can reach the blood supply through the GI tract. Therefore, it is desirable to have an anti-asthmatic drug which is potent and stable in the lungs but which is rapidly metabolized in the blood supply. Cromakalim has useful anti-asthmatic properties but has cardiovascular side-effects if it gets into the blood supply. The structures UK 143220 and UK 157147 were developed from cromakalim so that they would be quickly metabolized in the blood supply (Fig. 10.22). For example, UK 143220 contains an ester which is quickly hydrolysed by esterases to produce an inactive carboxylic acid, while UK 157147 contains a phenol group which is quickly conjugated and eliminated. Both these compounds are now being considered as clinical candidates.

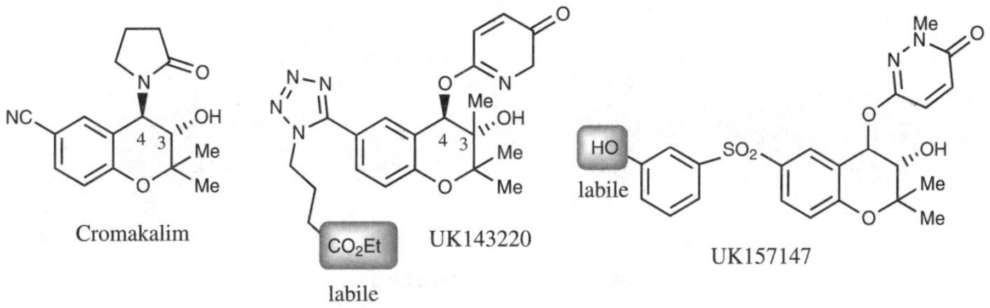

Fig. 10.22 Metabolically labile analogues of cromakalim.

10.6.2 Self-destruct drugs

A self-destruct drug is one which is chemically stable under one set of conditions but becomes unstable and spontaneously degrades under another set of conditions. The advantage of a self-destruct drug is that inactivation does not depend on the activity of metabolic enzymes which could vary from patient to patient. The best example of a self-destruct drug is the neuromuscular blocking agent atracurium (see Chapter 15), which is stable at acid pH but 'self-destructs' when it meets the slightly alkaline conditions of the blood.

10.7 Drug design – targetting drugs

One of the major goals in drug design is to find ways of targetting drugs to the exact locations in the body where they are most needed. In Chapters 8 and 9 we discussed how the introduction of receptor or enzyme specificity can help target drugs to particular tissues. Here, we discuss other tactics related to the distribution of drugs.

10.7.1 Targetting tumour cells – 'search and destroy' drugs

A major goal in cancer chemotherapy is to target drugs efficiently against tumour cells rather than normal cells. One method of achieving this is to design drugs which make use of specific molecular transport systems. The idea is to attach the active drug to an important 'building block' molecule which is needed in large amounts by the rapidly dividing tumour cells. One approach has been to attach the active drug to an amino acid or a nucleic acid base, e.g. uracil mustard (Fig. 10.23).

Of course, normal cells require these building blocks as well, but tumour cells often grow more quickly than normal cells and require the building blocks more urgently. Therefore, the uptake of these drugs should be greater in tumour cells than in normal cells.

A more recent idea has been to attach the active drug (or a poison such as ricin) to monoclonal antibodies which can recognize antigens unique to the tumour cell. Once

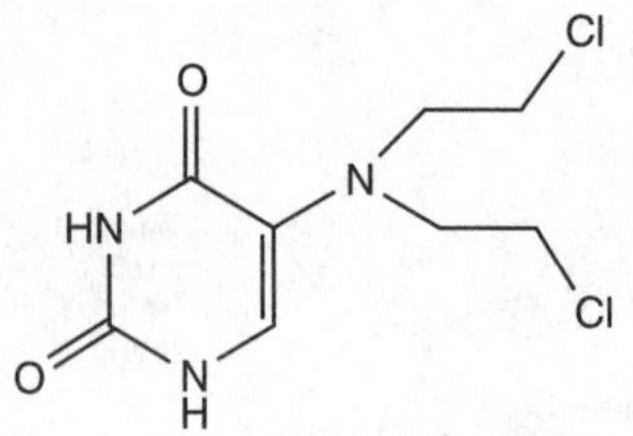

Fig. 10.23 Uracil mustard.

the antibody binds to the antigen, the drug or poison would be released to kill the cell. The difficulty in this approach is in identifying suitable antigens and producing the antibodies in significant quantity. So far the tactic has only been attempted on animals. However, the approach has great promise for the future.

10.7.2 Targetting infections of the gastrointestinal tract

If we want to target drugs against an infection of the GI tract, then we want to prevent the drugs being absorbed into the blood supply. This can easily be done by using a fully ionized drug which is incapable of crossing cell membranes. For example, highly ionized sulphonamides are used against GI infections. They are incapable of crossing the gut wall and are therefore directed efficiently against the infection.

10.7.3 Targeting peripheral regions over the central nervous system

It is often possible to target drugs such that they act peripherally (i.e. around the body) and not in the CNS. By increasing the polarity of drugs, they are less likely to cross the blood–brain barrier and this means they are less likely to have CNS side-effects. Achieving selectivity for the CNS over the peripheral regions of the body is not so straightforward.

10.8 Drug design – reducing toxicity

It is often found that a drug may fail clinical trials because of toxic side-effects. This may be due to toxic metabolites, in which case the drug should be made more resistant to metabolism as described earlier.

It is also worth checking to see whether there are any functional groups present in the molecule which are particularly prone to producing toxic metabolites. For example, it is known that functional groups such as aromatic nitro groups, aromatic amines, bromoarenes, hydrazines, hydroxylamines, or polyhalogenated groups are often metabolized to toxic products. Such groups should be removed or replaced.

Side-effects might also be reduced or eliminated by varying apparently harmless substituents. For example, the halogen substituents of the antifungal agent UK 47265 were varied to find a compound which was less toxic to the liver. This led to fluconazole which was introduced to the clinic under the trade name of Diflucan (Fig. 10.24).

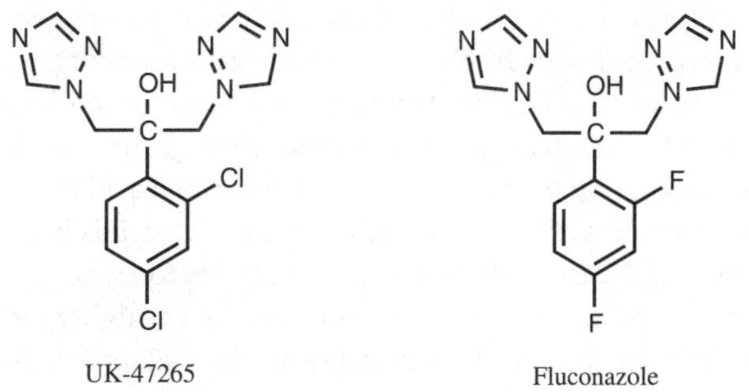

UK-47265   Fluconazole

Fig. 10.24 Varying aromatic substituents to reduce toxicity.

Varying the position of substituents can often reduce or eliminate side-effects. For example, the dopamine antagonist SB269652 inhibits P450 enzymes as a side-effect. Placing the cyano group at a different position prevented this inhibition (Fig. 10.25).

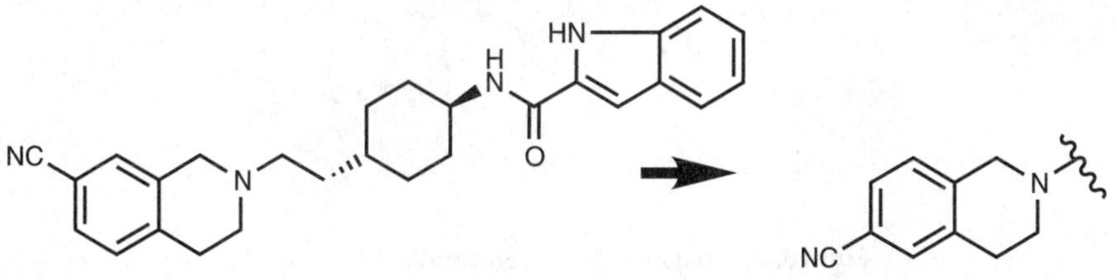

Fig. 10.25 SB269652.

10.9 Prodrugs

Prodrugs are compounds which are inactive in themselves but which are converted in the body to the active drug. They have been useful in tackling problems such as acid sensitivity, poor membrane permeability, drug toxicity, bad taste, and short duration of action. By using prodrugs, the medicinal chemist turns the reactions of drug metabolism to his or her benefit. When designing prodrugs, it is important to ensure that the prodrug is effectively converted to the active drug once it has been absorbed into the blood supply, but it is also important to ensure that any groups cleaved from the molecule are non-toxic.

10.9.1 Prodrugs to improve membrane permeability

10.9.1.1 Esters as prodrugs

Prodrugs have proved very useful in temporarily masking an 'awkward' functional group which is important to receptor binding, but which hinders the drug from crossing cell membranes. For example, a carboxylic acid functional group may have an important role to play in binding the drug to a receptor via ionic or hydrogen bonding. However, the very fact that it is an ionizable group may prevent it from crossing a fatty cell membrane. The answer is to protect the acid function as an ester. The less polar ester can cross fatty cell membranes and, once it is in the bloodstream, it is hydrolysed back to the free acid by esterases in the blood.

Examples of ester prodrugs used to aid membrane permeability include enalapril, which is the prodrug for the antihypertensive agent enalaprilate (Fig. 10.26), and pivampicillin, which is a penicillin prodrug (see Chapter 14).

Fig. 10.26 Enalapril (R=Et); Enalaprilate (R=H).

Not all esters are hydrolysed equally efficiently and a range of esters may need to be tried to find the best one. It is possible to make esters more susceptible to hydrolysis by introducing electron withdrawing groups to the alcohol moiety (e.g. OCH_2CF_3; OCH_2CO_2R; $OCONR_2$; OAr). The inductive effect of these groups aids the hydrolysis mechanism by making the alcohol a better leaving group (Fig. 10.27). However, care has to be taken not to make the ester too reactive or it becomes chemically unstable and hydrolyses before it reaches the blood supply.

The protease inhibitor candoxatrilat (Fig. 10.28) has to be given intravenously since it is too polar to be absorbed from the GI tract. Different esters were tried as prodrugs, revealing that an ethyl ester was inefficiently hydrolysed and that more activated esters were required, a 5-indanyl ester proving to be the best. The 5-indanol which is released on hydrolysis is non-toxic.

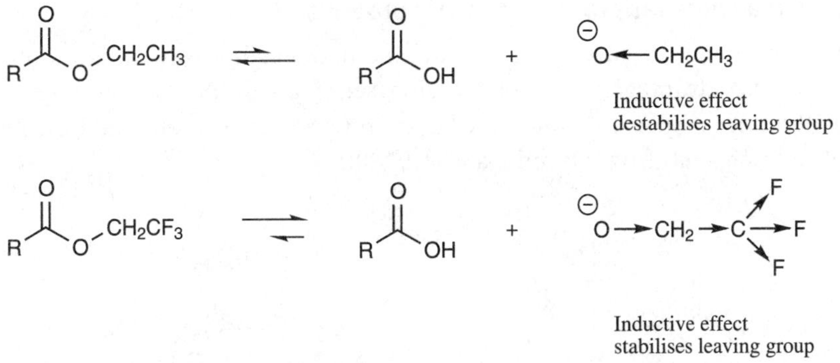

Fig. 10.27 Inductive effects on the stability of leaving groups.

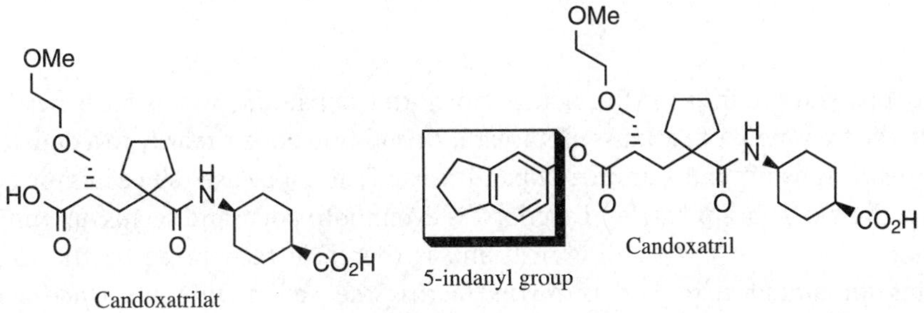

Candoxatrilat

5-indanyl group

Candoxatril

Fig. 10.28 Protease inhibitors.

10.9.1.2 *N-methylation*

Since *N*-demethylation is a common metabolic reaction in the liver, polar amines can be *N*-methylated to reduce polarity and improve membrane permeability. Several hypnotics and antiepileptics take advantage of this reaction (e.g. hexobarbitone; Fig. 10.29).

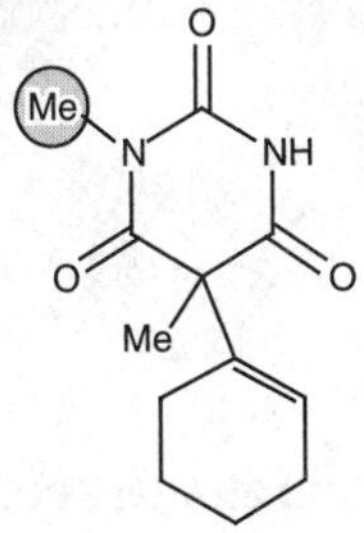

Fig. 10.29 Hexobarbitone.

10.9.1.3 Trojan horse approach for carrier proteins

Another way round the problem of membrane permeability is to design a prodrug which can take advantage of carrier proteins (see Chapter 3) in the cell membrane, such as the ones responsible for carrying amino acids into a cell. The best known example of such a prodrug is levodopa (Fig. 10.30).

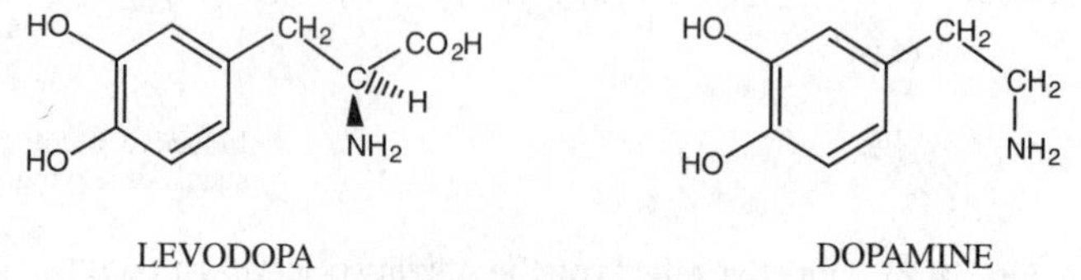

LEVODOPA   DOPAMINE

Fig. 10.30 Levodopa.

Levodopa is a prodrug for the neurotransmitter dopamine and as such has been used in the treatment of Parkinson's disease, a condition due primarily to a deficiency of the neurotransmitter dopamine. Dopamine itself cannot be used since it is too polar to cross the blood–brain barrier. Levodopa is even more polar and seems an unlikely prodrug. However, it is an amino acid and is therefore recognized by the carrier proteins for amino acids and is carried across the cell membrane. Once across the barrier, a decarboxylase enzyme removes the acid group and generates dopamine (Fig. 10.31).

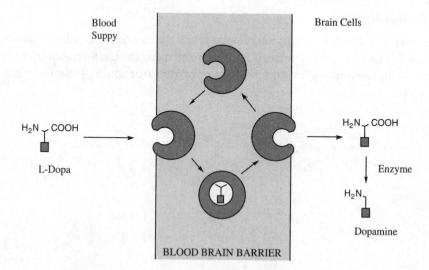

Fig. 10.31 Transport of levodopa across the blood–brain barrier.

10.9.2 Prodrugs to prolong drug activity

Sometimes prodrugs are designed to be slowly converted to the active drug, thus prolonging a drug's activity. For example, 6-mercaptopurine (Fig. 10.32) suppresses the body's immune response and is therefore useful in protecting donor grafts. However, the drug tends to be eliminated from the body too quickly. The prodrug azathioprine is slowly converted to 6-mercaptopurine, allowing a more sustained activity. Since the conversion is chemical and unaffected by enzymes, the rate of conversion can be altered, depending on the electron withdrawing ability of the heterocyclic group. The greater the electron withdrawing power, the faster the breakdown. The NO_2 group is therefore present to ensure an efficient conversion to 6-mercaptopurine, since it is strongly electron withdrawing.

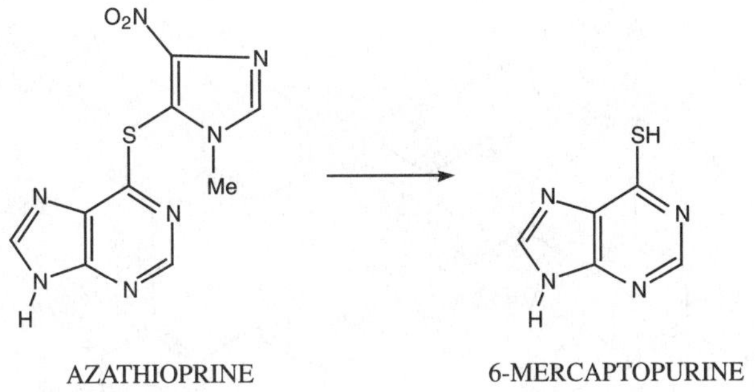

Fig. 10.32 Azathioprine acts as a prodrug for 6-mercaptopurine.

There is a belief that the well known sedatives Valium (Fig. 10.33) and Librium might be prodrugs and are only active because they are metabolized by *N*-demethylation to nordazepam. Nordazepam itself has been used as a sedative, but loses activity quite quickly because of metabolism and excretion. Valium, if it is a prodrug for nordazepam, demonstrates again how a prodrug can be used to lead to a more sustained action.

Another approach to maintaining a sustained level of drug over long periods is deliberately to associate a very lipophilic group to the drug. This means that the majority of the drug is stored in fat tissue and, if the lipophilic group is only slowly removed, the drug is steadily released into the bloodstream over a long period of time. The antimalarial agent cycloguanil pamoate (Fig. 10.34) is one such agent. The active drug is bound ionically to an anion with a large lipophilic group.

Similarly, lipophilic esters of the antipsychotic drug fluphenazine are used to prolong its action. The prodrug is given by intramuscular injection such that it slowly diffuses into the blood supply where it is rapidly hydrolysed (Fig 10.35).

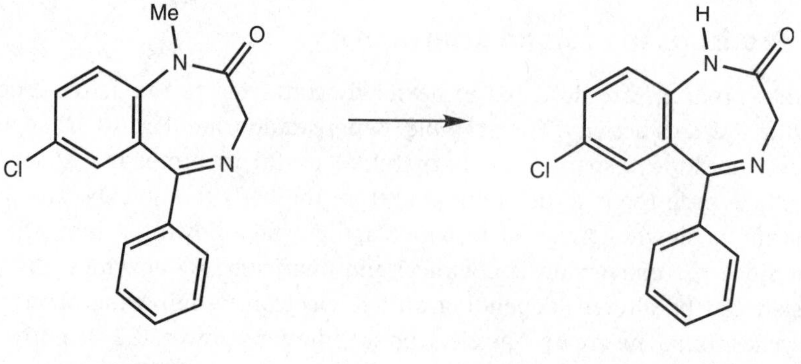

DIAZEPAM
(VALIUM)

NORDAZEPAM

Fig. 10.33 Valium as a possible prodrug for nordazepam.

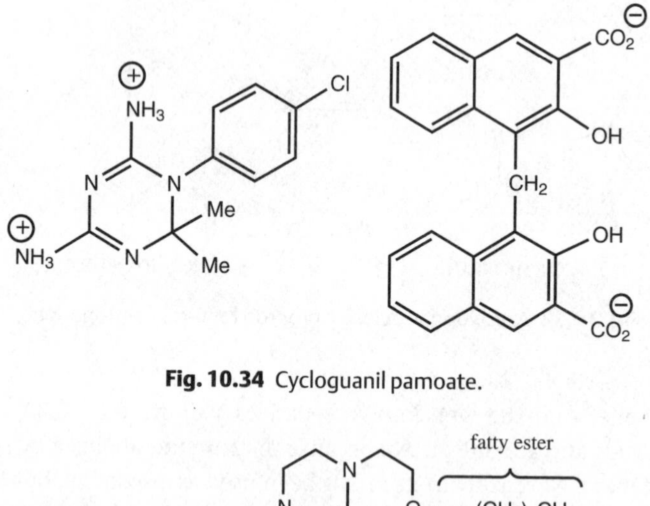

Fig. 10.34 Cycloguanil pamoate.

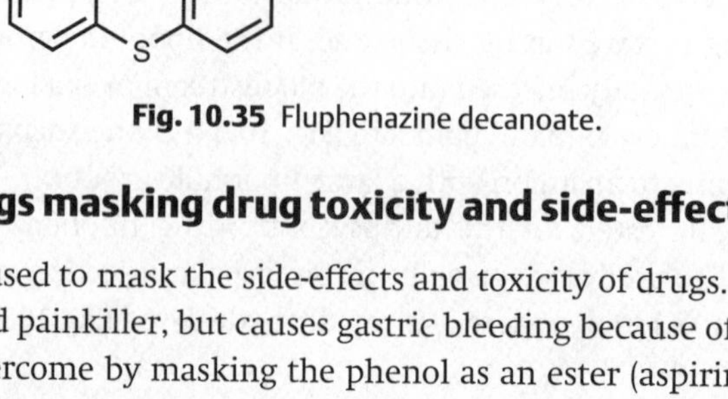

Fig. 10.35 Fluphenazine decanoate.

10.9.3 Prodrugs masking drug toxicity and side-effects

Prodrugs can be used to mask the side-effects and toxicity of drugs. For example, salicylic acid is a good painkiller, but causes gastric bleeding because of the free phenolic group. This is overcome by masking the phenol as an ester (aspirin) (Fig. 10.36). The ester is later hydrolysed by esterases to free the active drug.

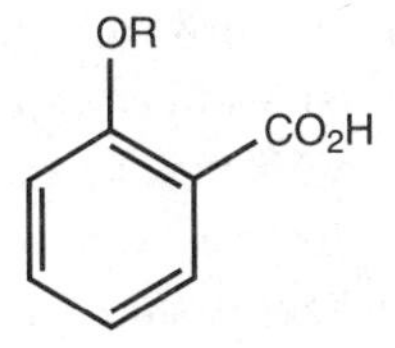

Fig. 10.36 Aspirin (R=Ac) and salicylic acid (R=H).

Prodrugs can be used to give a slow release of drugs which would be too toxic to give directly. Propiolaldehyde is useful in the aversion therapy of alcohol, but is not used itself since it is an irritant. However, the prodrug pargylene can be converted to propiolaldehyde by enzymes in the liver (Fig. 10.37).

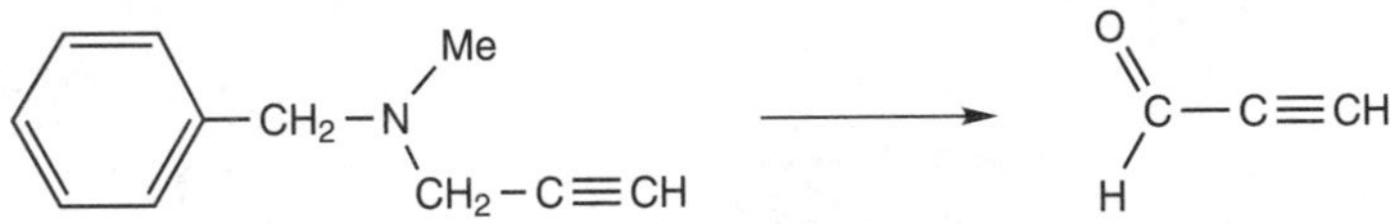

Fig. 10.37 Pargylene as a prodrug for propiolaldehyde.

An extension of this tactic is to design a prodrug that is converted to the active drug at the target site itself. If this can be achieved it will greatly reduce the side-effects of highly toxic drugs. Cyclophosphamide is a successful anticancer drug which is not toxic itself, but which becomes active after metabolism in the liver by being converted to a toxic alkylating agent. It can therefore be taken orally without causing damage to the gut wall (Fig. 10.38). It was also hoped that the high level of phosphoramidase enzyme present in some tumour cells would lead to a greater concentration of alkylating agent in these cells and result in some selectivity of action.

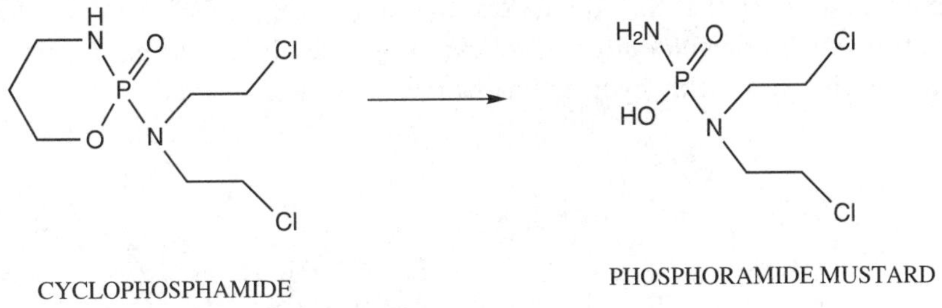

CYCLOPHOSPHAMIDE

PHOSPHORAMIDE MUSTARD

Fig. 10.38 Phosphoramide mustard from cyclophosphamide.

Many important antiviral drugs (e.g. zidovudine, acyclovir, and penciclovir) are non-toxic prodrugs which show selective toxicity towards virally infected cells because they are converted to toxic triphosphates by viral enzymes present in the infected cells. They serve both as competitive inhibitors and as chain terminators (see Chapter 7).

LDZ is an example of a diazepam prodrug which avoids the drowsiness side-effects associated with diazepam. These side-effects are associated with the high initial plasma levels of diazepam on administration, and the use of a prodrug avoids this problem. An aminopeptidase hydrolyses off a non-toxic lysine moiety and the resulting amine spontaneously cyclizes to diazepam (Fig. 10.39).

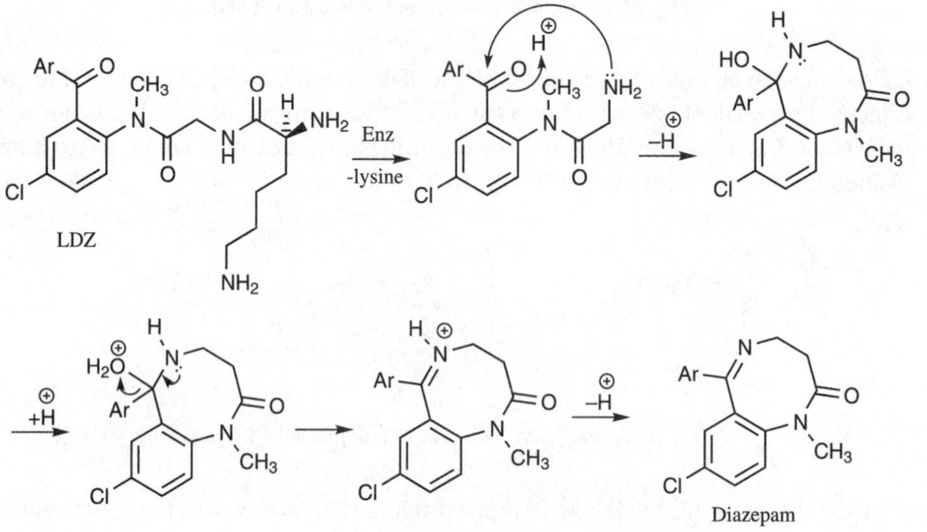

Fig. 10.39 Diazepam prodrug.

10.9.4 Prodrugs to lower solubility in water

Some drugs have a revolting taste. One way to avoid this problem is to reduce their solubility in water so that they do not dissolve on the tongue. For example, the bitter taste of the antibiotic chloramphenicol can be avoided by using the palmitate ester (Fig 10.40) which is quickly hydrolysed once swallowed.

R=H   Chloramphenicol
R= CO(CH$_2$)$_{14}$CH$_3$ Chloramphenicol palmitate
R= CO(CH$_2$)$_2$CO$_2$H   Chloramphenicol succinate

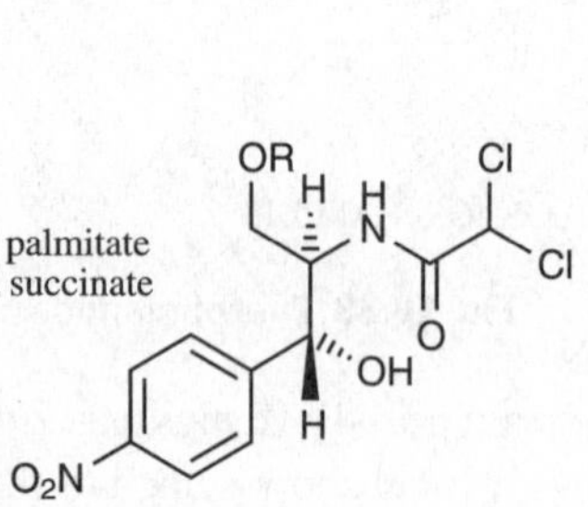

Fig. 10.40 Chloramphenicol prodrugs.

10.9.5 Prodrugs to improve solubility in water

Prodrugs have been used to increase the solubility in water of drugs. This is particularly useful for drugs given intravenously so that higher concentrations and smaller volumes can be used. For example, the succinate ester of chloramphenicol (Fig. 10.40) increases the water solubility of chloramphenicol because of the extra carboxylic acid. It is also worth noting that hydrolysis of the ester produces succinic acid which is present naturally in the body.

Prodrugs designed to increase solubility in water have proved useful in preventing the pain associated with some injections, caused by the poor solubility of the drug at the site of injection. For example, the antibacterial agent clindamycin is painful when injected, but using a phosphate ester prodrug improves solubility and prevents the pain (Fig. 10.41).

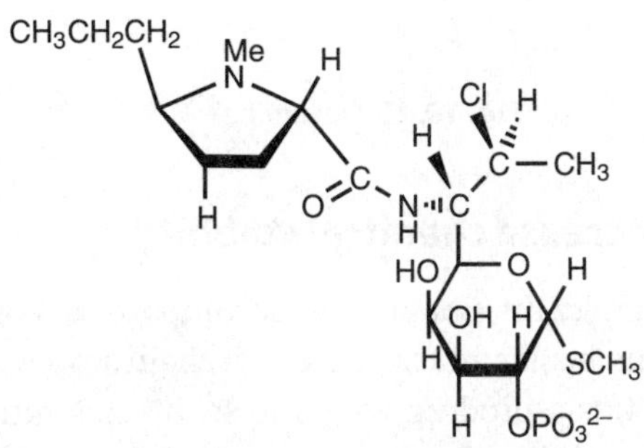

Fig. 10.41 Clindamycin phosphate.

Polar prodrugs have also been used to improve the absorption of non-polar drugs from the gut. Drugs have to have some solubilty in water if they are to be absorbed, otherwise they dissolve in fatty globules and fail to interact effectively with the gut wall. The steroid oestrone is one such drug. By using a lysine ester prodrug, water solubility and absorption is increased (Fig. 10.42).

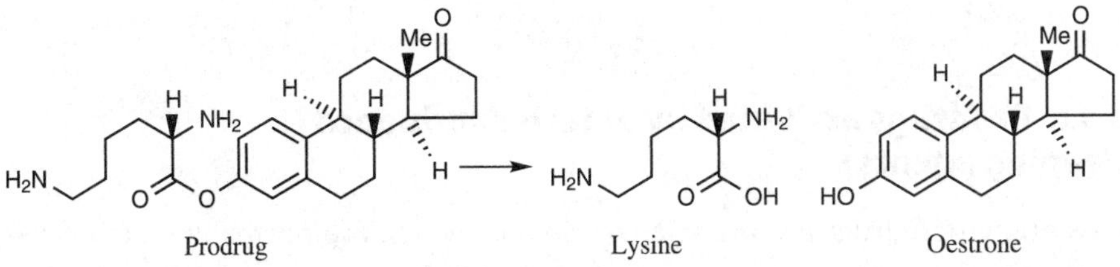

Fig. 10.42 Lysine ester of oestrone.

10.9.6 Prodrugs used in the targetting of drugs

Hexamine (Fig 10.43) is a stable inactive compound at a pH greater than 5. However at more acidic pH, the compound spontaneously degrades to generate formaldehyde which has antibacterial properties. This is useful in the treatment of urinary tract infections. The normal pH of blood is slightly alkaline and so hexamine passes round the body unchanged. However, once it is excreted into the urinary tract, it encounters urine which is acidic as a result of the bacterial infection. Consequently, hexamine degrades to generate formaldehyde just where it is needed.

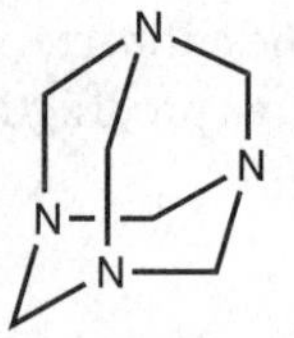

Fig. 10.43 Hexamine.

10.9.7 Prodrugs to increase chemical stability

The antibacterial agent ampicillin (Chapter 14) decomposes in concentrated aqueous solution because of intramolecular attack of the side chain amino group on the lactam ring. Hetacillin (Fig. 10.44) is a prodrug which locks up this nitrogen in a ring and prevents the reaction. Once the prodrug has been administered, the hetacillin slowly decomposes on its own to release ampicillin and acetone.

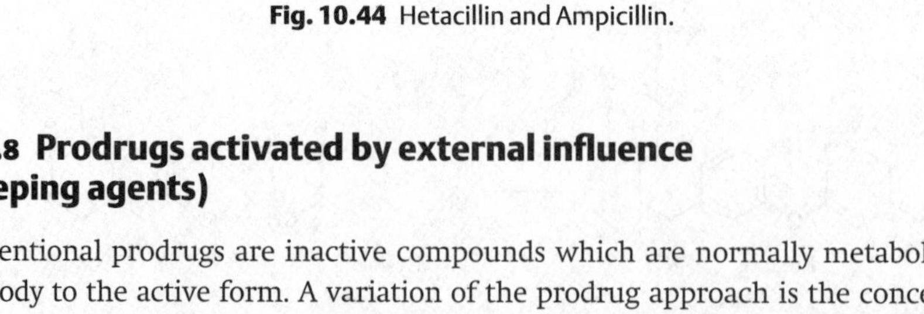

Fig. 10.44 Hetacillin and Ampicillin.

10.9.8 Prodrugs activated by external influence (sleeping agents)

Conventional prodrugs are inactive compounds which are normally metabolized in the body to the active form. A variation of the prodrug approach is the concept of a 'sleeping agent'. This is an inactive compound which is only converted to the active drug by some form of external influence. The best example of this approach is the use of photosensitizing agents such as porphyrins or chlorins in the treatment of cancers

– photodynamic therapy. By itself, the photosensitizing agent has little effect. Given intravenously, it accumulates within cells and has some selectivity for tumour cells. If these cells are now irradiated with light, the porphyrin is converted to an excited state and reacts with molecular oxygen to produce highly toxic singlet oxygen. Singlet oxygen can then attack proteins and unsaturated lipids in the cell membrane, leading to the formation of hydroxyl radicals which further react with DNA, leading to cell destruction. Foscan (temoporfin) (Fig. 10.45) is an example of a chlorin photosensitizing agent undergoing clinical trials for the treatment of a variety of cancers.

10.9.9 Prodrugs of prodrugs

Famciclovir (Famvir) (Fig. 10.46) is a prodrug of penciclovir which is itself a prodrug! Penciclovir is poorly absorbed from the gut owing to its polarity. Famciclovir is less polar and is absorbed more easily. It is then metabolized, mainly in the liver, to form

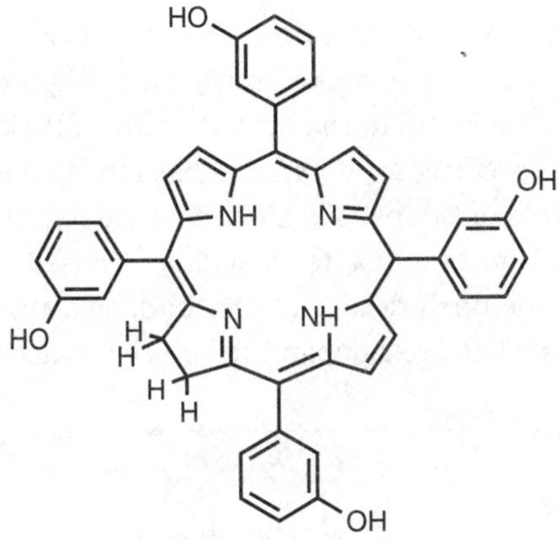

Fig. 10.45 Foscan.

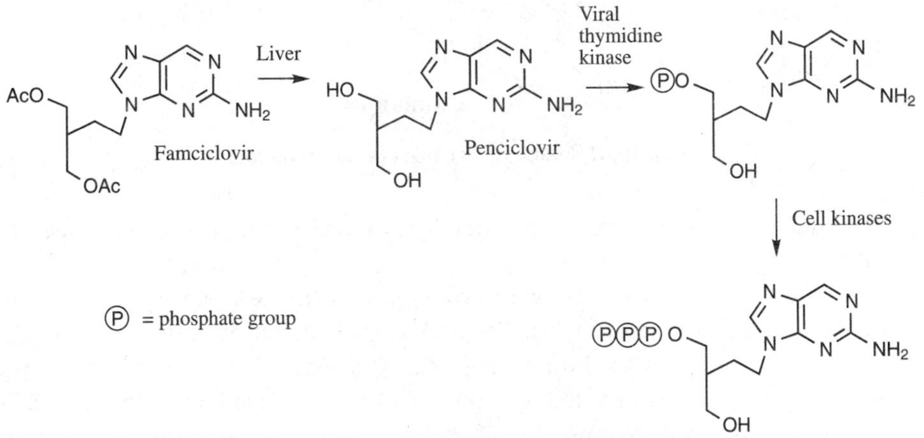

Fig. 10.46 Famvir as a prodrug.

penciclovir which is phosphorylated in virally infected cells as described previously (see Section 7.5).

10.10 Drug alliances – synergism

Some drugs are found to affect the activity or pharmacokinetic properties of other drugs, and this can be put to good use. The following are some examples.

10.10.1 'Sentry' drugs

In this approach, a second drug is administered along with the drug which is 'going into action'. The role of the second drug is to guard or assist the principal drug. Usually, the second drug antagonizes an enzyme which metabolizes the principal drug.

For example, clavulanic acid inhibits the enzyme β-lactamase and is therefore able to protect penicillins from that particular enzyme (see Chapter 14).

Another example is to be found in the drug therapy of Parkinson's disease. The use of L-dopa (levodopa) as a prodrug for dopamine has already been described. However, to be effective, large doses of L-dopa (3–8 g per day) are required, and over a period of time these dose levels lead to side-effects such as nausea and vomiting. L-Dopa is susceptible to the enzyme dopa decarboxylase and, as a result, much of the L-dopa administered is decarboxylated to L-dopamine before it reaches the CNS (Fig. 10.47).

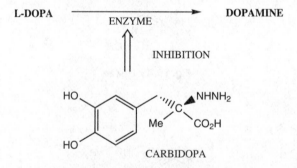

Fig. 10.47 Inhibition of L-dopa decarboxylation.

This build up of dopamine in the peripheral blood supply leads to nausea and vomiting.

An antagonist of dopa decarboxylase would inhibit the decarboxylation of L-dopa and allow smaller doses to be used. The drug carbidopa has been used successfully in this respect and effectively inhibits dopa decarboxylase. Furthermore, since it is a highly polar compound containing two phenolic groups, a hydrazine moiety, and an acidic group, it is unable to cross the blood–brain barrier and so cannot prevent the conversion of L-dopa to dopamine in the brain.

Another example concerns several important peptides and proteins which could be used as drugs if it were not for the fact that they are quickly broken down by protease enzymes. One way round this problem could be to inhibit the protease enzymes. Candoxatril (Fig. 10.28) is a protease inhibitor which has some potential in this respect and is under clinical evaluation.

10.10.2 Localizing a drug's area of activity

Adrenaline is an example of a drug which has been used alongside another drug to localize the area of activity. When injected with the local anaesthetic procaine, adrenaline constricts the blood vessels in the vicinity of the injection and so prevents procaine being 'washed away' by the blood supply.

10.10.3 Increasing absorption

Metoclopramide (Fig. 10.48) is administered along with analgesics in the treatment of migraine. Its function is to increase gastric motility, leading to faster absorption of the analgesic and quicker pain relief.

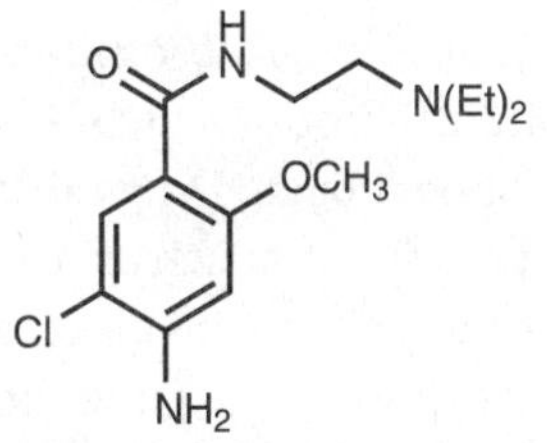

Fig. 10.48 Metoclopramide.

10.11 Methods of administration

There are a large variety of ways in which drugs can be administered and many of these avoid some of the problems associated with oral administration. The main routes are: oral, sublingual, rectal, epithelial, inhalation and injection.

10.11.1 Oral administration

Orally administered drugs are taken by mouth and are the preferred option for most patients. Care has to be taken if drugs interact with food. For example, tetracycline binds strongly to calcium ions which inhibits absorption. Therefore, foods such as milk should be avoided. Some drugs bind other drugs and prevent absorption. For example, cholestyramine (used to lower cholesterol levels) binds to warfarin and also to thyroxine, and so these drugs should be taken separately.

10.11.2 Sublingual administration

These are drugs which are applied under the tongue. The drug is rapidly absorbed into the blood supply of the oral cavity and this provides a means of avoiding the acids of the stomach or the enzymes of the liver, thus avoiding first pass metabolism. Glyceryl trinitrate is administered in this way and the Incas absorbed cocaine sublingually by chewing coca leaves.

10.11.3 Rectal administration

Drugs administered in this way are usually targeted for a local effect. Absorption into the blood supply is very efficient by this route. However, it is not the most popular of methods with patients!

10.11.4 Epithelial administration

There are several methods by which drugs can be applied to epithelial (or surface) cells.

Topical drugs are those which are applied to the skin. For example, steroids are applied topically to treat local skin irritations. It is also possible for some of the drug to be absorbed through the skin (transdermal absorption) and to enter the blood supply, especially if the drug is lipophilic. Nicotine patches work in this fashion, as do hormone replacement therapies for oestrogen. Drugs absorbed by this method do so at a steady rate and avoid the acidity of the stomach or the enzymes in the gut or gut wall.

Nasal sprays administer drugs directly to the mucosal surfaces of the nose and have been used to administer analogues of peptide hormones such as antidiuretic hormone. These drugs would be quickly degraded if taken orally.

Eye drops are used to administer drugs directly to the eye and thus reduce the possibility of side-effects elsewhere in the body. For example, glaucoma is treated in this way. Nevertheless, some absorption into the blood supply can still occur and some asthmatic patients suffer bronchospasms when taking timolol eye drops.

10.11.5 Inhalation

Inhalation sprays are used to direct drugs such as volatile and gaseous anaesthetics or anti-asthmatic agents to the airways. Inhalation results in a greater concentration of the drug in the airways than elsewhere in the body, thus reducing side-effects. Nevertheless, some of the drug is inevitably absorbed and can lead to side-effects. For example, salbutamol can lead to tremor.

Recreational drugs such as nicotine or cannabis can be absorbed by smoking. However, this is a particularly hazardous method of taking drugs. A normal cigarette is like a mini-furnace, resulting in the production of a complex mixture of potentially carcinogenic compounds, especially from the tars present in tobacco. The tars in cannabis are considerably more dangerous than those in tobacco. If cannabis is to be used in medicine, safer methods of administration are desirable (i.e. inhalers).

10.11.6 Injection

Drugs can be injected under the skin (subcutaneous), into muscle (intramuscular), into veins (intravenous) or into the spinal cord (intrathecal).

Intravenous injection is the fastest method of administering a drug and avoids the problems of absorption and first pass metabolism. An intravenous drip also allows the drug to be administered in a controlled manner such that there is a steady level of drug in the system. Drugs such as the local anaesthetic lignocaine are given by intravenous injection.

Drugs given by subcutaneous or intramuscular injection are more slowly absorbed into the blood supply, but are still absorbed faster than by oral administration. The rate of absorption depends on diffusion of the drug and local blood flow. The latter can be reduced by adding adrenaline to constrict blood vessels. The former can be slowed by using a poorly absorbed salt, ester or complex of the drug. The advantage of slowing the absorption of the drug is in prolonging its activity. For example, oily suspensions of steroid hormone esters are used to slow absorption.

Intrathecal injections are used to administer methotrexate in the treatment of childhood leukaemia to prevent relapse in the CNS. Antibacterial agents which normally do not cross the blood–brain barrier can also be given in this way.

10.11.7 Implants

Continuous osmotically driven minipumps have been developed which can release insulin at varying rates depending on blood glucose levels. These are implanted under the skin and appear to be the best answer to the problem of providing insulin at the correct levels at the correct times.

10.12 Formulation

The way in which a drug is formulated can sometimes avoid some of the problems associated with oral administration. Usually drugs are taken orally as tablets or capsules. A tablet is usually a compressed preparation that contains 5–10% of the drug, 80% of fillers, disintegrants, lubricants, glidants and binders, and 10% of compounds ensuring easy disintegration, disaggregation, and dissolution of the tablet in the stomach or intestine. The disintegration time can be modified for a rapid effect or sustained release. Special coatings can make the tablet resistant to the stomach acids such that it only disintegrates in the duodenum as a result of enzyme action or alkaline pH. Pills can also be coated with sugar, varnish or wax to disguise taste. Some tablets are designed with an osmotically active core surrounded by an impermeable membrane with a pore in it which allows the drug to exit in solution at a constant rate as the tablet moves through the digestive tract.

Capsules are a gelatinous envelope which enclose the active substance. Capsules can be designed to remain intact for some hours after ingestion to delay absorption. Capsules may contain a mixture of slow and fast release particles to produce rapid and sustained absorption in the same dose.

The drug itself needs to be soluble in aqueous solution and the dissolution rate can be varied by formulation. Such factors as particle size and crystal form can significantly affect dissolution. Fast dissolution is not always ideal. For example, slow dissolution rates can prolong the duration of action or avoid initial large plasma levels.

A physical way of protecting drugs from metabolic enzymes in the bloodstream is to inject small vesicles called liposomes filled with the drug. These vesicles or globules consist of a bilayer of fatty phospholipid molecules (similar to a cell membrane) and will travel round the circulation, slowly leaking their contents. Liposomes are found to be concentrated in malignant tumours and this provides a possible method of delivering antitumour drugs to these cells. Another future possibility for targetting liposomes is to incorporate antibodies into the liposome surface such that specific tissue antigens would be recognized.

Microspheres made up of biologically erodable polymers can be designed so that they stick to the gut wall. The drug is present within the sphere and, because the sphere sticks to the wall, absorption of the drug is increased. This has still to be used clinically, but has proven effective in enhancing the absorption of insulin and plasmid DNA in test animals.

Protein-based polymers are being developed as drug delivery systems for the controlled release of charged drugs. For example, the cationic drugs leu-enkephalin or naltrexone could be delivered using polymers with anionic carboxylate groups. Ionic interactions between the drug and the protein result in folding and assembly of the protein polymer to form a protein–drug complex and the drug is then released at a slow and constant rate. The amount of drug carried could be predetermined by the density of carboxylate binding sites present and the accessible surface area of the vehicle. The rate of release could be controlled by varying the number of hydrophobic amino acids present. The more hydrophobic amino acids present, the weaker the affinity betwen the carboxylate binding groups and the drug. Once the drug is released, the protein carrier would be metabolized like any normal protein.

10.13 Neurotransmitters and hormones as drugs?

Before finishing this chapter, let us consider the body's own neurotransmitters and hormones. Why do we not use these as drugs?

10.13.1 Neurotransmitters

Many non-peptide neurotransmitters are simple molecules which can be easily prepared in the laboratory, so why do we not use these as drugs? For example, if there is a shortage of dopamine in the brain, why not administer more dopamine to make up the balance?

Unfortunately, this is not possible for a number of reasons. Many neurotransmitters are not chemically stable enough to survive the acid of the stomach and would have to be injected. Even if they were injected, there is little chance that they would survive to reach their target receptors. The body has efficient mechanisms which inactivate neurotransmitters as soon as they have passed on their message from nerve to target cell. Therefore, any neurotransmitter injected into the blood supply would be swiftly inactivated by enzymes or by cell uptake.

Even if they were to survive, they would be poor drugs indeed, leading to many undesirable side-effects. For example, the shortage of neurotransmitter may only be at one small area in the brain and be normal elsewhere. If we gave the natural neurotransmitter, how would we stop it producing an overdose of transmitter at these other sites? This is, of course, a problem with all drugs, but in recent years it has been discovered that receptors for a specific neurotransmitter are slightly different depending on where they are in the body. The medicinal chemist can design synthetic drugs which take advantage of that difference such that they 'ignore' receptors which the natural neurotransmitter would not. In this respect, the medicinal chemist has actually improved on nature.

We cannot even assume that the body's own neurotransmitters are perfectly safe, and free from the horrors of tolerance and addiction associated with drugs such as heroin. It is quite possible to be addicted to one's own neurotransmitters and hormones. Some people are addicted to exercise and are compelled to exercise long hours each day to feel good. The very process of exercise leads to the release of hormones and neurotransmitters. This can produce a 'high' which drives susceptible people to exercise more and more. If they stop exercising, they suffer withdrawal symptoms such as deep depression.

The same phenomenon probably drives mountaineers into attempting feats which they know may quite well lead to their death. The thrill of danger produces hormones and neurotransmitters which in turn produce a 'high'. Perhaps this, too, is why war has such a fascination for mankind.

To conclude, many of the body's own neurotransmitters are known and can be easily synthesized, but they cannot be effectively used as medicines.

10.13.2 Natural hormones as drugs

Unlike natural neurotransmitters, natural hormones have potential in drug therapy since they normally circulate round the body and behave like drugs. Indeed, adrenaline is commonly used in medicine to treat (amongst other things) severe allergic reactions. Most hormones are peptides and proteins, and some naturally occurring peptide and protein hormones are already used in medicine (e.g. insulin, calcitonin, human growth factor, interferons, and colony stimulating factors). However, others have proved ineffective. This is because peptides and proteins suffer serious drawbacks because of their susceptibility to digestive and metabolic breakdown. Furthermore, proteins are large molecules which could possibly induce an adverse immunological response.

10.13.3 Peptides and proteins as drugs

Because of the difficulties already mentioned, there is often a reluctance to develop peptides and proteins as drugs, but this does not mean that peptide drugs have no role to play in medicinal chemistry. For example, we have already mentioned the immunosuppressant cyclosporin, which can be administered orally. Another important peptide drug is Zoladex (Fig. 10.49), which is adminstered as a subcutaneous implant and is used against breast and prostate cancers, earning 700 million dollars for its maker per year. Peptide drugs can be useful if one chooses the right disease and method of administration.

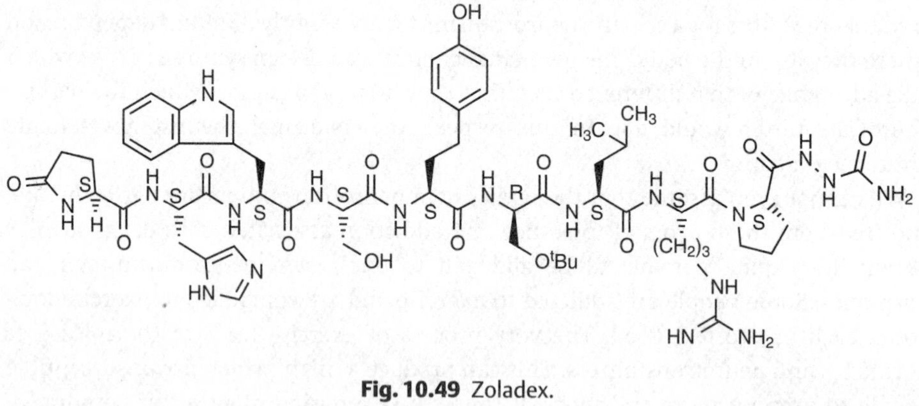

Fig. 10.49 Zoladex.

There are also methods which can be used to stabilize susceptible peptide bonds. One answer is to replace the peptide bond with another functional group which is stable to these hydrolytic enzymes. For example, a peptide bond might be replaced with a double bond. If the compound retains activity, then the double bond represents a bioisostere for the peptide link. Other methods of stabilizing peptide bonds include the replacement of an L-amino acid with the corresponding D-enantiomer or an unnatural amino acid. In both cases, the tactic is to make the peptide bond unrecognizable to the enzyme.

10.13.4 Oligonucleotides as drugs – antisense drugs

The therapeutic potential of antisense drugs was mentioned in Chapter 7. However, there are several disadvantages to the use of oligonucleotides as drugs. They are rapidly degraded by nucleases. They are large and highly charged and are not easily absorbed through cell membranes. Attempts to stabilize these molecules and/or reduce polarity have involved modifying the phosphate linkages in the sugar–phosphate backbone. For example, phosphorothioates and methylphosphonates have been extensively studied and oligonucleotides containing these linkages show promise as therapeutic agents (Fig. 10.50). Alterations to the sugar moiety have also been tried. For example, a methyl group at position 2′, or using the α-anomer of a deoxyribose sugar, increases resistance to nucleases. Bases have been modified to improve and increase the number of hydrogen bonding interactions with the target mRNAs.

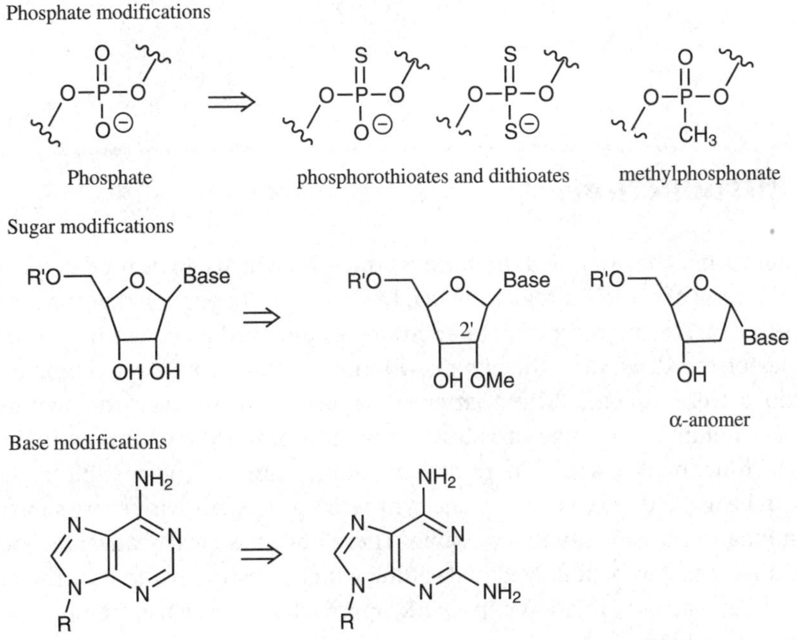

Phosphate modifications

Phosphate   phosphorothioates and dithioates   methylphosphonate

Sugar modifications

α-anomer

Base modifications

Fig. 10.50 Modifications on oligonucleotides.

11 Quantitative structure–activity relationships

11.1 Introduction

In Chapters 9 and 10 we studied the various strategies which can be used in the design of drugs. Several of these strategies involved a change in shape such that the new drug had a better 'fit' for its receptor. Other strategies involved a change in the physical properties of the drug such that its distribution, metabolism, or receptor binding interactions were affected. These latter strategies often involved the synthesis of analogues containing a range of substituents on aromatic/heteroaromatic rings or accessible functional groups. There are an infinite number of possible analogues which can be made if we were to try and synthesize analogues with every substituent and combination of substituents possible. Therefore, it is clearly advantageous if a rational approach can be followed in deciding which substituents to use. The quantitative structure–activity relationship (QSAR) approach has proved extremely useful in tackling this problem.

The QSAR approach attempts to identify and quantify the physicochemical properties of a drug and to see whether any of these properties has an effect on the drug's biological activity. If such a relationship holds true, an equation can be drawn up which quantifies the relationship and allows the medicinal chemist to say with some confidence that the property (or properties) has an important role in the distribution or mechanism of the drug. It also allows the medicinal chemist some level of prediction. By quantifying physicochemical properties, it should be possible to calculate in advance what the biological activity of a novel analogue might be. There are two advantages to this. First, it allows the medicinal chemist to target efforts on analogues which should have improved activity and thus cut down the number of analogues which have to be made. Second, if an analogue is discovered which does not fit the equation, it implies that some other feature is important and provides a lead for further development.

What are these physicochemical features which we have mentioned?

Essentially, they refer to any structural, physical, or chemical property of a drug. Clearly, any drug will have a large number of such properties and it would be a Herculean task to quantify and relate them all to biological activity at the same time. A simple, more practical approach is to consider one or two physicochemical properties of the drug and to vary these while attempting to keep other properties constant. This is not as simple as it sounds, since it is not always possible to vary one property without affecting another. Nevertheless, there have been numerous examples where the approach has worked.

11.2 Graphs and equations

In the simplest situation, a range of compounds is synthesized to vary one physicochemical property (e.g. $\log P$) and to test how this affects the biological activity ($\log 1/C$) (we will come to the meaning of $\log 1/C$ and $\log P$ in due course). A graph is then drawn to plot the biological activity on the y axis versus the physicochemical feature on the x axis (Fig. 11.1).

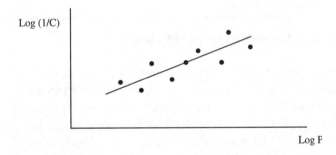

Fig. 11.1 Biological activity versus physicochemical property.

It is then necessary to draw the best possible line through the data points on the graph. This is done by a procedure known as 'linear regression analysis by the least squares method'. This is quite a mouthful and can produce a glazed expression on any chemist who is not mathematically orientated. In fact, the principle is quite straightforward.

If we draw a line through a set of data points, most of the points will be scattered on either side of the line. The best line will be the one closest to the data points. To measure how close the data points are, vertical lines are drawn from each point (Fig. 11.2). These verticals are measured and then squared to eliminate the negative values. The squares are then added up to give a total. The best line through the points will be the line where this total is a minimum.

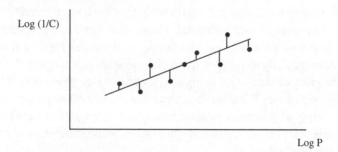

Fig. 11.2 Proximity of data points to line of best fit.

The equation of the straight line will be $y = k_1x + k_2$ where k_1 and k_2 are constants. By varying k_1 and k_2, different equations are obtained until the best line is derived. This whole process can be speedily done by computer.

The next stage in the process is to see whether the relationship is significant. We may have obtained a straight line through points which are so random that it means nothing. The significance of the equation is given by a term known as the **regression coefficient** (r) This coefficient can again be calculated by computer. For a perfect fit, $r^2 = 1$. Good fits generally have r^2 values of 0.95 or above.

11.3 Physicochemical properties

There are many physical, structural, and chemical properties which have been studied by the QSAR approach, but the most commonly studied are hydrophobic, electronic, and steric. This is because it is possible to quantify these effects relatively easily.

In particular, hydrophobic properties can be easily quantified for complete molecules or for individual substituents. On the other hand, electronic and steric properties are more difficult to quantify, and quantification is only really feasible for individual substituents.

Consequently, QSAR studies on a variety of totally different structures are relatively rare and are limited to studies on hydrophobicity. It is more common to find QSAR studies being carried out on compounds of the same general structure, in which substituents on aromatic rings or accessible functional groups are varied. The QSAR study then considers how the hydrophobic, electronic, and steric properties of the substituents affect biological activity.

The three most studied physicochemical properties will now be considered in some detail.

11.3.1 Hydrophobicity

The hydrophobic character of a drug is crucial to how easily it crosses cell membranes (see Section 10.2.3) and may also be important in receptor interactions. Changing substituents on a drug may well have significant effects on its hydrophobic character and hence its biological activity. Therefore, it is important to have a means of predicting this quantitatively.

The partition coefficient (P)

The hydrophobic character of a drug can be measured experimentally by testing the drug's relative distribution in an octanol/water mixture. Hydrophobic molecules prefer to dissolve in the octanol layer of this two-phase system, whereas hydrophilic molecules prefer the aqueous layer. The relative distribution is known as the partition coefficient (P) and is obtained from the following equation:

$$P = \frac{\text{Concentration of drug in octanol}}{\text{Concentration of drug in aqueous solution}}$$

Hydrophobic compounds will have a high P value, whereas hydrophilic compounds will have a low P value.

Varying substituents on the lead compound will produce a series of analogues having different hydrophobicities and therefore different P values. By plotting these P values against the biological activity of these drugs, it is possible to see if there is any relationship between the two properties. The biological activity is normally expressed as $1/C$, where C is the concentration of drug required to achieve a defined level of biological activity. (The reciprocal of the concentration $(1/C)$ is used, since more active drugs will achieve a defined biological activity at lower concentration.)

The graph is drawn by plotting $\log 1/C$ versus $\log P$. The scale of numbers involved in measuring C and P usually covers several factors of ten and so the use of logarithms allows the use of more manageable numbers.

In studies where the range of the $\log P$ values is restricted to a small range (e.g. $\log P = 1\text{--}4$), a straight-line graph is obtained (Fig. 11.1), showing that there is a relationship between hydrophobicity and biological activity. Such a line would have the following equation:

$$\log \left(\frac{1}{C}\right) = k_1 \log P + k_2.$$

For example, the binding of drugs to serum albumin is determined by their hydrophobicity, and a study of 40 compounds resulted in the following equation:

$$\log \left(\frac{1}{C}\right) = 0.75 \log P + 2.30.$$

The equation shows that serum albumin binding increases as $\log P$ increases. In other words, hydrophobic drugs bind more strongly to serum albumin than hydrophilic drugs. Knowing how strongly a drug binds to serum albumin can be important in estimating effective dose levels for that drug. When bound to serum albumin, the drug cannot bind to its receptor and so the dose levels for the drug should be based on the amount of unbound drug present in the circulation. The equation above allows us to calculate how strongly drugs of similar structure will bind to serum albumin and gives an indication of how 'available' they will be for receptor interactions.

Despite such factors as serum albumin binding, it is generally found that increasing the hydrophobicity of a lead compound results in an increase in biological activity. This reflects the fact that drugs have to cross hydrophobic barriers such as cell membranes to reach their target. Even if no barriers are to be crossed (e.g. *in vitro* studies), the drug has to interact with a target system such as an enzyme or receptor where the binding site is usually hydrophobic. Therefore, increasing hydrophobicity aids the drug in crossing hydrophobic barriers or in binding to its target site.

This might imply that increasing log P should increase the biological activity *ad infinitum*. In fact, this does not happen. There are several reasons for this. For example, the drug may become so hydrophobic that it is poorly soluble in the aqueous phase. Alternatively, it may be 'trapped' in fat depots and never reach the intended site. Finally, hydrophobic drugs are often more susceptible to metabolism and subsequent elimination.

A straight-line relationship between log P and biological activity is observed in many QSAR studies because the range of log P values studied is often relatively narrow. For example, the study carried out on serum albumin binding was restricted to compounds having log P values in the range 0.78–3.82. If these studies were to be extended to include compounds with very high log P values then we would see a different picture. The graph would be parabolic, as shown in Fig. 11.3. Here, the biological activity increases as log P increases until a maximum value is obtained. The value of log P at the maximum (log P^0) represents the optimum partition coefficient for biological activity. Beyond that point, an increase in log P results in a decrease in biological activity.

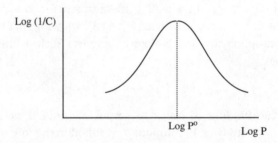

Fig. 11.3 Parabolic log (1/C) versus log P curve.

If the partition coefficient is the only factor influencing biological activity, the parabolic curve can be expressed by the mathematical equation:

$$\log\left(\frac{1}{C}\right) = -k_1 (\log P)^2 + k_2 \log P + k_3.$$

Note that the $(\log P)^2$ term has a negative sign in front of it. When P is small, the $(\log P)^2$ term is very small and the equation is dominated by the log P term. This represents the first part of the graph where activity increases with increasing P. When P is large, the $(\log P)^2$ term is more significant and eventually 'overwhelms' the $\log P$ term. This represents the last part of the graph where activity drops with increasing P. k_1, k_2, and k_3 are constants and can be determined by a suitable computer programme.

There are relatively few drugs where activity is related to the $\log P$ factor alone. Those that do tend to operate in cell membranes where hydrophobicity is the dominant feature controlling their action. The best examples of drugs which operate in cell membranes are the general anaesthetics. These are thought to function by entering the central nervous system (CNS) and 'dissolving' into cell membranes where they affect membrane structure and nerve function. In such a scenario, there are no specific drug–receptor interactions and the mechanism of the drug is controlled purely by its ability to enter cell membranes (i.e. its hydrophobic character). The general anaesthetic activity of a range of ethers was found to fit the parabolic equation:

$$\log \left(\frac{1}{C}\right) = -0.22 \, (\log P)^2 + 1.04 \log P + 2.16.$$

According to the equation, anaesthetic activity increases with increasing hydrophobicity (P), as determined by the $\log P$ factor. The negative $(\log P)^2$ factor shows that the relationship is parabolic and that there is an optimum value for $\log P$ ($\log P^0$), beyond which increasing hydrophobicity causes a decrease in anaesthetic activity.

With this equation, it is now possible to predict the anaesthetic activity of other ether structures, given their partition coefficients.

There are limitations to the use of this particular equation. For example, it is derived purely for anaesthetic ethers and is not applicable to other structural types of anaesthetics. This is generally true in QSAR studies. The procedure works best if it is applied to a series of compounds which have the same general structure.

QSAR studies have been carried out on other structural types of general anaesthetics and in each case a parabolic curve has been obtained. Although, the constants for each equation are different, it is significant that the optimum hydrophobicity (represented by $\log P^0$) for anaesthetic activity is close to 2.3, regardless of the class of anaesthetic being studied. This finding suggests that all general anaesthetics operate in a similar fashion, controlled by the hydrophobicity of the structure.

Since different anaesthetics have similar $\log P^0$ values, the $\log P$ value of any compound can give some idea of its potential potency as an anaesthetic. For example, the $\log P$ values of the gaseous anaesthetics ether, chloroform, and halothane are 0.98, 1.97, and 2.3 respectively. Their anaesthetic activity increases in the same order.

Since general anaesthetics have a simple mechanism of action based on the efficiency with which they enter the CNS, it implies that $\log P$ values should give an indication of how easily any compound can enter the CNS. In other words, compounds having a $\log P$ value close to 2 should be capable of entering the CNS efficiently. This is generally found to be true. For example, the most potent barbiturates for sedative and hypnotic activity are found to have $\log P$ values close to 2.

As a rule of thumb, drugs which are to be targeted for the CNS should have a $\log P$ value of approximately 2. Conversely, drugs which are designed to act elsewhere in the body should have log P values significantly different from 2 to avoid possible CNS side-effects (e.g. drowsiness).

As an example of this, the cardiotonic agent shown in Fig. 11.4(a) was found to produce 'bright visions' in some patients, which implied that it was entering the CNS. This was supported by the fact that the log P value of the drug was 2.59. To prevent the drug entering the CNS, the 4-OMe group was replaced with a 4-S(O)Me group. This particular group is approximately the same size as the methoxy group, but more hydrophilic. The log P value of the new drug (sulmazole; Fig. 11.4(b)) was found to be 1.17. The drug was now too hydrophilic to enter the CNS and was free of CNS side-effects.

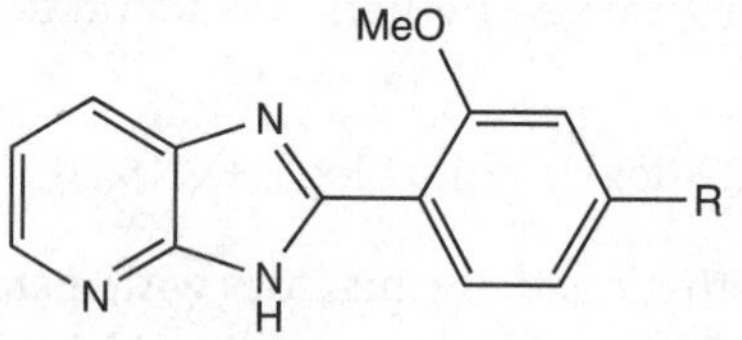

a) R= OMe
b) R= S(O)Me Sulmazole

Fig. 11.4 Cardiotonic agents.

The substituent hydrophobicity constant (π)

We have seen how the hydrophobicity of a compound can be quantified by using the partition coefficient P. However, to obtain P we have to measure it experimentally and that means that we have to synthesize the compounds. It would be much better if we could calculate P theoretically and decide in advance whether the compound is worth synthesizing. QSAR would then allow us to target the most promising looking structures. For example, if we were planning to synthesize a range of barbiturate structures, we could calculate log P values for them all and concentrate on the structures which had log P values closest to the optimum log P^0 value for barbiturates.

Fortunately, partition coefficients can be calculated by knowing the contribution that various substituents make to hydrophobicity. This contribution is known as the substituent hydrophobicity constant (π).

The substituent hydrophobicity constant is a measure of how hydrophobic a substituent is, relative to hydrogen. The value can be obtained as follows. Partition coefficients are measured experimentally for a standard compound with and without a substituent (X). The hydrophobicity constant (π_X) for the substituent (X) is then obtained using the following equation:

$$\pi_X = \log P_X - \log P_H$$

where P_H is the partition coefficient for the standard compound, and P_X is the partition coefficient for the standard compound with the substituent.

A positive value of π indicates that the substituent is more hydrophobic than hydrogen. A negative value indicates that the substituent is less hydrophobic. The π values for a range of substituents are shown in Table 11.1.

Table 11.1

Group	CH$_3$	t-Bu	OH	OCH$_3$	CF$_3$	Cl	Br	F
π (Aliphatic substituents)	0.50	1.68	-1.16	0.47	1.07	0.39	0.60	-0.17
π (Aromatic substituents)	0.52	1.68	-0.67	-0.02	1.16	0.71	0.86	0.14

These π values are characteristic for the substituent and can be used to calculate how the partition coefficient of a drug would be affected by adding these substituents. The P value for the lead compound would have to be measured experimentally but, once this is known, the P value for analogues can be calculated quite simply.

As an example, consider the log P values for benzene (log $P = 2.13$), chlorobenzene (log $P = 2.84$), and benzamide (log $P = 0.64$) (Fig. 11.5). Since benzene is the parent compound, the substituent constants for Cl and CONH$_2$ are 0.71 and -1.49 respectively. Having obtained these values, it is now possible to calculate the theoretical log P value for *meta* chlorobenzamide:

$$\log P_{\text{(chlorobenzamide)}} = \log P_{\text{(benzene)}} + \pi_{\text{Cl}} + \pi_{\text{CONH}_2}$$

$$= 2.13 + 0.71 + (-1.49)$$

$$= 1.35.$$

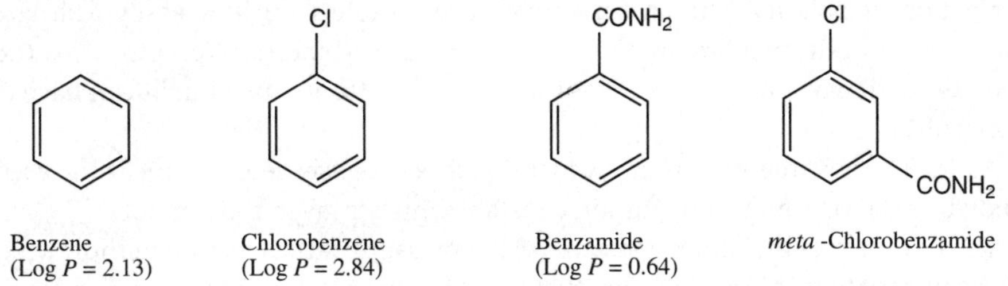

Benzene
(Log $P = 2.13$)

Chlorobenzene
(Log $P = 2.84$)

Benzamide
(Log $P = 0.64$)

meta -Chlorobenzamide

Fig. 11.5 Values for log P.

The observed log P value for this compound is 1.51.

It should be noted that π values for aromatic substituents are different from those used for aliphatic substituents. Furthermore, neither of these sets of π values are in fact true constants and are accurate only for the structures from which they were derived. They can be used as good approximations when studying other structures, but it is possible that the values will have to be adjusted to obtain accurate results.

P versus π

QSAR equations relating biological activity to the partition coefficient *P* have already been described, but there is no reason why the substituent hydrophobicity constant π cannot be used in place of *P* if only the substituents are being varied.

The equation obtained would be just as relevant as a study of how hydrophobicity affects biological activity. That is not to say that *P* and π are exactly equivalent – different equations would be obtained with different constants.

Apart from the fact that the constants would be different, the two factors have different emphases. The partition coefficient *P* is a measure of the drug's overall hydrophobicity and is therefore an important measure of how efficiently a drug is transported to its target site and bound to its receptor. The π factor measures the hydrophobicity of a specific region on the drug's skeleton. Thus, any hydrophobic bonding to a receptor involving that region will be more significant to the equation than the overall transport process. If the substituent is involved in hydrophobic bonding to a receptor, then the QSAR equation using the π factor will emphasize that contribution to biological activity more dramatically than the equation using *P*.

Most QSAR equations will have a contribution from *P* or from π or from both. However, there are examples of drugs which have only a slight contribution. For example, a study on antimalarial drugs showed very little relationship between anti-malarial activity and hydrophobic character. This finding lends support to the theory that these drugs are acting in red blood cells, since previous research has shown that the ease with which drugs enter red blood cells is not related to their hydrophobicity.

11.3.2 Electronic effects

The electronic effects of various substituents will clearly have an effect on a drug's ionization or polarity. This in turn may have an effect on how easily a drug can pass through cell membranes or how strongly it can bind to a receptor. It is therefore useful to have some measure of the electronic effect a substituent can have on a molecule.

As far as substituents on an aromatic ring are concerned, the measure used is known as the Hammett substitution constant which is given the symbol σ.

The Hammett substitution constant (σ) is a measure of the electron withdrawing or electron donating ability of a substituent and has been determined by measuring the dissociation of a series of substituted benzoic acids compared to the dissociation of benzoic acid itself.

Benzoic acid is a weak acid and only partially ionizes in water (Fig. 11.6).

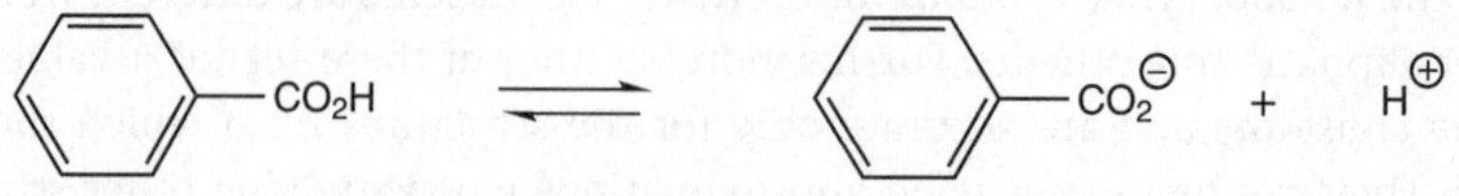

Fig. 11.6 Ionization of benzoic acid.

An equilibrium is set up between the ionized and non-ionized forms, in which the relative proportions of the species is known as the equilibrium or dissociation constant K_H (the subscript H signifies that there are no substituents on the aromatic ring).

$$K_H = \frac{[PhCO_2^-]}{[PhCO_2H]}$$

When a substituent is present on the aromatic ring, this equilibrium is affected. Electron withdrawing groups, such as a nitro group, result in the aromatic ring having a stronger electron withdrawing and stabilizing influence on the carboxylate anion. The equilibrium will therefore shift more to the ionized form such that the substituted benzoic acid is a stronger acid and has a larger K_X value (X represents the substituent on the aromatic ring) (Fig. 11.7).

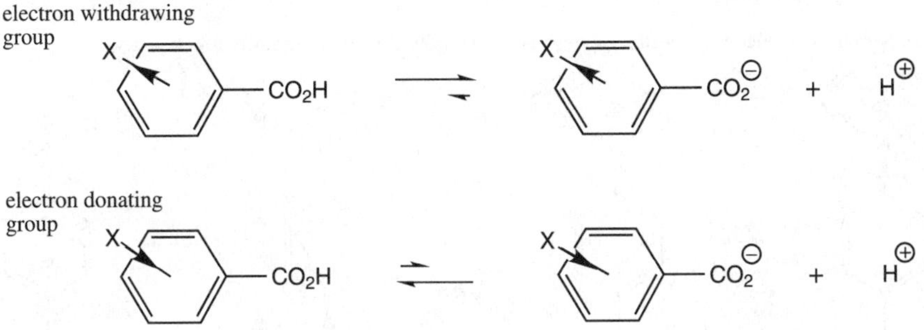

Fig. 11.7 Position of equilibrium dependent on substituent group X.

If the substituent X is an electron donating group such as an alkyl group, then the aromatic ring is less able to stabilize the carboxylate ion. The equilibrium shifts to the left and a weaker acid is obtained with a smaller K_X value (Fig. 11.7).

The Hammett substituent constant (σ_X) for a particular substituent X is defined by the following equation:

$$\sigma_X = \log \frac{K_X}{K_H} = \log K_X - \log K_H$$

Benzoic acids containing electron withdrawing substituents will have larger K_X values than benzoic acid itself (K_H) and therefore the value of σ_X for an electron withdrawing substituent will be positive. Substituents such as Cl, CN, or CF_3 have positive σ values.

Benzoic acids containing electron donating substituents will have smaller K_X values than benzoic acid itself and hence the value of σ_X for an electron donating substituent will be negative. Substituents such as Me, Et, and t-Bu have negative σ values. The Hammett substituent constant for H is zero.

The Hammett constant takes into account both resonance and inductive effects. Therefore, the value of σ for a particular substituent will depend on whether the substituent is *meta* or *para*. This is indicated by the subscript m or p after the σ symbol.

For example, the nitro substituent has $\sigma_p = 0.78$ and $\sigma_m = 0.71$. In the *meta* position, the electron withdrawing power is due to the inductive influence of the substituent, whereas at the *para* position inductive and resonance effects both play a part, and so the σ_p value is greater (Fig. 11.8).

meta Nitro group - electronic influence on R is inductive

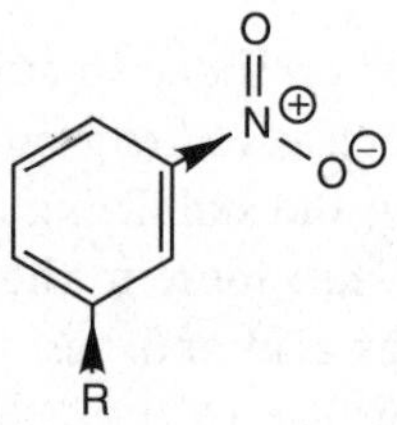

para Nitro group - electronic influence on R is due to inductive and resonance effects

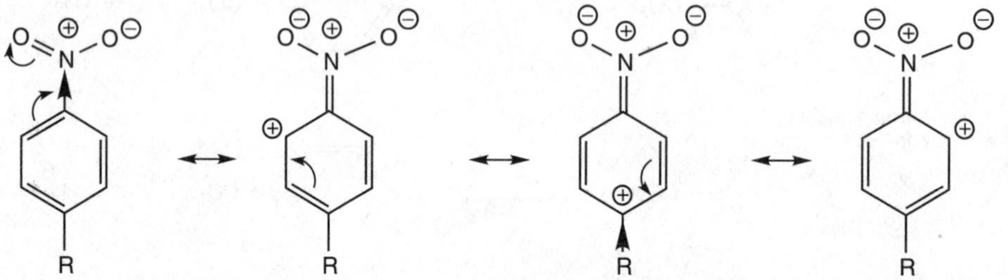

Fig. 11.8

For the OH group $\sigma_m = 0.12$ while $\sigma_p = -0.37$. At the *meta* position, the influence is inductive and electron withdrawing. At the *para* position, the electron donating influence due to resonance is more significant than the electron withdrawing influence due to induction (Fig. 11.9).

Most QSAR studies start off by considering σ and, if there is more than one substituent, the σ values are summed ($\Sigma\sigma$). However, as more compounds are synthesized, it is possible to refine or fine tune the QSAR equation. As mentioned above, σ is a measure of a substituent's inductive and resonance electronic effects. With more detailed studies, the inductive and resonance effects can be considered separately. Tables of constants are available which quantify a substituent's inductive effect (F) and its resonance effect (R). In some cases, it might be found that a substituent's effect on activity is due to F rather than R, and *vice versa*. It might also be found that a substituent has a more significant effect at a particular position on the ring and this can also be included in the equation.

There are limitations to the electronic constants which we have described so far. For example, Hammett substituent constants cannot be measured for *ortho* substituents since such substituents have an important steric, as well as electronic, effect.

meta Hydroxyl group - electronic influence on R is inductive

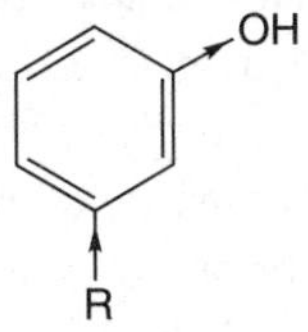

para Hydroxyl group - electronic influence on R dominated by resonance effects

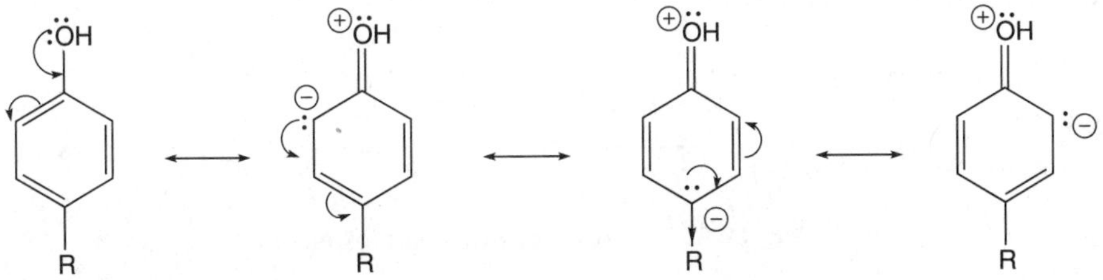

Fig. 11.9

There are very few drugs whose activities are solely influenced by a substituent's electronic effect, since hydrophobicity usually has to be considered as well. Those that do are generally operating by a mechanism whereby they do not have to cross any cell membranes. Alternatively, *in vitro* studies on isolated enzymes may result in QSAR equations lacking the hydrophobicity factor, since there are no cell membranes to be considered.

The insecticidal activity of diethyl phenyl phosphates (Fig. 11.10) is one of the few examples where activity is related to electronic factors alone:

$$\log \left(\frac{1}{C} \right) = 2.282\sigma - 0.348.$$

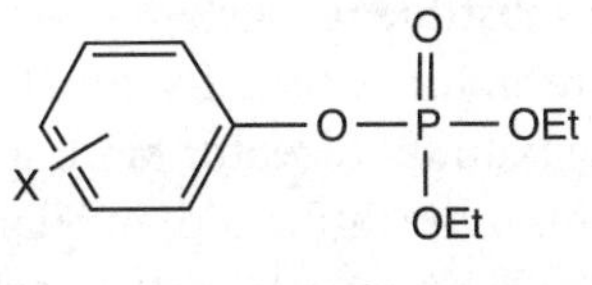

Fig. 11.10 Diethyl phenyl phosphate.

The equation reveals that substituents with a positive value for σ (i.e. electron withdrawing groups) will increase activity. The fact that the π parameter is not significant is a good indication that the drugs do not have to pass into or through a cell membrane to have activity. In fact, these drugs are known to act against an enzyme called acetylcholinesterase which is situated on the outside of cell membranes (see Chapter 15).

The above constants (σ, R, and F) can only be used for aromatic substituents and are therefore only suitable for drugs containing aromatic rings. However, a series of aliphatic electronic substituent constants are available. These were obtained by measuring the rates of hydrolysis for a series of aliphatic esters (Fig. 11.11). Methyl ethanoate is the parent ester and it is found that the rate of hydrolysis is affected by the substituent X. The extent to which the rate of hydrolysis is affected is a measure of the substituent's electronic effect at the site of reaction (i.e. the ester group). The electronic effect is purely inductive and is given the symbol σ_I. Electron donating groups reduce the rate of hydrolysis and therefore have negative values. For example, σ_I values for methyl, ethyl, and propyl are -0.04, -0.07, and -0.36 respectively. Electron withdrawing groups increase the rate of hydrolysis and have positive values. The σ_I values for $^+NMe_3$ and CN are 0.93 and 0.53 respectively.

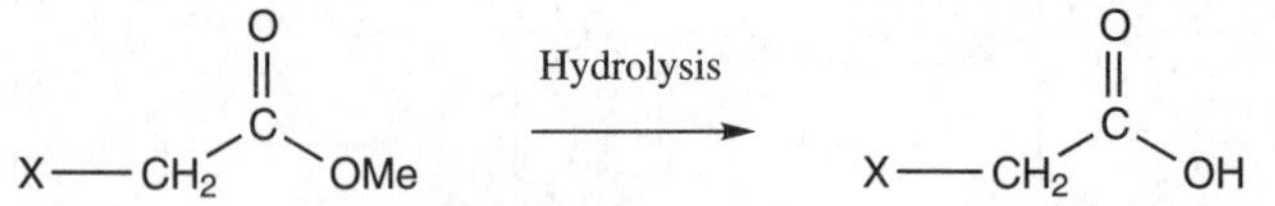

Fig. 11.11 Hydrolysis of an aliphatic ester.

It should be noted that the inductive effect is not the only factor affecting the rate of hydrolysis. The substituent may also have a steric effect. For example, a bulky substituent may 'shield' the ester from attack and lower the rate of hydrolysis. It is therefore necessary to separate out these two effects. This can be done by measuring hydrolysis rates under basic conditions and also under acidic conditions. Under basic conditions, steric and electronic factors are important, whereas under acidic conditions only steric factors are important. By comparing the rates, values for the electronic effect (σ_I), and for the steric effect (E_S) (see below) can be determined.

11.3.3 Steric factors

For a drug to interact with an enzyme or a receptor, it has to approach, then bind to a binding site. The bulk, size, and shape of the drug may have an influence on this process. For example, a bulky substituent may act like a shield and hinder the ideal interaction between drug and receptor. Alternatively, a bulky substituent may help to orientate a drug properly for maximum receptor binding and increase activity.

Quantifying steric properties is more difficult than quantifying hydrophobic or electronic properties. Several methods have been tried and three are described here. It is highly unlikely that a drug's biological activity will be affected by steric factors alone, but these factors are frequently to be found in Hansch equations (see Section 11.4.).

Taft's steric factor (E_s)

Attempts have been made to quantify the steric features of substituents by using Taft's steric factor (E_s). The value for E_s can be obtained as described in Section 11.3.2.

However, the number of substituents which can be studied by this method is restricted.

Molar refractivity

Another measure of the steric factor is provided by a parameter known as molar refractivity (MR). This is a measure of the volume occupied by an atom or group of atoms. The MR is obtained from the following equation:

$$MR = \frac{(n^2 - 1)}{(n^2 + 2)} \times \frac{MW}{d}$$

where n is the index of refraction, MW is the molecular weight, and d is the density. The term MW/d defines a volume, while the $(n^2 - 1)/(n^2 + 2)$ term provides a correction factor by defining how easily the substituent can be polarized. This is particularly significant if the substituent has π electrons or lone pairs of electrons.

Verloop steric parameter

Another approach to measuring the steric factor involves a computer programme called sterimol which calculates steric substituent values (Verloop steric parameters) from standard bond angles, van der Waals radii, bond lengths, and possible conformations for the substituent. Unlike E_s, the Verloop steric parameters can be measured for any substituent. For example, the Verloop steric parameters for a carboxylic acid group are demonstrated in Fig. 11.12. L is the length of the substituent while B_1–B_4 are the radii of the group.

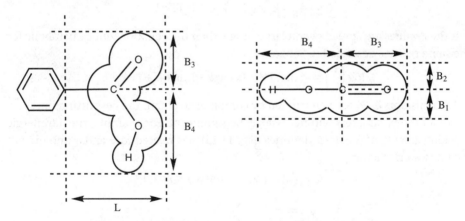

Fig. 11.12 Verloop parameters for a carboxylic acid group.

11.3.4 Other physicochemical parameters

The physicochemical properties most commonly studied by the QSAR approach have been described above, but other properties have also been studied. These include dipole moments, hydrogen bonding, conformation, and interatomic distances. However, difficulties in quantifying these properties limit the use of these parameters. Several QSAR formulae have also been developed based on the highest occupied and/or the lowest unoccupied molecular orbitals of the test compounds. The calculation of these orbitals can be carried out using semiempirical quantum mechanical methods. This used to be the province of specialists in quantum chemistry but advances in computers and computer power have allowed medicinal chemists to carry out these studies with normal laboratory computers.

11.4 Hansch equation

In Section 11.3 we looked at the physicochemical properties commonly used in QSAR studies and how it is possible to quantify them. In a simple situation where biological activity is related to only one such property, a simple equation can be drawn up. However, the biological activity of most drugs is related to a combination of physicochemical properties. In such cases, simple equations involving only one parameter are relevant only if the other parameters are kept constant. In reality, this is not easy to achieve and equations which relate biological activity to more than one parameter are more common. These equations are known as Hansch equations and they usually relate biological activity to the most commonly used physicochemical properties (P and/or π, σ, and a steric factor). If the hydrophobicity values are limited to a small range then the equation will be linear, as follows:

$$\log\left(\frac{1}{C}\right) = k_1 \log P + k_2\, \sigma + k_3\, E_s + k_4.$$

If the P values are spread over a large range, then the equation will be parabolic for the same reasons described in Section 11.3.1.

$$\log\left(\frac{1}{C}\right) = -k\,(\log P)^2 + k_2 \log P + k_3\, \sigma + k_4\, E_s + k_5.$$

The constants k_1–k_5 are determined by computer to obtain the best fitting line.

Not all the parameters will necessarily be significant. For example, the adrenergic blocking activity of β-halo-arylamines (Fig. 11.13) was related to π and σ and did not include a steric factor:

$$\log\left(\frac{1}{C}\right) = 1.22\pi - 1.59\sigma + 7.89.$$

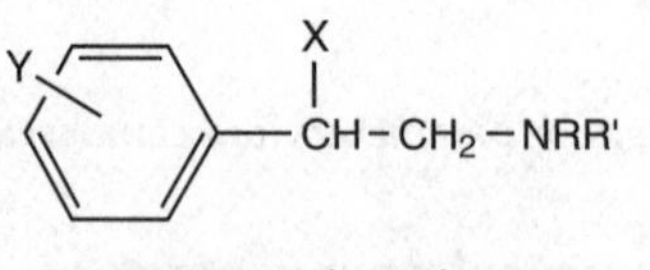

Fig. 11.13 β-Halo-arylamines.

This equation tells us that biological activity increases if the substituents have a positive π value and a negative σ value. In other words, the substituents should be hydrophobic and electron donating.

Since the P value and the π factor are not necessarily correlated, it is possible to have Hansch equations containing both of these factors. For example, a series of 102 phenanthrene aminocarbinols (Fig. 11.14) was tested for antimalarial activity and found to fit the following equation:

$$\log\left(\frac{1}{C}\right) = -0.015\,(\log P)^2 + 0.14\log P + 0.27\Sigma\,\pi_X + 0.40\,\Sigma\pi_Y$$
$$+ 0.65\,\Sigma\sigma_X + 0.88\,\Sigma\sigma_Y + 2.34.$$

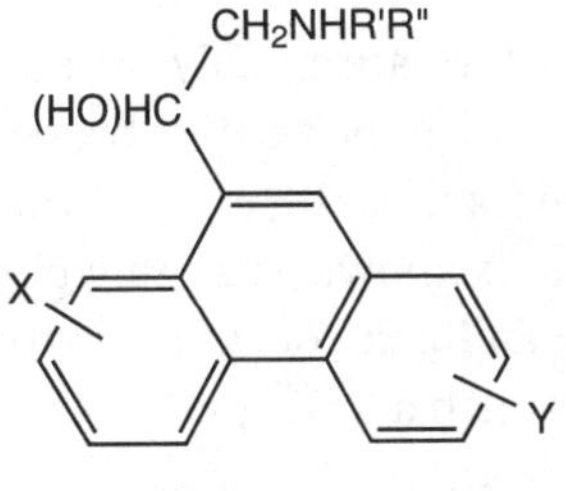

Fig. 11.14 Phenanthrene aminocarbinol structure.

This equation tells us that antimalarial activity increases very slightly as the hydrophobicity of the molecule (P) increases. The constant of 0.14 is low and shows that the increase is slight. The $(\log P)^2$ term shows that there is an optimum P value for activity. The equation also shows that activity increases significantly if hydrophobic substituents are present on ring X and in particular on ring Y. This could be taken to imply that some form of hydrophobic interaction is involved at these sites. Electron withdrawing substituents on both rings are also beneficial to activity, more so on ring Y than ring X.

When carrying out a Hansch analysis, it is important to choose the substituents carefully to ensure that the change in biological activity can be attributed to a particular parameter. There are plenty of traps for the unwary. Take, for example, drugs which contain an amine group. One of the most frequently carried out studies on amines is to synthesize analogues containing a homologous series of alkyl substituents on the nitrogen atom (i.e. Me, Et, n-Pr, n-Bu). If activity increases with the chain length of the substituent, is it due to increasing hydrophobicity or to increasing size or to both? If we look at the π and MR values of these substituents, then we find that both increase in a similar fashion across the series and we would not be able to distinguish between them (Table 11.2).

Table 11.2

Substituent	H	Me	Et	n-Pr	n-Bu	OMe	NHCONH$_2$	I	CN
π	0.00	0.56	1.02	1.50	2.13	-0.02	-1.30	1.12	-0.57
MR	0.10	0.56	1.03	1.55	1.96	0.79	1.37	1.39	0.63

In this example, a series of substituents would have to be chosen where π and MR are not related. The substituents H, Me, OMe, NHCOCH$_2$, I, and CN would be more suitable.

11.5 The Craig plot

Although tables of π and σ factors are readily available for a large range of substituents, it is often easier to visualize the relative properties of different substituents by considering a plot where the y axis is the value of the σ factor and the x axis is the value of the π factor. Such a plot is known as a Craig plot. The example shown in Fig. 11.15 is the Craig plot for the σ and π factors of *para* aromatic substituents. There are several advantages to the use of such a Craig plot.

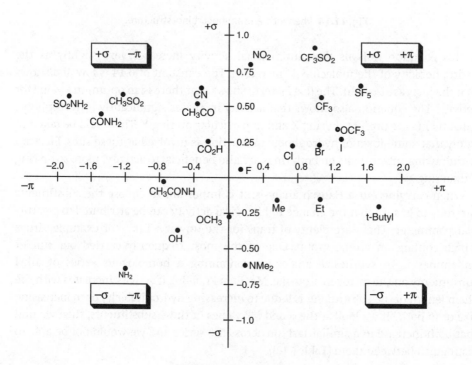

Fig. 11.15 Craig plot.

• The plot shows clearly that there is no overall relationship between π and σ. The various substituents are scattered around all four quadrants of the plot.

• It is possible to tell at a glance which substituents have positive π and σ parameters, which substituents have negative π and σ parameters, and which substituents have one positive and one negative parameter.

- It is easy to see which substituents have similar π values. For example, the ethyl, bromo, trifluoromethyl, and trifluoromethylsulphonyl groups are all approximately on the same vertical line on the plot. In theory, these groups could be interchangeable on drugs where the principal factor affecting biological activity is the π factor. Similarly, groups which form a horizontal line can be identified as being iso-electronic or having similar σ values (e.g. CO_2H, Cl, Br, I).

- The Craig plot is useful in planning which substituents should be used in a QSAR study. To derive the most accurate equation involving π and σ, analogues should be synthesized with substituents from each quadrant. For example, halide substituents are useful representatives of substituents with increased hydrophobicity and electron withdrawing properties (positive π and positive σ), whereas an OH substituent has more hydrophilic and electron donating properties (negative π and negative σ). Alkyl groups are examples of substituents with positive π and negative σ values, whereas acyl groups have negative π and positive σ values.

- Once the Hansch equation has been derived, it will show whether π or σ should be negative or positive to obtain good biological activity. Further developments would then concentrate on substituents from the relevant quadrant. For example, if the equation shows that positive π and positive σ values are necessary, then further substituents should only be taken from the top right quadrant.

Craig plots can also be drawn up to compare other sets of physicochemical parameters, such as hydrophobicity and *MR*.

11.6 The Topliss scheme

In certain situations, it might not be feasible to make the large range of structures required for a Hansch equation. For example, the synthetic route involved might be difficult and only a few structures can be made in a limited time. In these circumstances, it would be useful to test compounds for biological activity as they are synthesized and to use these results to determine the next analogue to be synthesized.

A Topliss scheme is a 'flow diagram' which allows such a procedure to be followed. There are two Topliss schemes, one for aromatic substituents (Fig. 11.16) and one for aliphatic side chain substituents (Fig. 11.17). The schemes were drawn up by considering the hydrophobicity and electronic factors of various substituents and are designed such that the optimum substituent can be found as efficiently as possible. However, they are not meant to be a replacement for a full Hansch analysis. Such an analysis would be carried out in due course, once a suitable number of structures have been synthesized.

The Topliss scheme for aromatic substituents (Fig. 11.16) assumes that the lead compound has been tested for biological activity and contains a monosubstituted aromatic ring. The first analogue in the scheme is the 4-chloro derivative, since this

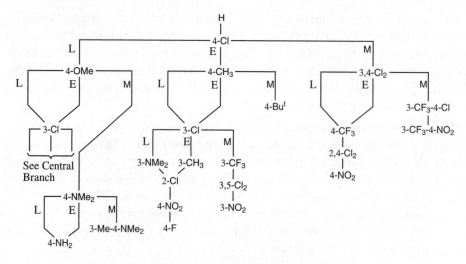

Fig. 11.16 Topliss scheme for aromatic substituents.

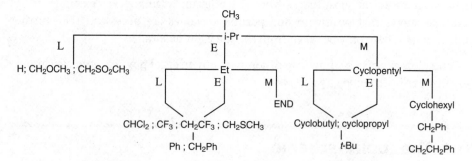

Fig. 11.17 Topliss scheme for aliphatic side chain substituents.

derivative is usually easy to synthesize. The chloro substituent is more hydrophobic and electron withdrawing than hydrogen and, therefore, π and σ are positive.

Once the chloro analogue has been synthesized, the biological activity is measured. There are three possibilities. The analogue will have less activity (L), equal activity (E), or more activity (M). The type of activity observed will determine which branch of the Topliss scheme is followed next.

If the biological activity increases, then the (M) branch is followed and the next analogue to be synthesized is the 3,4-dichloro-substituted analogue. If, on the other hand, the activity stays the same, then the (E) branch is followed and the 4-methyl analogue is synthesized. Finally, if activity drops, the (L) branch is followed and the next analogue is the 4-methoxy analogue.

Biological results from the second analogue now determine the next branch to be followed in the scheme.

What is the rationale behind this?

Let us consider the situation where the 4-chloro derivative increases in biological activity. Since the chloro substituent has positive π and σ values, it implies that one or both of these properties are important to biological activity. If both are important, then adding a second chloro group should increase biological activity yet further. If it does, substituents are varied to increase the π and σ values even further. If it does not, then an unfavourable steric interaction or excessive hydrophobicity is indicated. Further modifications then test the relative importance of π and steric factors.

We shall now consider the situation in which the 4-chloro analogue drops in activity. This suggests either that negative π and/or σ values are important to activity or that a *para* substituent is sterically unfavourable. It is assumed that an unfavourable σ effect is the most likely reason for the reduced activity and so the next substituent is one with a negative σ factor (i.e. 4-OMe). If activity improves, further changes are suggested to test the relative importance of the σ and π factors. If, on the other hand, the 4-OMe group does not improve activity, it is assumed that an unfavourable steric factor is at work and the next substituent is a 3-chloro group. Modifications of this group would then be carried out in the same way as shown in the centre branch of Fig. 11.16.

The last scenario is that in which the activity of the 4-chloro analogue is little changed from the lead compound. This could arise from the drug requiring a positive π value and a negative σ value. Since both values for the chloro group are positive, the beneficial effect of the positive π value might be cancelled out by the detrimental effects of a positive σ value. The next substituent to try in that case is the 4-methyl group; this has the necessary positive π value and negative σ value. If this still has no beneficial effect, then it is assumed that there is an unfavourable steric interaction at the *para* position and the 3-chloro substituent is chosen next. Further changes continue to vary the relative values of the π and σ factors.

The validity of the Topliss scheme was tested by looking at structure–activity results for various drugs which had been reported in the literature. For example, the biological activities of 19 substituted benzenesulphonamides (Fig. 11.18) have been reported. The second most active compound was the nitro-substituted analogue, which would have been the fifth compound synthesized if the Topliss scheme had been followed.

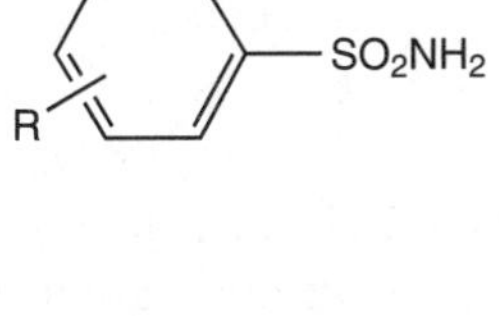

	Order of Synthesis	R	Biological Activity	High Potency
R—⟨benzene⟩—SO$_2$NH$_2$	1	H	-	
	2	4-Cl	M	
	3	3,4-Cl$_2$	L	
	4	4-Br	E	
	5	4-NO$_2$	M	*

M= More Activity
L= Less Activity
E = Equal Activity

Fig. 11.18 Biological activity of substituted benzenesulphonamides.

Another example comes from the anti-inflammatory activities of substituted aryl-tetrazolylalkanoic acids (Fig. 11.19). Twenty-eight of these were synthesized. Using the Topliss scheme, three of the four most active structures would have been prepared from the first eight compounds synthesized.

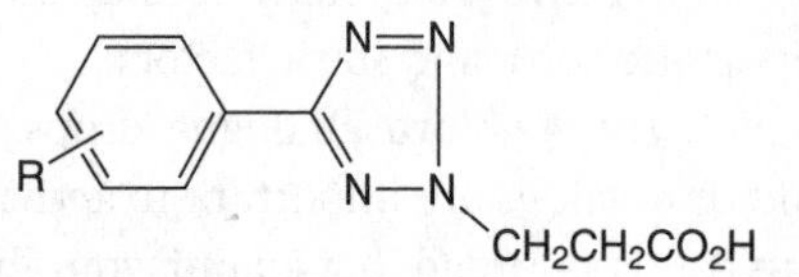

Order of Synthesis	R	Biological Activity	High Potency
1	H	-	
2	4-Cl	L	
3	4-MeO	L	
4	3-Cl	M	*
5	3-CF$_3$	L	
6	3-Br	M	*
7	3-I	L	
8	3,5-Cl$_2$	M	*

M= More Activity
L= Less Activity
E = Equal Activity

Fig. 11.19 Anti-inflammatory activities of substituted aryltetrazolylalkanoic acids.

The Topliss scheme for aliphatic side chains (Fig. 11.17) was set up following a similar rationale to the aromatic scheme, and is used in the same way for side groups attached to a carbonyl, amino, amide, or similar functional group. The scheme only attempts to differentiate between the hydrophobic and electronic effects of substituents and not the steric properties. Thus, the substituents involved have been chosen to try and minimize any steric differences. It is assumed that the lead compound has a methyl group. The first analogue suggested is the isopropyl analogue. This has an increased π value and in most cases would be expected to increase activity, since it has been found from experience that the hydrophobicity of most lead compounds is less than the optimum hydrophobicity required for activity.

Let us concentrate first on the situation in which activity rises. Following this branch, a cyclopentyl group is now used. A cyclic structure is used since it has a larger π value, but keeps any increase in steric factor to a minimum. If activity rises again, more hydrophobic substituents are tried. If activity does not rise, then there could be two explanations. Either the optimum hydrophobicity has been passed or there is an electronic effect (σ_I) at work. Further substituents are then used to determine which is the correct explanation.

Let us now look at the situation where the activity of the isopropyl analogue stays much the same. The most likely explanation is that the methyl and isopropyl groups are on either side of the hydrophobic optimum. Therefore, an ethyl group is used next, since it has an intermediate π value. If this does not lead to an improvement, it is possible that there is an unfavourable electronic effect. The groups used have been electron donating, and so electron withdrawing groups with similar π values are now suggested.

Finally, we shall look at the case where activity drops for the isopropyl group. In this case, hydrophobic and/or electron donating groups could be bad for activity and the groups suggested are suitable choices for further development.

11.7 Bioisosteres

Tables of substituent constants are available for various physicochemical properties. A knowledge of these constants allows the medicinal chemist to identify substituents which may be potential bioisosteres. Thus, the substituents CN, NO_2, and COMe have similar hydrophobic, electronic, and steric factors, and might be interchangeable. Such interchangeability was observed in the development of cimetidine (see Chapter 18). The important thing to notice is that groups can be bioisosteric in some situations, but not others. Consider for example the table shown in Fig. 11.20.

Substituent	$\overset{O}{\underset{\parallel}{-C}}-CH_3$	$\overset{NC\diagdown\diagup CN}{\underset{\parallel}{-C}}-CH_3$	$\overset{O}{\underset{\parallel}{-S}}-CH_3$	$\overset{O}{\underset{\underset{O}{\parallel}}{-S}}-CH_3$	$\overset{O}{\underset{\underset{O}{\parallel}}{-S}}-NHCH_3$	$\overset{O}{\underset{\parallel}{-C}}-NMe_2$
π	−0.55	0.40	−1.58	−1.63	−1.82	−1.51
σ_p	0.50	0.84	0.49	0.72	0.57	0.36
σ_m	0.38	0.66	0.52	0.60	0.46	0.35
MR	11.2	21.5	13.7	13.5	16.9	19.2

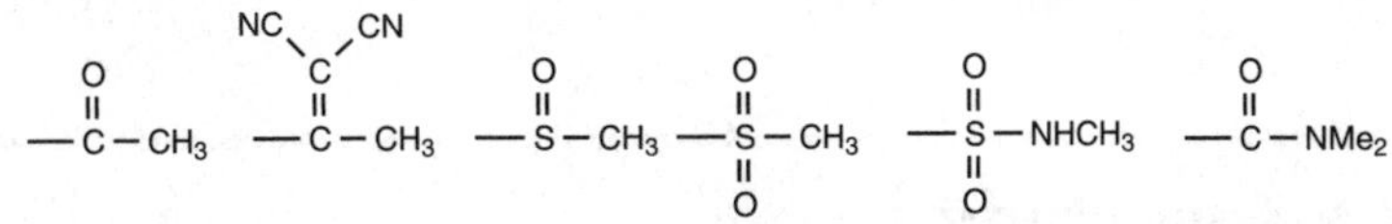

Fig. 11.20 Physicochemical parameters for six substituents.

This table shows some physicochemical parameters for six different substituents. If the most important physicochemical parameter for biological activity is σ_p, then the $COCH_3$ group (0.50) would be a reasonable bioisostere for the $SOCH_3$ group (0.49). If, on the other hand, the dominant parameter is π, then a more suitable bioisostere for $SOCH_3$ (−1.58) would be SO_2CH_3 (−1.63).

11.8 Planning a QSAR study

When starting a QSAR study it is important to decide which physicochemical parameters are going to be studied and to plan the analogues such that the parameters under study are suitably varied. For example, it would be pointless to synthesize analogues where the hydrophobicity and steric volume of the substituents are correlated if these two parameters are to go into the equation.

It is also important to synthesize enough structures to make the results statistically meaningful. As a rule of thumb, five structures should be made for every parameter studied. Typically, the initial QSAR study would involve the two parameters π and σ, and possibly E_s. Craig plots could be used to choose suitable substituents.

Certain substituents are worth avoiding in the initial study since they may have properties other than those being studied. For example, substituents which might ionize (CO_2H, NH_2, SO_2H) should be avoided. Groups which might easily be metabolized should be avoided if possible (e.g. esters or nitro groups).

If there are two or more substituents, then the initial equation usually considers the total π and σ contribution.

As more analogues are made, it is often possible to consider the hydrophobic and electronic effect of substituents at specific positions of the molecule. Furthermore, the electronic parameter σ can be split into its inductive and resonance components (F and R). Such detailed equations may show up a particular localized requirement for activity. For example, a hydrophobic substituent may be favoured in one part of the skeleton, while an electron withdrawing substituent is favoured at another. This in turn gives clues about the binding interactions involved between drug and receptor.

11.9 Case study

An example of how the QSAR equation can change and become more specific as a study develops is demonstrated from a study carried out by workers at Smith Kline & French on the anti-allergic activity of a series of pyranenamines (Fig. 11.21). In this study, substituents were varied on the aromatic ring, and the remainder of the molecule was kept constant. Nineteen compounds were synthesized and the first QSAR equation was obtained by only considering π and σ:

$$\log\left(\frac{1}{C}\right) = -0.14 \, \Sigma\pi - 1.35 \, (\Sigma\sigma)^2 - 0.72$$

where $\Sigma\pi$ and $\Sigma\sigma$ are the total π and σ values for all substituents present.

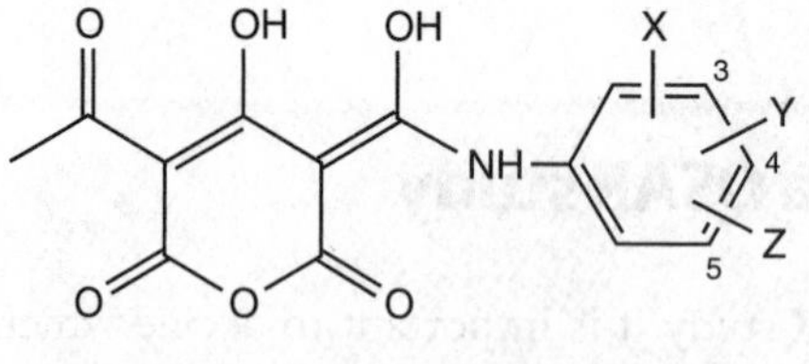

Fig. 11.21 Structure of pyranenamine.

The negative coefficient for the π term shows that activity is inversely proportional to hydrophobicity, which is quite unusual. The $(\Sigma\sigma)^2$ term is also quite unusual. It was chosen since there was no simple relationship between activity and σ. In fact, it was observed that activity dropped if the substituent was electron withdrawing or electron donating. Activity was best with neutral substituents. To take account of this, the $(\Sigma\sigma)^2$ term was introduced. Since the coefficient in the equation is negative, activity is lowered if σ is anything other than zero.

A further range of compounds was synthesized with hydrophilic substituents to test this equation, making a total of 61 structures. This resulted in the following inconsistencies.

• The activities for the substituents 3-NHCOMe, 3-NHCOEt, and 3-NHCOPr were all similar. However, according to the equation, the activities should have dropped as the alkyl group became larger because of increasing hydrophobicity.

• Activity was greater than expected if there was a substituent such as OH, SH, NH_2, or NHCOR at position 3, 4, or 5.

• The substituent $NHSO_2R$ was bad for activity.

• The substituents $3,5\text{-}(CF_3)_2$ and $3,5\text{-}(NHCOMe)_2$ had much greater activity than expected.

• An acyloxy group at the 4-position resulted in an activity five times greater than predicted by the equation.

These results implied that the initial equation was too simple and that properties other than π and σ were important to activity. At this stage, the following theories were proposed to explain the above results.

• The similar activities for 3-NHCOMe, 3-NHCOEt, and 3-NHCOPr could be because of a steric factor. The substituents had increasing hydrophobicity which is bad for activity, but were also increasing in size and it was proposed that this was good for activity. The most likely explanation is that the size of the substituent forces the drug into the correct orientation for optimum receptor interaction.

• The substituents which unexpectedly increased activity when they were at positions 3, 4, or 5 are all capable of hydrogen bonding. This suggests an important hydrogen bonding interaction with the receptor. For some reason, the $NHSO_2R$ group is an exception, which implies there is some other unfavourable steric or electronic factor peculiar to this group.

• The increased activity for 4-acyloxy groups was explained by suggesting that these analogues are acting as prodrugs. The acyloxy group is less polar than the hydroxyl group and so these analogues would be expected to cross cell membranes and reach the receptor more efficiently than analogues bearing a free hydroxyl group. At the receptor, the ester group could be hydrolysed to reveal the hydroxyl group which would then take part in hydrogen bonding with the receptor.

• The structures having substituents $3,5\text{-}(CF_3)_2$ and $3,5\text{-}(NHCOMe)_2$ are the only disubstituted structures where a substituent at position 5 has an electron withdrawing effect, so this feature was also introduced into the next equation.

The revised QSAR equation was as follows:

$$\log\left(\frac{1}{C}\right) = -0.30 \ \Sigma\pi - 1.5 \ (\Sigma\sigma)^2 + 2.0 \ (F\text{-}5) + 0.39 \ (345\text{-}HBD) - 0.63 \ (NHSO_2)$$
$$+ \ 0.78 \ (M\text{-}V) + 0.72 \ (4\text{-}OCO) - 0.75.$$

The π and σ parameters are still present, but a number of new parameters have now been introduced.

• The *F*-5 term represents the inductive effect of a substituent at position 5. Since the coefficient is positive and large, it shows that an electron withdrawing group substantially increases activity. However, since only two compounds in the 61 synthesized had a 5-substituent, there might be quite an error in this result.

• The advantage of having hydrogen bonding substituents at position 3, 4, or 5 is accounted for by including a hydrogen bonding term (*345-HBD*). The value of this term depends on the number of hydrogen bonding substituents present. If one such group is present, the *345-HBD* term is 1. If two such groups were present, the parameter is 2. Therefore, for each hydrogen bonding substituent present at positions 3, 4, or 5, log (1/*C*) increases by 0.39.

• The $NHSO_2$ term was introduced since this group was poor for activity despite being capable of hydrogen bonding. The negative coefficient indicates the drop in activity. A figure of 1 is used for any $NHSO_2R$ substituent present.

• The *M-V* term represents the volume of any *meta* substituent, and since the coefficient is positive, it indicates that substituents with a large volume at the *meta* position increase activity.

• The *4-OCO* term is either 0 or 1 and is only present if an acyloxy group is present at position 4, and so log (1/*C*) is increased by 0.72 if the acyl group is present.

The most important parameters in the above equation are the hydrophobic parameter and the *4-OCO* parameter.

A further 37 structures were synthesized to test steric and *F*-5 parameters as well as exploring further groups capable of hydrogen bonding. Since hydrophilic substituents were good for activity, a range of very hydrophilic substituents were also tested to see if there was an optimum value for hydrophilicity. The results obtained highlighted one more anomaly in that two hydrogen bonding groups *ortho* to each other were bad for activity. This was attributed to the groups hydrogen bonding with each other rather than to the receptor.

A revised equation was obtained as follows:

$$\log\left(\frac{1}{C}\right) = -0.034 \ (\Sigma\pi)^2 - 0.33 \ (\Sigma\pi) + 4.3 \ (F\text{-}5) + 1.3 \ (R\text{-}5) - 1.7 \ (\Sigma\sigma)^2 + 0.73 \ (345\text{-}HBD)$$
$$- \ 0.86 \ (HB\text{-}INTRA) - 0.69 \ (NHSO_2) + 0.72 \ (4\text{-}OCO) - 0.59.$$

The main points of interest from this equation are as follows.

• Increasing the hydrophilicity of substituents allowed the identification of an optimum value for hydrophobicity ($\Sigma\pi = -5$) and introduced the $(\Sigma\pi)^2$ parameter into the equation. The value of –5 is remarkably low and indicates that the receptor site is hydrophilic

• As far as electronic effects are concerned, it is revealed that the resonance effects of substituents at the 5-position also have an influence on activity.

• The unfavourable situation where two hydrogen bonding groups are *ortho* to each other is represented by the *HB-INTRA* parameter. This parameter is given the value 1 if such an interaction is possible and the negative constant (–0.86) shows that such interactions decrease activity.

• It is interesting to note that the steric parameter is no longer significant and has disappeared from the equation.

The compound having the greatest activity has two NHCOCH(OH)CH$_2$OH substituents at the 3- and 5-positions and is 1000 times more active than the original lead compound. The substituents are very polar and are not ones which would normally be used. They satisfy all the requirements determined by the QSAR study. They are highly polar groups which can take part in hydrogen bonding. They are *meta* with respect to each other, rather than *ortho*, to avoid undesirable intramolecular hydrogen bonding. One of the groups is at the 5-position and has a favourable F-5 parameter. Together the two groups have a negligible $(\Sigma\sigma)^2$ value. Such an analogue would certainly not have been obtained by trial and error and this example demonstrates the strengths of the QSAR approach.

All the evidence from this study suggests that the aromatic ring of this series of compounds fits into a hydrophilic pocket in the receptor which contains polar groups capable of hydrogen bonding.

It is further proposed that a positively charged residue such as arginine, lysine, or histidine might be present in the pocket which could interact with an electronegative substituent at position 5 of the aromatic ring (Fig. 11.22).

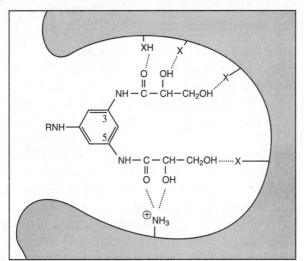

Fig. 11.22 Hypothetical receptor binding interactions of a pyranenamine.

This example demonstrates that QSAR studies and computers are powerful tools in medicinal chemistry. However, it also shows that the QSAR approach is a long way from replacing the human factor. One cannot put a series of facts and figures into a computer and expect it to magically produce an instant explanation of how a drug works. The medicinal chemist still has to interpret results, propose theories, and test those theories by incorporating the correct parameters into the QSAR equation. Imagination and experience still count for a great deal.

11.10 3D QSAR

11.10.1 Introduction

In recent years, a method known as 3D QSAR has been developed whereby the 3D properties of a molecule are considered as a whole rather than considering individual substituents or moieties. This has proved remarkably useful in the design of new drugs. Moreover, the necessary software and hardware are readily affordable and easy to use. Specialist skills and apparatus are no longer required. The philosophy of 3D QSAR revolves around the assumption that the most important features about a molecule are its overall size and shape, and its electronic properties (electrostatic fields).

If these features can be defined, then it is possible to study how they affect biological properties. There are several approaches to 3D QSAR but the method which has gained ascendency is known as CoMFA (comparative molecular field analysis). CoMFA methodology is based on the assumption that drug receptor interactions are not covalent and that changes in biological activity correlate with the changes in the steric and/or electrostatic fields of the drug molecules.

11.10.2 Defining steric and electrostatic fields

To define the necessary steric and electrostatic fields, a molecule is constructed on the computer using molecular modelling software (see Chapter 13). The molecule is then energy minimized to ensure that the bond angles and bond lengths are energetically favourable. Once the molecule has been constructed, the next stage is to build a lattice or grid around the structure (Fig. 11.23). We shall call the intersections of this lattice the lattice (or grid) points.

Each lattice point defines a position in space relative to the molecule.

The next stage is to add a 'probe atom' (usually a proton or an sp^3 hybridized carbon atom with a positive charge) at one of the lattice points. Using the software, it is then possible to measure the steric and electrostatic interactions between the probe atom and the test molecule. This process is then repeated for all the lattice points. By doing this, it is possible to define the shape and electrostatic properties of the molecule in a quantitative way and in a way which the computer can understand. For example, the closer the lattice points are to the molecule, the higher the steric interaction between

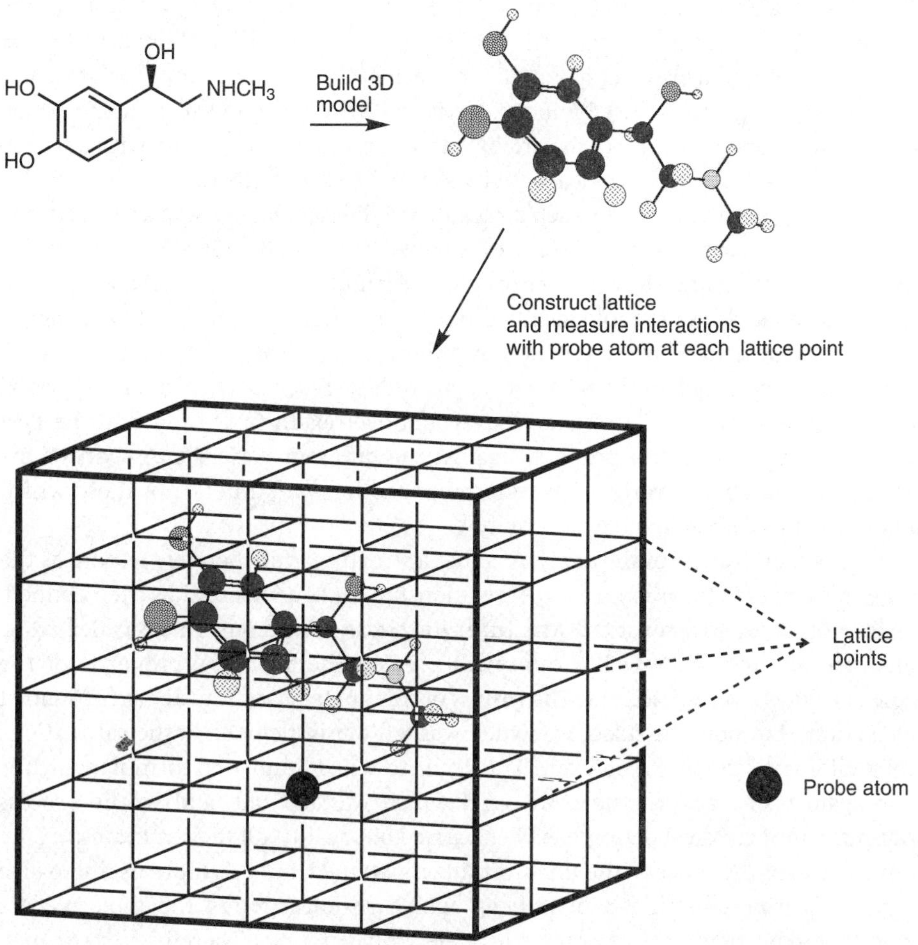

Fig. 11.23 3D QSAR.

the probe atom and the molecule. A series of countours can then be drawn linking lattice points of similar value and, by defining a 'cut-off' value, the shape and size of the molecule can be defined numerically by means of the closest contour line.

A similar process can be carried out to measure the electronic distribution of the molecule. This is done by measuring the electrostatic interactions between the charged probe atom at each lattice point and the test molecule. Again, the results are shown as a contour map surrounding the molecule.

11.10.3 Relating shape and electronic distribution with biological activity

Defining the size, shape, and electronic distribution of a series of molecules is relatively straightforward and is carried out automatically by the software programme. The next stage is to relate these properties to the biological activity of the molecules.

This is less straightforward and differs significantly from 2D QSAR. In 2D QSAR, there are relatively few variables involved. For example, if we consider lipophilicity, a π factor, σ factor and a size factor for each molecule, then we have four variables per molecule to compare against biological activity. With 100 molecules in the study, there are far more molecules than variables and it is possible to come up with an equation relating variables to biological activity as previously described.

In 3D QSAR, the variables for each molecule are the calculated steric and electronic interactions at a couple of thousand lattice points. With a 100 molecules under study, the number of variables now far outweighs the number of structures, and it is not possible to relate these to biological potency by the standard multiple linear regression analysis described in Section 11.2. A different statistical procedure is followed, using a technique called partial least squares (PLS). Essentially, it is an analytical computing process which is repeated over and over again to try and find the best formula relating biological property against the different variables. As part of the process, the number of variables is reduced as the software 'filters' out those which are clearly unrelated to biological activity.

An important feature of the analysis is that a structure is deliberately left out as the computer strives to form some form of relationship. Once a formula has been defined, the formula is tested against the structure which was missed out. This is called cross-validation and tests how well the formula predicts the biological property for the molecule which was missed out. The results of this are fed back into another round of calculations, but now the structure which was left out is included in the calculations and a different structure is left out. This leads to a new improved formula which is once again tested against the compound which was left out, and so the process continues until cross-validation has been carried out against all the structures.

At the end of the process, the final formula is obtained, but it is more useful to give a graphical representation which shows which regions around the molecule are important to biological activity on steric or electronic grounds. Therefore a steric map would show a series of coloured contours indicating beneficial and detrimental steric interactions around a representative molecule from the set of molecules tested (Fig. 11.24). A similar contour map would be created to illustrate the beneficial electrostatic interactions.

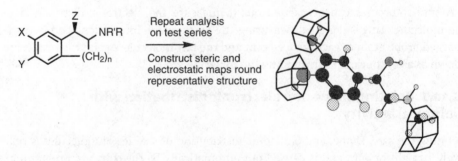

Fig. 11.24 Contour map.

11.10.4 Hydrophobic potential

In 2D QSAR, the hydrophobic factor is very important and most equations include a log P value. Surprisingly, the calculation of a hydrophobic factor in 3D QSAR is not so crucial and most successful studies have been based on steric and electrostatic factors alone. Nevertheless, 3D QSAR studies can be carried out using a hydrophobic factor if the probe atom is replaced with a water molecule to calculate the hydrophobic potential at each lattice point.

11.10.5 Advantages of 3D QSAR over 2D QSAR

Some of the problems involved with a 2D QSAR study include the following:

• Only molecules of similar structure can be studied.

• The validity of the numerical descriptors is open to doubt. These descriptors are obtained by measuring reaction rates and equilibria constants in model reactions and are listed in tables. However, separating one property from another is not always possible in experimental measurement. For example, the Taft steric factor is not purely a measure of the steric factor since the measured reaction rates used to define it are also affected by electronic factors. Also, the octanol:water partition coefficients which are used to measure log P are known to be affected by the hydrogen bonding character of molecules.

• The tabulated descriptors may not include entries for unusual substituents.

• It is necessary to synthesize a range of molecules where substituents are varied to test a particular property (e.g. hydrophobicity). However, synthesizing such a range of compounds may not be straightforward or feasible.

• 2D QSAR equations do not directly suggest new compounds to synthesize.

These problems are avoided with 3D QSAR which offers the following advantages:

• Favourable and unfavourable interactions are represented graphically by 3D contours around a representative molecule. A graphical picture such as this is easier to visualize than a mathematical formula.

• In 3D QSAR the properties of the test molecules are calculated individually by computer programme. There is no reliance on experimental or tabulated factors. There is no need to confine the study to molecules of similar structure. As long as one is confident that all the compounds in the study share same pharmacophore and interact in the same way with the target, they can all be analysed in a 3D QSAR study.

• The graphical representation of beneficial and non-beneficial interactions allows medicinal chemists to design new structures. For example, if a contour map shows a favourable steric effect at one particular location, then this implies that the target receptor or enzyme has space for further extension at that location. This may lead to further favourable receptor–drug interactions.

- Both 2D and 3D QSAR can be used without needing to know the structure of the biological target.

11.10.6 Potential problems of 3D QSAR

There are several pitfalls which have to be avoided when carrying out 3D QSAR. The main ones are:

- Care has to be taken to ensure that each molecule is in the active conformation or shape when it is built on the computer.

- Each molecule must be properly aligned on the screen with respect to the others such that their pharmacophores match up.

Knowing the active conformation is possible in rigid structures such as steroids. However, it is more difficult with flexible molecules which are capable of several bond rotations. Therefore, it is useful to have a conformationally restrained analogue which is biologically active and which can act as a guide to the likely active conformation. More flexible molecules can then be constructed on the computer with the conformation most closely matching that of the more rigid analogue. If the structure of the target receptor or enzyme is known, this can be useful in deciding the likely active conformation and alignment of the molecule before the 3D QSAR analysis.

It is also crucial to identify the likely pharmacophore such that important atoms or groups of atoms are positioned in the same area of space for each molecule. However, there may be some difficulty identifying the pharmacophore in some molecules. In that case, a pharmacophore mapping exercise can be carried out by computer (see Chapter 13). This is likely to be successful if there are some rigid active compounds available, allowing a restriction on the number of possible conformations.

One has to be careful to ensure that all the compounds in the study interact with the target in similar ways. For example, a QSAR study on all possible acetylcholinesterase inhibitors is doomed to failure. In the first place, the great diversity of structures involved, ranging from choline to decamethonium, makes it impossible to align these structures in an unbiased way or to generate a 3D pharmacophore. Second, the various inhibitors do not interact with the target enzyme in the same way. X-ray studies of enzyme inhibitor complexes show that the inhibitors tacrine, edrophonium, and decamethonium all have different binding orientations in the active site.

3D QSAR provides a summary of how structural changes in a drug affect biological activity. However, it is dangerous to assume too much. For example, a 3D QSAR model may show that increasing the bulk of the molecule at a particular location increases activity. This might suggest that there is an accessible hydrophobic pocket allowing extra binding interactions. However, it is also possible that the extra steric bulk causes the molecule to bind in a different orientation from the other molecules in the analysis, and that this is the reason for the increased activity.

12 Combinatorial synthesis

12.1 Introduction

Combinatorial chemistry has been one of the most rapidly developing fields in the pharmaceutical industry in recent years and is now seen as an essential tool, both in the discovery and the development of new drugs. So what is combinatorial synthesis and why is it important?

Put at its simplest, combinatorial synthesis is a means of producing a large number of compounds in a short time period using a defined reaction route and a large variety of starting materials and reagents. Usually, this is done on a very small scale using solid phase synthesis so that the process can be automated or semi-automated. This allows each reaction of the synthetic route to be performed in several reaction vessels at the same time and under identical conditions.

Combinatorial synthesis can be carried out such that a single product is obtained in each different reaction flask – a process known as **parallel synthesis**. This is useful for drug optimization. Alternatively, the process can be designed such that *mixtures* of compounds are produced in each reaction vessel. This is frequently used for drug discovery, i.e. finding a lead compound (Fig. 12.1).

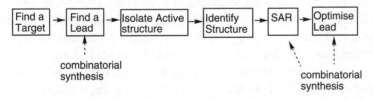

Fig. 12.1 The role of combinatorial synthesis in drug discovery and drug optimization.

12.2 Combinatorial synthesis for drug optimization

To carry out drug optimization, the synthesis of a large number of analogues is required to test structure–activity relationships and/or to improve the activity of the lead compound. Parallel synthesis allows the rapid synthesis of analogues which vary only slightly in structure from the lead compound. The emphasis is on producing individual compounds (one to each reaction flask) which can be isolated, identified, and tested. In the optimization phase, the techniques of combinatorial chemistry can be used alongside rational design.

12.3 Combinatorial synthesis for drug discovery

The requirement to find new lead compounds in drug discovery has been the major driving force in the development of combinatorial synthesis. There has been a rapid explosion in the number of new drug targets discovered by genomic projects across the world, many of which provide the opportunity of developing new treatments for old diseases. With so many new targets being discovered, pharmaceutical companies are faced with the problem of identifying the function of each target, finding a lead compound for the target and optimizing that lead structure as quickly as possible. Herein lies the real need for combinatorial synthesis. Whereas in the past the driving force in medicinal chemistry has been the discovery of a lead compound, the driving force now is the discovery of new drug targets. It has been stated that a pharmaceutical company might expect to set up and carry out lead discovery programmes against about 100 targets per year and will need to screen over a million compounds if it is to find a lead compound quickly and efficiently. Combinatorial chemistry provides a means of producing this quantity of compounds, but it is important to remember that structural diversity is also important to increase the chances of finding a 'hit'. To increase the number and diversity of compounds produced, combinatorial synthesis is carried out in such a way that mixtures of compounds are produced in each reaction flask, allowing a single chemist to produce thousands and even millions of novel structures in the time that he or she would take to synthesize a few dozen by conventional means. This method of synthesis 'goes against the grain' of conventional organic synthesis. Usually chemists set out to produce a single identifiable structure which can be purified and characterized. In combinatorial synthesis, the emphasis is on producing mixtures. The structures of the compound in the mixture are not known with certainty. Neither are the components separated and purified. Instead, each mixture is tested for biological activity as a whole. If there is no activity, then there is no need to study that mixture any more and it is stored. If activity *is* observed, the challenge is now to identify which component (or components) of the mixture is the active compound. We shall discuss this in more detail later on. Overall, there is an economy of effort since a negative result for a mixture of 100 compounds saves the effort of synthesizing, purifying, and identifying each component of that mixture.

In a sense, combinatorial synthesis can be looked upon as the synthetic equivalent of nature's chemical pool. Through evolution, nature has produced a huge number and variety of chemical structures, some of which are biologically active. Traditional medicinal chemistry dips into that pool to pick out the active principles and develop them. Combinatorial synthesis is now producing pools of purely synthetic structures which can be dipped into for active compounds.

Comparing the advantages and disadvantages of both pools, the diversity of structures from the natural pool is far greater than that likely to be achieved by combinatorial synthesis. However, isolating, purifying, and identifying new agents from natural sources is a relatively slow process and there is no guarantee that a lead compound will be discovered against a specific drug target. The advantage of combinatorial chemistry is the fact that it produces new compounds faster than those derived from natural sources and can produce a diversity not found in the banks of compounds held by pharmaceutical companies.

Combinatorial synthesis is not without its critics, and cynics have described its use in drug discovery as nothing more than a technologically advanced form of trial and error. Admittedly, there is some truth in this, but it would be wrong to say that combinatorial synthesis removes the intellectual challenge from drug discovery and development. There is more to combinatorial chemistry than devising a reaction sequence and programming a machine to churn out thousands upon thousands of compounds. A certain amount of thought has to go into the design of the process to ensure that it will be as efficient as possible in producing a 'hit'. Careful thought also has to go into what types of novel structure are likely to be pharmaceutically active before synthesizing them. The future will see an increasing use of robots and machines to synthesize new structures, but that does not mean we have to think like them.

12.4 Combinatorial synthesis – solid phase techniques

Although some combinatorial experiments have been performed in solution, the majority have been achieved using solid phase techniques where the reaction is carried out on a solid support such as a resin bead. There are several advantages to this.

• A range of different starting materials can be bound to separate beads. The beads can then be mixed together such that all the starting materials can be treated with another reagent in a single experiment. The starting materials and products are still physically distinct since they are bound to separate beads. In most cases, mixing all the starting materials together in solution chemistry is a recipe for disaster, with polymerizations and side reactions producing a tarry mess.

- Since the starting materials and products are bound to a solid support, excess reagents or unbound by-products can easily be removed by washing the resin.

- Large excesses of reagents can be used to drive the reactions to completion (greater than 99%) because of the ease with which excess reagent can be removed.

- If one uses low loadings (less than 0.8 mmol per g support), undesired side reactions (such as cross-linking) can be suppressed.

- Intermediates in a reaction sequence are bound to the bead and do not need to be purified.

- The individual beads can be separated at the end of the experiment to give individual products.

- The polymeric support can be regenerated and reused if appropriate cleavage conditions and suitable anchor/linker groups are chosen (see later).

- Automation is possible with solid phase synthesis.

Solid phase synthesis was pioneered by Merrifield for the synthesis of peptides. As a result, most of the early work carried out on combinatorial synthesis was performed on peptides. However, peptides have serious disadvantages as drugs (see Chapters 9 and 10) and so a large amount of research has been carried out to extend solid phase synthetic methods to the synthesis of small, non-peptide molecules. The essential requirements for solid phase synthesis are:

- a cross-linked insoluble polymeric support which is inert to the synthetic conditions (e.g. a resin bead)

- an anchor or linker covalently linked to the resin. The anchor will have a reactive functional group such that substrates can be attached to it

- a bond linking the substrate to the linker which will be stable to the reaction conditions used in the synthesis

- a means of cleaving the product or the intermediates from the linker

- chemical protecting groups for functional groups not involved in the synthetic route.

12.4.1 The solid support

The earliest form of resin used by Merrifield was partially cross-linked polystyrene beads where the styrene was cross-linked with 1% divinylbenzene. The beads are derivatized with a chloromethyl group (the anchor/linker), to which amino acids can be coupled via an ester group (Fig. 12.2). This ester group is stable to the reaction conditions used in peptide synthesis but can be cleaved at the end of the synthesis using vigorous acidic conditions.

One disadvantage of the polystyrene bead is the fact that it is hydrophobic and the growing peptide chain is hydrophilic. As a result, the growing peptide chain is not solvated and often folds in on itself, forming internal hydrogen bonds. This in turn

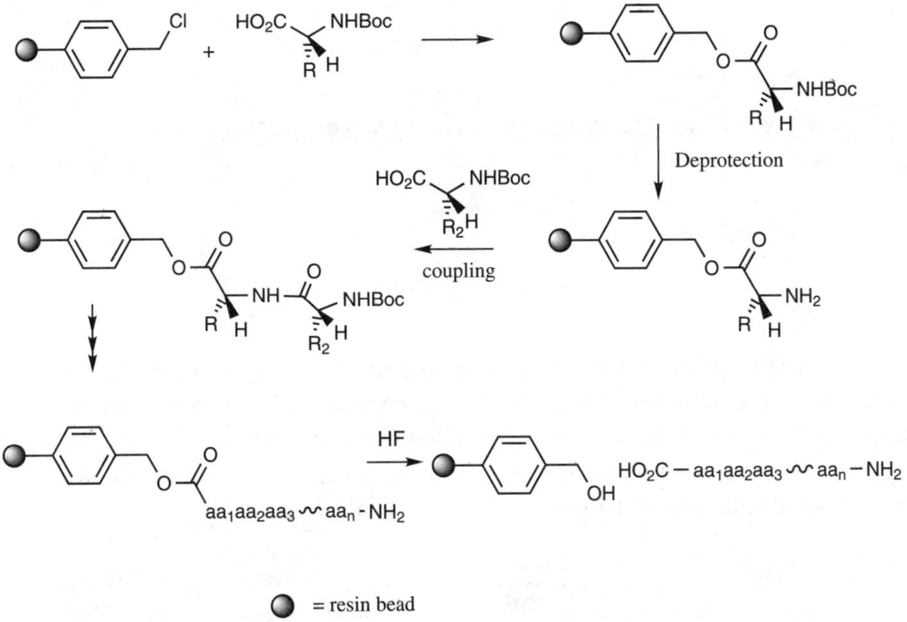

Fig. 12.2 Peptide synthesis.

hinders the access of further amino acids to the exposed end of the growing chain. To address this, more polar solid phases were developed, such as Sheppard's polyamide resin. Other resins have been developed to be more suitable for the combinatorial synthesis of non-peptides. For example, Tentagel resin is 80% polyethylene glycol grafted to cross-linked polystyrene and provides an environment similar to ether or tetrahydrofuran.

Although beads are the common shape for the solid support, a range of other shapes has been designed (e.g. pins) to maximize the surface area available for reaction and hence maximize the amount of compound linked to the solid support. Functionalized glass surfaces have also been used and are suitable for oligonucleotide synthesis.

12.4.2 The anchor/linker

The anchor/linker is a molecular unit covalently attached to the solid support. It contains a reactive functional group with which the first of the reagents in the proposed synthesis can react and hence become attached to the resin. The resulting link must be stable to the reaction conditions used throughout the synthesis, but be easily cleaved to release the final compound once the synthesis is complete (Fig. 12.3).

Different linkers are used depending on the functional group which will be present on the substrate, and the functional group which is desired on the final product once it is released.

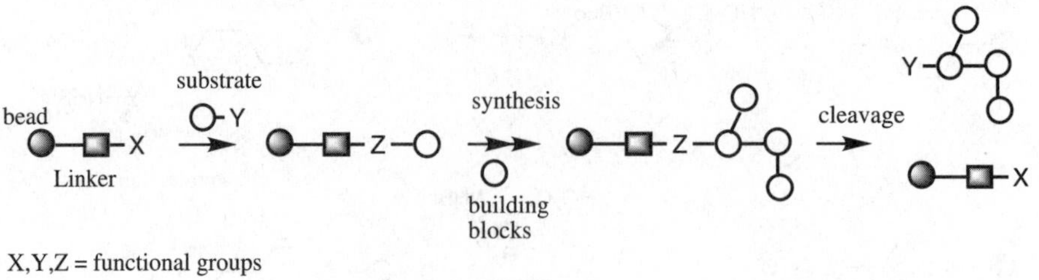

Fig. 12.3 Anchor/linker.

Resins having different linkers are given different names. For example, the Wang resin has a linker which is suitable for the attachment and release of carboxylic acids while the Rink resin is suitable for the attachment of carboxylic acids and the release of carboxamides (the linkage point is circled) (Fig. 12.4). The dihydropyran derivatized resin is suitable for the attachment and release of alcohols.

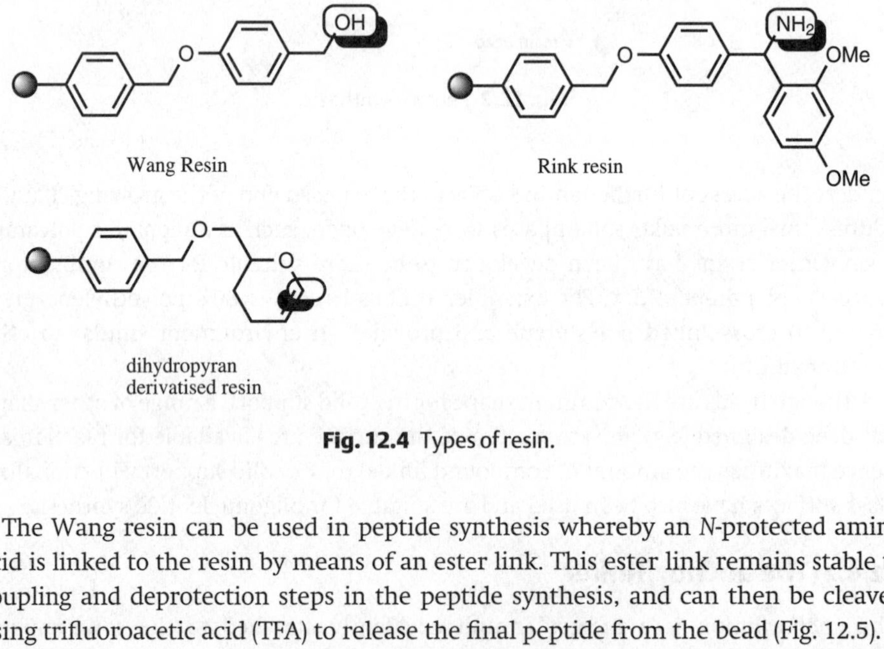

Fig. 12.4 Types of resin.

The Wang resin can be used in peptide synthesis whereby an *N*-protected amino acid is linked to the resin by means of an ester link. This ester link remains stable to coupling and deprotection steps in the peptide synthesis, and can then be cleaved using trifluoroacetic acid (TFA) to release the final peptide from the bead (Fig. 12.5).

Substrates with a carboxylic acid (RCO_2H) can be linked to the Rink resin via an amide link. Once the reaction sequence is complete, treatment with TFA releases the product with a primary amide group rather than the original carboxylic acid (**R'**$CONH_2$; Fig. 12.6).

Primary and secondary alcohols (ROH) can be linked to a dihydropyran-functionalized resin. Linking the alcohol is performed in the presence of pyridinium 4-toluenesulphonate (PPTS) in dichloromethane. Once the reaction sequence has been completed, cleavage can be carried out using trifluoroacetic acid (Fig. 12.7).

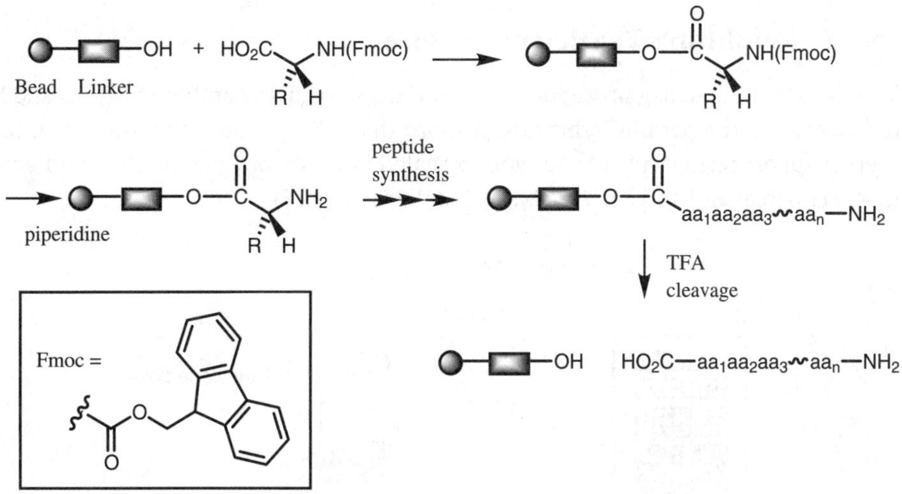

Fig. 12.5 Peptide synthesis with Wang resin.

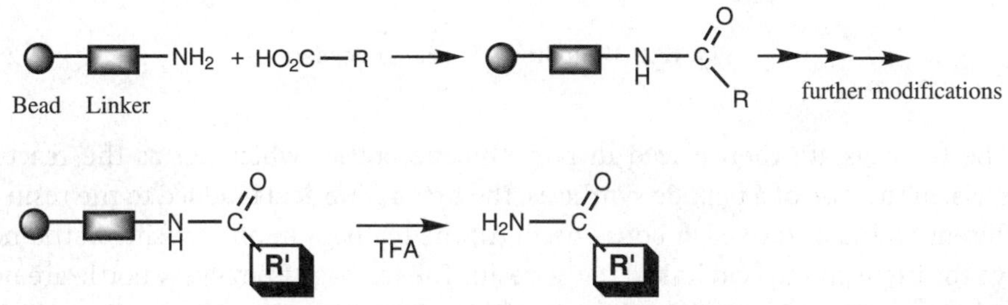

Fig. 12.6 Combinatorial synthesis with a Rink resin.

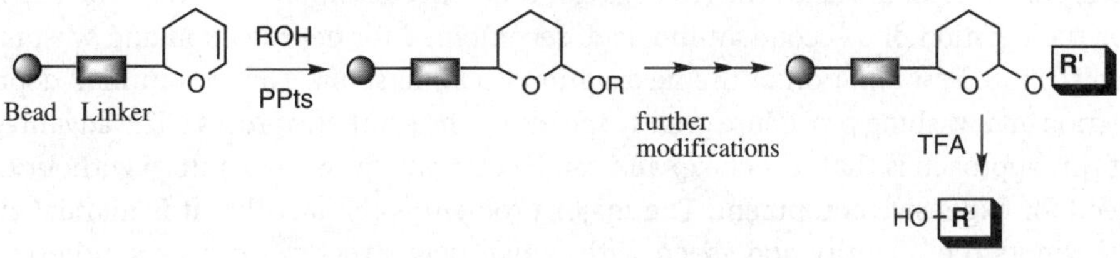

Fig. 12.7 Combinatorial synthesis with a dihydropyran functionalized resin.

12.5 Methods of parallel synthesis

Having looked at the basics of solid phase synthesis, we shall now look at some methods of combinatorial synthesis. The simplest examples involve parallel syntheses where a single reaction product is produced in each reaction vessel.

12.5.1 Houghton's Tea Bag procedure

The Houghton's Tea Bag procedure is a manual approach to parallel synthesis and has been used for the parallel synthesis of more than 150 peptides at a time. The polymeric support resin (typically 100 mg) is sealed in polypropylene meshed containers (3 × 4 cm) (teabags) and each tea bag is labelled (Fig. 12.8).

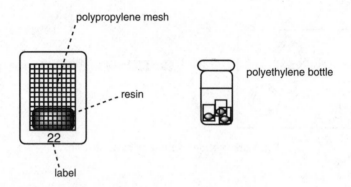

Fig. 12.8 Houghton's Tea Bag procedure.

The tea bags are then placed in polyethylene bottles which act as the reaction vessels. In the case of a peptide synthesis, the first amino acid is added to the resin – a different amino acid to each bottle used. All the tea bags in one specific bottle now have the same amino acid linked to the resin. The tea bags from every bottle are now combined in the one vessel for deprotection and washing. This allows all the amino acids to be deprotected at one time, avoiding the need to carry out deprotection separately on each amino acid. The tea bags can then be redistributed between the bottles for the addition of a second amino acid, recombined for deprotection and washing, redistributed for addition of the next amino acid and so on. The 'communal' deprotection and washing procedure greatly speeds up the synthetic process. The advantage of this approach is that it is cheap and can be carried out in a laboratory without the need for expensive equipment. The major problem is the fact that it is manual and this limits the quantity and speed with which new structures can be synthesized. Thus, pharmaceutical industries now use automation or semi-automation for the parallel synthesis of structures.

12.5.2 Automated parallel synthesis

Automated synthesizers (or laboratory robots) can cope with the parallel synthesis of 42, 96 or 144 structures, depending on the size of reaction tubes used. The solid phase support is in the form of sticks or pins (4 mm diameter, 40 mm length) which can be dipped into each reaction tube or well (Fig. 12.9). In the case of peptide synthesis, common operations such as the washing and deprotection of peptides is done by

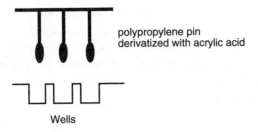

polypropylene pin
derivatized with acrylic acid

Wells

Fig. 12.9 Automated synthesis.

dipping the rods into large baths, but the coupling is done in the wells such that each well has a unique amino acid. About 80–300 nmol of peptides can be synthesized per rod. Other synthesizers use small reaction vials, and beads as the solid phase.

The addition of reagents and the removal of excess reagents can be carried out automatically. Reactions can be carried out under inert atmospheres, and the reactions can be heated or cooled as required.

These reactors are suitable both for multiple parallel synthesis of individual compounds where one substance is prepared per reaction flask, and multiple parallel syntheses where each reaction flask contains a mixture of different but structurally similar compounds (see Section 12.6).

12.6 Methods in mixed combinatorial synthesis

12.6.1 General principles

Combinatorial synthesis is often designed to produce a mixture of products in each reaction vessel, starting with a wide range of starting materials and reagents. This does not mean that all possible starting materials are thrown together in one reaction flask. Planning has to go into designing a combinatorial synthesis to minimize the effort involved and to maximize the number of different structures obtained.

As an example, suppose you wish to synthesize all the possible dipeptides of five different amino acids. Using orthodox chemistry, you would synthesize these one at a time. There are 25 possible dipeptides and so you would have to carry out 25 separate experiments (Fig. 12.10).

Glycine (Gly)	25 separate	Gly-Gly	Ala-Gly	Phe-Gly	Val-Gly	Ser-Gly
Alanine (Ala)	experiments	Gly-Ala	Ala-Ala	Phe-Ala	Val-Ala	Ser-Ala
Phenylalanine (Phe)	⟶	Gly-Phe	Ala-Phe	Phe-Phe	Val-Phe	Ser-Phe
Valine (Val)		Gly-Val	Ala-Val	Phe-Val	Val-Val	Ser-Val
Serine (Ser)		Gly-Ser	Ala-Ser	Phe-Ser	Val-Ser	Ser-Ser

Fig. 12.10 Traditional synthesis of dipeptides.

However, using combinatorial chemistry the same products could be obtained with far less effort. If all five different amino acids are separately bound to resin beads, the beads can then be mixed together and treated with a second amino acid to produce all possible dipeptides in five experiments. For example, in one experiment the five different amino acids could be combined with glycine to produce five of the 25 possible dipeptides (Fig. 12.11).

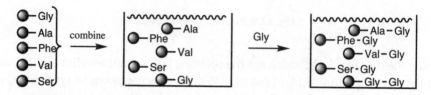

Fig. 12.11 Synthesis of five different dipeptides.

This mixture could then be tested for activity. If the results were positive, the emphasis would be on identifying which of the dipeptides was active. If there was no activity present, then the mixture could be ignored and stored.

In studies such as these, one can generate large numbers of mixtures, many of which are inactive. However, these mixtures are not discarded. Although they may not contain a lead compound on this particular occasion, they may provide the necessary lead compound for a different field of medicinal chemistry. Therefore, all the mixtures (both active and inactive) resulting from a combinatorial synthesis are stored and are referred to as Combinatorial or Compound Libraries.

The example we gave above produced 25 compounds in five mixtures. However, using combinatorial synthesis, it is possible to produce different structures numbering in the order of thousands or millions. Since the quantities involved are extremely small (perhaps only five beads containing one individual structure), even huge numbers like these can be stored and used for later studies. Because of the vast number and the small quantities of compounds present in these mixtures, the exact structure of each component in a mixture is not known, although the chemist has a general idea of the type of structure likely to be present based on the type of synthesis carried out. The library acts as a source of potential new leads in the same way as a plant or herb would.

12.6.2 The mix and split method

When generating large quantities of different structures, it is important to minimize the effort involved, and the mix and split method is a popular way of doing this. An example best illustrates the principle. Let us assume that we want to make all the possible tripeptides of three different amino acids (e.g. Gly, Val, Ala). The mix and split method would work like this.

Stage 1

Link each amino acid to a solid support (Fig. 12.12). (To clarify the diagrams, each sphere represents a resin bead with suitable linker unit.)

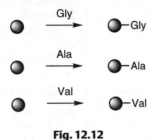

Fig. 12.12

Stage 2

Mix the beads together and separate into three equal portions (Fig. 12.13).

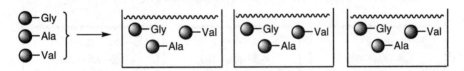

Fig. 12.13

Stage 3

React each portion with a different amino acid (Fig. 12.14).

All possible nine dipeptides have now been synthesized in three separate experiments. Samples of each portion could be retained for **recursive deconvolution** (see later).

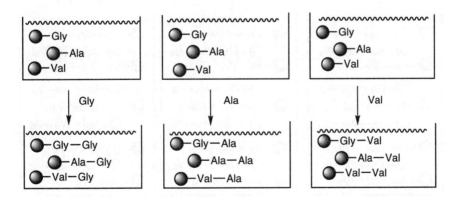

Fig. 12.14

Stage 4

Isolate all the beads, mix them all together and split into three equal portions. Each portion will now have all nine possible dipeptides (Fig. 12.15).

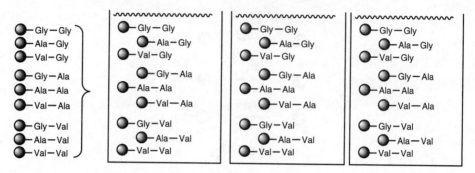

Fig. 12.15

Stage 5

React each portion with the three amino acids (Fig. 12.16).

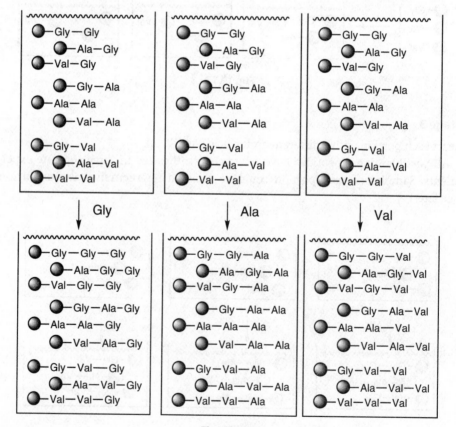

Fig. 12.16

All 27 possible tripeptides have now been synthesized in another three experiments. Thus 27 compounds have been obtained in six experiments as opposed to 27.

In this example, amino acids were linked together, but any monomer unit or combination of chemical structures could be linked together using the same strategy, and so the process is not limited to peptide synthesis.

12.6.3 Mix and split in the production of positional scanning libraries

A variation of the mix and split method allows the creation of positional scanning libraries. In this method, the same library compounds are prepared several times and in each library a different residue in the sequence is held constant. For example a series of hexapeptide libraries totalling 34 million compounds was produced, each library consisting of six sets of mixtures, and with each mixture containing 1 889 568 peptides. In each of the mixtures one of the amino acid positions was held constant.

The first library consisted of six mixtures in which the first amino acid was constant *within* each individual mixture but was different *between* mixtures. The most active mixture was then identified and, since the amino acid at position 1 was constant in that mixture, it could be identified. The second library consisted of a series of six mixtures in which the second amino acid was constant within each mixture but different between the mixtures. Thus it was known which amino acid was present at position 2 in the most active of these mixtures.

Testing all the mixtures in all the libraries revealed the preferred residues at each of the six residue positions.

However, one has to be careful here. Although the most active amino acid at each position can be identified, it does not mean that the most active structure is the one linking each of these amino acids. After all, the activity being measured is the combined effect of several different hexapeptide structures in the active mixtures. Therefore, it is quite possible that the most active hexapeptide is in a mixture which has a lower overall activity than another mixture, since the latter might contain a large number of compounds with moderate activity.

One way around this is to identify the top three to four amino acids at each position rather than just the most active one. Once these have been identified, all the possible variations of these amino acids could be synthesized.

For example, several libraries of a hexapeptide were prepared in a search for structures which would bind to the μ opiate receptor. The most active amino acids at each position were found to be Try (Y), Gly (G), Phe (F), Phe (F), Leu (L), and Arg (R). Linking these together gave a hexapeptide (YGFFLR) which was only weakly active. However, the results showed that mixtures containing Gly or Phe at position 3; Phe, Tyr, Met or Leu at position 5; and Phe, Tyr, or Arg at position 6 were also active.

Therefore all hexapeptides having the sequence Y,G,G/F, F, F/M/L, F/Y/R were made, with the most active being YGGFMY.

12.7 Isolating the active component in a mixture – deconvolution

Assuming that a compound mixture proves to be biologically active, the tricky job of identifying the active component (or components) now needs to be carried out. Isolation and identification of the most active compound in a mixture is known as **deconvoluting** the mixture. There are several methods of doing this.

12.7.1 Micromanipulation

Each bead in a mixture contains only one type of structural product. Therefore, the individual beads can be separated, the product cleaved and then tested. This procedure can be aided by colorimetric analysis in which products are tested for activity while they are still bound to the beads. The active beads are distinguished by a colour reaction and can then be 'picked out' by micromanipulation.

12.7.2 Recursive deconvolution

Micromanipulation is tedious and has serious drawbacks when handling large quantities of beads. A method known as recursive deconvolution can be useful in cutting down the amount of work involved and can be illustrated by considering the library of tripeptides described in Section 12.6.2. Here we synthesized three mixtures. Let us assume that one of these three mixtures shows activity. How do we find out which of the nine possible tripeptides is the active component? We could synthesize all nine possible tripeptides separately and test each one. However, we could cut down the work if samples of the dimer mixtures produced during the combinatorial synthesis were retained. The procedure involved is known as recursive deconvolution (Fig. 12.17).

Let us assume that the third tripeptide mixture in Section 12.6.2 showed activity. This means that the active tripeptide has valine at the *N*-terminus. The next stage is to take the three dipeptide mixtures which were retained and to link valine to each mixture. This gives us the nine tripeptides we need in three separate mixtures.

Now the second and third amino acids are the same in each mixture. The three mixtures can now be tested. If one of these mixtures is active then we can identify the second and third amino acids. Let us assume that the mixture containing Ala and Val is the active mixture. The three component tripeptides in this mixture can now be individually synthesized and tested.

In this example, we looked at tripeptides involving three different amino acids, but typical combinatorial syntheses involve a larger variety of monomeric units, and the larger the number of monomers the larger the number of compounds produced. For example, a series of 34 million hexapeptides was synthesized by this method using 18 amino acids and 324 mixtures.

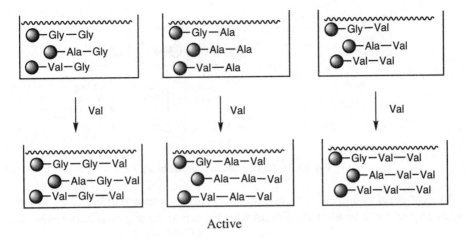

Active

Fig. 12.17 Recursive deconvolution.

Appropriate use of the mix and split method, and the retention of intermediate mixtures for recursive deconvolution, is crucial in economizing the effort involved.

12.7.3 Sequential release

Linkers have been devised which allow release of a certain percentage of the product from the bead. The process can be repeated to release another percentage of product and so the product is released sequentially rather than all at once. Therefore, a mixture of beads can be treated to release some of their bound product, which is then tested in solution. If the mixture is active, the same beads are split into smaller mixtures and further product is released from the beads and tested (Fig. 12.18).

This process can be repeated several times until the active bead is identified (Fig. 12.19).

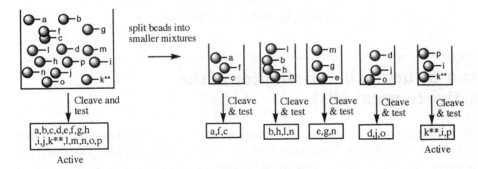

Fig. 12.18 Sequential release.

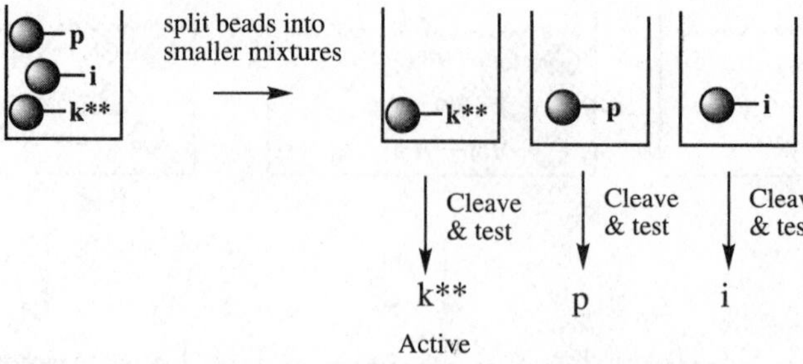

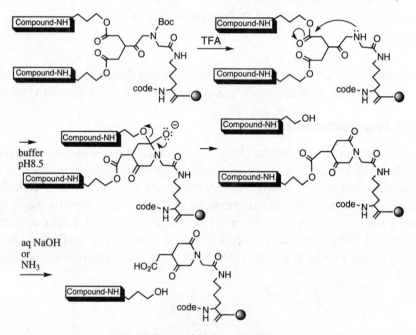

Fig. 12.19 Identification of active bead.

An example of a double cleavable linker for peptides is shown in Fig. 12.20. The first cleavage is initiated by addition of neutral buffer, while the second is initiated by base.

Fig. 12.20 Double cleavable linker.

12.8 Structure determination of the active compound(s)

The direct structural determination of components in a compound mixture is no easy task. However, advances have been made in obtaining interpretable mass, nuclear magnetic resonance (NMR), Raman, infrared and ultraviolet spectra on products attached to a single resin bead. In the case of peptides, peptide sequencing can be used to determine the sequence, with the peptide still attached to the bead. Each 100-micrometer bead contains approximately 100 picomoles of peptide which is sufficient for microsequencing. With non-peptides, the structural determination of an active compound can be achieved by a systematic iterative resynthesis as described in Section 12.6. However, this can be tedious.

Alternatively, **tagging** procedures can be used during the synthesis.

12.8.1 Tagging

In this process, two molecules are built up on the same bead. One of these is the new structure to be tested while the other is a molecular tag (usually a peptide or oligonucleotide). This tag will act as a code for each step of the synthesis. For this to work, the bead must have a multiple linker capable of linking both the structure being synthesized and the molecular tag. A reagent is added to one part of the linker, and an encoding amino acid (or nucleotide) to another part of the linker. After each subsequent stage of the combinatorial synthesis, an amino acid (or nucleotide) is added to the tag to indicate what reagent was used.

One example of a multiple linker is called the safety catch linker (SCAL) (Fig. 12.21), which includes lysine and tryptophan. Both these amino acids have a free amino group.

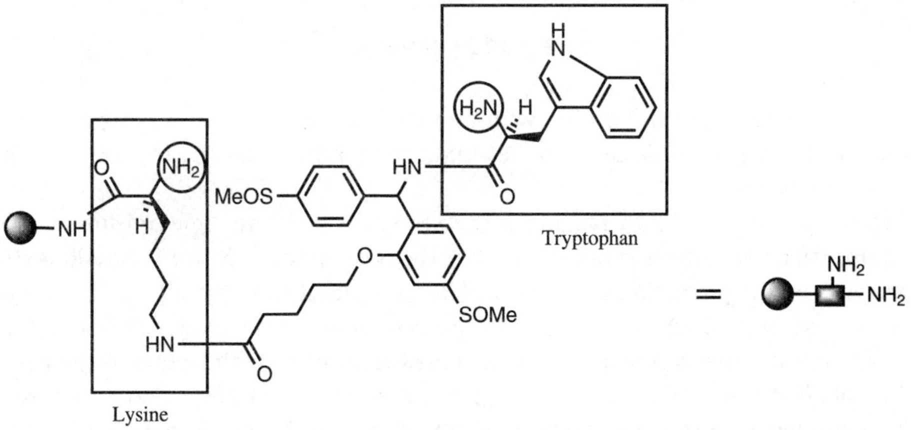

Fig. 12.21 Safety catch linker.

The compound to be synthesized is constructed on the tryptophan moiety and, after each stage of the synthesis, a tagging amino acid is built onto the lysine moiety so that by the end of the synthesis there is a tripeptide present in which each amino acid defines the identity of the variable groups R, R' and R" in the non-peptide structure (Fig. 12.22).

The non-peptide structure can be cleaved by reducing the two sulphoxide groups in the safety catch linker, then treating with acid. Under these conditions, the tripeptide sequence remains attached to the bead and can be sequenced on the bead to identify the structure of the compound which has been released.

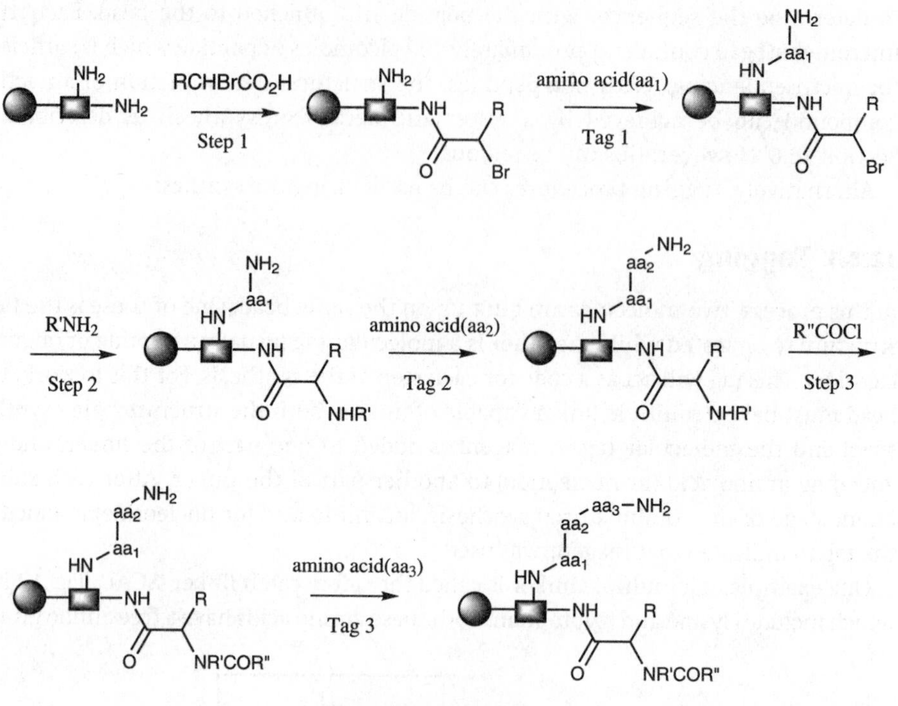

Fig. 12.22 Tagging.

The same strategy can be used with an oligonucleotide as the tagging molecule instead. This tagging code can then be amplified by replication and the code read by DNA sequencing.

There are drawbacks to tagging processes since they are time-consuming and require elaborate instrumentation. Building the coding structure itself also adds extra restraints on the protection strategies which can be used and may impose limitations on the reactions which can be used. In the case of oligonucleotides, their inherent instability can prove a problem. Another possible problem with tagging is the possibility of an unexpected reaction taking place, resulting in a different structure from that expected. Nevertheless, the tagging procedure is still valid since it identifies the starting materials and the reaction conditions and, when these are repeated on larger scale, any unusual reactions would be discovered.

12.8.2 Encoded sheets

Pfizer have developed a method which allows ready separation of the individual solid phase products and which includes an inbuilt code to determine the synthetic history. Resin beads are sandwiched between two woven sheets of inert polypropylene and the sheets are fused together so that the beads are immobile. These sheets can then be marked into squares and each square is given a three letter code. For example, in Fig. 12.23, three such sheets measuring 6 cm × 6 cm are marked with nine squares and each square is given a three letter code. The three sheets are separated and each sheet is treated with an amino acid. Thus all the beads in the top sheet now have leucine attached. All the beads in the second sheet have serine attached and all the beads in the bottom sheets have glycine attached. The sheets can be washed, dried, and treated with piperidine to remove the Fmoc protecting group.[1]

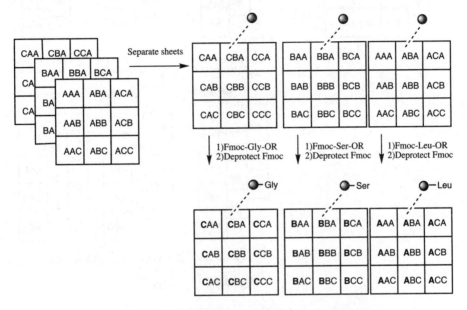

Fig. 12.23 Pfizer's encoded sheets.

In stage 2 (Fig. 12.24), the sheets are restacked as before, then cut into strips of three to give three sets of columns. Each set of columns is treated with another activated Fmoc amino acid to generate a unique dipeptide on each strip of laminar material.

The strips can be treated as before and the Fmoc protecting group removed.

The strips are restacked and cut into individual squares (Fig. 12.25). These are divided by rows this time to form three sets of squares. Each set of squares is now treated with the third amino acid. In this way, all 27 possible tripeptides are synthesized, with each square containing a unique tripeptide sequence which is identified by the code present on the square.

12.8.3 Photolithography

Photolithography is a technique which permits miniaturization and spatial resolution such that specific products are synthesized on a plate of immobilized solid support. In the synthesis of peptides, the solid support surface contains an amino group protected

---

[1] The structure of the Fmoc protecting group is shown in Fig. 12.5.

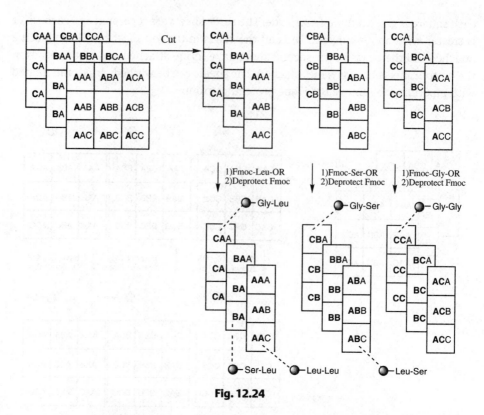

Fig. 12.24

by the photolabile protecting group nitroveratryloxycarbonyl (NVOC) (Fig. 12.26). Using a mask, part of the surface is exposed to light, resulting in deprotection of the exposed region. The plate is then treated with a protected amino acid and the coupling reaction takes place only on the region of the plate which has been deprotected. The plate is then washed to remove excess amino acid. The process can be repeated on a different region using a different mask and so different peptide chains can be built on different parts of the plate; the sequences are known based on the record of masks used.

Incubation of the plate with a protein receptor can then be carried out to detect active compounds which bind to the binding site. A convenient method to assess such interactions is to incubate the plate with a fluorescently tagged receptor. Only those regions of the plate which contain active compounds will bind to the receptor and fluoresce. The fluorescence intensity can be measured using fluorescence microscopy and is a measure of the affinity of the compound for the receptor. Alternatively, testing can be carried out such that active compounds are detected by radioactivity or chemiluminescence.

The photodeprotection described above can be achieved in high resolution. At a 20-micron resolution, plates can be prepared with 250 000 separate compounds per square centimetre.

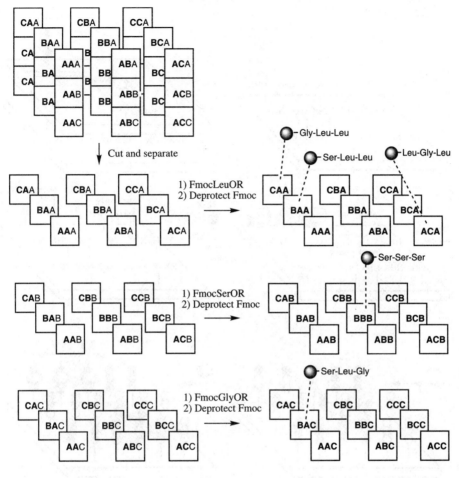

Fig. 12.25

12.9 Limitations of combinatorial synthesis

In principle, combinatorial chemistry could be used to synthesize all the 10 240 billion possible decapeptides. However, there are limitations since one has to consider the practical details of weight and volume. First, how many beads will be needed for a combinatorial synthesis? For statistical reasons, the number of beads should exceed the number of target molecules by a factor of 10. Otherwise, one might not sample all the possible structures present. For example, if there are only five beads representing each of the 3.2 million components of a pentapeptide library and one fifth of the whole mixture is removed as a sample, then the probability of finding all the peptides present in that sample is only 63%.

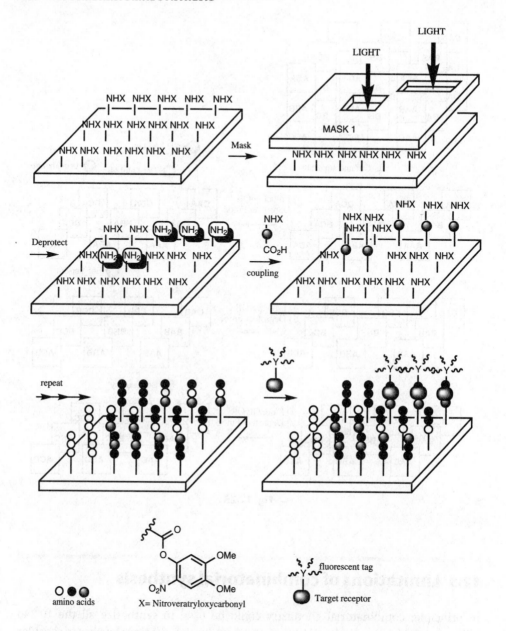

Fig. 12.26 Photolithography.

Assuming that one uses the required excess of beads, the weight of beads required to make a complete library of dipeptides would be 8.4 mg. To make a complete library of tetrapeptides you would need 3.4 g which is still practical. However, to make a complete library of decapeptides you would need 215.3 tons!

12.10 Examples of combinatorial chemistry

Combinatorial chemistry has proved its worth in throwing up new lead compounds in a variety of fields. Much of the early work in combinatorial chemistry was carried out on peptides since the solid phase procedures had already been developed. This resulted in the discovery of new HIV protease inhibitors, antimicrobial agents, opiate receptor ligands, and aspartic acid protease inhibitors. However, peptides are not ideal drug candidates since they usually have poor oral activity because of metabolism by digestive enzymes.

The first move away from peptides was to use the same peptide coupling procedures but using non-natural amino acids. Peptides could also be modified once they were built by reactions such as N-methylation. Peptides have also been built linking N-substituted glycine units to produce structures known as **peptoids** where the side chain is attached to the nitrogen rather than the α-carbon. Some of these have been shown to be ligands for various important receptors and show increased metabolic stability.

However, there is currently more interest in the creation of heterocyclic combinatorial libraries. One of the earliest examples was the synthesis of 1,4-benzodiazepines – an ideal synthesis since three distinct units are brought together. The final product has five variable substituents, two of which can be positionally varied on the aromatic ring (Fig. 12.27). Piperazinediones, 2,5-disubstituted tetrahydrofurans, and thiazolidines have also been synthesized, and many more examples are published each year.

The sort of reactions which can be carried out on solid phase has also been extended such that most common reactions are now feasible, and even include moisture-sensitive and organometallic reactions (e.g. Aldol, Dibal reduction, Wittig, LDA reduction,[2] Heck coupling, Stille coupling, Mitsunobu reaction).

When moving into the field of heterocyclic synthesis, it is important to spend some time optimizing the reaction conditions before launching into a full scale library synthesis. Otherwise, the reactions might proceed in a totally different way from expected or might not take place at all. Since there is no easy way of analysing what is happening on the beads, this could lead to millions of useless compounds being synthesized before the problem is identified. Therefore, it is a good idea to carry out some model syntheses using a couple of 'worst case' scenarios – structures which might not be expected to react well because of steric or electronic factors. These model studies can be used to optimize conditions or to avoid using monomers which are particularly unreactive. When the library synthesis is being carried out, a parallel synthesis of a single compound by solid phase should be carried out as a further check.

---

[2] LDA = Lithium diisopropylamide.

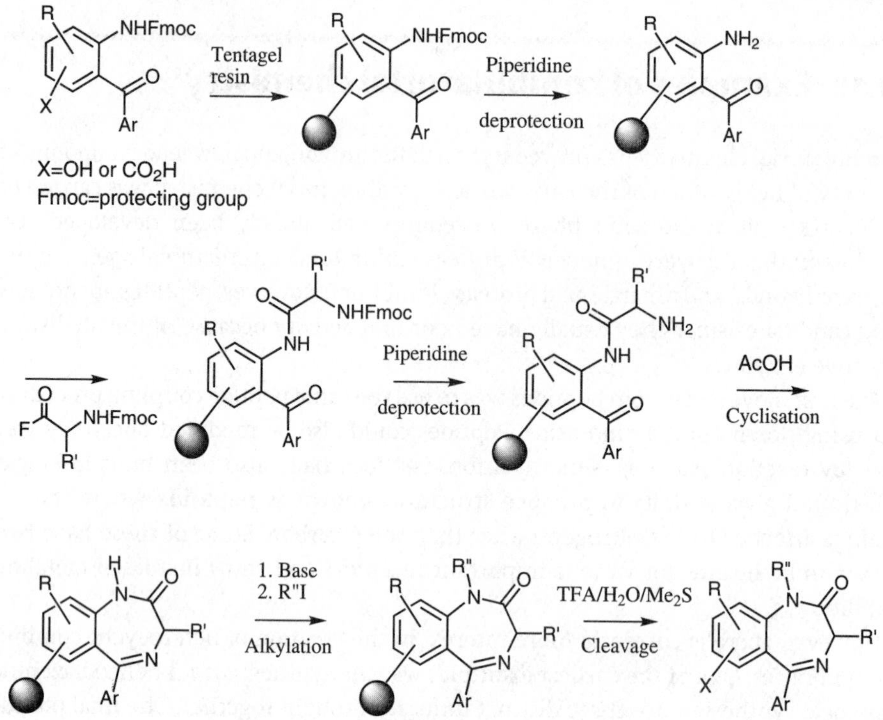

Fig. 12.27 1,4-Benzodiazepine synthesis.

12.11 Planning and designing a combinatorial synthesis

12.11.1 'Spider-like' scaffolds

To find a new lead compound from combinatorial synthesis, we need to generate a large number of different structures, but we also need to ensure that these are as diverse in structure as possible. This may not seem very likely if we are restricted to using a single reaction sequence. However, if we are careful in the type of molecule we synthesize and the method by which we synthesize it, then such diversity *is* possible. In general, it is best to synthesize 'spider-like' molecules, so called because they consist of a central body (called the centroid or scaffold) from which various 'arms' (substituents) are placed.

These 'arms' contain different functional groups which are used to probe the binding site for binding regions once the spider-like molecule has entered (Fig. 12.29). The chances of success are greater if the arms are evenly spread around the scaffold since this allows a more thorough exploration of the 3D space (conformational space) around the molecule. The molecules made in the synthesis would also be planned in

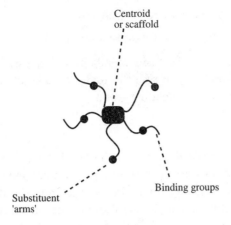

Fig. 12.28 'Spider-like' molecule.

advance to ensure that they would contain different functional groups on their 'arms', and at different distances from the central scaffold.

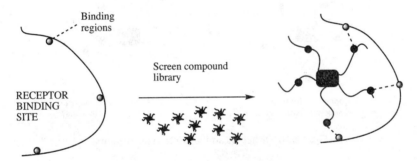

Fig. 12.29 Probing for an interaction.

12.11.2 Designing 'drug-like' molecules

The 'spider-like' approach increases the chances of finding a lead compound which will interact with the target receptor or enzyme, but it is also worth remembering that compounds with good binding interactions do not necessarily make good medicines. There are also the pharmacokinetic issues to be taken into account (see Chapter 11). It is worthwhile, therefore, introducing certain restrictions to the types of molecule which will be produced by the combinatorial synthesis to increase the chance that the lead compound will be orally active. In general, the chances of oral activity are increased if the structure has the following:

• a molecular weight less than 500
• a calculated log P value of less than $+5$

- no more than five hydrogen bond donating groups
- no more than 10 hydrogen bond accepting groups.[3]

Groups which are liable to be easily metabolized (e.g. esters) should be avoided. Scaffolds or substituents likely to result in toxic compounds should also be avoided (e.g. aromatic nitro groups or alkylating groups).

12.11.3 Scaffolds

Most scaffolds are synthesized by the synthetic route used for the combinatorial synthesis, and the synthesis used determines the number and variety of substituents which can be attached to the scaffold. The ideal scaffold should be small to allow a wide variation of substituents. It should also have its substituents widely dispersed round its structure (spider-like) rather than restricted to one part of the structure (tadpole-like) if the conformational space around it is to be fully explored (Fig. 12.30). Finally, the synthesis should allow each of the substituents to be varied independently of the other.

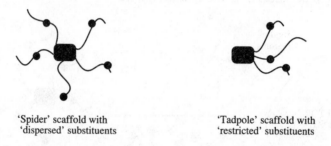

'Spider' scaffold with
'dispersed' substituents

'Tadpole' scaffold with
'restricted' substituents

Fig. 12.30 Dispersed and restricted substituents.

Scaffolds can be flexible (e.g. a peptide backbone) or rigid (a cyclic system). They may contain groups which are capable of forming useful bonding interactions with the binding site or they may not. Some scaffolds are already common in medicinal chemistry (e.g. benzodiazepine, hydantoin, tetrahydroisoquinoline, and benzenesulphonamide), and are associated with a diverse range of activities. Such scaffolds are termed 'privileged' scaffolds.

The following examples of scaffolds (Fig 12.31) illustrate some of the principles described above. Benzodiazepines, hydantoins, β-lactams and pyridines are examples of extremely good scaffolds. They all have small molecular weights and there are various synthetic routes available which allow the substitution patterns needed to explore the conformational space fully. For example, the synthesis in Fig. 12.26 produces a bicyclic benzodiazepine scaffold. The conformational space around this scaffold can be explored by varying the five different substituents (R, X, R', R'' and Ar).

---

[3] These requirements are known as the 'Rule of Five' since the numbers involved are multiples of five.

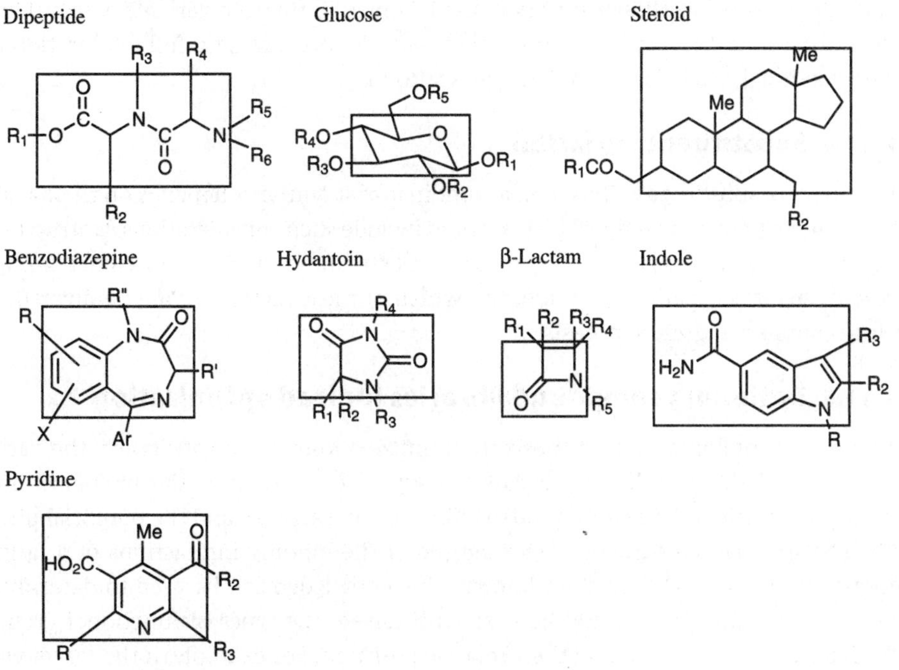

Fig. 12.31 Examples of scaffolds.

Peptide scaffolds are flexible scaffolds which have the capacity to form hydrogen bonds with target binding sites. They are easy to synthesize and a large variety of different substituents is possible by using the amino acid building blocks. Further substitution is possible both on the terminal amino and carboxylic acid functions. The substituents are widely distributed along the peptide chain, allowing a thorough exploration of conformational space. However, if we consider our 'rules', then the peptide scaffold should ideally be restricted to dipeptides and tripeptides to keep the molecular weight below 500. It is interesting to note that the antihypertensive agents captopril and enalapril are dipeptide-like and are orally active, whereas larger peptides such as the enkephalins are not orally active.

The remaining scaffolds shown have various disadvantages. Glucose might be considered a potential scaffold. It has a small molecular weight and has the possibility of having five substituents around the ring. However, glucose suffers a severe disadvantage in that it contains multiple hydroxyl groups. Attaching different substituents to similar groups would therefore require complex protection and deprotection strategies.

Steroids might also appear attractive as scaffolds. However, the molecular weight of the steroid skeleton itself (314) limits the size of the substituents which can be added if we wish to keep the overall molecular weight below 500. Furthermore, there are relatively few positions in which substituents can be easily attached. On both accounts, it is not possible to explore fully the conformational space around the steroid scaffold.

The indole scaffold shown suffers a disadvantage in that the variable substituents are all located in the same region of the molecule, preventing a full exploration of conformational space (i.e. it is a 'tadpole' scaffold).

12.11.4 Substituent variation

The variety of substituents chosen in a combinatorial synthesis depends on their availability and the diversity required. This would include such considerations as structure, size, shape, lipophilicity, dipole moment, electrostatic charge, functional groups present, etc. It is usually best to identify which of these factors should be diversified before commencing the synthesis.

12.11.5 Designing compound libraries for lead optimization

When using combinatorial synthesis to optimize a known lead structure, the variations planned should take into account several factors such as the biological and physical properties of the compound, its binding interactions, and the potential problems of particular substituents. For example, if the binding interactions of a target receptor with its usual ligand are known, this knowledge can be used to determine what size of compounds would be best synthesized, the types of functional groups which ought to be present, and their relative positions. For example, if the target was a zinc-containing protease (e.g. angiotensin converting enzyme), a library of compounds containing a carboxylic acid or thiol group would be relevant.

12.11.6 Computer-designed libraries

It has been claimed that half of all known drugs involve only 32 scaffolds. Furthermore, it has been stated that a relatively small number of moieties account for the large majority of side chains in known drugs. This may imply that it is possible to define 'drug-like molecules' and so use computer software programmes to design more focused combinatorial compound libraries. Descriptors used in this approach include log P, molecular weight, number of hydrogen bond donors, number of hydrogen bond acceptors, number of rotatable bonds, aromatic density, the degree of branching in the bonding, and the presence or absence of specific functional groups. One can also choose to filter out compounds which do not obey the rules mentioned in 12.11.2.

Computer programmes can also be used to identify the structures which should be synthesized to maximize the number of different pharmacophores produced (see Chapter 13).

12.12 Testing for activity

We shall now look in more detail at how the products from combinatorial synthesis are tested for biological activity.

12.12.1 High throughput screening

Since combinatorial synthesis produces a large quantity of structures in a very short time period, biological testing has to be carried out quickly and automatically. The process is known as high throughput screening (HTS) and was developed in advance of combinatorial synthesis. Indeed, one of the pressures to develop combinatorial synthesis was the existence of HTS. Since biological testing was so rapid and efficient, the pharmaceutical companies soon ran out of novel structures to test, and the synthesis of new structures became the 'logjam' in the whole process of drug discovery. Combinatorial synthesis solved that problem and the number of new compounds synthesized each year has increased dramatically by several factors. In fact, the issue has now come full circle. There are now so many compounds being produced that the focus is on making HTS even more efficient. Traditionally, compounds are automatically tested and analysed on a plate containing 96 small wells with a capacity of 0.1 ml. There is now a move to test plates of similar size but which contain 1536 wells, where the test volumes are reduced to 1–10 μl. Moreover, methods such as fluorescence and chemiluminescence are being developed which will allow the simultaneous identification of active wells. Further miniaturization of open systems is unlikely because of the problems of evaporation of small volumes (less than 1 μl). However, miniaturization using closed systems is on the horizon. The next major advance will involve the science of microfluidics, which involves the manipulation of tiny volumes of liquids in confined space. Microfluidic circuits on a chip can be used to control fluids electronically, allowing separation of an analytical sample using capillary electrophoresis. Companies are now developing machines which combine ultra small scale synthesis and miniaturized analysis. A single 10×10 cm silicon wafer can be microfabricated to support 10^5 separate syntheses/bioassays on a nanolitre scale!

12.12.2 Screening 'on-bead' or 'off-bead'

Sometimes structures can be tested for biological activity when they are still attached to the solid phase. 'On-bead' screening assays involve interactions with targets which are tagged with an enzyme, fluorescent probe, radionuclide, or a chromophore. A positive interaction results in a recognizable effect such as fluorescence or a colour change. These screening assays are rapid and 10^8 beads can be readily screened. Active beads could then be 'picked' out by micromanipulation and the structure of the active compound determined.

However, a false-negative might be obtained if the solid phase sterically interferes with the assay. If such interference is suspected, it is better to release the drug from the solid phase before testing. This avoids the uncertainty of false-negatives. However, there are cases where the compounds released prove to be insoluble in the test assay and give a negative result, whereas they give a positive result when attached to the bead.

13 | Computers in medicinal chemistry

13.1 Introduction

Computers are an essential tool in modern medicinal chemistry and are important both in drug discovery and development. Rapid advances in computer hardware and software have meant that many of the operations which were once the exclusive province of the expert can now be carried out on ordinary laboratory computers with little specialist expertise in the molecular or quantum mechanics involved. In the next few sections, we shall look at some examples of how computers are used in medicinal chemistry. However, it has to be appreciated that it is not possible to do full justice to this subject in a single chapter, and the author has been fairly selective in what has been included. Moreover, the pace of change is such that the material reported here could well be out of date within a few months of publication, if not before!

13.2 Molecular and quantum mechanics

The various operations carried out in molecular modelling involve the use of programmes or algorithms which calculate the structure and property data for the molecule in question. For example, it is possible to calculate the energy of a particular arrangement of atoms (conformation), modify the structure to create an energy minimum and calculate properties such as charge, dipole moment, and heat of formation. The mathematical details of these operations are too involved to be included in an introductory text, but it is important to appreciate a few general principles about how these processes are carried out. The computational methods used to calculate structure and property data can be split into two categories – molecular mechanics and quantum mechanics.

13.2.1 Molecular mechanics

In molecular mechanics, equations are used which follow the laws of classical physics and apply them to molecular nuclei without consideration of the electrons. In essence, the molecule is treated as a series of spheres (the atoms) connected by springs (the bonds). Equations derived from classical mechanics are used to calculate the different interactions and energies (force fields) resulting from bond stretching, angle bending, torsional energies, and non-bonded interactions. These calculations require data or parameters which are stored in tables within the programme and which describe interactions between different sets of atoms. The energies calculated by molecular mechanics have no meaning as absolute quantities but are useful when comparing different conformations of the same molecule. Molecular mechanics is fast and less intensive on computer time than quantum mechanics. However, it cannot calculate electronic properties since electrons are not included in the calculations. Operations involving the use of the MM2 programme in molecular modelling software involve molecular mechanics.

13.2.2 Quantum mechanics

Quantum mechanics uses quantum physics to calculate the property of a molecule by considering the interactions between the electrons and nuclei of the molecule. Unlike molecular mechanics, atoms are not treated as solid spheres. To make the calculations feasible, various approximations have to be made. First, nuclei are regarded as motionless. This is reasonable since the motion of the electrons is much faster in comparison. Since electrons are considered to be moving around fixed nuclei, it is possible to describe electronic energy separately from nuclear energy. Second, it is assumed that the electrons move independently of each other and so the influence of other electrons and nuclei is taken as an average.

Quantum mechanical methods can be subdivided into two broad methods – *ab initio* or semi-empirical. The former is more rigorous and does not require any stored parameters or data. However, it is expensive on computer time and is restricted to small molecules. Semi-empirical methods are quicker though less accurate, and can be carried out on larger molecules. There are various forms of semi-empirical programme (i.e. programmes such as MINDO/3, MNDO, MNDO-d, AM1 and PM3). These methods are quicker since they use further approximations and make use of stored parameters.

13.2.3 Choice of method

The method of calculation chosen depends on what calculation needs to be done, as well as the size of the molecule. As far as size of molecule is concerned, *ab initio* calculations are limited to molecules containing tens of atoms, semi-empirical calculations on molecules containing hundreds of atoms, and molecular mechanics on molecules containing thousands of atoms.

Molecular mechanics is useful for the following operations or calculations:

- energy minimization
- identifying stable conformations
- energy calculations for specific conformations
- generating different conformations
- studying molecular motion.

Quantum mechanical methods are suitable for calculating the following:

- molecular orbital energies and coefficients
- heat of formation for specific conformations
- partial atomic charges calculated from molecular orbital coefficients
- electrostatic potentials
- dipole moments
- transition state geometries and energies
- bond dissociation energies.

13.3 Drawing chemical structures

Chemical drawing packages do not require the calculations described in Section 13.2, but they are often integrated into molecular modelling programmes. In the not so distant past, drawing chemical structures for a report or a scientific paper was a rather tedious business which involved tracing the skeleton of the molecule with templates, then using a typewriter to add elements and substituents. Positioning the paper in the typewriter to get the substituent at the correct position was quite an art! Various software packages, such as ChemDraw, ChemWindow, and Isis Draw, are now available which can be used to construct diagrams quickly and with a professional result. For example, the diagrams in this book have all been prepared using the ChemDraw package.

Some drawing packages are linked to other items of software which allow quick calculations of various molecular properties. For example, once the structure of adrenaline was created in ChemDraw Ultra, the structure's correct IUPAC chemical name was easily obtained, as well as its molecular formula, molecular weight, exact mass, and theoretical elemental analysis. It was also possible to get calculated predictions of the compound's ^{1}H and ^{13}C NMR chemical shifts, melting point, freezing point, log P value, molar refractivity, and heat of formation (Fig. 13.1).

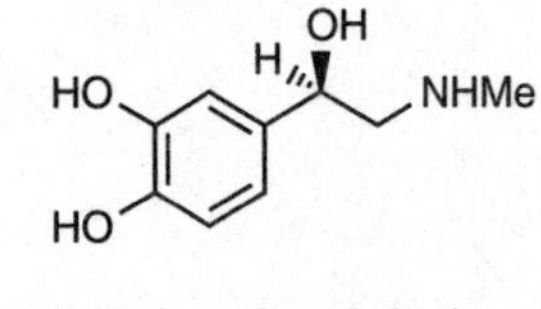

Calculated Properties
$C_9H_{13}NO_3$
Exact Mass: 183.09
Mol. Wt.: 183.20
C, 59.00; H, 7.15;
N, 7.65; O, 26.20

Predicted properties
LogP = -0.61 - 0.63
Molar refractivity
  48.66-49.08 [cm.cm.cm/mol]
b.pt. 618.55K;
Freezing point 539.03 K
Heat of formation
  -451.22 kJ/mol

4-(1-Hydroxy-2-methylamino-ethyl)-
benzene-1,2-diol

Predicted ^{13}C nmr

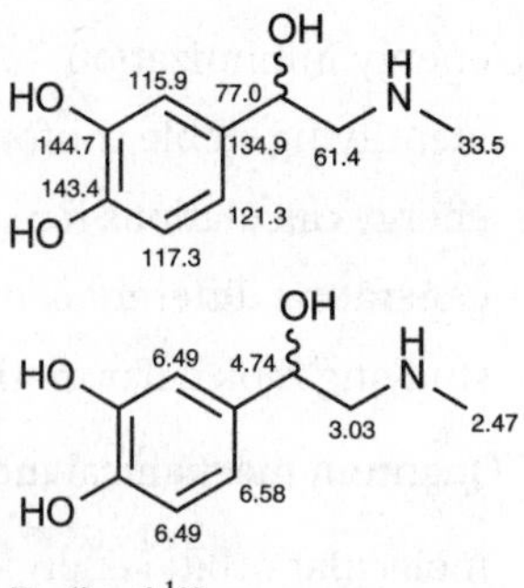

Predicted ^{1}H nmr

Fig. 13.1 Drawing chemical structures.

13.4 3D structures

Molecular modelling software allows the chemist to construct a 3D molecular structure on the computer. There are several software packages available, such as Chem3D, Alchemy, Sybyl, Hyperchem, ChemX, and CAChe. The 3D model can be made by constructing the molecule atom by atom, and bond by bond. However, it is also possible to convert a 2D drawing automatically into a 3D structure, and most molecular modelling packages will have this facility. For example, the 2D structure of adrenaline below was drawn in ChemDraw, then copied and pasted into Chem3D, resulting in the automatic construction of the 3D model shown (Fig. 13.2).

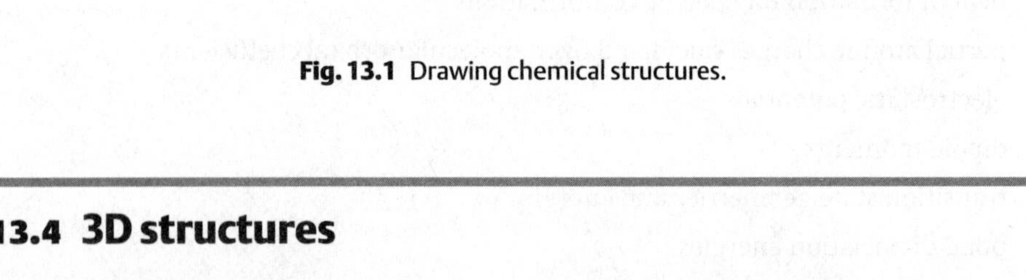

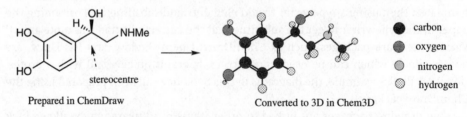

stereocentre

Prepared in ChemDraw

Converted to 3D in Chem3D

carbon
oxygen
nitrogen
hydrogen

Fig. 13.2 Conversion of a 2D drawing to a 3D model.

13.5 Energy minimization

Whichever software programme is used to create a 3D structure, a process called energy minimization should be carried out once the structure is built. This is because the construction process may have resulted in unfavourable bond lengths, bond angles, or torsion angles. Unfavourable non-bonded interactions may also be present (i.e. atoms from different parts of the molecule occupying the same region of space). The energy minimization process is usually carried out by a molecular mechanics programme which calculates the energy of the starting molecule, then varies the bond lengths, bond angles, and torsion angles to create a new structure. The energy of the new structure is calculated to see whether it is energetically more stable or not. If the starting structure is inherently unstable, a slight alteration in bond angle or bond length will have a large effect on the overall energy of the molecule, resulting in a large energy difference (ΔE; Fig. 13.3). The programme will recognize this and carry out more changes, recognizing those which lead to stabilization and those which do not. Eventually, a structure will be found where structural variations result in only slight changes in energy – an energy minimum. The programme will interpret this as the most stable structure and will stop at that stage.

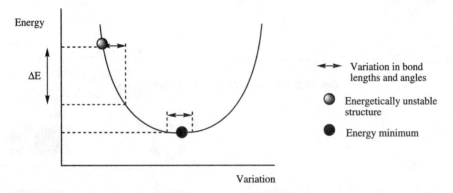

Fig. 13.3 Energy minimization.

For example, a 2D structure of aporphine was converted to a 3D structure using Chem3D. However, the catechol ring was found to be non-planar with different lengths of C–C bond. Energy minimization corrected the deformed aromatic ring, resulting in the desired planarity and the correct length of bonds (Fig. 13.4).

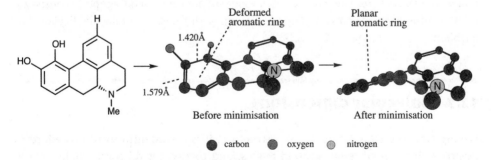

Fig. 13.4 Energy minimization carried out on aporphine.

13.6 Viewing 3D molecules

Once a structure has undergone energy minimization, it can be rotated in various axes to study its shape from different angles. For example, the 3D structure of adrenaline is shown from different aspects in Fig. 13.5. It is also possible to display the structure in different formats (i.e. cylindrical bonds, wire frame, ball and stick, space filling; Fig. 13.6).

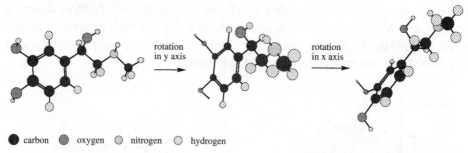

● carbon   ◍ oxygen   ◯ nitrogen   ◌ hydrogen

Fig. 13.5 Viewing a 3D model in different axes.

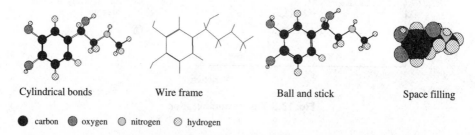

Cylindrical bonds   Wire frame   Ball and stick   Space filling

● carbon   ◍ oxygen   ◯ nitrogen   ◌ hydrogen

Fig. 13.6 Different methods of visualizing molecules.

There is another format, known as the ribbon format, which is suitable for protraying α helical regions within protein structures. This often simplifies the highly complex looking structure of a protein, allowing easier visualization of its secondary and tertiary structure. The ball and stick model of an α helical decapeptide consisting of 10 alanine units is shown in Fig. 13.7, along with the same molecule displayed as a ribbon.

13.7 Molecular dimensions

Having constructed a 3D model of a structure, it is a straightforward procedure to measure all its bond lengths, bond angles, and torsion (or dihedral) angles. These values can be read from relevant tables or by highlighting the relevant atoms and

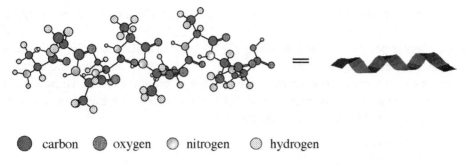

● carbon ● oxygen ○ nitrogen ◎ hydrogen

Fig. 13.7 Ribbon representation of a helical decapeptide (Chem3D).

bonds on the structure itself. The various bond lengths, bond angles, and torsion angles measured for noradrenaline are illustrated in Fig. 13.8. It is also a straightforward process to measure the separation between any two atoms in a molecule.

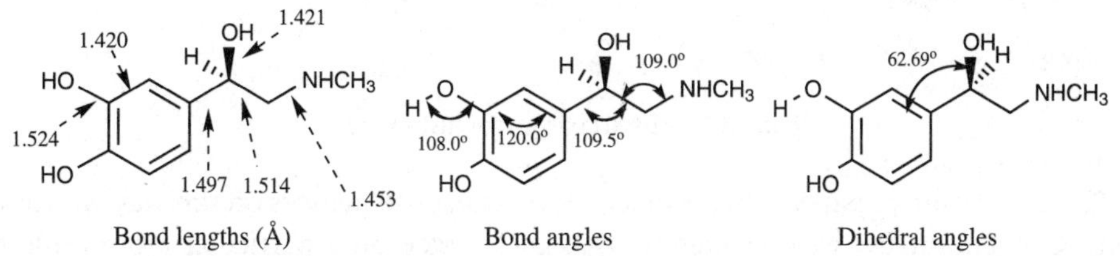

Bond lengths (Å)   Bond angles   Dihedral angles

Fig. 13.8 Molecular dimensions for noradrenaline (Chem3D).

13.8 Molecular properties

Various properties of the 3D structure can be calculated once it has been built and minimized. For example, the steric energy is automatically measured as part of the minimization process and takes into account the various strain energies within the molecule, such as bond stretching or bond compression, deformed bond angles, deformed torsion angles, non-bonded interactions arising from atoms too close to each other in space, and unfavourable dipole–dipole interactions. The steric energy is useful when comparing different conformations of the same structure, but should not be compared with the steric energies of different molecules.

Other properties for the structure can be calculated, such as the predicted heat of formation, dipole moment, charge density, electrostatic potential, electron spin density, hyperfine coupling constants, partial charges, polarizability, and infrared vibrational frequencies. Some of these will be described below.

13.8.1 Partial charges

It is important to realize that the valence electrons in molecules are not fixed to any one particular atom and can move around the molecule as a whole. Since the electrons

are likely to spend more of their time nearer electronegative atoms than electroposi-
tive atoms, this distribution is not uniform and results in some parts of the molecule
being slightly positive and others being slightly negative. For example, the partial
charges for histamine are shown in Fig. 13.9. The numbers represent the average
charge on each atom subtracted from the number of protons present. These figures
demonstrate that two of the nitrogens and two of the carbons are slightly negative in
nature, while the third nitrogen is slightly positive.

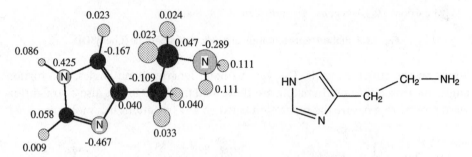

Fig. 13.9 Partial charges for histamine.

The calculation of partial charges has important consequences on the way we view
ions. Conventionally, we consider charges to be fixed on a particular atom (unless
delocalization is possible). For example, we would consider the positive charge on the
histamine ion to reside fully on the nitrogen atom (Fig. 13.10).

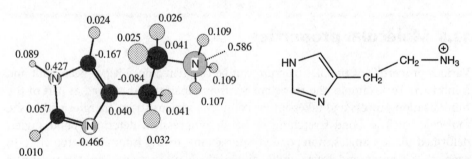

Fig. 13.10 Charge distribution on the histamine ion.

However, calculation of partial charges shows that only about half of the positive
charge is localized on the terminal nitrogen and that the remainder is spread over the
molecule. This has important consequences for the way we think of ionic interactions
between drugs and receptors. It implies that charged areas both on the receptor and
the drug are more diffuse than originally thought. This in turn suggests that we have
wider scope in designing novel drugs. For example, in the classical viewpoint of
charge distribution, a certain molecule might be considered to have its charged centre
too far away from the corresponding 'centre' in the receptor binding site. If these
charged areas are actually more diffuse, then this is not necessarily true (Fig. 13.11).

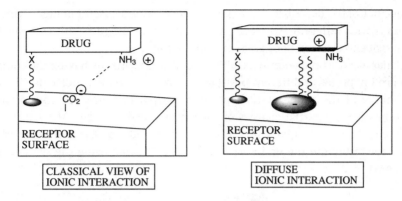

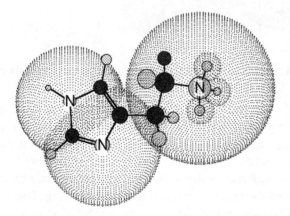

Fig. 13.11 Ionic interactions.

It is worth pointing out, however, that such calculations are carried out on structures in isolation from their environment. In the body, histamine is in an aqueous environment and would be surrounded by water molecules which would solvate the charge and consequently have an effect on charge distribution.

Partial charges can also be represented by dot clouds. The size of each cloud represents the amount of charge and the clouds can be coloured red or blue to show what sort of charge it is. The dot clouds shown in Fig. 13.12 represent the relative sizes of partial charges on the histamine ion, with the majority of the charge focused on the three nitrogens.

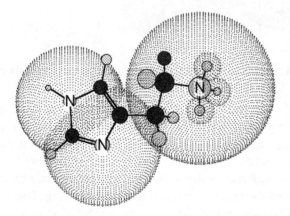

Fig. 13.12 Dot cloud representation of partial charge.

13.8.2 Molecular electrostatic potentials

Another way to consider charge distribution is to view the molecule as a whole rather than as individual atoms and bonds, such that one can identify areas of the molecule which are electron rich or electron poor. This is particularly important in the 3D QSAR technique of CoMFA described in Chapter 11. It can also be useful in identifying how compounds with different structures might line up to interact with corresponding electron rich and electron poor areas in a binding site.

Molecular electrostatic potentials (MEPs) can be calculated by having the computer place a proton 'probe' at different positions in space around the molecule. The interaction energy of the probe at each position is then measured by considering its interaction with the partial charges of each atom 'within range'. Alternatively, the MEP can be calculated by quantum mechanics by considering the molecular orbitals. The MEP for histamine shown in Fig. 13.13 was calculated using the semiempirical method AM1.

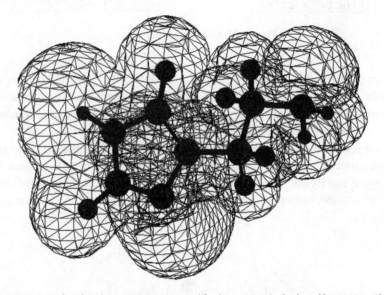

Fig. 13.13 Molecular electrostatic potential for histamine (calculated by AM1 in Chem3D).

An example of how electrostatic potentials have been used in drug design can be seen in the design of the cromakalim analogue (II) (Fig. 13.14), where the cyanoaromatic ring was replaced with a pyridine ring. This was part of a study looking into analogues of cromakalim which would have similar antihypertensive properties but which might have different pharmacokinetics. To retain activity, it was important that the heteroaromatic ring which was introduced was as similar in character to the original aromatic ring as possible. Consequently, the MEP's of various bicyclic systems were calculated and compared with the parent bicyclic system (IV) (Fig. 13.15). To simplify the analysis, the study was carried out in 2D within the plane of the bicyclic systems, and maps were created showing areas of negative potential (Fig. 13.16). The contours represented the various levels of the MEP and can be taken to indicate possible hydrogen bonding regions around each molecule. The analysis demonstrated that the bicyclic system (III) had similar electrostatic properties to (IV), resulting in the choice of structure (II) as an analogue.

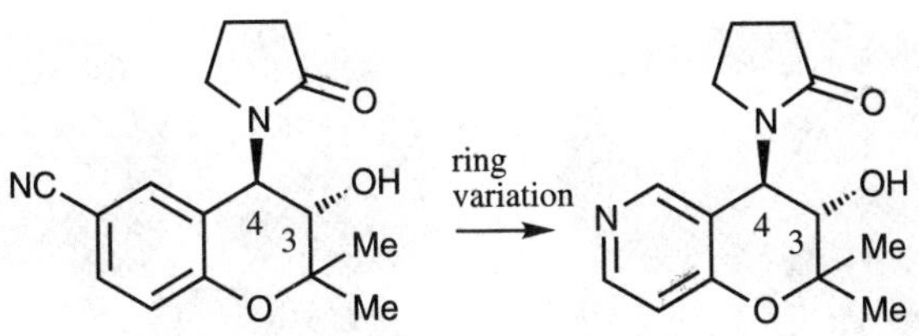

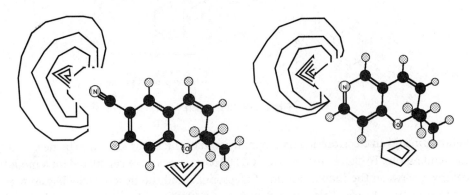

Fig. 13.14 Ring variation on cromakalim.

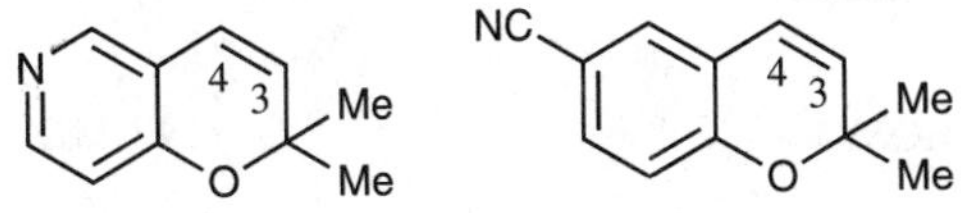

Fig. 13.15 Bicyclic models in cromakalim study.

Fig. 13.16 MEP's of cromakalim and analogue.

13.8.3 Molecular orbitals

The molecular orbitals of a compound can be calculated using quantum mechanics. For example, ethene can be shown to have 12 molecular orbitals. The highest occupied molecular orbital (HOMO) and lowest unoccupied molecular orbital (LUMO) are shown in Fig. 13.17 .

A study of the HOMO and LUMO orbitals is particularly useful since frontier molecular orbital theory states that these orbitals are the most important in terms of a molecule's reactivity. An example of the use of HOMO and LUMO orbitals in explaining drug–receptor interactions has been demonstrated with ketanserin (Fig. 13.18). Ketanserin is an antagonist at serotonin receptors, but has a greater binding affinity than would be expected from normal bonding interactions.

To explain this greater binding affinity, it was proposed that a charge transfer interaction was taking place between the electron deficient fluorobenzoyl ring system of

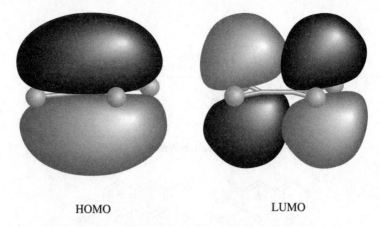

HOMO   LUMO

Fig. 13.17 HOMO and LUMO molecular orbitals for ethene.

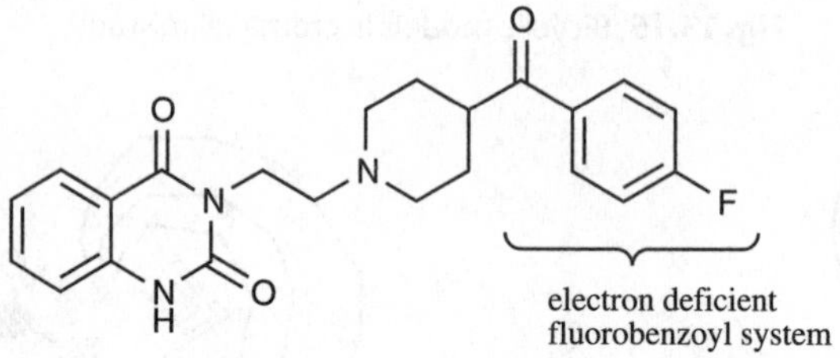

Fig. 13.18 Ketanserin.

ketanserin and an electron rich tryptophan residue which was known to be nearby in the binding site. To check this, HOMO and LUMO energies were calculated for a model complex between the indole system of tryptophan and the fluorobenzoyl system of ketanserin (Fig. 13.19). This showed that the HOMO for the indole–fluorobenzoyl complex resided on the indole structure while the LUMO was on the fluorobenzoyl moiety, indicating that charge transfer is possible. With other antagonists, there was not this same clearcut separation between the HOMO and LUMO orbitals, with the indole system being involved in both orbitals.

13.8.4 Spectroscopic transitions

It is possible to calculate the infrared or ultraviolet transitions for a molecule. As far as the infrared is concerned, a theoretical spectrum can be generated, but it is highly unlikely that it will accurately match the actual infrared spectrum. Nevertheless, the position and identification of specific absorptions can be identified and can be useful in the design of drugs.

For example, it is found that the activity of penicillins is related to the position of the β-lactam carbonyl stretching vibration in the infrared. Calculating the theoretical wave number for a range of β-lactam structures can be useful in identifying which ones are likely to have useful activity before synthesizing them.

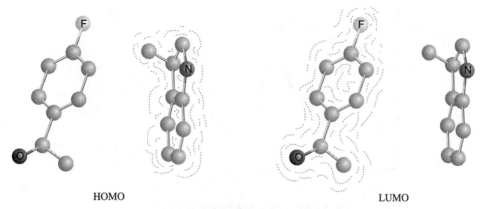

HOMO   LUMO

Fig. 13.19 HOMO and LUMO molecular orbitals (dot surfaces) for the indole–fluorobenzoyl complex.

13.9 Conformational analysis

13.9.1 Local and global energy minima

In Section 13.5, we saw how energy minimization is carried out to produce a stable conformation for a 3D structure. However, the structure obtained is not necessarily the most stable conformation. This is because energy minimization stops as soon as it reaches the first stable conformation it finds, and that will be the one closest in structure to the starting structure. This can be illustrated in Fig. 13.20 in which the most stable conformation is separated from another higher energy conformation by an energy saddle. If the 3D structure initially created is on the energy curve at the position shown, energy minimization will stop when it reaches the first stable conformation it encounters – a local energy minimum. At this point, variations in structure result in low energy changes and so the minimization will stop. To cross the saddle to the more stable conformation, structural variations would have to be carried out which increase the strain energy of the structure and these will be rejected by the programme. The minimization programme works as if it were blind and when it reaches a hollow it assumes that it is at the bottom of the hill and has no way of knowing that there is a more stable conformation (a global energy miminum) beyond the energy saddle.

Therefore, to identify the most stable conformation, it is necessary to generate different conformations of the molecule and to compare their steric energies. There are two methods of doing this: molecular dynamics or stepwise rotation of bonds.

13.9.2 Molecular dynamics

Using molecular mechanics (MM2), it is possible to generate a variety of different conformations by using a molecular dynamics programme which 'heats' the molecule

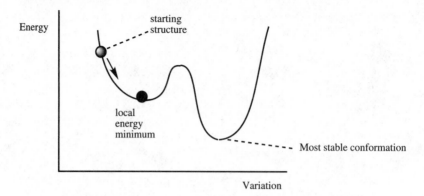

Fig. 13.20 Local and global energy minima.

to 800–900 K. Of course, this does not mean that the inside of your computer is about to melt. It means that the programme allows the structure to undergo bond stretching and bond rotation as if it was being heated. As a result, energy barriers between different conformations are overcome, allowing the crossing of energy saddles. In the process, the molecule is 'heated' at a high T (900 K) for a certain period (e.g. 5 pico-seconds), then 'cooled' to 300 K for another period (e.g. 10 ps) to give a final structure. The process can be repeated automatically as many times as wished to give as many different structures as required. Each of these structures can then be recovered, energy minimized and its steric energy measured. By carrying out this procedure, it is usually possible to identify distinct conformations, some of which might be more stable than the initial conformation.

For example, the 2D drawing of butane shown in Fig. 13.21 was imported into Chem3D and energy minimized. Because of the way the molecule was represented, energy minimization stopped at the first local energy minimum it found, which was the gauche conformation having a steric energy of 3.038 kcal/mol. The molecular dynamics programme was run to generate other conformations and successfully produced the fully staggered *trans* conformation which, after optimization, had a steric energy of 2.175 kcal/mol, showing that the latter was more stable by about 1 kcal/mol.

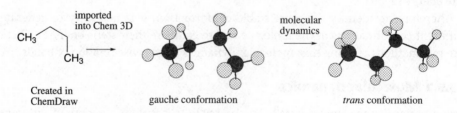

Fig. 13.21 Generations of butane conformations.

In fact, this particular problem could be solved more efficiently by the stepwise rotation of bonds described below (see Section 13.9.3). Molecular dynamics is more useful for creating different conformations of molecule which are not conducive to stepwise bond rotation (e.g. cyclic systems), or which would take too long to analyse by that process (large molecules).

For example, the twist boat conformation of cyclohexane remains as the twist boat when energy minimization is carried out. 'Heating' the molecule by molecular dynamics in Chem3D produces a variety of different conformations, including the more stable chair conformation (Fig. 13.22).

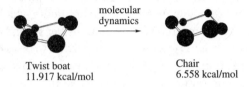

molecular
dynamics

Twist boat
11.917 kcal/mol

Chair
6.558 kcal/mol

Fig. 13.22 Generation of the cyclohexane chair conformation by molecular dynamics.

13.9.3 **Stepwise bond rotation**

Although molecular dynamics can be used to generate different conformations, there is no guarantee that it will identify all the possible conformations of a structure. A more systematic process is to generate different conformations by automatically rotating every single bond by a set number of degrees. For example, 13 different conformations of butane were generated by automatically rotating the central bond in 30° steps. The steric energy of each conformation was calculated and graphed (Fig. 13.23), revealing the most stable conformation to be fully staggered and the least stable to be the eclipsed. In this operation, energy minimization is not carried out on each structure, since one wants to identify both stable and unstable conformations.

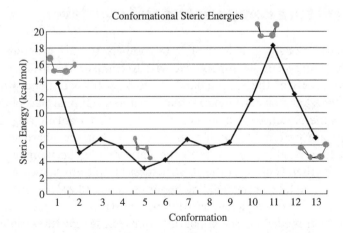

Fig. 13.23 Graph showing relative stabilities of various butane conformations.

Some modelling software packages (e.g. ChemX) can automatically identify all the rotatable single bonds in a structure. Bonds to hydrogen or to methyl groups are excluded in this analysis since rotations of these bonds do not generate significantly different conformations. Once the rotatable bonds have been identified, the programme generates all the possible conformations which can arise from rotating these bonds by a set amount determined by the operator. The number of conformations generated will depend on the number of rotatable bonds present and the set amount of rotation.

For example, a structure with three rotatable bonds could be analysed for conformations resulting from 10° increments at each bond to generate 46 656 conformations. With four rotatable bonds, 30° increments would generate 20 736 conformations. In general, about 1000 conformations per second can be processed on a standard bench top computer. However, it is important to be as efficient as possible, and care should be taken in deciding how much each bond should be rotated at a time to ensure that a representative but manageable number of conformations are created.

It is also possible to make the process more efficient depending on the information desired. For example, if you are only interested in identifying stable conformations, the programme can automatically filter out conformations which are eclipsed or near eclipsed. It is also possible to filter out 'nonsense' conformations. These are conformations where atoms occupy the same position in space. Such conformations can arise since bond rotations are being carried out by the programme without analysing what is happening elsewhere in the molecule.

Once a series of conformations has been generated, they can be tabulated and sorted into their order of stability. The most stable conformations can then be energy minimized and their structures compared.

13.10 Structure comparisons and overlays

Using molecular modelling, it is possible to compare the 3D structures of two or more molecules. For example, suppose we wish to compare the structure of cocaine with procaine. Both of these compounds have a local anaesthetic property, and structure–activity relationships indicate that the important pharmacophore for local anaesthesia is the presence of an amine, an ester, and an aromatic ring. These functional groups are present in cocaine and procaine, but the pharmacophore also requires the functional groups to be in the same relative positions in space with respect to each other. Looking at the 2D structures of procaine and cocaine, it would be tempting to match up corresponding bonds as in Fig. 13.24, but this would place the nitrogen atoms one bond length apart in the overlay.

Using molecular modelling, the important atoms of the structures can be matched up (e.g. the nitrogens and the aromatic rings in both structures). The software then

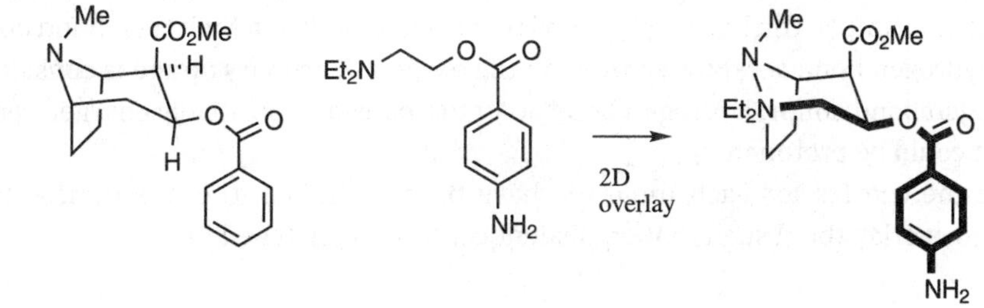

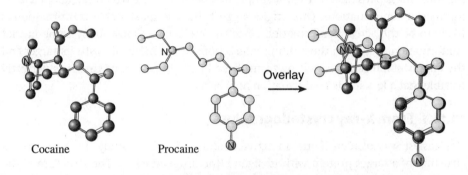

Fig. 13.24 2D overlay of cocaine and procaine.

strives to find the best fit, resulting in the overlay shown in Fig. 13.25. Here the procaine molecule has been laid across the centre of the bicylic system in cocaine such that both the aromatic rings and nitrogen atoms overlap.

Cocaine   Procaine   Overlay

Fig. 13.25 Overlay of cocaine and procaine using Chem3D.

It is important to appreciate that the fitting process is carried out on a rigid basis (i.e. the molecules are locked in one conformation and no bond rotations are permitted). This means that it is important to be sure that each molecule is in the active conformation before carrying out the fitting process.

Some modelling software (e.g. ChemX) has the capacity to overlay two molecules automatically without the operator having to define the centres or how they should be matched up. The programme searches each molecule for what it considers to be important centres. These are centres which are normally involved in binding interactions (i.e. aromatic rings, hydrogen bond donors, hydrogen bond acceptors, positively charge centres, acids, and bases). As far as aromatic rings are concerned, the centre of the ring is defined as the centre. For hydrogen bond donors or acceptors (X–H), the heteroatom (X) is defined as the centre. To be precise, the centre for a hydrogen bond donor should be the hydrogen atom, but there is a large amount of uncertainty as to where this atom is located because of bond rotation, and so the heteroatom is defined as the centre of an 'available volume' within which the hydrogen atom is located. Certain functional groups can be defined as being more than one type of centre. For example, the hydroxyl group is considered both as a hydrogen bond donor and a hydrogen bond acceptor centred on the oxygen. A primary amine is considered a hydrogen bond donor, hydrogen bond acceptor, base, and positively charged centre (since it could be protonated).

Once the centres for each molecule have been identified, the programme then strives to overlay them such that equivalent centres are matched up.

13.11 Identifying the active conformation

A frequently encountered problem in drug design is trying to decide what shape or conformation a molecule is in when it fits its target binding site – the active conformation. This is particularly true for simple flexible molecules which can adopt a large number of conformations. One might suggest that the most stable conformation is likely to be the active conformation since the molecule is most likely to be in that conformation. However, the binding interactions of the molecule with its target and the energy stabilization resulting from these interactions means that it is perfectly feasible that a less stable conformation may be used.

13.11.1 From X-ray crystallography

The easiest way of identifying an active conformation is to study the X-ray crystal structure of a target protein with its ligand (the drug) attached. The structure of the protein–ligand complex can then be studied by computer and the conformation of the ligand identified. However, not all proteins can be easily crystallized and so other methods of identifying active conformations may have to be used.

13.11.2 Comparison of rigid and non-rigid ligands

Identification of active conformations is made much easier if one of the active compounds is a rigid molecule which only has one possible conformation. The geometry of the pharmacophore (the important binding centres) can be determined for the rigid molecule. More flexible molecules can then be compared with the rigid molecule to find a conformation which will place the important binding groups in the same relative geometry. This can be done to some extent with solid models, but a molecular modelling programme is a much more effective method.

For example, the neuromuscular blocking agent tubocurarine is a fairly rigid structure in which the important pharmacophore is the two quaternary nitrogen atoms. When built on the computer, the distance between these atoms is measured as 11.527 Å (Fig. 13.26). Decamethonium also acts as a neuromuscular blocking agent, but is extremely flexible, which means that a large number of conformations is possible. The most stable conformation is the extended one where the quaternary nitrogens are

14.004 Å apart (Fig. 13.27). Using molecular dynamics, a variety of different conformations for decamethonium can be generated, as described in Section 13.9.2, resulting in the identification of a conformation where the quaternary nitrogens are 11.375 Å apart. This could then be proposed as a likely candidate for the active conformation.

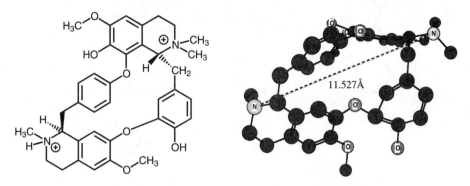

Fig. 13.26 Computer-generated model of tubocurarine (Chem3D).

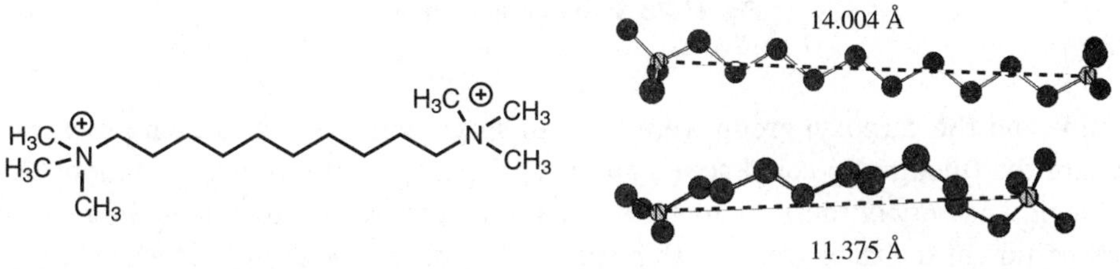

Fig. 13.27 Computer-generated conformations of decamethonium (Chem3D).

If a fully rigid molecule is not available to act as a template, it may be possible to match up different structures which have an element of rigidity somewhere in their skeleton. For example, structures I–III are all antagonists of the 5-HT$_{2A}$ (serotonin) receptor. An active conformation for structure I could be proposed which matched the rigid moieties of structures II and III (Fig.13.28). Structure II gives the conformation for the top half of the molecule while structure III gives the conformation of the bottom half.

Another method for finding active conformations is to consider all the reasonable conformations for a range of active compounds, and then to determine the common volume or space into which the various important binding groups can be placed to interact with the receptor. A study such as this was carried out to determine the active conformation of the antihypertensive agent captopril (Fig. 13.29). Since captopril is flexible, the exact 3D relationship of the important binding groups (i.e. the carboxyl, amide, and thiol groups) in the active conformation is not known. There was also no X-ray crystallographic data to study how captopril was bound to its target enzyme. To address this problem, a variety of rigid analogues (I–III) were synthesized in which the

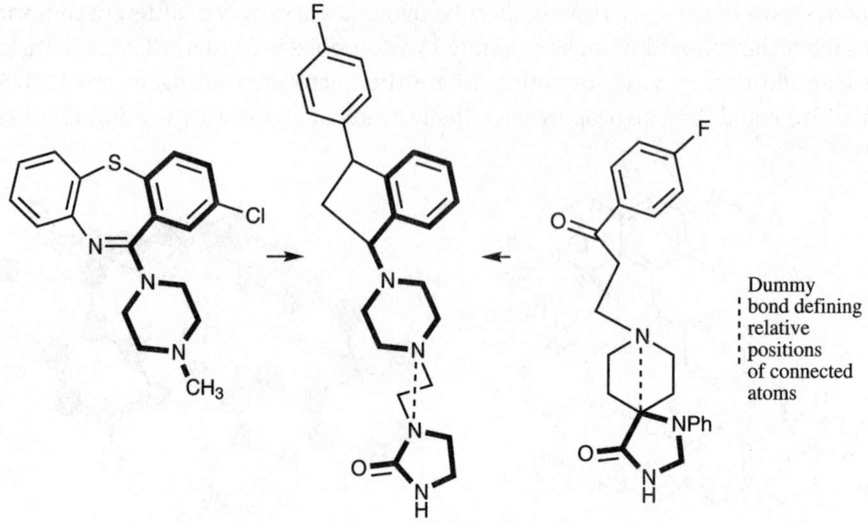

Fig. 13.28 Serotonin antagonists.

II   I   III

Dummy
bond defining
relative
positions
of connected
atoms

amide and the carboxyl group were fixed in space with respect to each other, but where the thiol group could access an area of space because of bond rotation. The biological activity of these compounds was then measured to study how the relative orientation of the thiol group, with respect to the other two groups, affected biological activity.

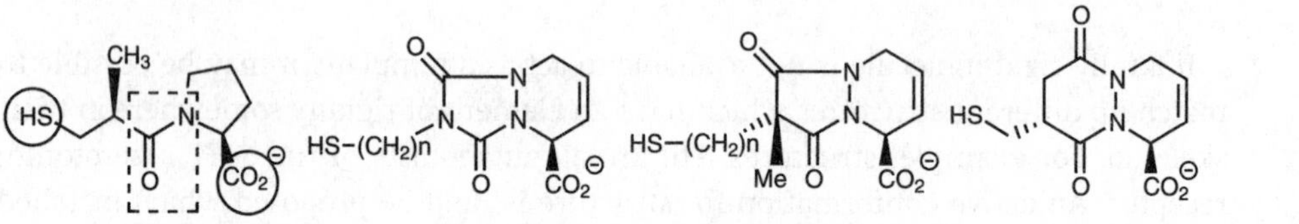

Captopril   I n=1,2,3   II n=1,2   III

Fig. 13.29 Captopril and rigid analogues.

The possible conformations for captropril arising from bond rotation around the two bonds shown in Fig. 13.30 were determined using molecular modelling, and a spatial map (A) was generated to show the possible regions in space which were accessible to the thiol group (Fig. 13.31).

However, some of the conformations involved in this analysis are high energy eclipsed conformations and are unlikely to be important. These conformations were filtered out from the analysis by programming the software to reject conformations with steric energy greater than 50 kcal/mol, thus leaving only stable conformations.

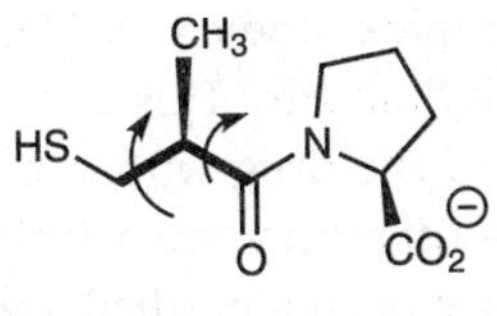

Fig. 13.30 Bond rotations in captopril.

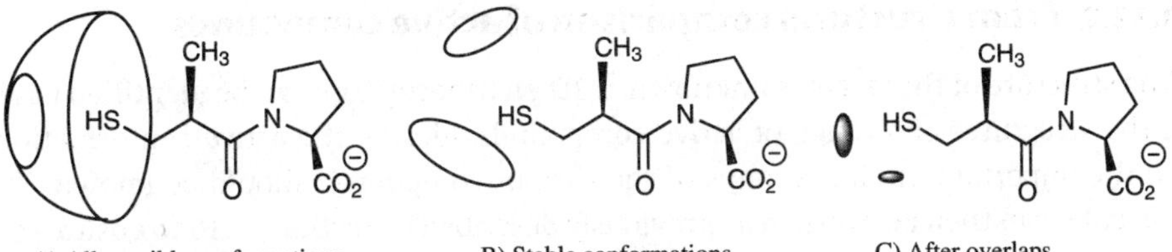

A) All possible conformations   B) Stable conformations   C) After overlaps

Fig. 13.31 Generating spatial maps in conformational analysis.

When this was done, the spatial map (B) showed that the thiol group was restricted to two main regions in space with respect to the other two binding groups.

A spatial map for one of the active rigid analogues was now generated and compared with the one generated by captopril. The overlap between the maps was considered to be the most likely location for the thiol group. The process was then repeated for the other rigid analogues, narrowing down the possible area which would be occupied by the thiol group yet further. The study identified two 'hot spots' of space (C) where the thiol was likely to be positioned. Only those conformations of captopril which could place the thiol group into these 'hot spots' were likely to be active conformations.

13.12 3D pharmacophore identification

A 3D pharmacophore represents the relative positions of important binding groups in space and disregards the molecular skeleton which holds them in place. Thus, the 3D pharmacophore for a particular binding site should be common to all the various ligands which bind to it. Once the 3D pharmacophore has been identified, structures can be analysed to see whether they can adopt a stable conformation which will contain the required pharmacophore. If they do (and there are no steric clashes with the binding site) the structure should be active.

13.12.1 From X-ray crystallography

The crystal structure of a target protein with its ligand bound to the binding site can be used to identify the 3D pharmacophore. The protein–ligand structure can be fed into the computer and the complex studied to identify the bonding interactions which hold the ligand in the binding site. This is done by measuring the distances between likely binding groups in the drug with complimentary binding groups on nearby amino acids to see whether they are within bonding distance. Once the binding groups on the ligand have been identified, their positions can be mapped to produce the pharmacophore.

13.12.2 From structural comparison of active compounds

If the structure of the target is unknown, a 3D pharmacophore can be identified based on the structures of a range of active compounds. Ideally, the active conformations and the important binding groups of the various compounds should be known. The molecules can then be overlain as previously described in Section 13.10 to ensure that the important binding groups are matched up as closely as possible. It would be rare for the binding groups to match up exactly and so an allowed region in space for each important binding group can be identified for the 3D pharmacophore.

13.12.3 Automatic identification of pharmacophores

It is possible to identify possible 3D pharmacophores for a range of active compounds using some software programmes (e.g. ChemX), even if the important binding groups are unknown or uncertain. First, the programme identifies potential binding centres in a particular molecule. These are the hydrogen bond donors, hydrogen bond acceptors, aromatic rings, acidic groups, basic groups, and positively charged centres[1]. It is also possible to search for hydrophobic centres involving hydrocarbon skeletons of three or more carbon atoms. Here, the hydrophobic centre is calculated as the midpoint of the carbon atoms in question.

Let us assume that dopamine is the structure being analysed. ChemX would identify four important binding centres for dopamine – the centre of the aromatic ring, both phenolic oxygens (hydrogen bond donors and acceptors), and the amine nitrogen (hydrogen bond donor or acceptor, base, positively charged centre; Fig. 13.32).

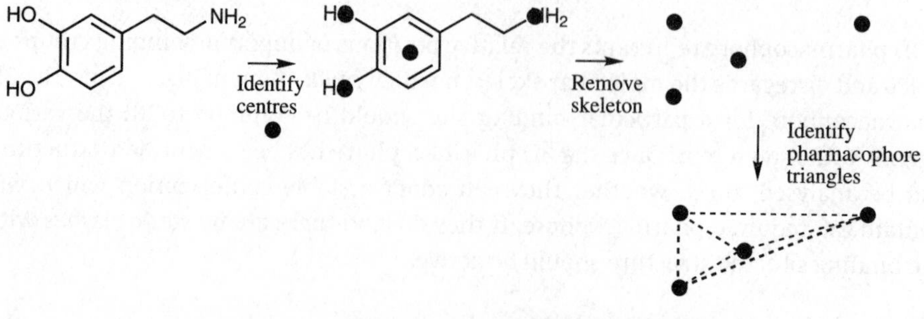

Fig. 13.32 Pharmacophore identification in dopamine.

[1] The programme does not normally search for negatively charged centres since these are not usually important binding groups in a drug.

The programme now identifies the various triangles which can connect up the important centres. In the case of dopamine, there are four such triangles. Each one is defined by the length of each side and the type of binding centres present, resulting in a set of pharmacophore triangles[2]. Of course, this analysis has only been carried out on one conformation of dopamine. The programme can now be used to generate a range of different conformations, as described in Section 13.9.3 and, for each conformation, another set of pharmacophore triangles can be defined. Adding all these together gives the total number of possible pharmacophore triangles for dopamine in all the conformations created.

Another structure with dopamine-like activity can now be analysed. Once all its pharmacophore triangles have been determined, they can be compared with those for dopamine and the common pharmacophores identified. The process is then repeated for all the active compounds until pharmacophore triangles common to all the structures have been identified. These are then plotted on a 3D plot where the x, y and z axes correspond to the lengths of the three sides of each triangle. This produces a visual display which allows easy identification of distinct pharmacophores. Very similar pharmacophores can be quickly spotted since they are clustered close together in specific regions of the plot.

For example, the three structures in Fig. 13.33 were analysed and found to have 38 common pharmacophores. When these were plotted, seven distinct groups of pharmacophore were identified (Fig. 13.34). Each pharmacophore present in the grid can be highlighted and revealed. For example, one of the possible active pharmacophores consists of two hydrogen bond acceptors and an aromatic centre.

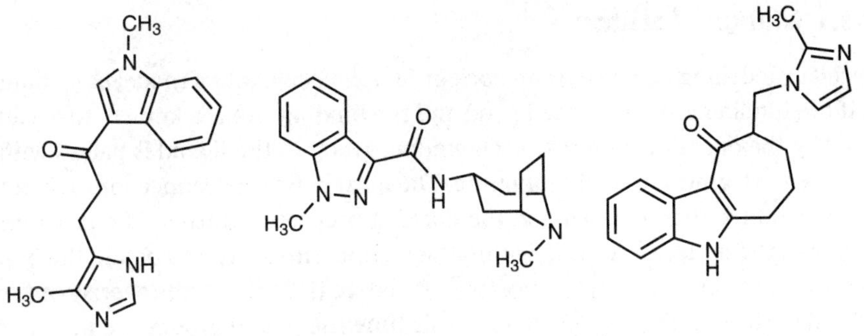

Fig. 13.33 Test structures.

Note that it is advisable to begin this exercise with the most active compound and then proceed through the structures in order of activity.

---

[2] Since some of the points specified represent more than one type of binding centre, this means that the number of pharmacophore triangles will be greater than four. For example, if one of the points is a phenol then it represents a hydrogen bond donor or acceptor. Therefore, any triangle including this point must result in two pharmacophore triangles – one for the donor and one for the acceptor.

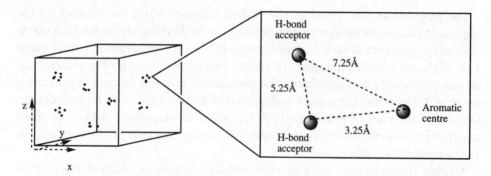

Fig. 13.34 Pharmacophore plot.

The above analysis can be simplified enormously if certain groups are known to be essential for binding. The programme can then be run such that only triangles containing these centres are included. For example, if the nitrogen atom of dopamine is known to be an essential bonding centre, then the number of possible triangles for each conformation is cut from four to three (missing out the triangle connecting the phenolic oxygens and the aromatic ring).

13.13 Docking procedures

13.13.1 Manual docking

Molecular modelling can be used to dock or fit a molecule into a model of its binding site. If the binding groups on the ligand and the binding site are known, they can be defined by the operator such that each binding group in the ligand is paired with its complimentary group in the binding site. The ideal bonding distance for each potential interaction is then defined and the docking procedure started. The programme then moves the molecule around within the binding site to try and obtain the best fit as defined by the operator. In essence, the procedure is similar to the overlay or fitting process described in Section 13.10, only this time the paired groups are not directly overlapped but fitted such that the groups are within preferred bonding distances of each other. Both the ligand and the protein remain in the same conformation throughout the process and so this is a rigid fit. Once a molecule has been successfully docked, fit optimization is carried out. This is essentially the same as energy minimization, but carried out on the ligand–target protein complex. Different conformations of the molecule can be docked in the same way and the interaction energies measured to identify which conformation fits the best.

13.13.2 Automatic docking

Automatic docking can be carried out by some programmes (e.g. ChemX) where the programme itself decides how it will dock the ligand. First, an X-ray crystal structure of the target protein is defined and loaded into the computer. The amino acids in the binding pocket are then left displayed while the remainder of the protein is hidden. (The structure is still present, but hiding the atoms makes analysis and visualization easier.)

It helps to know which amino acids are involved in binding, but if this information is not available, an analysis of potentially important binding centres in the binding site can be carried out (i.e. hydrogen bond donors and acceptors, acids, bases, positive centres, and aromatic rings). Once all the possible centres have been identified, a location is defined in the binding site where a complimentary group should be positioned for a bonding interaction (Fig. 13.35).

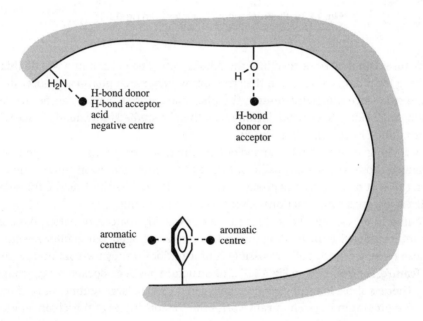

Fig. 13.35 Identifying binding centres and locations for complimentary groups.

In this process, it is assumed that aromatic rings interact face on, but face to edge interactions could be included if necessary. The complimentary centre for a hydrogen bond donor or acceptor is placed along the line of the carbon heteroatom bond. This is an unlikely position for a hydrogen bond since the interaction is more likely to be at an angle to allow interaction with the lone pair of the heteroatom. However, the position defined is midway between the most likely positions for hydrogen bonding, and a suitable tolerance value is chosen to ensure that the ideal positions are included (i.e. the complimentary centres are treated as volumes of space rather than distinct points, with the size of the volume being determined by the operator). Having identified the potential complimentary centres, all the possible pharmacophore triangles involving these centres are calculated (Fig. 13.36). A large number of triangles may be generated and it is a good idea to simplify the situation before carrying out this step. For example, some centres can be rejected if they are on the margins of the binding site. On the other hand, other binding centres which are known to be important could be defined such that they must be included in pharmacophoric triangles.

Fig. 13.36 Pharmacophore triangles from Fig. 13.35.

Structures can now be docked into the binding site. The programme will do this by searching for pharmacophores in the structure which match pharmacophores in the binding site. Once a match is found, the pharmacophore triangles can be overlaid, resulting in docking. Successful dockings can then be studied individually to see what conformation or conformations can fit.

This docking analysis can be carried out automatically on a range of compounds to distinguish those which are capable of fitting the binding site from those which are not. In this sort of analysis, the process can be accelerated by aborting the full search for all a structure's conformations as soon as one fit is found.

A 'bump filter' can also be included to assess quickly if a conformation has a bad steric interaction with the binding site and then reject it. Such 'bump filter rejections' may also reject a range of other possible conformations if they too contain the undesired feature. For example, in Fig. 13.37, the structure has been docked using groups X and Y. There is also a long alkyl chain which can take up a large number of conformations. The programme systematically varies the torsion angles of this chain, working outwards from the heart of the molecule. If a conformation having specific torsion angles χ_1, χ_2, and χ_3 is rejected because of a bad steric interaction with the receptor, the programme will not bother varying χ_4–χ_6.

Another way to identify the location of important centres in the binding site is to place a potential energy grid into the binding site and to put probe atoms at every grid point to calculate interaction energies between the probe and the receptor. The grid points with the highest interaction energies are then retained as the centres for the binding site. The type of centre identified (i.e. hydrogen bonding, hydrophobic, etc.) would depend on the type of probe atom used. When it comes to docking, positive scores are generated when the correct centre on the ligand is matched up with the

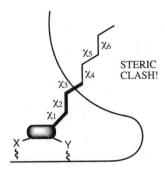

Fig. 13.37 'Bump' filter.

corresponding centre in the binding site. The molecule would be moved around the binding site until a maximum score was achieved and this would be taken as the best fit. A 'goodness of fit' can then be measured, taking into account the shape fit (a measure of how well the ligand makes contact with the surface of the binding site), and the chemical fit (checking whether interactions such as hydrogen bonding, polar, electrostatic, or hydrophobic are good or bad. Good interactions are scored positively and bad ones negatively.

A potential energy grid can be used for the calculation of ligand–receptor interaction energies. The idea is to precalculate interaction energies with the receptor for various atom types placed on the grid points. A 3D table of these interaction energies is then stored and, when a ligand is docked, the atoms closest to each grid point are identified and the interaction energies automatically read off the table and summed to give a score for the goodness of fit.

13.14 Automated screening of databases for lead compounds

The automated docking procedure described in Section 13.13.2 can be used to screen a variety of different 3D structures to see whether they will fit the binding site of a particular target (electronic screening or database mining). This would be useful for a pharmaceutical company wishing to screen its own or other chemical stocks (libraries) for suitable lead compounds.

Screening of databases can also be done purely by searching for suitable pharmacophores. The process is accelerated by a quick filter which eliminates any structure which does not contain the necessary centres. The operator has the ability to vary the tolerances involved in the search to find pharmacophores which nearly match the desired pharmacophores.

13.15 Receptor mapping

Drug design is made far easier if the structure of the target protein and its binding site are known. The best way of obtaining this information is from X-ray crystallography of protein crystals, preferably with a ligand bound to the binding site. Unfortunately, not all proteins are easily crystallized (e.g. membrane proteins). In cases like this, model receptors and binding sites may be constructed to aid the drug design process.

13.15.1 Constructing a model receptor

A model of a protein can be created using molecular modelling if the primary amino acid sequence is known and the X-ray structure of a related protein has been determined.

Of particular interest in this respect is the protein bacteriorhodopsin, which has been crystallized and its structure determined by X-ray crystallography. Bacteriorhodopsin is an example of a G-protein-coupled receptor consisting of seven transmembrane helices (see Chapter 6). Many of the important receptors in medicinal chemistry belong to this same family of proteins and so the structure of bacteriorhodopsin has been a vital template in constructing models of these membrane-bound receptors. By identifying the primary amino acid sequence of the target receptor and looking for suitable stretches of hydrophobic amino acids, it is possible to identify the seven transmembrane helices and then to use bacteriorhodopsin as a template to construct the helices in a similar position relative to each other. The linking loops can then be modelled in to give the total 3D structure.

If a new protein has been discovered, its primary structure is first determined. Suitable software is then used to compare its primary sequence with the primary sequences of other proteins to find a closely related protein. This involves comparing the sequences with respect to conserved amino acids, hydrophobic regions, and secondary structure. Once a reference protein of similar structure has been identified, it is used as a template to build the peptide backbone of the new protein. First of all, regions which are similar in the new protein and the template protein are identified. The backbone for the new protein is constructed to match the corresponding region in the template protein. This leaves connecting regions whose structure cannot be determined from the template. A suitable conformation for these intervening sequences might be found by searching the protein databases for a similar sequence in another protein. Alternatively, a loop may be generated to connect two known regions. Once the backbone has been constructed, the side chains are added in energetically favourable conformations. Energy minimization is carried out and the structure is refined with molecular dynamics in the absence and presence of ligand. Once the model has been constructed, it is tested experimentally. For example, the model would indicate that certain amino acids might be important in the binding site.

These could then be mutated to see if this has an effect on ligand binding. Studies such as this have identified amino acids which are important in binding neurotransmitters in a range of G-protein receptors (Fig. 13.38).

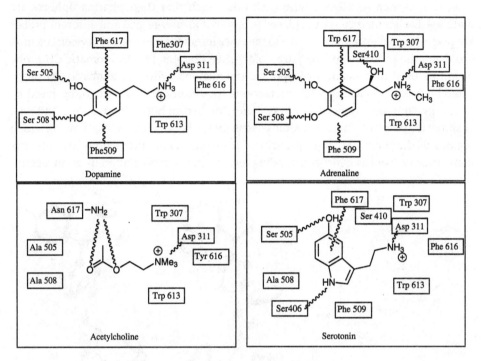

Fig. 13.38 Amino acids in binding sites.

This study shows interesting similarities and differences between the four receptors. For example, all four binding sites interact with ligands having a charged nitrogen group and contain a hydrophobic pocket to receive it. There are several conserved aromatic residues in this pocket at positions 307, 613 and 616. An aspartate residue at position 311 is also present in all cases and is capable of forming an ionic interaction.

However, there are also differences between the different binding sites which account for the different ligand selectivities. For example, the amino acids at positions 505 and 508 in the catecholamine receptors are serine, whereas the corresponding amino acids in the cholinergic receptors are alanine. The amino acid at position 617 in the catecholamine receptors is phenylalanine (allowing an interaction with the aromatic portion of the catecholamines), whereas in the cholinergic receptor this amino acid is arginine (allowing a hydrogen bonding interaction with the ester group of acetylcholine).

13.15.2 Constructing a binding site

Rather than construct a complete model protein, it is possible to use molecular modelling to design a model binding site based on the structures of the compounds which bind to it. To do this effectively, a range of structurally different compounds with a range of activities should be chosen. The active conformations should be identified as far as possible and a 3D pharmacophore identified as described previously. The molecules should then be aligned with each other such that their pharmacophores are matched. Each molecule is then placed in a potential energy grid and different probes are placed at each grid point in turn to measure interaction energies between the molecule and the probe atom (compare CoMFA; see Chapter 11). An aromatic CH probe would be used to measure hydrophobic interactions, while an aliphatic OH probe could be used to measure polar interactions. The interactions are then displayed by isoenergy contours (typically –1.5 kcal/mol for hydrophobic interactions and –4.0 kcal/mol for polar interactions). Altanserin is shown below (Fig. 13.39), showing which regions of the molecule could participate in hydrophobic interactions (e.g. the aromatic rings) as well as hydrogen bonding regions (e.g. carbonyl oxygens or nitrogen).

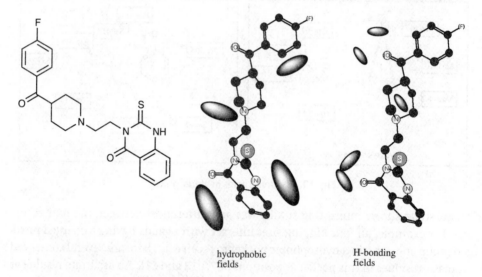

hydrophobic
fields

H-bonding
fields

Fig. 13.39 Analysis of hydrophobic and hydrogen bonding fields around altanserin.

The fields for all the molecules in the study can then be compared to identify common fields. Once these have been identified, suitable amino acids can be positioned to allow the required interaction. For example, an aspartate residue could be used to allow an ionic interaction, and amino acids such as a phenylalanine, tryptophan, isoleucine, leucine or valine could be used for a hydrophobic interaction.

A range of structures, including altanserin and ketanserin, was used to construct the receptor map for the 5-HT$_{2a}$-receptor. Taking ketanserin as the representative structure for these compounds, various hydrogen, ionic, and hydrophobic bonding interactions were identified. Structure–activity relationships (SARs) were then used to identify whether any of these proposed interactions were important or not. In this case, SARs indicated that the two carbonyl groups were not important, and so the hydrogen bonding regions derived from these groups probably do not exist in the receptor. Suitable amino acids can now be placed in the relevant positions. The choice of which amino acids should be used is helped by knowing the amino acid sequence of the target protein and the structure of a comparable protein. The 5-HT$_{2a}$-receptor belongs to a superfamily of proteins which include bacteriorhodopsin, the structure of which is known. Allying this information with the primary amino acid sequence of the receptor led to the choice of amino acids shown in Fig. 13.40.

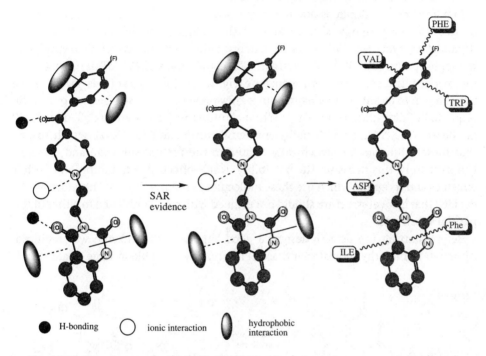

Fig. 13.40 Receptor map for the 5-HT$_{2a}$-receptor using ketanserin as representative ligand.

Once built, known compounds can be docked to the model receptor binding site, the complex minimized, and binding energies calculated. These can then be compared with experimental binding affinities to see how well the model agrees with experiment. If the results make sense, the model can then be used for the design and synthesis of new agents.

13.16 *De novo* design

De novo design involves the design of novel structures based on the structure of the binding site with which they are meant to interact.

13.16.1 Thymidylate synthase inhibitors

In theory, it should be possible to design a drug for a particular target if one knows the structure of the binding site. The process sounds quite easy. A molecular skeleton could be designed which would fit the binding site and fill the available space. Suitable functional groups could then be incorporated into the structure to ensure binding interactions with nearby amino acid residues.

In reality, *de novo* design is not as straightforward as it seems and success has been limited. However, one good illustration is the design of inhibitors for the enzyme thymidylate synthase. This enzyme catalyses the methylation of deoxyuridylate monophosphate (dUMP) to deoxythymidylate monophosphate (dTMP) using 5,10-methylene tetrahydrofolate as a coenzyme (Fig. 13.41). Inhibitors of this enzyme have been shown to be antitumour agents since they prevent the biosynthesis of one of the required building blocks for DNA. Traditional inhibitors have been modelled on dUMP or the enzyme cofactor 5,10-methylene tetrahydrofolate (Fig. 13.42), which means that these inhibitors are structurally related to the natural substrate and cofactor. Unfortunately, this increases the possibility of side-effects due to inhibition of other enzymes and receptors which use these molecules as natural ligands. Therefore, it was decided that a novel structure should be designed which was unrelated to either of the natural substrates.

Before starting the *de novo* design, a good supply of the enzyme was required. Although human thymidylate synthase is not readily available in large quantities, it

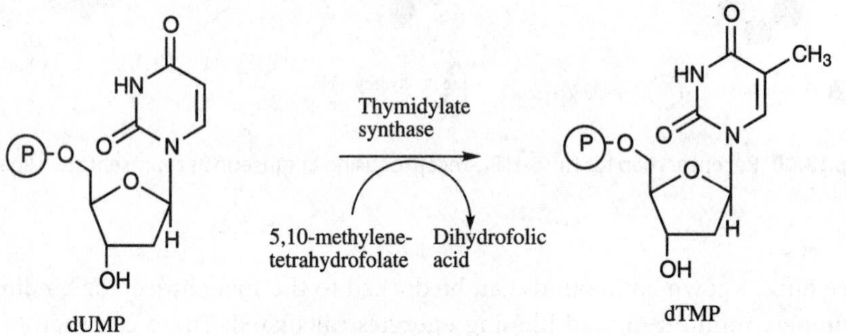

Fig. 13.41 Thymidylate synthase.

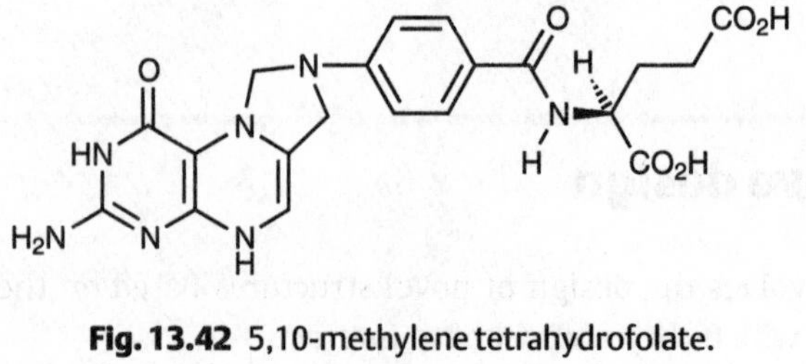

Fig. 13.42 5,10-methylene tetrahydrofolate.

was possible to obtain good quantities of the bacterial version from *E. coli* by using recombinant DNA technology to clone the gene and then expressing it in fast growing cells. The bacterial enzyme is not identical to the human version, but it is very similar and so it was considered to be a reasonable analogue.

The enzyme was crystallized along with the known inhibitors 5-fluorodeoxyuridylate and CB3717 (Fig. 13.43). These structures mimic the substrate and the coenzyme, respectively, and bind to the sites normally inhabited by these structures. The structure of the enzyme–inhibitor complex was then determined by X-ray crystallography and loaded into the computer.

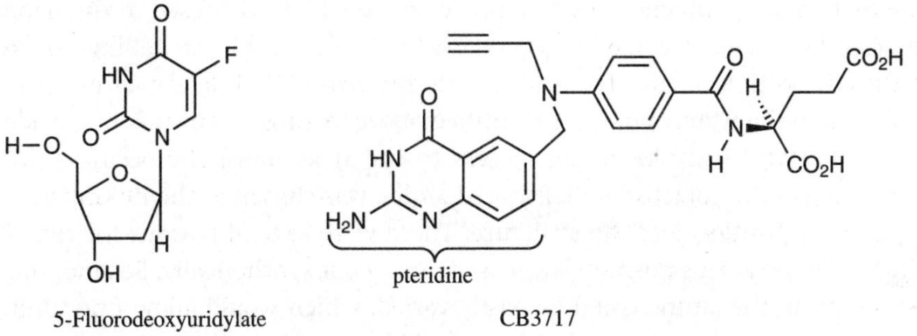

Fig. 13.43 Inhibitors of thymidylate synthase.

A study of the enzyme–inhibitor complex revealed where the inhibitors were bound and also the binding interactions involved. For CB3717, the binding interactions around the pteridine portion of the inhibitor were identified as involving hydrogen bonding interactions to two amino acids (the carboxylate ion of Asp-169 and the main chain peptide link next to Ala-263). There was also a hydrogen bonding interaction to a water molecule which acted as a hydrogen bonding bridge to Arg-21 (Fig. 13.44).

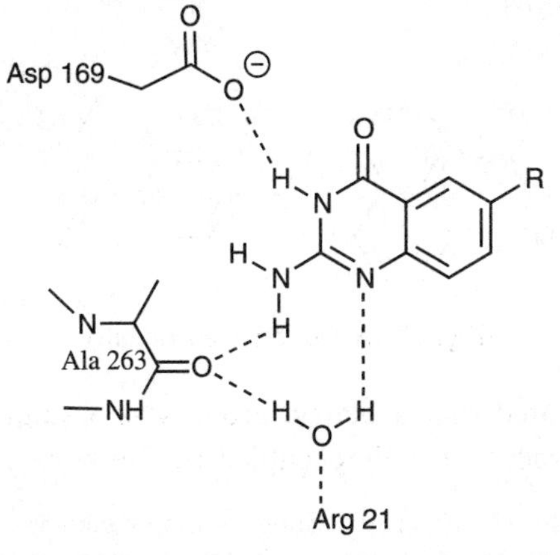

Fig. 13.44 Binding interactions in active site.

Using molecular modelling, the inhibitor was deleted from the binding site to allow further analysis of the empty binding site[3].

A grid was set up within the binding site, and an aromatic CH probe was placed at each grid point to measure hydrophobic interactions and thus identify hydrophobic regions. From this analysis it was discovered that the pteridine portion of CB3717 was positioned in a hydrophobic pocket despite the presence of the hydrogen bonding interactions which held it there. The boundaries of this hydrophobic region were determined and a naphthalene ring was found to be a suitable hydrophobic molecule to fit the pocket, yet still leave room for the addition of a functional group which would be capable of forming the important hydrogen bonds.

The functional group chosen was a cyclic amide which was fused to the naphthalene scaffold to create a naphthostyryl scaffold (Fig. 13.45). Modelling suggested that the NH portion of the amide would bind to Asp-169 while the carbonyl group would bind to the water molecule identified above. A substituent was now added to the naphthostyryl scaffold to gain access to the space normally occupied by the benzene ring of the cofactor. A dialkylated amine was chosen as the linking unit and was placed at position 5 of the structure. There were several reasons for this. First, adding an amine at this position was easy to carry out synthetically. Second, the two substituents on the amine could be easily varied, which would allow fine tuning of the compound. Last, by using an amine it would be possible to have a branching point which could have different substituents without adding an asymmetric centre. If a carbon atom had been added instead, two different substituents would have inevitably led to an asymmetric centre.

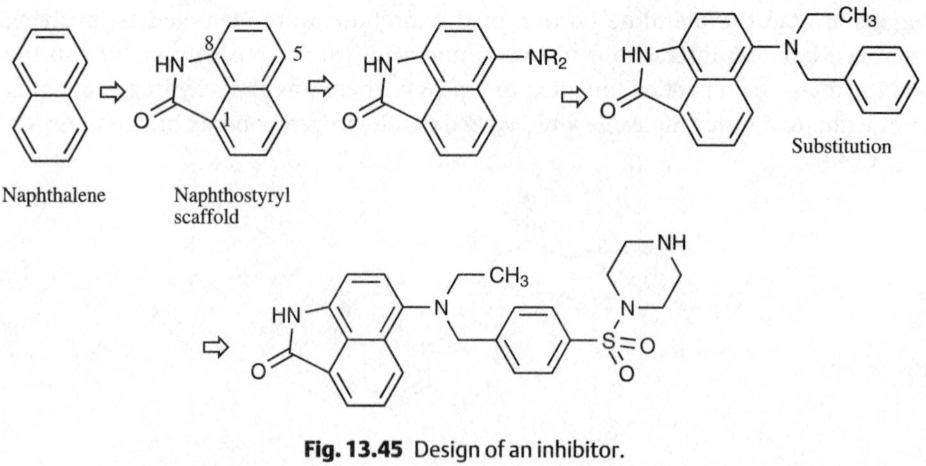

Fig. 13.45 Design of an inhibitor.

Modelling demonstrated that a benzyl group was a suitable substituent for the amine to access the space normally occupied by the benzene ring of the cofactor.

[3] It should be stated at this point that generating the empty binding site from the enzyme–ligand complex is better than studying the empty binding site from the pure enzyme. This is because the latter does not take into account the induced fit which occurs on ligand binding.

A (phenylsulphonyl)piperazine group was then added to the *para* position of the aromatic ring to make the molecule more water-soluble – a necessary property if the synthesized structures were to be bound to the enzyme and crystallized for further X-ray studies.

This structure was now synthesized and tested for inhibition against both the bacterial and human versions of the enzyme, and was found to be active, with higher activity for the human enzyme. A crystal structure of the novel inhibitor bound to the bacterial enzyme was successfully obtained and studied to see whether the inhibitor had fitted the binding site as expected. In fact, it was found that the naphthalene ring of the inhibitor was wedged deeper into the pocket than expected because of more favourable hydrophobic interactions. As a result, the cyclic amide was failing to form the direct hydrogen bond interaction to Asp-169 which had been planned, and was hydrogen bonding to a bridging water molecule instead. The lactam carbonyl oxygen was also too close to Ala-263 and this had caused this residue to shift 1 Å from its usual position. This in turn had displaced the water molecule that had been the intended target for hydrogen bonding (Fig. 13.46).

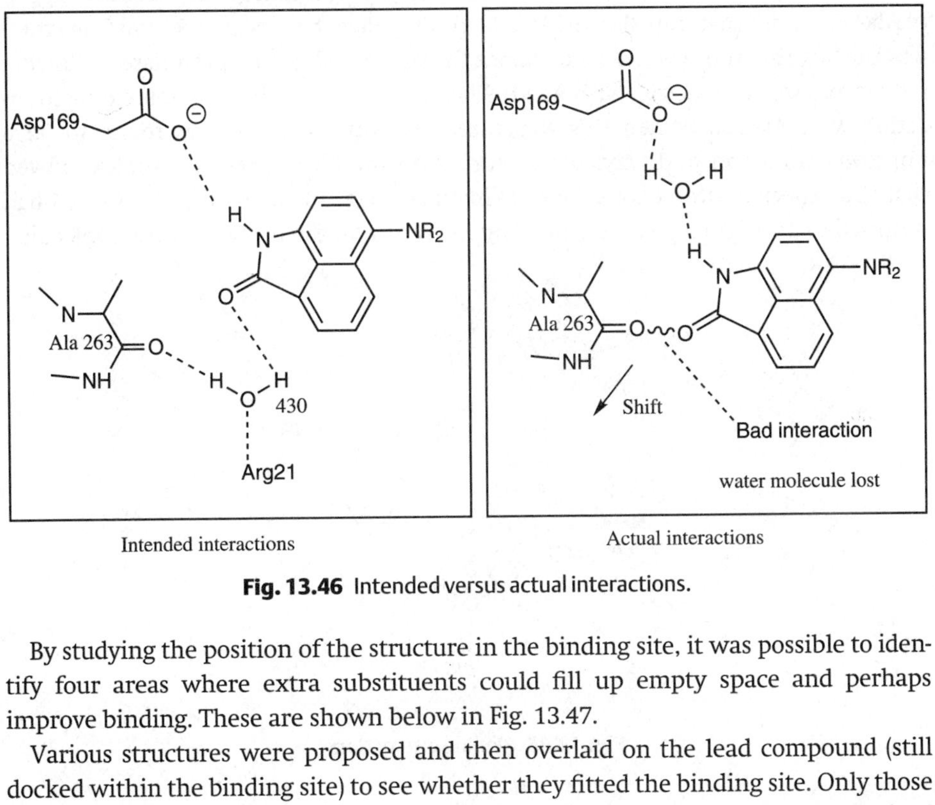

Intended interactions   Actual interactions

Fig. 13.46 Intended versus actual interactions.

By studying the position of the structure in the binding site, it was possible to identify four areas where extra substituents could fill up empty space and perhaps improve binding. These are shown below in Fig. 13.47.

Various structures were proposed and then overlaid on the lead compound (still docked within the binding site) to see whether they fitted the binding site. Only those which passed this test and which were in stable conformations were synthesized (41 in total) and tested for activity. The optimum substituent at each position was then identified.

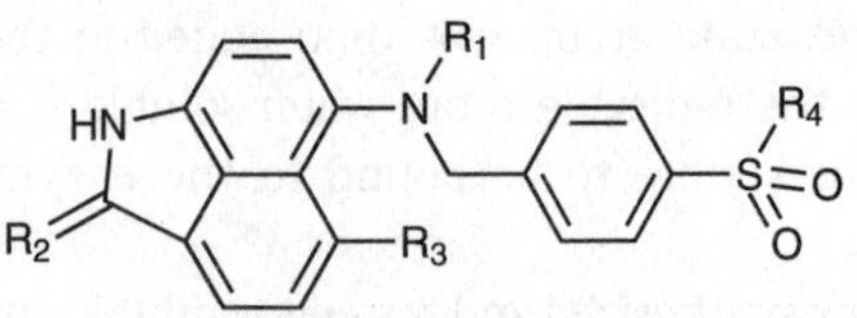

Fig. 13.47 Variable positions (R).

As far as region 1 (R$_1$) was concerned, modelling showed that this substituent fitted into a hydrophobic pocket which became hydrophilic the deeper one got. This suggested that a hydrogen bonding substituent at the end of an alkyl chain might be worth trying (*extension*) and, indeed, a CH$_2$CH$_2$OH group led to an improvement in binding affinity. It was also found that a methyl group was better than the original ethyl group.

In region 3 (R$_3$), there was room for a small group such as a chlorine atom or a methyl group, and both of these substituents led to an increase in activity.

As far as region 2 was concerned (R$_2$), the carbonyl oxygen was replaced with an amidine group which would be capable of hydrogen bonding to the carbonyl oxygen of Ala-263 rather than repelling it. An added advantage in using a basic amidine group was the fact that there was a good chance that it would become protonated, allowing a stronger ionic interaction with Asp-169, as well as a better hydrogen bonding inter-action with Ala-263. When this structure was synthesized, it was found to have improved inhibition, and a crystal structure of the inhibitor–enzyme complex showed that the expected interactions were taking place (Fig. 13.48). Moreover, Ala-263 had returned to its original position, allowing the return of the bridging water molecule.

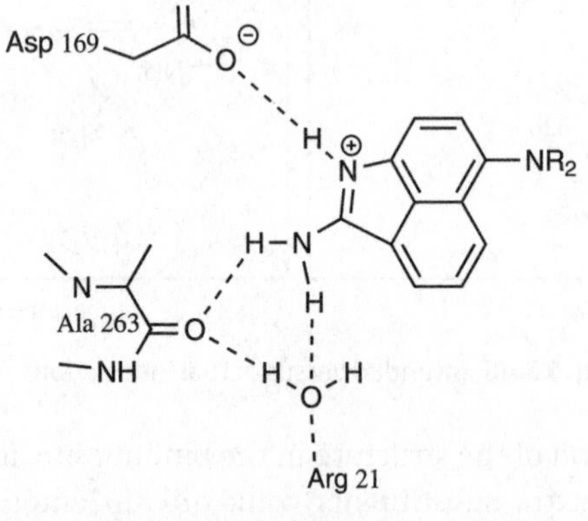

Fig. 13.48 Binding interactions.

Region 4 (R$_4$) was relatively unimportant for inhibitory activity since groups at this position protruded out of the active site into the surrounding solvent and had only minimal contact with the enzyme. Nevertheless, the piperazine ring was replaced with a morpholine group since the latter had some advantages with respect to selectivity and pharmacological properties.

Having identified the optimum groups at each position, structures were synthesized combining some or all of these groups. The presence of the amidine resulted in the best improvement in activity and so the presence of this group was mandatory. Interestingly, adding all the optimum groups is not as beneficial as adding some of them.

The modified structure (Fig. 13.49) was synthesized and was found to be a potent inhibitor which was 500 times more active than the original amide. A crystal structure of the enzyme–inhibitor complex showed a much better fit and the compound was put forward for clinical trials as an antitumour agent.

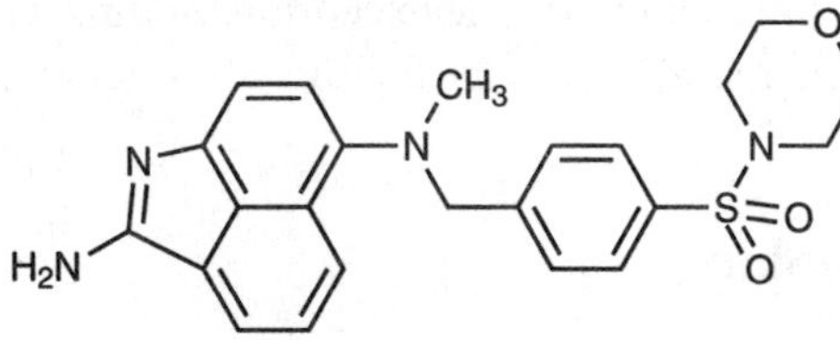

Fig. 13.49 Modified inhibitor.

A couple of general points are worth mentioning about the strategies involved in *de novo* design. First of all, it may be tempting to design a molecule which completely fills the available space in the binding site. However, this would not be a good idea for the following reasons:

• The position of atoms in the crystal structure is accurate only to 0.2–0.4 Å and allowance should be made for that.

• It is possible that the designed molecule may not bind to the binding site exactly as predicted. If the intended fit is too 'tight', a slight alteration in the binding mode may prevent the molecule binding at all. It would be better to have a loose-fitting structure in the first instance and to check whether it binds as predicted. If it does not, the loose fit gives the molecule a chance to bind in an alternative fashion.

• It is worth leaving scope for variation and elaboration in the molecule, to study how fine tuning affects the molecule's properties with respect to binding affinity and pharmacokinetics.

Other points to take into consideration in *de novo* design are the following:

• Flexible molecules are better than rigid molecules since the former are more likely to find an alternative binding conformation should they fail to bind as expected. This allows modifications to be carried out based on the actual binding mode. If a rigid molecule fails to bind as predicted, it may not bind at all.

• It is pointless designing molecules which are difficult or impossible to synthesize.

• Similarly, it is pointless designing molecules which need to adopt an unstable conformation in order to bind.

• Consideration of the energy losses involved in water desolvation should be taken into account.

• There may be subtle differences in structure between receptors and enzymes from different species. For example, the computer modelling studies described above were carried out on bacterial thymidylate synthase enzyme rather than the human version. Fortunately, the activities of the designed inhibitors were actually greater for the human enzyme than for the bacterial enzyme and this was put down to the fact that the hydrophobic space available for the naphthalene ring was larger in the human enzyme than for the bacterial one. Fortunately, most changes carried out had beneficial effects for both enzymes, but with one exception. Adding a methyl group at R_3 to the amidine led to an increase in activity for the bacterial enzyme but not for the human enzyme.

13.16.2 Fragment linking

Computer programmes have also been written which automatically design novel structures to fit known binding sites. For example, ChemX has a *de novo* capability which can be used if a pharmacophore triangle has been defined for the binding site. The programme creates simple structures to match this pharmacophore by accessing a databank of drug-like fragments (e.g. aromatic rings, amino groups, etc.), and then automatically linking them together to create novel structures which contain the required pharmacophore. Each new structure is vetted for its ease of synthesis and a score is given. Scores are determined by the sort of linking bond involved. For example, a carbon–carbon bond would be more favoured than, say, an oxygen–nitrogen bond.

13.17 Planning combinatorial syntheses

Combinatorial synthesis (see Chapter 12) is a method of creating a large number of compounds in small scale and in a small time scale. The combinatorial synthesis could be carried out to synthesize as many possible compounds as possible from the starting materials and reagents available. However, molecular modelling can help to focus the study such that a smaller number of structures are made whilst retaining the chances of success.

One method of doing this is based on the identification of pharmacophore triangles. Let us assume that the combinatorial synthesis is being carried out to generate 1000 compounds with as diverse a range of structures as possible to find an active compound. The number of different pharmacophores generated from the 1000 compounds would be an indication of the structural diversity. Therefore, a library of compounds which generated 100 000 different pharmacophores would be superior to a library of similar size which produced only 100 different pharmacophores. Pharmacophore searching can be done on all the possible target structures which can be generated from the synthesis to select those structures which will give the widest diversity. These would then be the compounds synthesized.

First, all the possible target structures are automatically ranked on their level of rigidity. This can be achieved by identifying the number of rotatable bonds. Pharmacophore searching then starts with the most rigid structure. All possible pharmacophore triangles are identified for the structure. If different conformations are possible, these are generated and the various pharmacophore triangles arising from these are added to the total. The next structure is then analysed for all of its pharmacophore triangles. Again, triangles would be identified for all the conformations which are possible. The pharmacophores from the first and second structures are then compared. If more than 10% of the pharmacophores from the second structure are different to those from the first, the structure is added to the list of those structures which should be synthesized for the library. Both sets of pharmacophores are combined and the next structure is analysed for all of its pharmacophores. These are compared with the total number of pharmacophores from structures 1 and 2 and are only accepted if more than 10% new pharmacophores can be generated. This process is repeated throughout all the target structures, eliminating all compounds which generate less than 10% of new pharmacophores. In this way, it is possible to cut the number of structures which need to be synthesized by 80–90%, with only a 10% drop in the number of pharmacophores generated.

There is a good reason for starting this analysis with a rigid structure. A rigid structure only has a few conformations and there is a good chance that most of these will be presented when the structure interacts with its target. Therefore, one can be confident that these conformations and their associated pharmacophores are fairly represented. If the analysis was started with a highly flexible molecule having a large number of conformations and pharmacophore triangles, there is less chance that these conformations will be fairly represented when the structure meets the target protein. Rigid structures which express specific conformations more clearly may be rejected later in the analysis since the pharmacophores are already represented in theory. As a result, some pharmacophores which should be present are actually left untested.

It is also possible to use modelling software to carry out a substituent search when planning a combinatorial synthetic library. Here, one defines the common scaffold created in the synthesis as well as the number of substituents which are attached and their point of attachment. Next, the general structures of the starting materials used to introduce these substituents are defined. The substituents which can be added to the structure can then be identified by having the computer search databanks for commercially available starting materials. The programme then generates all the possible structures which can be included in the library, based on the starting materials which are available. Once these have been identified, they can be analysed for pharmacophore diversity as described above.

Alternatively, the various substituents which are possible can be clustered into similar groups based on their structural similarity. This allows starting materials to be preselected, choosing a representative compound from each group. The structural similarity of different substituents would be based on a number of criteria, such as the distance between important binding centres, the types of centre present, particular bonding patterns, and functional groups.

13.18 Database handling

The development of a drug requires the analysis of a large amount of data. For example, activity against a range of targets has to be measured to ensure that the compounds not only have good activity against the target protein, but also have good selectivity with respect to a range of other targets. When it comes to rationalizing results, many other parameters have to be considered, such as molecular weight, log P, pKa, etc. The handling of such large amounts of data requires dedicated software.

There are several software programmes available for the handling of data which allow medicinal chemists to assess quickly biological activity versus physical properties, or to compare the activities of a series of compounds at two different targets. Such programmes allow results to be presented in a visual qualitative fashion, allowing quick identification of any likely correlations between different sets of data.

For example, if one wanted to see whether the log P value of a series of compounds was related to their α- and/or β-adrenergic activity, then a 2D plot could be drawn up comparing α-adrenergic activity versus β-adrenergic activity. The log P value of each compound could then be indicated by a colour code for the various points on the plot. In this way, it would be easy to see whether these three properties were related. Such an analysis might show, for example, that log P is related to compounds having low α- and high β-activity.

Some programmes (e.g. DIVA) can be used to assess the biological results from a combinatorial synthetic study. The scaffold used in the synthesis is defined first, then the substituents are defined. Once the biological test results are obtained, a tree diagram can be drawn up to assess which substitution point is most important for activity. For example, suppose there were three substitution points on the scaffold, the programme could analyse the data to identify which of the substitution points was the most important in controlling the activity. The data relevant for this particular substituent could then be split into three groups corresponding to good, average, and poor activity. For each of these groups, the programme could be used to identify the next most important substitution point, and so on.

13.19 Case study

It was once claimed that drugs could be designed using molecular modelling alone. However, to date, no drug has been designed purely by molecular modelling. Even the thymidylate synthase inhibitors were modified in the light of experimental results. Molecular modelling should be seen as one of the powerful weapons which the medicinal chemist has to hand, but it is not the only weapon. The following case study illustrates how computer modelling has been used as an aid in drug design alongside many of the other tools and strategies described in previous chapters. The project in question was carried out by SmithKline Beecham and was aimed at finding novel anxiolytic and antidepressant agents.

13.19.1 The target

First, a suitable target had to be identified and it was decided to design an antagonist for a serotonin receptor. Serotonin (or 5-hydroxytryptamine; 5-HT) (Fig. 13.50) is an important neurotransmitter in the central nervous system (CNS), and abnormalities in serotonin levels are thought to be involved in a variety of disorders such as anxiety, depression, and migraine. Antagonists are therefore useful in treating these diseases. However, these agents have various problems and side-effects associated with them. The discovery of various types of serotonin receptor allowed the possibility of designing antagonists which might be more selective in their action and have fewer side-effects as a result.

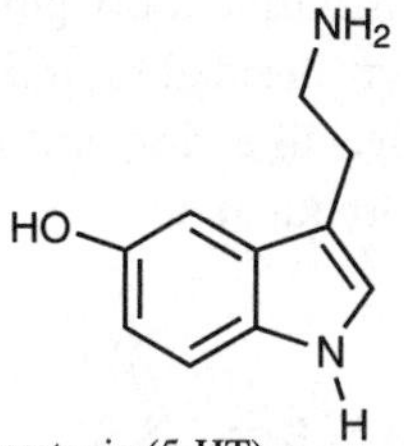

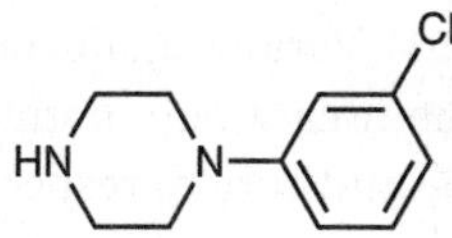

Serotonin (5-HT)   *meta*-Chlorophenylpiperazine (*m*CPP)

Fig. 13.50 Serotonin and *m*CPP.

There are seven main types of serotonin receptor (5-HT_1 to 5-HT_7) and several subtypes of these. For example, the 5-HT_2-receptor consists of three subtypes (5-HT_{2A}, 5-HT_{2B}, and 5-HT_{2C}). Of particular interest was the 5-HT_{2C}-receptor since there was some evidence that this receptor might be involved in anxiety. For example, *meta* chlorophenylpiperazine (*m*CPP) is an agonist which shows some selectivity for the 5-HT_{2B}- and 5-HT_{2C}-receptors, and has been shown to cause anxiety in animal and human studies. Since the 5-HT_{2B}-receptor is mainly in the peripheral nervous system, it is likely that the CNS effects are caused by the agonist acting on the 5-HT_{2C}-receptor which is only present in the CNS. The serotonin antagonists which were on the market at the start of this project did not show any selectivity for the 5-HT$_{2C}$- over the 5-HT$_{2A}$-receptor and so it was argued that a selective antagonist might have improved properties over established drugs.

Having identified a likely target, the aim was now to identify a suitable profile of activity for the intended drug. Clearly, the antagonist should be as selective as possible for the 5-HT$_{2C}$-receptor to reduce the possibility of side-effects. In particular, the challenge was to obtain selectivity with respect to the 5-HT$_{2A}$-receptor. Furthermore, the new drug should not affect metabolic enzymes (e.g. cytochrome P450 enzymes), again to reduce possible side-effects and to avoid the possibility of drug–drug interactions. The ideal drug should also be non-sedating, have no interaction with alcohol, have a fast onset of action, high response rate, and show no withdrawal effects.

13.19.2 Testing procedures

Both *in vitro* and *in vivo* tests were required for the study. The *in vitro* tests involved radioligand binding studies carried out on human 5-HT$_{2A}$-, 5-HT$_{2B}$-, and 5-HT$_{2C}$-receptors which had been cloned and expressed in fast growing cells. Testing for any inhibitory activity against cytochrome P450 enzymes was also performed *in vitro*.

In vivo activity was measured by the ability of compounds to block hypoactivity in rats, brought on by administering *m*CPP.

13.19.3 Lead compound to SB200646

The lead compound for the project was a drug which had been produced by Lilly Pharmaceuticals. One of the problems with this compound was its insolubility in water and so it was decided to replace the phenyl ring with a more polar pyridine ring (*ring variation*) (Fig. 13.51). Various analogues were synthesized to find the best substitution positions for the urea group, both for the pyridine ring and the indole ring, revealing that 3- and 5-substitution, respectively, were ideal.

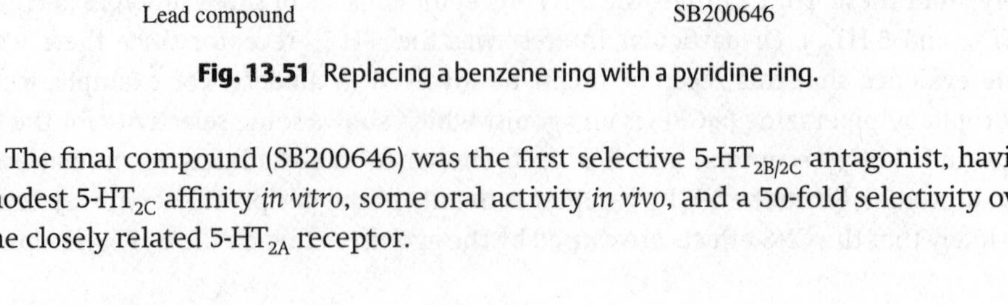

Fig. 13.51 Replacing a benzene ring with a pyridine ring.

The final compound (SB200646) was the first selective 5-HT$_{2B/2C}$ antagonist, having modest 5-HT$_{2C}$ affinity *in vitro*, some oral activity *in vivo*, and a 50-fold selectivity over the closely related 5-HT$_{2A}$ receptor.

13.19.4 **SB200646 to SB206553**

The urea functional group was the most flexible region of the molecule and so the next tactic was to reduce this flexibility by locking one of the bonds into a ring (*rigidification*).

Four conformationally restrained structures (Fig. 13.52) were prepared where $n = 1,2$. In both cases, the introduction of a six-membered ring was bad for affinity, whereas the five-membered ring increased affinity.

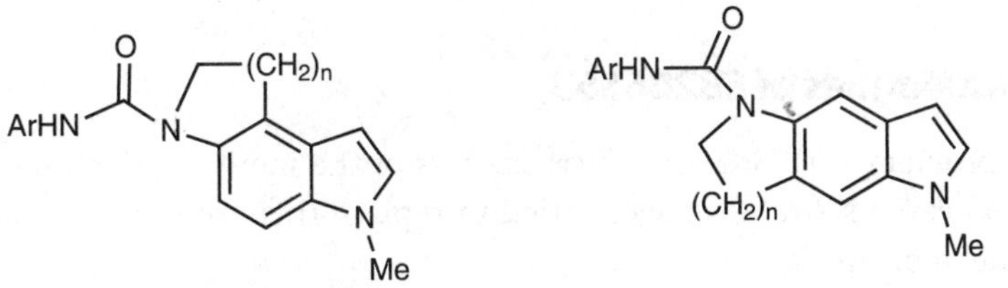

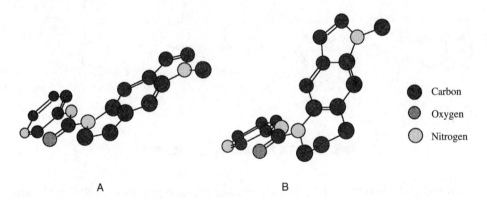

Fig. 13.52 Rigidification.

The difference in affinity between the six-membered and five-membered rings was rationalized by carrying out molecular modelling on these structures. This showed a marked difference in geometry between the two ring sytems (Fig. 13.53). With a five-membered ring (A), the overall structure is roughly planar. This is not the case with the six-membered ring (B). Here, the overall structure is more twisted. Clearly, the planar molecule interacts with the receptor more strongly than the twisted molecule.

● Carbon

◕ Oxygen

○ Nitrogen

A   B

Fig. 13.53 Relative geometries of the rigidified analogues.

Therefore, rigidification was carried out using the five-membered ring, leading to SB206553 (Fig. 13.54) which resulted in a 10-fold increase in *in vitro* affinity and a 160-fold selectivity over the 5-HT$_{2A}$-receptor. The compound was also shown to relieve anxiety *in vivo* with a fourfold increase in potency in the rat hypolocomotion assay. Unfortunately, this compound was found to undergo metabolism in the body, resulting in demethylation and the production of a non-selective active metabolite.

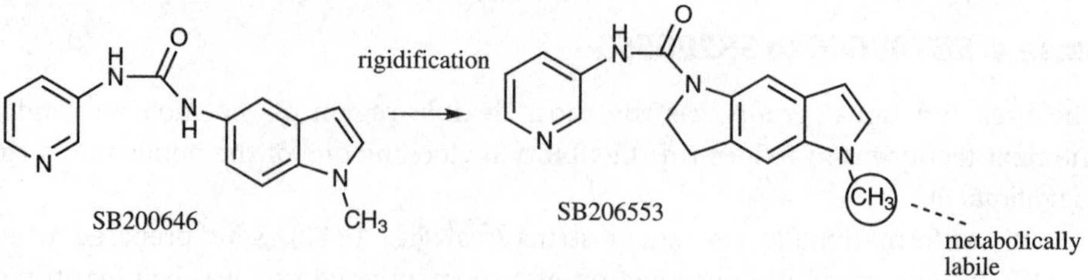

Fig. 13.54 SB206553.

13.19.5 Analogues of SB206553

The main problem with SB206553 was the loss of the methyl group due to metabolism, and so various strategies were tried to replace this group or to make it more stable to metabolism.

First, a number of tricyclic analogues of SB206553 were synthesized to explore the effects of fusing the five-membered ring at different positions of the aromatic ring (*ring variation*) (Fig. 13.55). Such a variation shifts the methyl group relative to the benzene ring, and this might be sufficient to make it unrecognizable to metabolic enzymes, especially if the benzene ring is crucial to how the molecule binds to the active site. However, all of these compounds were less active and less selective, demonstrating that such a tactic also makes the methyl group less recognizable to the target receptor.

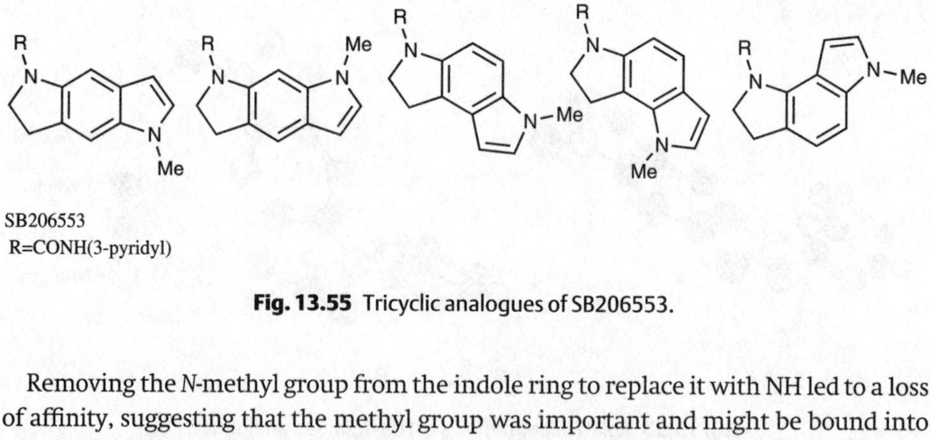

SB206553
R=CONH(3-pyridyl)

Fig. 13.55 Tricyclic analogues of SB206553.

Removing the N-methyl group from the indole ring to replace it with NH led to a loss of affinity, suggesting that the methyl group was important and might be bound into a hydrophobic pocket in the binding site. To explore the size of this proposed pocket, the methyl group was replaced with larger alkyl groups (*varying substituents*) to see whether this would lead to increased hydrophobic interactions and an increased affinity. From this, it was found that groups such as Et, Pr, and ^{i}Pr resulted in a similar 5-HT_{2C}-affinity and slightly increased selectivity. However, an N-benzyl group was bad, both for affinity and for selectivity. Substitution elsewhere in the pyrrole ring was also bad for affinity (Fig 13.56).

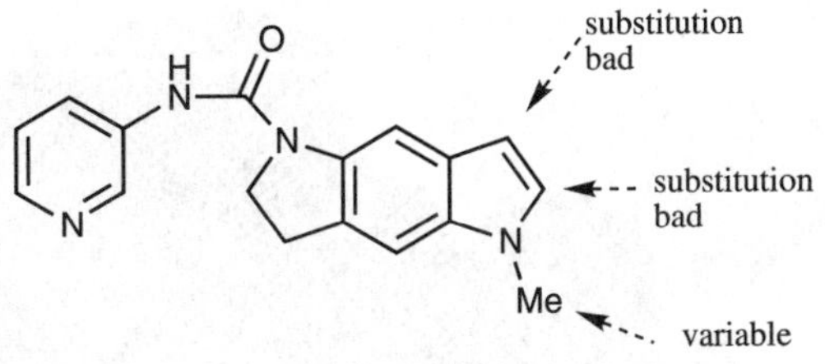

Fig. 13.56 Varying substituents on SB206553.

An isostere for the pyrrole ring was now sought by replacing it with alternative five-membered heterocycles (Fig. 13.57) (*ring variation*). A thiophene led to a loss in selectivity and this was attributed to the loss of the methyl group, which could no longer fit the proposed hydrophobic pocket. A substituted furan ring was used instead, with a methyl substituent positioned to fit the proposed pocket. Since this group was now attached to carbon rather than nitrogen it was expected that it would be more stable to metabolism. The resulting compound retained good *in vitro* affinity and selectivity, but had poor oral activity *in vivo*. However, *in vivo* activity was better if the compound was administered intravenously, suggesting that the compound was either not absorbed from the gastrointestinal tract or was rapidly metabolized.

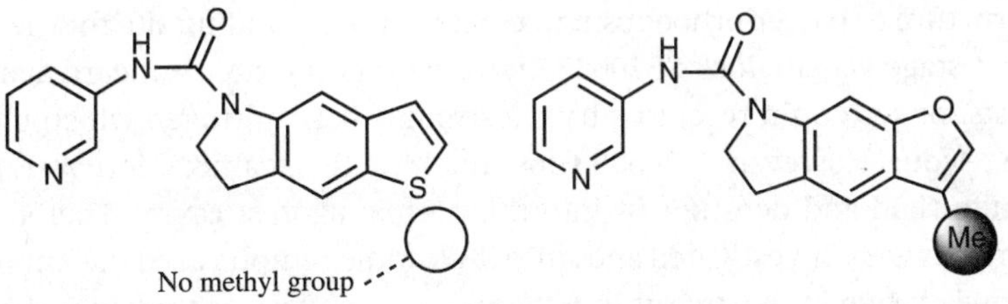

Fig. 13.57 Ring variation of SB206553.

13.19.6 Molecular modelling studies on SB206553 and analogues

Having synthesized several analogues, molecular modelling was carried out to overlay both active and inactive compounds, using the urea group as the common feature. The space occupied by each molecule was then analysed and compared with its affinity for both the 5-HT$_{2A}$- and 5-HT$_{2C}$-receptors. It was then assumed that compounds which showed poor activity must have substituents in sterically disallowed areas of the relevant binding site. This allowed a map to be created (Fig. 13.58) showing a volume in space (coloured dark grey) which could be accessed by molecules having 5-HT$_{2C}$-affinity, and which included a volume (coloured light grey) which was disallowed for 5-HT$_{2A}$-affinity. (The pyridine ring was not included in the calculations because of the large number of conformations it can adopt.) This crucial light grey area which was allowed for 5-HT$_{2C}$-activity but not for 5-HT$_{2A}$-activity was in the region of the *N*-methyl group of SB206553.

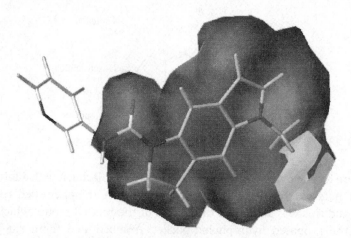

Fig. 13.58 5-HT$_{2C}$ allowed volume (dark grey) includes a 5-HT$_{2A}$ disallowed volume (light grey) crucial for selectivity (shown around the structure of SB206553).

A model 5-HT$_{2C}$-receptor was now built using molecular modelling software. The amino acid sequence for the 5-HT$_{2C}$-receptor was obtained and compared with other related serotonin receptors to identify the likely transmembrane regions (the receptor is a member of the 7-TM receptor family). A model was built by analogy with the crystal structure of bacteriorhodopsin to create a model binding site for the receptor.

The next stage was to dock SB206553 into the binding site. Standard agonists and antagonists for serotonin receptors have a strongly basic nitrogen which binds to an aspartate residue. However, SB206553 does not have this feature. Moreover, the molecule is quite rigid and does not fit into the normal agonist cavity. Therefore, other binding modes were investigated and since SAR studies emphasized the importance of the carbonyl oxygen in the urea group, the receptor model was studied to see whether there were any amino acid residues which could act as possible hydrogen bonding donors. Two serine residues were identified fairly close to the important aspartate residue, one of which (Ser-312) is unique to the 5-HT$_2$ class of receptors. Since these compounds do not bind to 5-HT$_1$-receptors, it was presumed that this was an important binding group.

SB206553 was positioned into the pocket such that it interacted with Ser-312 and it was found that a second hydrogen bonding interaction from the urea carbonyl group was possible to another serine residue (Ser-315). This binding interaction also allowed the two aromatic regions (the pyridine and tricyclic ring) to fit into hydrophobic pockets, one of which was dominated by aromatic amino acids, and one which was made up of both aliphatic and aromatic amino acids. However, it could not be determined which ring fitted which pocket and so both possibilities were considered. Once docked, energy minimizations were carried out for both binding modes, and interaction energies calculated. The more stable binding mode was the one in which the pyridine ring was placed in the more aromatic of the hydrophobic pockets (Fig. 13.59).

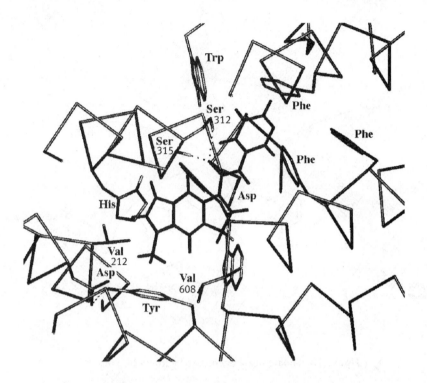

Fig. 13.59 Stable binding mode for SB206553.

In this binding mode, the carbonyl oxygen of the urea interacts through hydrogen bonds with two serine residues, while the indole *N*-methyl group is placed in a hydrophobic pocket adjacent to two valine residues (Val-608 and Val-212; Fig. 13.60). The receptor–ligand model was now studied to identify methods of achieving better selectivity for the 5-HT$_{2C}$-receptor over the 5-HT$_{2A}$-receptor. These studies revealed a difference between the two receptors in the pocket occupied by the *N*-methyl group. In the 5-HT$_{2C}$-receptor there are two valine residues in this pocket. However, in the 5-HT$_{2A}$-receptor, these amino acids are replaced with leucine. Since leucine is bulkier than valine it was proposed that the hydrophobic pocket is smaller in the 5-HT$_{2A}$ receptor, making it more difficult for the *N*-methyl group of SB206553 to fit and thus accounting for the observed selectivity.

13.19.7 SB206553 to SB221284

The next stage was to try and take advantage of these findings. The metabolically labile *N*-methyl group had to be removed, but it was important to replace it with something which could fit the identified hydrophobic pocket to retain selectivity. It was decided to remove the pyrrole ring containing the *N*-methyl group and add substituents to the remaining bicyclic indoline system (A) to act as isosteres for the lost ring (Fig. 13.61). However, since these structures were fairly difficult to synthesize, an

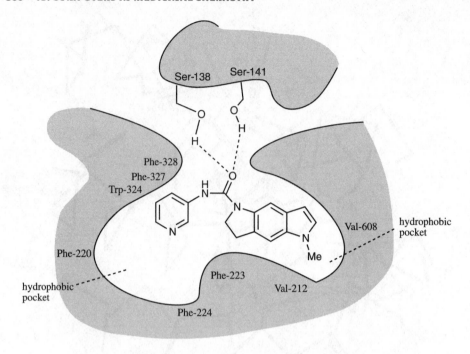

Fig. 13.60 Binding site showing two hydrophobic pockets.

initial study was carried out on the simpler phenyl urea system (B), and a large number of analogues (86) were prepared by combinatorial parallel synthesis to find the best substituents for that system.

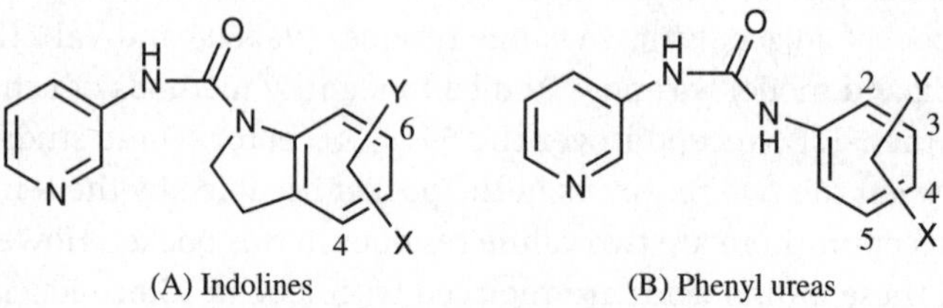

(A) Indolines   (B) Phenyl ureas

Fig. 13.61

Mono-, di- and tri-substituted structures were prepared at various positions and it was found that the preferred substitution pattern consisted of two substituents, *ortho* to each other at positions 3 and 4. Substitution at position 2 was bad for affinity and it was rationalized that this was because of a steric effect between the substituent and the urea group which caused the urea group to go out of plane relative to the aromatic ring. This would result in a twist in the molecule and poorer binding as a consequence.

Having identified positions 3 and 4 as the ideal substitution points, the nature of the substituents was studied to see whether good affinity was related to particular substituents. At position 3, small hydrophobic, electron withdrawing substituents were preferred (e.g. Cl or CF_3). In contrast, small hydrophobic electron donating groups were preferred at position 4 (e.g. SMe or OMe). QSAR studies showed a good correlation between the lipophilicity (π) of small substituents at position 4 and 5-HT$_{2C}$-binding affinity (pK_i) (Fig. 13.62).

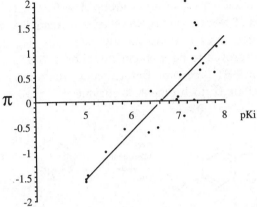

Fig. 13.62 QSAR plot relating lipophilicity of 4-substituents versus binding affinity.

Having established the preferred substituents in the phenyl urea series, these substituents were now tried out in the indoline series. However, since these are different structural systems, it was important to check that the results were valid for both series and so a selection of indolines was prepared and their affinities checked against the corresponding phenyl ureas. The graph (Fig. 13.63) showed the same substituent effect for the indolines as for the phenyl ureas. As expected, the indolines were more potent than the corresponding phenyl ureas.[1]

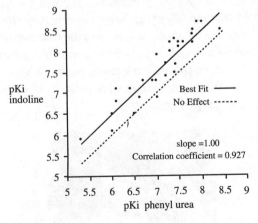

Fig. 13.63 Comparison of pKi valus for indolines and phenyl ureas

[1] The dotted line on the graph is what would be expected if the indolines and phenyl ureas had equal activity.

In the indoline series (Fig. 13.64), an electron withdrawing group at position 6 was preferred (e.g. Cl or CF_3). The substituent at position 5 was the one expected to fit the hydrophobic pocket which was so crucial to selectivity. Increasing the bulk of this substituent using different alkyl groups (*varying substituents*) increased the selectivity but also led to a fall in affinity. The best balance of affinity versus selectivity was achieved with a thioether or an ether group. The best selectivity was found with groups such as SEt , S(n-Pr) or O(i-Pr). However, *in vivo* activity was better with smaller groups such as SMe or OMe. Consequently, SB221284 (Fig. 13.64) was chosen for further study and proved to be a potent inhibitor of $5\text{-HT}_{2B/2C}$-receptors with good selectivity over the 5-HT_{2A}-receptor. Unfortunately, the compound had a major drawback in that it inhibited cytochrome P450 enzymes.

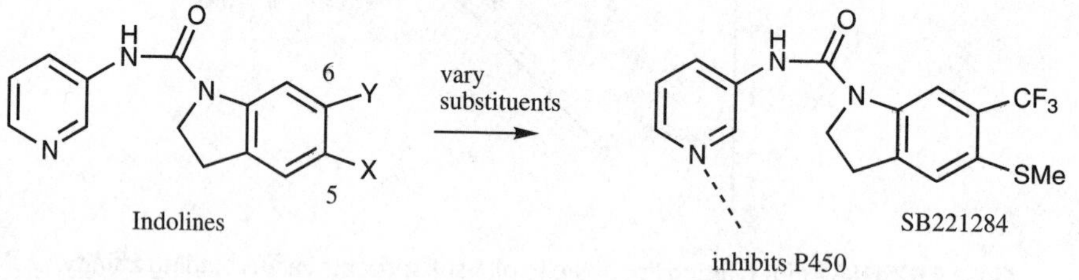

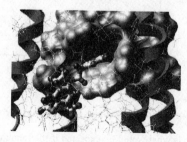

Fig. 13.64 Development of SB221284.

13.19.8 Modelling studies on SB221284

Ligand docking studies were now carried out with SB221284 on the model 5-HT_{2C}-receptor (Figs 13.65 and 13.66). The resulting complex showed that the pyridine and indoline rings were in the same hydrophobic pockets as before. The hydrogen bonding interactions to the two serine residues were present, and the *S*-methyl group fitted into the hydrophobic pocket previously occupied by the *N*-methyl group. Moreover, it was noted that the CF_3 group helped to orientate the thioether correctly for this interaction. Since the CF_3 group was *ortho* to the thiomethyl group, it restricted the latter's rotation (*conformational blocker*), thus favouring conformations where the thiomethyl group was directed towards its binding pocket. It was also found that both the pyridine ring and the thiomethyl group were twisted out of the plane of the indoline ring and were almost perpendicular to it.

Fig. 13.65 Ligand docking studies on SB221284.

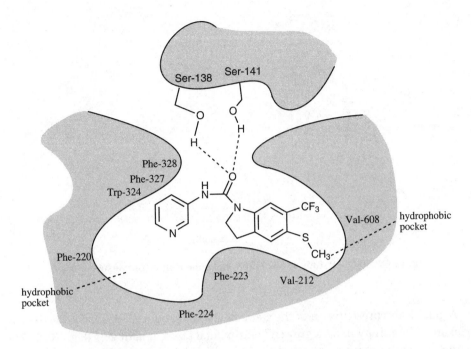

Fig. 13.66 Binding site interactions for SB221284.

Attention was now turned to the hydrophobic pocket occupied by the pyridine ring and it was observed that this pocket was quite deep, suggesting that further aromatic substituents could be added to increase the binding interactions.

Adding an extra substituent to the pyridine ring might also be expected to have a separate beneficial effect. Previous work had shown that the pyridine nitrogen in this class of compounds was responsible for the P450 inhibitory activity and that this activity could be prevented if substituents were placed close to the pyridine nitrogen to act as steric shields.

13.19.9 3D QSAR studies on analogues of SB221284

As a check on the authenticity of the receptor model, a series of 55 substituted analogues of SB221284 was synthesized and a 3D QSAR study (CoMFA) carried out. Each structure was docked into the model receptor and the receptor–ligand complex minimized. The ligands were then removed from the receptor and subjected to CoMFA analysis, leaving out eight structures to serve as a test set. Therefore 47 compounds were analysed and a QSAR equation derived which was checked against the test set of eight structures and found to be a good predictor for their affinities. Affinity values for the 47 structures used in the analysis were calculated and a graph of predicted affinity versus actual affinity showed a good relationship (Fig. 13.67). Further analysis showed that the steric fields derived for the CoMFA analysis were of more importance to the equation than the electrostatic fields.

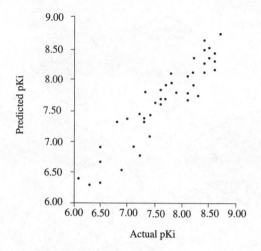

Fig. 13.67 3D QSAR study on SB221284 analogues – predicted versus actual pK_i.

A graphical representation of the steric fields was produced which showed beneficial areas for affinity in dark grey and detrimental areas in light grey (Fig. 13.68). The large number of detrimental areas present suggested that the indoline ring and its substituents were in a tight pocket, allowing little scope for further variation. These results were in agreement with what had been expected from the proposed receptor binding model.

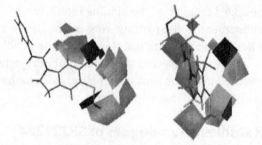

Fig. 13.68 3D QSAR study on SB221284. Steric fields (detrimental in light grey; beneficial in dark grey).

13.19.10 SB221284 to SB228357

Work was now carried out to determine whether adding further aromatic groups to the pyridine ring would have the expected beneficial effects on affinity as well as removing P450 inhibition. For this work, compounds were prepared with a methoxy group at position 5 instead of a methylthio group (Fig. 13.69). However, it had previously been shown that the methoxy and methylthio groups were essentially equivalent as far as affinity and selectivity were concerned.

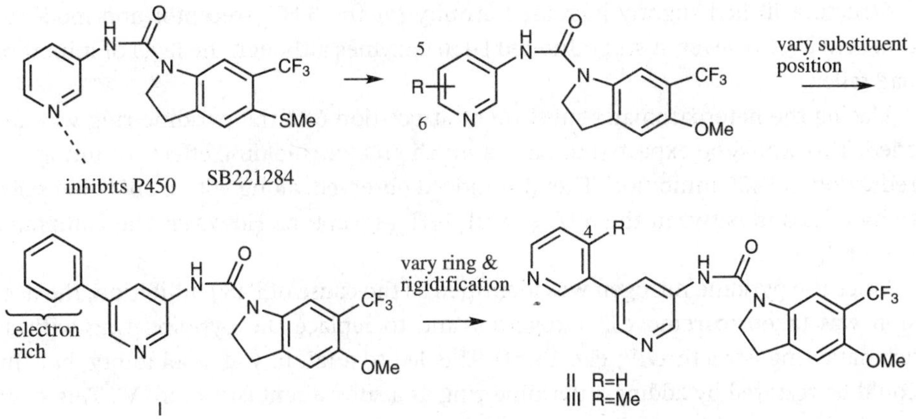

Fig. 13.69 Substitution on the pyridine ring.

Substitutions at various positions of the pyridine ring were carried out, with position 5 being the most promising. If the substituent was an aromatic ring (structure I), this resulted in slightly increased affinity, increased selectivity and also led to a drop in P450 inhibition by 100-fold. However, the level of P450 inhibition was still unacceptable and the structures had poor oral activity. It was also proposed that the aromatic ring might be susceptible to metabolism since it is relatively electron rich. Fluorine substituents were added to make the aromatic ring less electron rich. However, this was detrimental both to affinity and selectivity.

A pyridine ring was used in place of the aromatic ring to increase water solubility and metabolic stability (structure II). This resulted in a tenfold increase in affinity which suggested that an additional binding interaction might be taking place with the nitrogen. Unfortunately, the selectivity fell since affinity for the 5-HT_{2A}-receptor increased even more.

Fortunately, selectivity was recovered by placing a methyl group at position 4 (structure III). It was rationalized that the methyl group in this position forced the two heteroaromatic rings out of plane with each other, because of a steric clash with the *ortho* proton of the neighbouring ring (Fig. 13.70). The methyl group had therefore introduced a level of conformational constraint or rigidification and had forced the molecule into a conformation which favoured the 5-HT_{2C}-receptor over the 5-HT_{2A}-receptor.

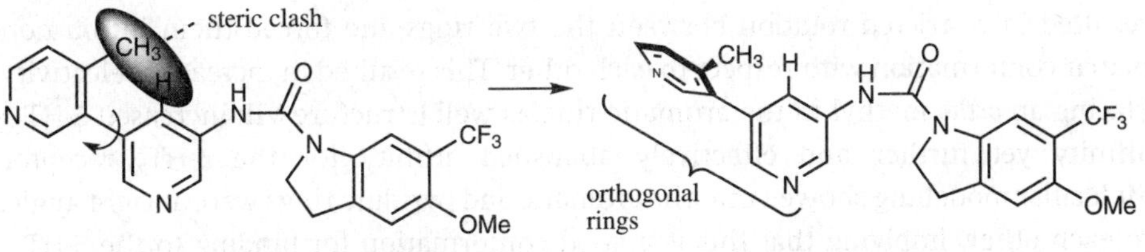

Fig. 13.70 Conformation constraint because of the *ortho* methyl group.

Structure III had slightly increased affinity for the 5-HT$_{2C}$-receptor and moderate oral activity. However, it still inhibited P450 enzymes although the level of inhibition had fallen.

Placing the heteroaromatic substituent at position 6 of the pyridine ring was also tried. This would be expected to have a much greater shielding effect, resulting in a reduction in P450 inhibition. This was indeed observed, along with a 2000-fold selectivity of action between the 5-HT$_{2C}$- and 5-HT$_{2A}$-receptors. However, the compound was orally inactive.

Since the pyridine nitrogen was identified as the cause of P450 inhibition, the decision was taken to remove it altogether and to replace the pyridine ring with an aromatic ring (structure IV; Fig. 13.71). This led to a fall in water solubility, but this could be restored by adding a pyridine ring as a substituent (structure V). This structure had improved selectivity, higher 5-HT$_{2C}$-affinity, potent oral activity, but was still observed to inhibit P450 enzymes.

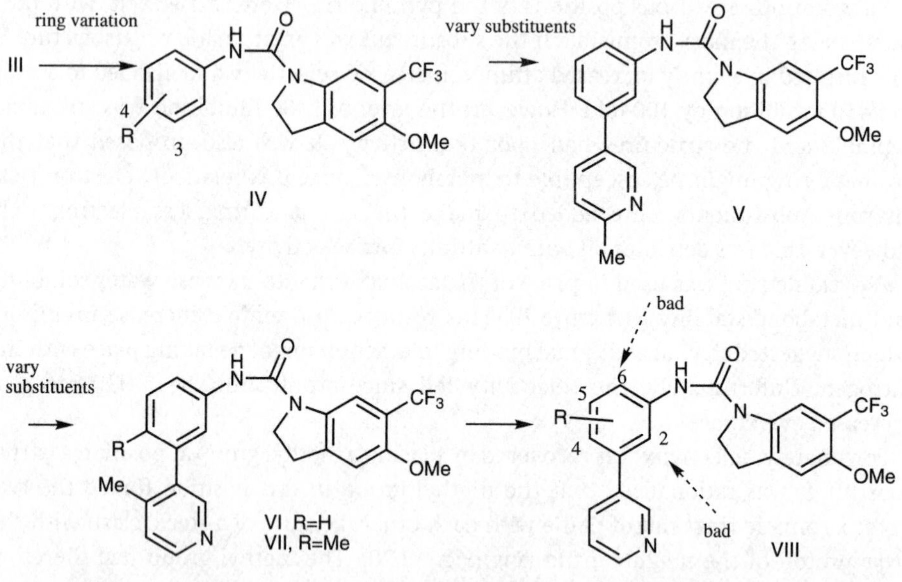

Fig. 13.71 Development of an agent with 5-HT$_{2C}$ affinity and no 5-HT$_{2A}$ affinity (structure VII).

Moving the methyl group in the substituent to the *ortho* position (structure VI) resulted in restricted rotation between the two rings and forced them into a non-planar conformation with respect to each other. This resulted in increased selectivity. Placing an *ortho* methyl in the aromatic ring as well (structure VII) increased 5-HT$_{2C}$-affinity yet further and effectively abolished affinity for the 5-HT$_{2A}$-receptor. Molecular modelling showed that the aromatic and pyridine rings were at right angles to each other, implying that this is a good conformation for binding to the 5-HT$_{2C}$-receptor but not the 5-HT$_{2A}$-receptor. Structure VII also had good oral activity but still retained some P450 activity. Nevertheless, structure VI had the better overall *in vitro* profile, with high *in vivo* potency and was chosen for further modification.

One problem with structure VI was its short duration of action – an indication that it was being metabolized, presumably at the electron rich phenyl ring. A series of analogues was therefore prepared, with electron withdrawing substituents on the aromatic ring (structure VIII). Substituents at positions 2 and 6 were found to be bad for affinity and selectivity (substitutions at this position are likely to twist the ring out of plane with the urea group). Substituents were better tolerated at positions 4 and 5, resulting in good affinity and selectivity. As mentioned previously, substitution at position 4 is particularly good for selectivity since it forces the neighbouring rings out of the same plane. However, the crucial feature at this stage was to increase the duration of activity, and SB228357 was found to have a duration of activity lasting 6 hours (Fig. 13.72).

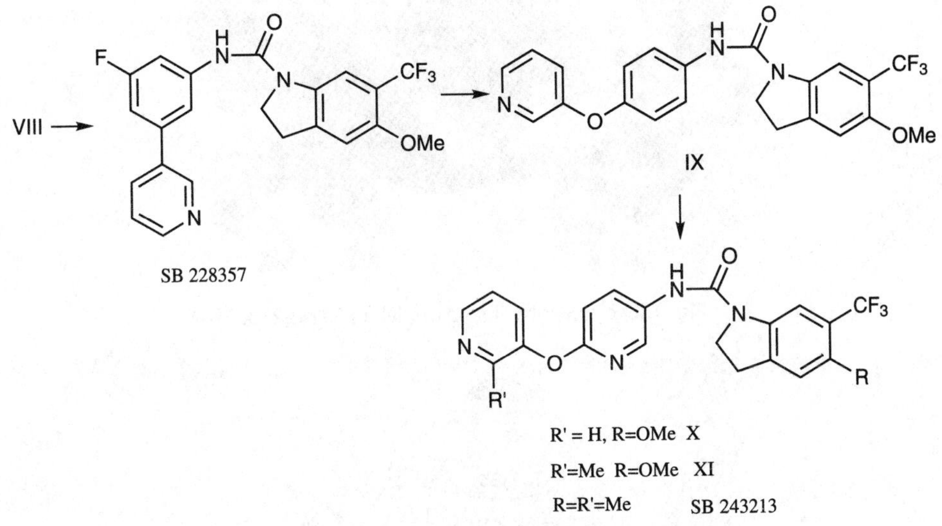

Fig. 13.72 Development of SB228357 and SB243213.

Further receptor ligand modelling suggested that there was still space in the hydrophobic pocket which would allow further extension. A linker group was therefore placed between the pyridine and the phenyl rings to allow a deeper occupation of the pocket (structure IX; Figure 13.72). However, this structure had poor oral activity, probably due to reduced solubility. As a result, the phenyl ring was replaced with the original pyridine ring (structure X). An *ortho* methyl group was added (structure XI) to increase the torsional angle between the two rings, resulting in increased selectivity and affinity for the 5-HT$_{2C}$-receptor. Moreover, selectivity over the 5-HT$_{2B}$-receptor was found for the first time (80-fold). Structure XI had low P450 activity and good *in vivo* activity, but there were worries that the 5-methoxy group on the indoline ring would be metabolically labile and so this was replaced with a methyl group. This compound

(SB243213) had a good profile with negligible P450 activity. The receptor ligand model also showed that the pyridine ring substituent is well inserted into the hydrophobic pocket (Figs 13.73 and 13.74).

SB243213 and SB228357 were both tested to ensure that they were inactive against a range of other receptors, ion channels, and enzymes. Long term studies were also carried out to see whether chronic dosing would affect receptor density or sensitivity. One of the key advantages of SB-243213 over SB-228357 was its much improved solubility and so the former was put forward for Phase 1 clinical trials as a non sedating antidepressant/anxiolytic.

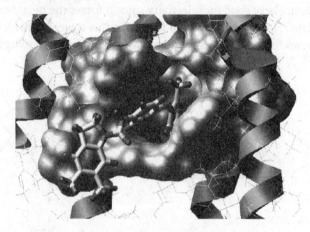

Fig. 13.73 Receptor ligand model for SB243213.

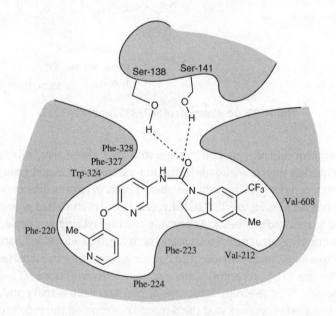

Fig. 13.74 Binding of SB243213.

14 Antibacterial agents

The fight against bacterial infection is one of the great success stories of medicinal chemistry. The topic is a large one and there are terms used in this chapter which are unique to this particular field. Rather than clutter the text with explanations and definitions, Appendix 3 and the Glossary contain explanations of such terms as antibacterial, antibiotic, Gram-positive, Gram-negative, cocci, bacilli, streptococci, and staphylococci.

14.1 The history of antibacterial agents

Bacteria were first identified in the 1670s by van Leeuwenhoek, following his invention of the microscope. However, it was not until the nineteenth century that their link with disease was appreciated. This followed the elegant experiments carried out by the French scientist Pasteur, who demonstrated that specific bacterial strains were crucial to fermentation, and that these and other microorganisms were far more widespread than was previously thought. The possibility that these microorganisms might be responsible for disease began to take hold.

An early advocate of a 'germ theory of disease' was the Edinburgh surgeon Lister. Despite the protests of several colleagues who took offence at the suggestion that they might be infecting their own patients, Lister introduced carbolic acid as an antiseptic and sterilizing agent for operating theatres and wards. The improvement in surgical survival rates was significant.

During the latter half of the nineteenth century, scientists such as Koch were able to identify the microorganisms responsible for diseases such as tuberculosis, cholera, and typhoid. Methods such as vaccination for fighting infections were studied. Research was also carried out to try and find effective antibacterial agents or antibiotics. However, the scientist who can lay claim to be the father of chemotherapy – the use of chemicals against infection – was Paul Ehrlich. Ehrlich spent much of his career studying histology, then immunochemistry, and won a Nobel prize for his contributions to immunology. However, in 1904 he switched direction and entered a field which he defined as chemotherapy. Ehrlich's 'Principle of Chemotherapy' was that a chemical could directly interfere with the proliferation of microorganisms at concentrations tolerated by the host. This concept was popularly known as the 'magic bullet', where the chemical was seen as a bullet which could search out and destroy the invading microorganism without adversely affecting the host. The process is one of selective toxicity, where the chemical shows greater toxicity to the target microorganism than to the host cells. Such selectivity can be represented by a 'chemotherapeutic index', which compares the minimum effective dose of a drug with the maximum dose which can be tolerated by the host. This measure of selectivity was eventually replaced by the currently used **therapeutic index** (see Glossary).

By 1910, Ehrlich had successfully developed the first example of a purely synthetic antimicrobial drug. This was the arsenic-containing compound salvarsan (Fig. 14.1). Although it was not effective against a wide range of bacterial infections, it did prove effective against the protozoal disease sleeping sickness (trypanosomiasis), and the spirochaete disease of syphilis. The drug was used until 1945 when it was replaced by penicillin.

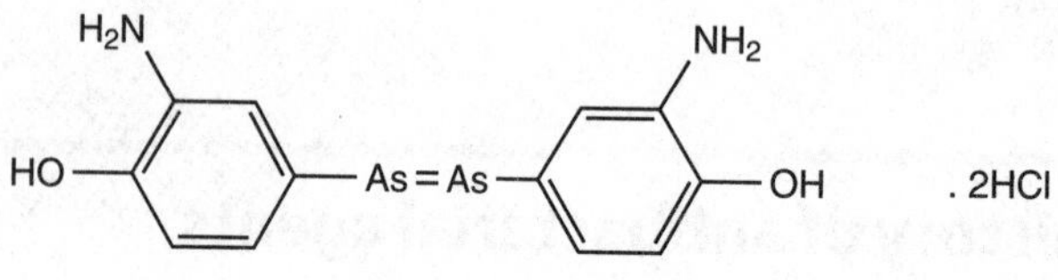

Fig. 14.1 Salvarsan.

Over the next 20 years, progress was made against a variety of protozoal diseases, but little progress was made in finding antibacterial agents until the introduction of proflavine in 1934.

Proflavine (Fig. 14.2) is a yellow-coloured aminoacridine structure which is particularly effective against bacterial infections in deep surface wounds, and was used to great effect during the Second World War. It is an interesting drug since it targets bacterial DNA rather than protein (see Chapter 7). Despite the success of this drug, it was not effective against bacterial infections in the bloodstream and there was still an urgent need for agents which would fight these infections.

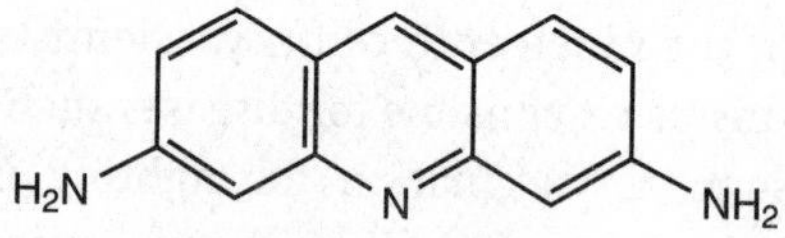

Fig. 14.2 Proflavine.

This need was met in 1935 with the discovery that a red dye called prontosil (Fig. 14.3) was effective against streptococcal infections *in vivo*. As discussed later, prontosil was eventually recognized as a prodrug for a new class of antibacterial agents – the sulpha drugs (sulphonamides). The discovery of these drugs was a real breakthrough since they represented the first drugs to be effective against bacterial infections carried in the bloodstream. They were the only effective drugs until penicillin became available in the early 1940s.

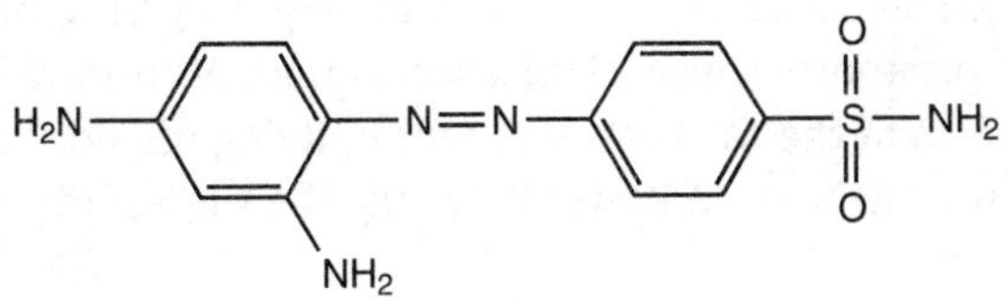

Fig. 14.3 Prontosil.

Although penicillin (see Fig. 14.18) was discovered in 1928, it was not until 1940 that effective means of isolating it were developed by Florey and Chain. Society was then rewarded with a drug which revolutionized the fight against bacterial infection and proved even more effective than the sulphonamides.

Despite penicillin's success, it was not effective against all types of infection and the need for new antibacterial agents still remained. Penicillin is an example of a toxic chemical produced by a fungus to kill bacteria which might otherwise compete with it for nutrients. The realization that fungi might be a source for novel antibiotics spurred scientists into a huge investigation of microbial cultures, both known and unknown.

In 1944, the antibiotic streptomycin (see Fig. 14.74) was discovered from a systematic search of soil organisms. It extended the range of chemotherapy to *Tubercle bacillus* and a variety of Gram-negative bacteria. This compound was the first example of a series of antibiotics known as the aminoglycoside antibiotics.

After the Second World War, the effort continued to find other novel antibiotic structures. This led to the discovery of the peptide antibiotics (e.g. bacitracin; 1945), chloramphenicol (see Fig. 14.76; 1947), the tetracycline antibiotics (e.g. chlortetracycline (see Fig. 14.75; 1948)), the macrolide antibiotics (e.g. erythromycin (see Fig. 14.77; 1952)), the cyclic peptide antibiotics (e.g. cycloserine; (Fig. 14.67; 1955), and, in 1955, the first example of a second major group of β-lactam antibiotics, cephalosporin C (see Fig. 14.42).

As far as synthetic agents were concerned, isoniazid (a pyridine hydrazide structure) was found to be effective against human tuberculosis in 1952, and, in 1962, nalidixic acid (Fig. 14.78) (the first of the quinolone antibacterial agents) was discovered. A second generation of this class of drugs was introduced in 1987 with ciprofloxacin (Fig. 14.78).

Many antibacterial agents are now available and the vast majority of bacterial diseases have been brought under control (e.g. syphilis, tuberculosis, typhoid, bubonic plague, leprosy, diphtheria, gas gangrene, tetanus, gonorrhoea).

This represents a great achievement for medicinal chemistry and it is perhaps sobering to consider the hazards which society faced in the days before penicillin.

Septicaemia was a risk faced by mothers during childbirth and could lead to death. Ear infections were common, especially in children, and could lead to deafness.

Pneumonia was a frequent cause of death in hospital wards. Tuberculosis was a major problem, requiring special isolation hospitals built away from populated centres. A simple cut or wound could lead to severe infection requiring the amputation of a limb, while the threat of peritonitis lowered the success rates of surgical operations.

These were the days of the thirties – still within living memory for many. Perhaps those of us born since the Second World War take the success of antibacterial agents too much for granted.

14.2 The bacterial cell

The success of antibacterial agents owes much to the fact that they can act selectively against bacterial cells rather than animal cells. This is largely due to the fact that bacterial and animal cells differ both in their structure and in the biosynthetic pathways which proceed inside them. Let us consider some of the differences between the bacterial cell (Fig. 14.4) and the animal cell.

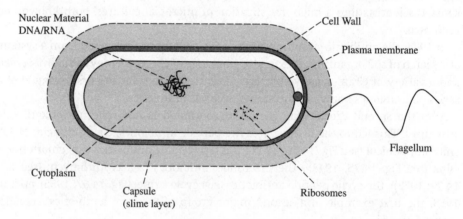

Nuclear Material
DNA/RNA

Cell Wall

Plasma membrane

Flagellum

Cytoplasm

Capsule
(slime layer)

Ribosomes

Fig. 14.4 The bacterial cell.

Differences between bacterial and animal cells

• The bacterial cell has a cell wall as well as a cell membrane, whereas the animal cell has only a cell membrane. The cell wall is crucial to the bacterial cell's survival. Bacteria have to survive a wide range of environments and osmotic pressures, whereas animal cells do not. If a bacterial cell lacking a cell wall was placed in an aqueous environment containing a low concentration of salts, water would freely enter the cell due to osmotic pressure. This would cause the cell to swell and eventually 'burst'. The cell wall does not stop water flowing into the cell directly, but it does prevent the cell from swelling and so indirectly prevents water entering the cell.

- The bacterial cell does not have a defined nucleus, whereas the animal cell does.

- Animal cells contain a variety of structures called organelles (e.g. mitochondria, etc.), whereas the bacterial cell is relatively simple.

- The biochemistry of a bacterial cell differs significantly from that of an animal cell. For example, bacteria may have to synthesize essential vitamins which animal cells can acquire intact from food. The bacterial cells must have the enzymes to catalyse these reactions. Animal cells do not, since the reactions are not required.

14.3 Mechanisms of antibacterial action

There are five main mechanisms by which antibacterial agents act (Fig. 4.5).

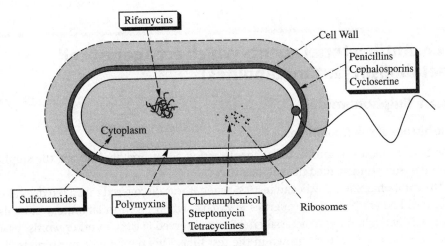

Fig. 14.5 Sites of antibacterial action.

Inhibition of cell metabolism

Antibacterial agents which inhibit cell metabolism are called antimetabolites. These compounds inhibit the metabolism of a microorganism, but not the metabolism of the host. They do this by inhibiting an enzyme-catalysed reaction which is present in the bacterial cell but not in animal cells. The best known examples of antibacterial agents acting in this way are the sulphonamides.

Inhibition of bacterial cell wall synthesis

Inhibition of cell wall synthesis leads to bacterial cell lysis (bursting) and death. Agents operating in this way include penicillins and cephalosporins. Since animal cells do not have a cell wall, they are unaffected by such agents.

Interactions with the plasma membrane

Some antibacterial agents interact with the plasma membrane of bacterial cells to affect membrane permeability. This has fatal results for the cell. Polymyxins and tyrothricin operate in this way.

Disruption of protein synthesis

Disruption of protein synthesis means that essential enzymes required for the cell's survival can no longer be made. Agents which disrupt protein synthesis include the rifamycins, aminoglycosides, tetracyclines, and chloramphenicol.

Inhibition of nucleic acid transcription and replication

Inhibition of nucleic acid function prevents cell division and/or the synthesis of essential enzymes. Agents acting in this way include nalidixic acid and proflavine.

We shall now consider these mechanisms in more detail.

14.4 Antibacterial agents which act against cell metabolism (antimetabolites)

14.4.1 Sulphonamides

The history of sulphonamides

The best examples of antibacterial agents acting as antimetabolites are the sulphonamides (sometimes called the sulpha drugs).

The sulphonamide story began in 1935 when it was discovered that a red dye called prontosil had antibacterial properties *in vivo* (i.e. when given to laboratory animals). Strangely enough, no antibacterial effect was observed *in vitro*. In other words, prontosil could not kill bacteria grown in the test tube. This remained a mystery until it was discovered that prontosil was not in fact the antibacterial agent.

Instead, it was found that the dye was metabolized by bacteria present in the small intestine of the test animal, and broken down to give a product called sulphanilamide (Fig. 14.6). It was this compound which was the true antibacterial agent. Thus, prontosil was the first example of a prodrug (see Chapter 10). Sulphanilamide was synthesized in the laboratory and became the first synthetic antibacterial agent active against a wide range of infections. Further developments led to a range of sulphonamides which proved effective against Gram-positive organisms, especially pneumococci and meningococci.

Despite their undoubted benefits, sulpha drugs have proved ineffective against infections such as Salmonella – the organism responsible for typhoid. Other problems have resulted from the way these drugs are metabolized, since toxic products are frequently obtained. This led to the sulphonamides mainly being superseded by penicillin.

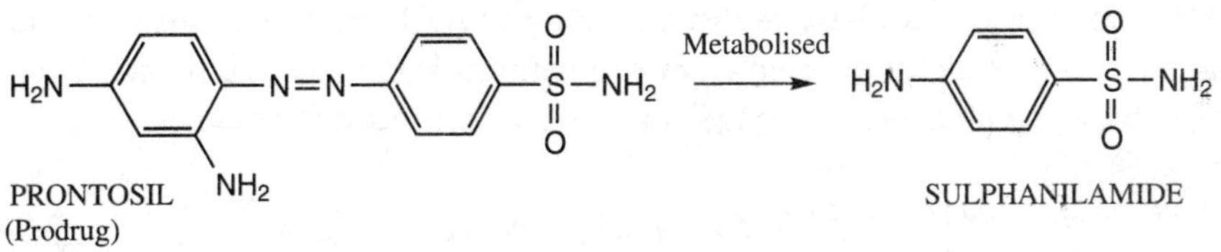

Fig. 14.6 Metabolism of prontosil.

Structure–activity relationships (SAR)

The synthesis of a large number of sulphonamide analogues (Fig. 14.7) led to the following conclusions.

- The *para* amino group is essential for activity and must be unsubstituted (i.e. R=H). The only exception is when R is an acyl group (i.e. amides). The amides themselves are inactive but can be metabolized in the body to regenerate the active compound (Fig. 14.8). Thus amides can be used as sulphonamide prodrugs (see later).
- The aromatic ring and the sulphonamide functional group are both required.
- The aromatic ring must be *para* substituted only.
- The sulphonamide nitrogen must be secondary.
- R″ is the only possible site that can be varied in sulphonamides.

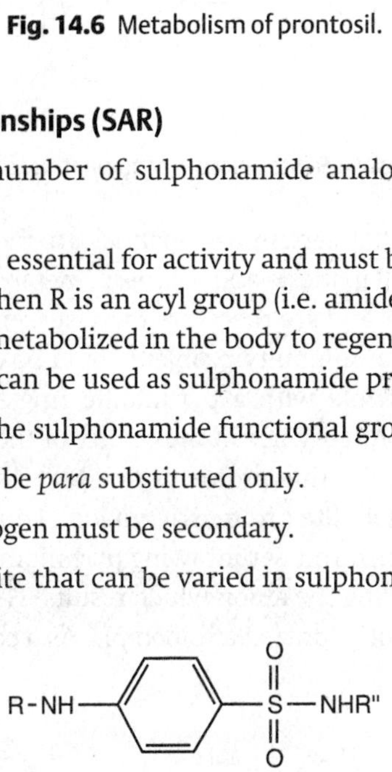

Fig. 14.7 Sulphonamide analogues.

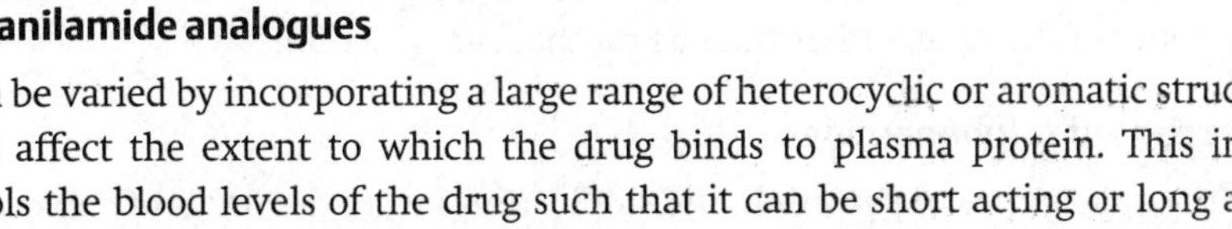

Fig. 14.8 Metabolism of acyl group to regenerate active compound.

Sulphanilamide analogues

R″ can be varied by incorporating a large range of heterocyclic or aromatic structures, which affect the extent to which the drug binds to plasma protein. This in turn controls the blood levels of the drug such that it can be short acting or long acting. Thus, a drug which binds strongly to plasma protein will be slowly released into the blood circulation and will be longer lasting.

Changing the nature of the group R″ has also helped to reduce the toxicity of some sulphonamides. The primary amino group of sulphonamides is acetylated in the body and the resulting amides have reduced solubility, which can lead to toxic effects. For example, the metabolite formed from sulphathiazole (an early sulphonamide) (Fig. 14.9) is poorly soluble and can prove fatal if it blocks the kidney tubules.

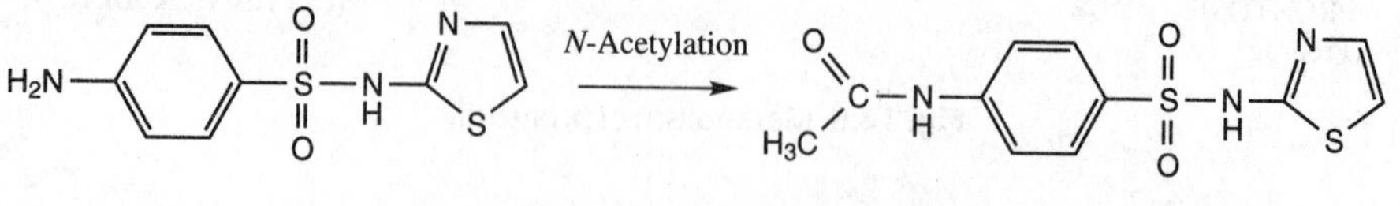

Fig. 14.9 Metabolism of sulphathiazole.

It is interesting to note that certain nationalities are more susceptible to this than others. For example, the Japanese and Chinese metabolize sulphathiazole more quickly than the Americans and are therefore more susceptible to its toxic effects.

It was discovered that the solubility problem could be overcome by replacing the thiazole ring in sulphathiazole with a pyrimidine ring to give sulphadiazine. The reason for the improved solubility lies in the acidity of the sulphonamide NH proton (Fig. 14.10). In sulphathiazole, this proton is not very acidic (high pK_a). Therefore, sulphathiazole and its metabolite are mostly un-ionized at blood pH. Replacing the thiazole ring with a more electron withdrawing pyrimidine ring increases the acidity of the NH proton by stabilizing the anion which results. Therefore, sulphadiazine and its metabolite are significantly ionized at blood pH. As a consequence, they are more soluble and less toxic.

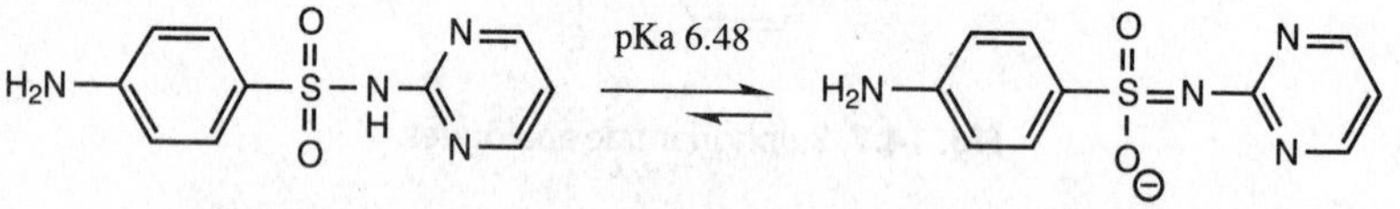

Fig. 14.10 Sulphadiazine.

Sulphadiazine was also found to be more active than sulphathiazole and soon replaced it in therapy.

To conclude, varying R" can affect the solubility of sulphonamides or the extent to which they bind to plasma protein. These variations are therefore affecting the pharmacokinetics of the drug, rather than its mechanism of action.

Applications of sulphonamides

Before the appearance of penicillin, the sulpha drugs were the drugs of choice in the treatment of infectious diseases. Indeed, they played a significant part in world history by saving Winston Churchill's life during the Second World War. Whilst visiting North Africa, Churchill became ill with a serious infection and was bedridden for several weeks. At one point, his condition was deemed so serious that his daughter was flown out from Britain to be at his side. Fortunately, he responded to the novel sulphonamide drugs of the day.

Penicillins largely superseded sulphonamides in the fight against bacterial infections and, for a long time, sulphonamides were relegated backstage. However, there has been a revival of interest with the discovery of a new 'breed' of longer lasting sulphonamides. One example of this new generation is sulphamethoxine (Fig. 14.11), which is so stable in the body that it need only be taken once a week.

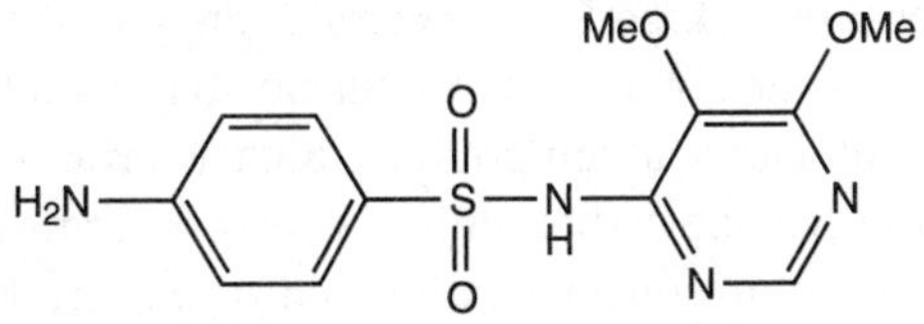

Fig. 14.11 Sulphamethoxine.

The sulpha drugs presently have the following applications in medicine:

- treatment of urinary tract infections
- eye lotions
- treatment of infections of mucous membranes
- treatment of gut infections.

Sulphonamides have been particularly useful against infections of the intestine and can be targeted specifically to that site by the use of prodrugs. For example, succinyl sulphathiazole (Fig. 14.12) is a prodrug of sulphathiazole. The succinyl group converts the basic sulphathiazole into an acid, which means that the prodrug is ionized in the slightly alkaline conditions of the intestine. As a result, it is not absorbed into the bloodstream and is retained in the intestine. Slow enzymatic hydrolysis of the succinyl group then releases the active sulphathiazole where it is needed.

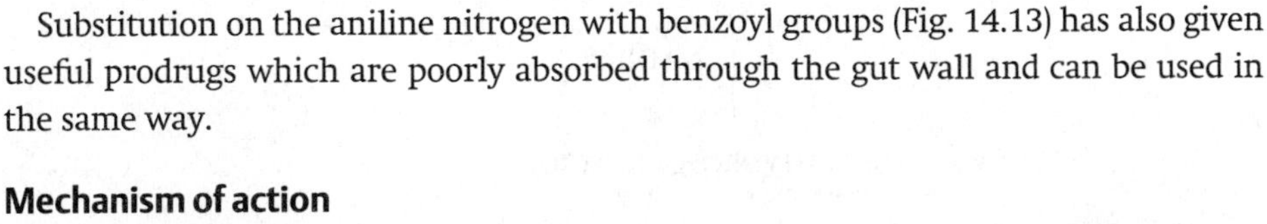

SUCCINYL SULFATHIAZOLE   SUCCINIC ACID   SULFATHIAZOLE

Fig. 14.12 Succinyl sulphathiazole is a prodrug of sulphathiazole.

Substitution on the aniline nitrogen with benzoyl groups (Fig. 14.13) has also given useful prodrugs which are poorly absorbed through the gut wall and can be used in the same way.

Mechanism of action

The sulphonamides act as competitive enzyme inhibitors and block the biosynthesis of the vitamin folic acid in bacterial cells (Fig. 14.14). They do this by inhibiting the

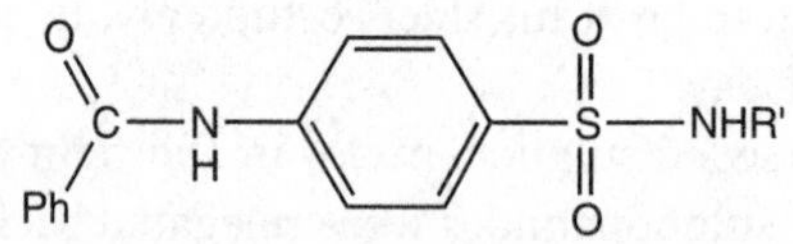

Fig. 14.13 Substitution on the aniline nitrogen with benzoyl groups.

enzyme responsible for linking together the component parts of folic acid. The consequences of this are disastrous for the cell. Under normal conditions, folic acid is the precursor for tetrahydrofolate – a compound which is crucial to cell biochemistry since it acts as the carrier for one-carbon units, necessary for many biosynthetic pathways. If tetrahydrofolate is no longer synthesized, then any biosynthetic pathway requiring one-carbon fragments is disrupted. The biosynthesis of nucleic acids is particularly disrupted and this leads to the cessation of cell growth and division.

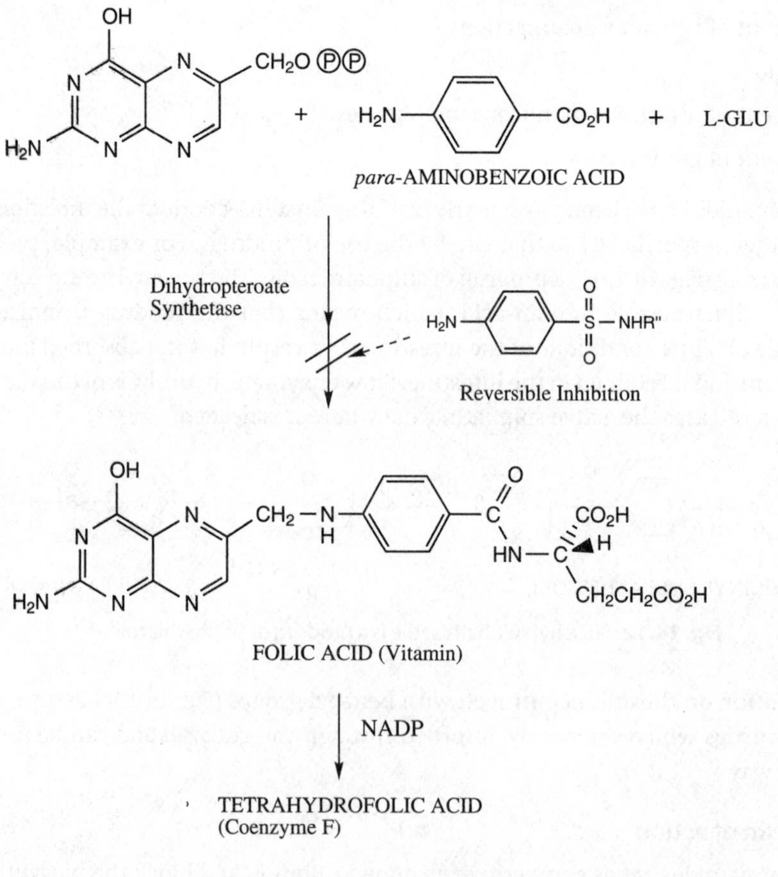

Fig. 14.14 Mechanism of action of sulphonamides.

Note that sulphonamides do not actively kill bacterial cells. They do, however, prevent the cells dividing and spreading. This gives the body's own defence systems enough time to gather their resources and wipe out the invader. Antibacterial agents which inhibit cell growth are classed as **bacteriostatic**, whereas agents which can actively kill bacterial cells (e.g. penicillin) are classed as **bactericidal**.

Sulphonamides act as inhibitors by mimicking *para* aminobenzoic acid (PABA) (Fig. 14.14) – one of the normal constituents of folic acid. The sulphonamide molecule is similar enough in structure to PABA that the enzyme is fooled into accepting it into its active site (Fig. 14.15). Once it is bound, the sulphonamide prevents PABA from binding. As a result, folic acid is no longer synthesized. Since folic acid is essential to cell growth, the cell will stop dividing.

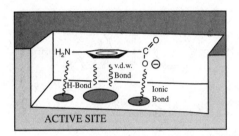 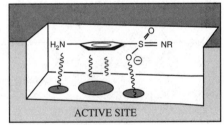

Fig. 14.15 Sulphonamide prevents PABA from binding by mimicking PABA.

One might ask why the enzyme does not join the sulphonamide to the other two components of folic acid to give a folic acid analogue containing the sulphonamide skeleton. This can in fact occur, but it does the cell no good at all since the analogue is not accepted by the next enzyme in the biosynthetic pathway.

Sulphonamides are competitive enzyme inhibitors and, as such, the effect can be reversible. This is demonstrated by certain organisms such as staphylococci, pneumococci, and gonococci which can acquire resistance by synthesizing more PABA. The more PABA there is in the cell, the more effectively it can compete with the sulphonamide inhibitor to reach the enzyme's active site. In such cases, the dose levels of sulphonamide have to be increased to bring back the same level of inhibition.

Folic acid is clearly necessary for the survival of bacterial cells. However, folic acid is also vital for the survival of human cells, so why do the sulpha drugs not affect human cells as well? The answer lies in the fact that human cells cannot make folic acid. They lack the necessary enzymes and so there is no enzyme for the sulphonamides to attack. Human cells acquire folic acid as a vitamin from the diet. Folic acid is brought through the cell membrane by a transport protein and this process is totally unaffected by sulphonamides.

We could now ask, 'If human cells can acquire folic acid from the diet, why can't bacterial cells infecting the human body do the same?' In fact, it is found that bacterial cells are unable to acquire folic acid since they lack the necessary transport protein required to carry it across the cell membrane. Therefore, they are forced to make it from scratch.

To sum up, the success of sulphonamides is because of two metabolic differences between mammalian and bacterial cells. In the first place, bacteria have a susceptible enzyme which is not present in mammalian cells. In the second place, bacteria lack the transport protein which would allow them to acquire folic acid from outside the cell.

14.4.2 Examples of other antimetabolites

There are other antimetabolites in medical use apart from the sulphonamides. Two examples are trimethoprim and a group of compounds known as sulfones (Fig. 14.16).

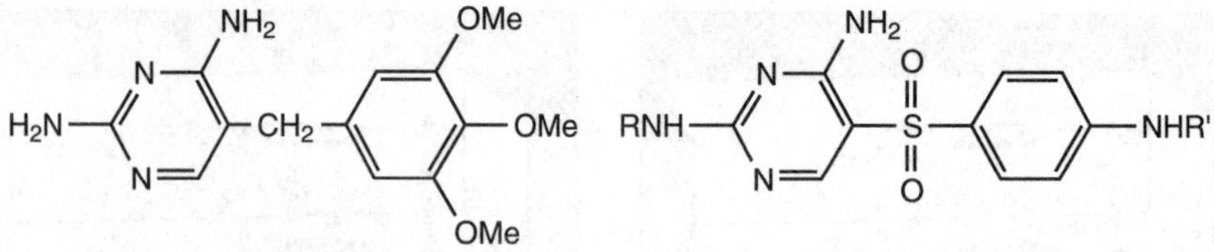

TRIMETHOPRIM
(Antimalarial)

SULPHONES
(Anti leprosy)

Fig. 14.16 Examples of antimetabolites in medical use.

Trimethoprim

Trimethoprim is a diaminopyrimidine structure which has proved to be a highly selective, orally active, antibacterial, and antimalarial agent. Unlike the sulphonamides, it acts against dihydrofolate reductase – the enzyme which carries out the conversion of folic acid to tetrahydrofolate. The overall effect, however, is the same as with sulphonamides – the inhibition of DNA synthesis and cell growth.

Dihydrofolate reductase is present in mammalian cells as well as bacterial cells, so we might wonder why trimethoprim does not affect our own cells. The answer is that trimethoprim is able to distinguish between the enzymes in either cell. Although this enzyme is present in both types of cell and carries out the same reaction, mutations over millions of years have resulted in a significant difference in structure between the two enzymes such that trimethoprim recognizes and inhibits the bacterial enzyme, but does not recognize the mammalian enzyme.

Trimethoprim is often given in conjunction with the sulphonamide sulphamethoxazole (Fig. 14.17). The latter inhibits the incorporation of PABA into folic acid, while the former inhibits dihydrofolate reductase. Therefore, two enzymes in the one biosynthetic route are inhibited. This is a very effective method of inhibiting a biosynthetic route and has the advantage that the doses of both drugs can be kept down to safe levels. To get the same level of inhibition using a single drug, the dose level of that drug would have to be much higher, leading to possible side-effects. This approach has been described as 'sequential blocking'.

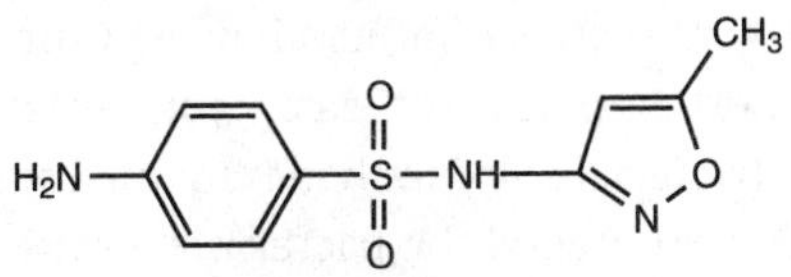

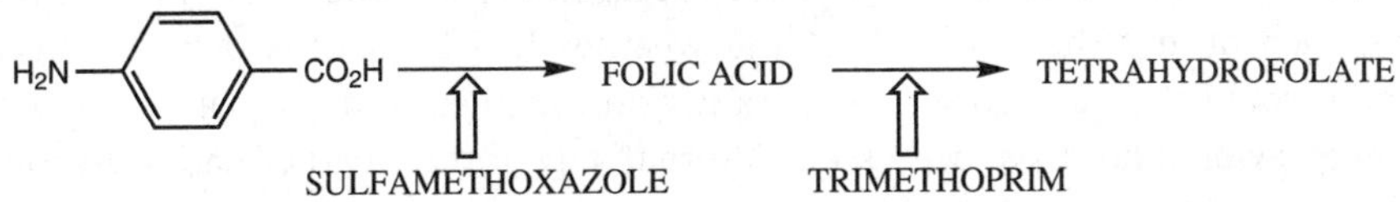

Fig. 14.17 Use of sulphamethoxazole and trimethoprim in 'sequential blocking'.

Sulphones

The sulphones (Fig. 14.16) are the most important drugs used in the treatment of leprosy. It is believed that they inhibit the same bacterial enzyme inhibited by the sulphonamides, i.e. dihydropteroate synthetase.

14.5 Antibacterial agents which inhibit cell wall synthesis

There are two major classes of drug which act in this fashion – penicillins and cephalosporins. We shall consider penicillins first.

14.5.1 Penicillins

History of penicillins

In 1877, Pasteur and Joubert discovered that certain moulds could produce toxic substances which killed bacteria. Unfortunately, these substances were also toxic to humans and of no clinical value. However, they did demonstrate that moulds could be a potential source of antibacterial agents.

In 1928, Fleming noted that a bacterial culture which had been left open to the air for several weeks had become infected by a fungal colony. Of more interest was the fact that there was an area surrounding the fungal colony where the bacterial colonies were dying. He correctly concluded that the fungal colony was producing an anti-bacterial agent which was spreading into the surrounding area. Recognizing the significance of this, he set out to culture and identify the fungus, and showed it to be a relatively rare species of *Penicillium*. It has since been suggested that the *Penicillium* spore responsible for the fungal colony originated from another laboratory in the building and that the spore was carried by air currents and was eventually blown through the window of Fleming's laboratory. This in itself appears a remarkable stroke of good fortune. However, a series of other chance events were involved in the story – not least the weather! A period of early cold weather had encouraged the fungus to grow while the bacterial colonies had remained static. A period of warm weather then followed which encouraged the bacteria to grow. These weather conditions were the ideal experimental conditions required for (a) the fungus to produce penicillin during the cold spell and (b) for the antibacterial properties of penicillin to be revealed during the hot spell. If the weather had been consistently cold, the bacteria would not have grown significantly and the death of cell colonies close to the fungus would not have been seen. Alternatively, if the weather had been consistently warm, the bacteria would have outgrown the fungus and little penicillin would have been produced. As a final twist to the story, the crucial agar plate had been stacked in a bowl of disinfectant prior to washing up, but was actually placed above the surface of the disinfectant. It says much for Fleming's observational powers that he bothered to take any notice of a culture plate which had been so discarded and that he spotted the crucial area of inhibition.

Fleming spent several years investigating the novel antibacterial substance and showed it to have significant antibacterial properties and to be remarkably non-toxic to humans. Unfortunately, the substance was also unstable and Fleming was unable to isolate and purify the compound. He therefore came to the conclusion that penicillin was too unstable to be used clinically.

The problem of isolating penicillin was eventually solved in 1938 by Florey and Chain by using a process known as freeze-drying, which allowed isolation of the antibiotic under much milder conditions than had previously been available. By 1941, Florey and Chain were able to carry out the first clinical trials on crude extracts of penicillin and achieved spectacular success. Further developments aimed at producing the new agent in large quantities were developed in the United States such that by 1944 there was enough penicillin for casualties arising from the D-Day landings.

Although the use of penicillin was now widespread, the structure of the compound was still not settled and was proving to be a source of furious debate because of the unusual structures being proposed. The issue was finally settled in 1945 when Dorothy Hodgkins established the structure by X-ray analysis (Fig. 14.18).

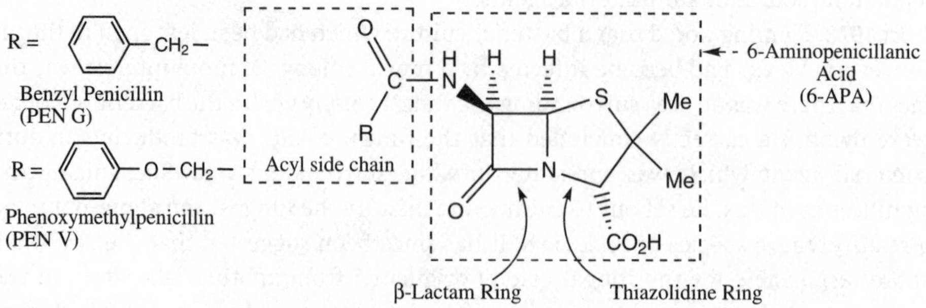

Fig. 14.18 The structure of penicillin.

The synthesis of such a highly strained molecule presented a huge challenge – a challenge which was met successfully by Sheehan, who completed a full synthesis of penicillin by 1957. The full synthesis was too involved to be of commercial use, but the following year Beechams isolated a biosynthetic intermediate of penicillin called 6-aminopenicillanic acid (6-APA) (Fig. 14.18), which provided a readily accessible biosynthetic intermediate of penicillin. This revolutionized the field of penicillins by providing the starting material for a huge range of semisynthetic penicillins.

Penicillins were used widely and often carelessly, so that the evolution of penicillin-resistant bacteria became more and more of a problem. The fight against these penicillin-resistant bacteria was promoted greatly when, in 1976, Beechams discovered a natural product called clavulanic acid (Fig. 14.55) which has proved highly effective in protecting penicillins from the bacterial enzymes which attack them.

Structure of penicillin

As mentioned above, the structure of penicillin (Fig. 14.18) is so unusual that many scientists remained sceptical until an X-ray analysis was carried out. Penicillin contains a highly unstable-looking bicyclic system consisting of a four-membered β-lactam ring fused to a five-membered thiazolidine ring. The skeleton of the molecule suggests that it is derived from the amino acids cysteine and valine (Fig. 14.19), and this has been established. The overall shape of the molecule is like a half-open book, as shown in Fig. 14.20.

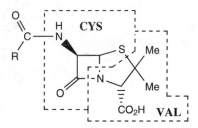

Fig. 14.19 Penicillin appears to be derived from cysteine and valine.

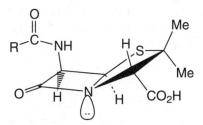

Fig. 14.20 Shape of penicillin.

The acyl side chain (R) varies depending on the make-up of the fermentation media. For example, corn steep liquor was used as the medium when penicillin was first mass-produced in the United States and this gave penicillin G (R=benzyl). This was due to high levels of phenylacetic acid (PhCH$_2$CO$_2$H) present in the medium.

Penicillin analogues

One method of varying the side chain is to add different carboxylic acids to the fermentation medium; for example, adding phenoxyacetic acid (PhOCH$_2$CO$_2$H) gives penicillin V (Fig. 14.18).

However, there is a limitation to the sort of carboxylic acid one can add to the medium (i.e. only acids of general formula RCH$_2$CO$_2$H), and this in turn restricts the variety of analogues which can be obtained. The other major disadvantage in obtaining analogues in this way is that it is a tedious and time-consuming business.

In 1957, Sheehan succeeded in synthesizing penicillin, and obtained a 1% yield of penicillin V using a multistep synthetic route. Clearly, a full synthesis was not an efficient way of making penicillin analogues.

In 1958–1960, Beechams managed to isolate a biosynthetic intermediate of penicillin which was also one of Sheehan's synthetic intermediates. The compound was 6-APA and it allowed the synthesis of a huge number of analogues by a semisynthetic method; thus, fermentation yielded 6-APA which could then be treated synthetically to give penicillin analogues. This was achieved by acylating the 6-APA with a range of acid chlorides (Fig. 14.21).

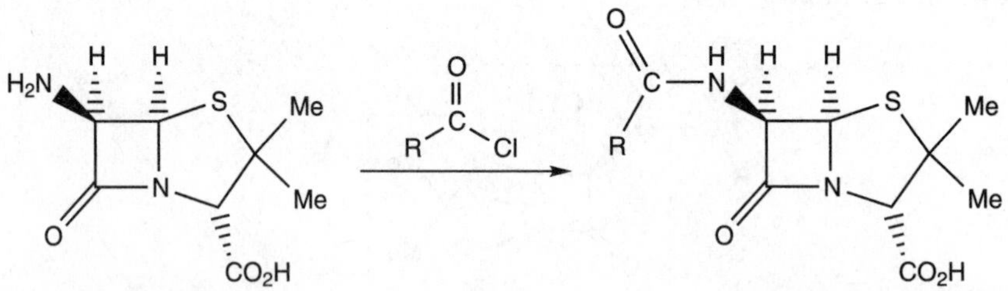

Fig. 14.21 Penicillin analogues achieved by acylating 6-APA.

6-APA is now produced by hydrolysing penicillin G or penicillin V with an enzyme (penicillin acylase) (Fig. 14.22) or by chemical methods (see later). These are more efficient procedures than fermentation.

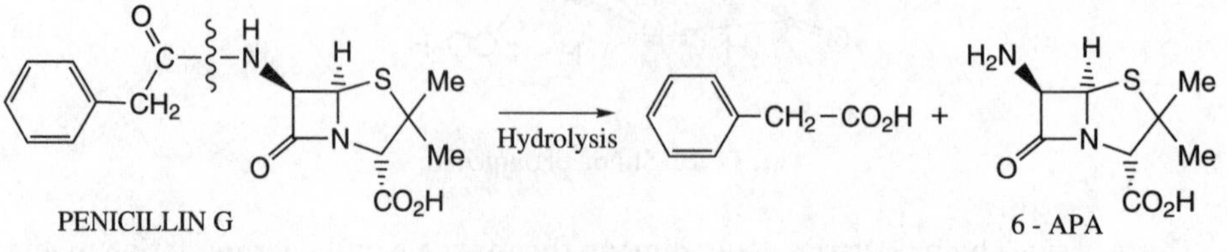

Fig. 14.22 Production of 6-APA.

We have emphasized the drive to make penicillin analogues with varying acyl side chains. No doubt the question could be asked – why bother? Is penicillin not good enough? Furthermore, what is so special about the acyl side chain? Could changes not be made elsewhere in the molecule as well?

To answer these questions we need to look at penicillin G (the first penicillin to be isolated) in more detail and to consider its properties. Just how good an antibiotic is penicillin G?

Properties of penicillin G

The properties of benzyl penicillin are summarized below.

- Active versus Gram-positive bacilli (e.g. staphylococci, meningitis, and gonorrhoea) and many (but not all) Gram-negative cocci.
- Non-toxic! This point is worth emphasizing. The penicillins are amongst the safest drugs known to medicine.
- Not active over a wide range (or spectrum) of bacteria.
- Ineffective when taken orally. Penicillin G can only be administered by injection. It is ineffective orally since it breaks down in the acid conditions of the stomach.
- Sensitive to all known β-lactamases. These are enzymes produced by penicillin-resistant bacteria which catalyse the degradation of penicillins.
- Allergic reactions are suffered by some individuals.

Clearly, there are several problems associated with the use of penicillin G, the most serious being acid sensitivity, sensitivity to penicillinase, and a narrow spectrum of activity. The purpose of making semisynthetic penicillin analogues is therefore to find compounds which do not suffer from these disadvantages. However, before launching into such a programme, a structure–activity study is needed to find out what features of the penicillin molecule are important to its activity. These features would then be retained in any analogues which are made.

Structure–activity relationships of penicillins

A large number of penicillin analogues have been synthesized and studied. The results of these studies led to the following conclusions (Fig. 14.23):

- The strained β-lactam ring is essential.
- The free carboxylic acid is essential.
- The bicyclic system is important (confers strain on the β-lactam ring – the greater the strain, the greater the activity, but the greater the instability of the molecule to other factors).
- The acylamino side chain is essential (except for thienamycin, see later).
- Sulfur is usual but not essential.
- The stereochemistry of the bicyclic ring with respect to the acylamino side chain is important.

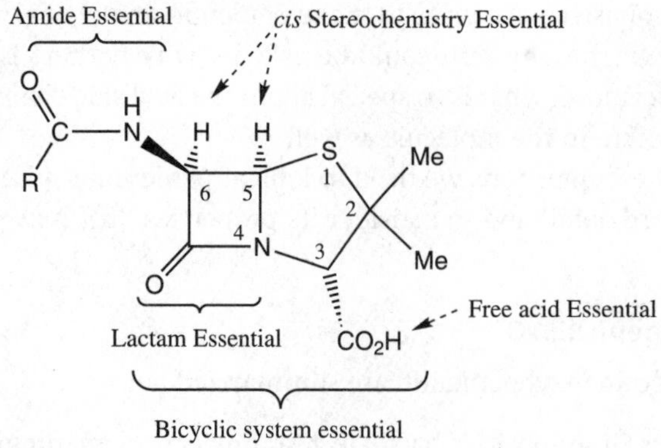

Amide Essential
cis Stereochemistry Essential
Free acid Essential
Lactam Essential
Bicyclic system essential

Fig. 14.23 Structure–activity relationships of penicillins.

The results of this analysis led to the inevitable conclusion that very little variation is tolerated by the penicillin nucleus and that any variation which can be made is restricted to the acylamino side chain.

We can now look at the three problems mentioned earlier and see how they can be tackled.

The acid sensitivity of penicillins

Why is penicillin G acid sensitive? If we know the answer to that question, we might be able to plan how to solve the problem. There are three reasons for the acid sensitivity of penicillin G.

Ring strain

The bicyclic system in penicillin consists of a four-membered ring and a five-membered ring. As a result, penicillin suffers large angle and torsional strains. Acid-catalysed ring opening relieves these strains by breaking open the more highly strained four-membered lactam ring (Fig. 14.24).

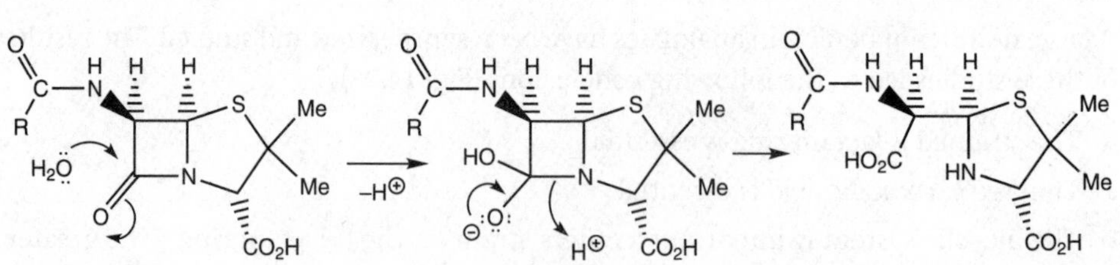

Fig. 14.24 Ring opening.

A highly reactive β-lactam carbonyl group

The carbonyl group in the β-lactam ring is highly susceptible to nucleophiles and as such does not behave like a normal tertiary amide which is usually quite resistant to nucleophilic attack. This difference in reactivity is due mainly to the fact that stabilization of the carbonyl is possible in the tertiary amide, but impossible in the β-lactam ring (Fig. 14.25). The β-lactam nitrogen is unable to feed its lone pair of electrons into the carbonyl group since this would require the bicyclic rings to adopt an impossibly strained flat system. As a result, the lone pair is localized on the nitrogen atom and the carbonyl group is far more electrophilic than one would expect for a tertiary amide. A normal tertiary amide is far less susceptible to nucleophiles since the resonance structures shown in Fig. 14.25 reduce the electrophilic character of the carbonyl group.

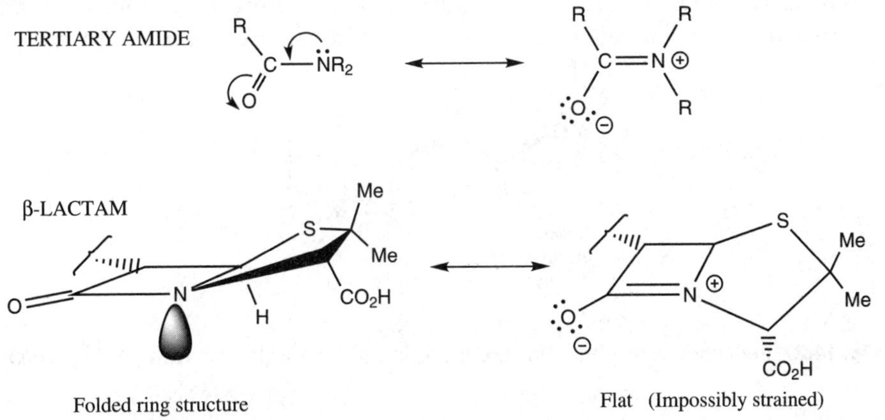

Fig. 14.25 Highly reactive β-lactam carbonyl group.

Influence of the acyl side chain (neighbouring group participation)

Fig. 14.26 demonstrates how the neighbouring acyl group can actively participate in a mechanism to open up the lactam ring. Thus, penicillin G has a self-destruct mechanism built into its structure.

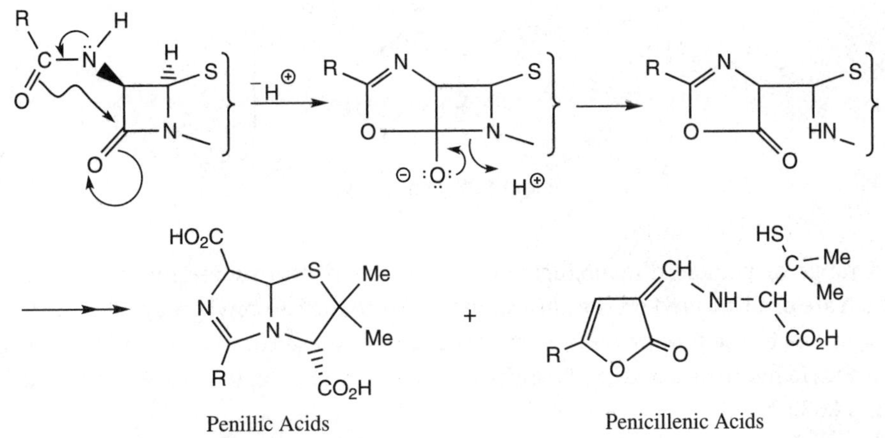

Fig. 14.26 Influence of the acyl side chain on acid sensitivity.

Tackling the problem of acid sensitivity

It can be seen that countering acid sensitivity is a difficult task. Nothing can be done about the first two factors since the β-lactam ring is vital for antibacterial activity. Without it, the molecule has no useful biological activity at all.

Therefore, only the third factor can be tackled. The task then becomes one of reducing the amount of neighbouring group participation to make it difficult, if not impossible, for the acyl carbonyl group to attack the β-lactam ring. Fortunately, such an objective is feasible. If a good electron withdrawing group is attached to the carbonyl group, then the inductive pulling effect should draw electrons away from the carbonyl oxygen and reduce its tendency to act as a nucleophile (Fig. 14.27).

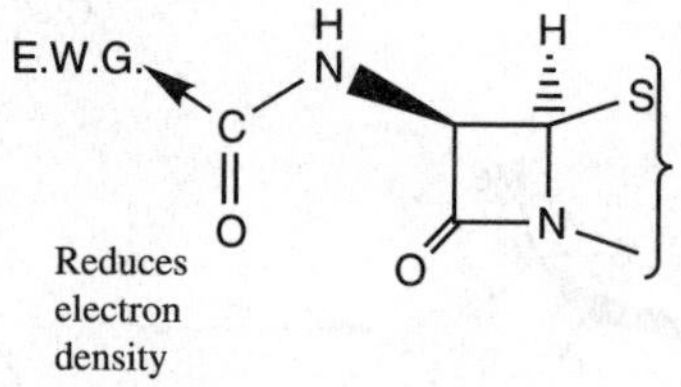

Fig. 14.27 Reduction of neighbouring group participation with electron withdrawing group.

Penicillin V (Fig. 14.28) has an electronegative oxygen on the acyl side chain with the electron withdrawing effect required. The molecule has better acid stability than penicillin G and is stable enough to survive the acid in the stomach. Thus, it can be given orally. However, penicillin V is still sensitive to penicillinases and is slightly less active than penicillin G. It also shares with penicillin G the problem of allergic sensitivity in some individuals.

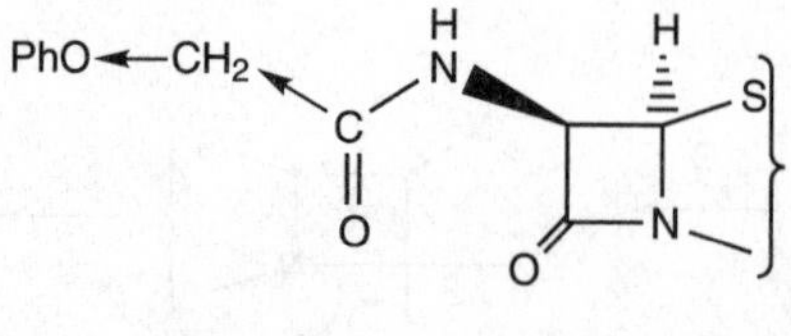

Fig. 14.28 Penicillin V.

A range of penicillin analogues which have been very successful are penicillins which are disubstituted on the *alpha* carbon next to the carbonyl group (Fig. 14.29). As long as one of the groups is electron withdrawing, these compounds are more resistant to acid hydrolysis and can be given orally (e.g. ampicillin (Fig. 14.36) and oxacillin (Fig. 14.33)).

To conclude, the problem of acid sensitivity is fairly easily solved by having an electron withdrawing group on the acyl side chain.

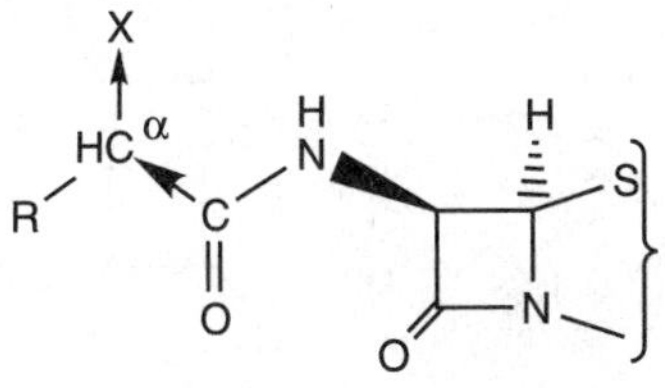

X = NH_2, Cl, PhOCONH,
Heterocycles

Fig. 14.29 Penicillin analogues.

Penicillin sensitivity to β-lactamases

β-Lactamases are enzymes produced by penicillin-resistant bacteria which can catalyse the reaction shown in Fig. 14.30 – i.e. the same ring opening and deactivation of penicillin which occurred with acid hydrolysis.

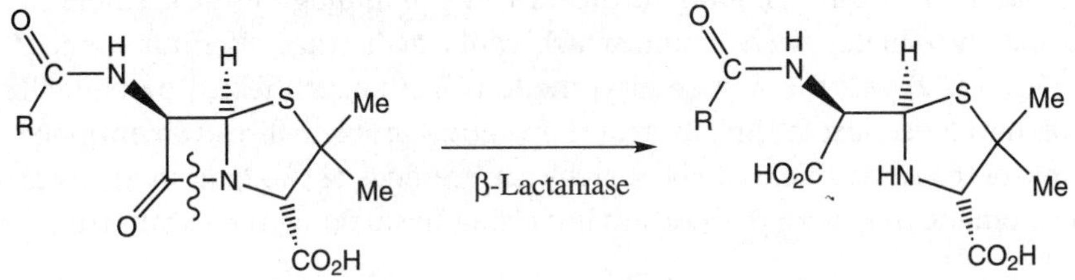

Fig. 14.30 β-Lactamase deactivation of penicillin.

The problem of β-lactamases became critical in 1960 when the widespread use of penicillin G led to an alarming increase of *Staphylococcus aureus* infections. These problem strains had gained the lactamase enzyme and had thus gained resistance to the drug. At one point, 80% of all *S. aureus* infections in hospitals were due to virulent, penicillin-resistant strains. Alarmingly, these strains were also resistant to all other available antibiotics.

Fortunately, a solution to the problem was just around the corner – the design of penicillinase-resistant penicillins. We say design, which implies that some sort of plan was used to counter the effects of the penicillinase enzyme. How then does one tackle a problem of this sort?

Tackling the problem of β-lactamase sensitivity

The strategy is to block the penicillin from reaching the penicillinase active site. One way of doing this is to place a bulky group on the side chain. This bulky group can then act as a 'shield' to ward off the penicillinase and therefore prevent binding (Fig. 14.31).

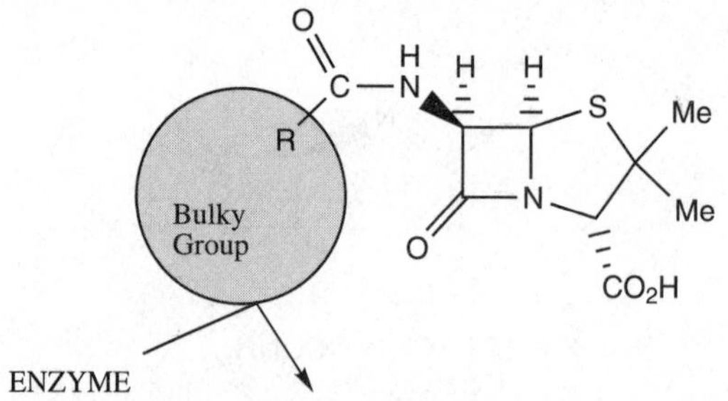

Fig. 14.31 Blocking penicillin from reaching the penicillinase active site.

Several analogues were made and the strategy was found to work. However, there was a problem. If the side chain was made too bulky, then the steric shield also prevented the penicillin from attacking the enzyme responsible for bacterial cell wall synthesis. Therefore, a great deal of work had to be done to find the ideal 'shield' that would be large enough to ward off the lactamase enzyme, but would be small enough to allow the penicillin to do its duty. The fact that it is the β-lactam ring which is interacting with both enzymes highlights the difficulty in finding the ideal 'shield'.

Fortunately, 'shields' were found which could make that discrimination. Methicillin (Fig. 14.32) was the first semisynthetic penicillin unaffected by penicillinase and was developed just in time to treat the *S. aureus* problem already mentioned. The principle of the steric shield can be seen by the presence of two *ortho* methoxy groups on the aromatic ring. Both of these are important in shielding the lactam ring.

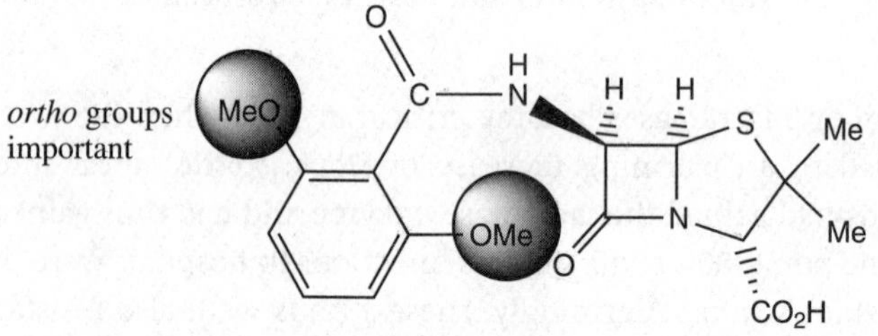

Fig. 14.32 Methicillin.

However, methicillin is by no means an ideal drug. Since there is no electron withdrawing group on the side chain, it is acid sensitive, and so has to be injected. It has only one-fiftieth of the activity of penicillin G against organisms sensitive to penicillin G, it shows poor activity against some streptococci, and it is inactive against Gram-negative bacteria.

Further work eventually solved the problem of acid sensitivity by incorporating into the side chain a five-membered heterocycle which was designed to act as a steric shield and also to be electron withdrawing (Fig. 14.33).

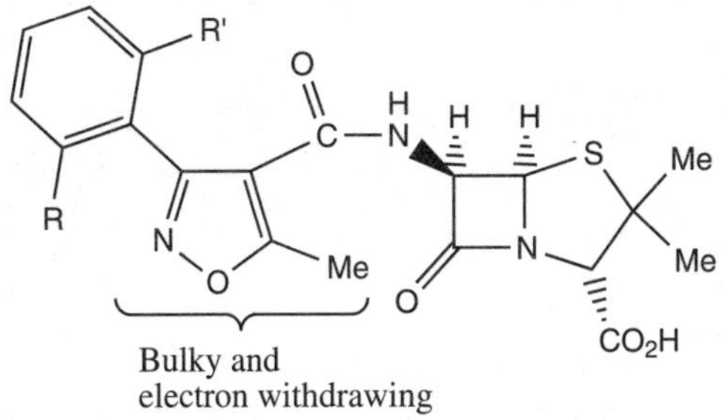

Oxacillin   R = R' = H
Cloxacillin   R = Cl, R' = H
Flucloxacillin R = Cl, R' = F

Fig. 14.33 Incorporation of a five-membered heterocycle.

These compounds (oxacillin, cloxacillin, and flucloxacillin) are resistant to acid and penicillinase, and are also useful against *S. aureus* infections.

The only difference between the above three compounds is the type of halogen substitution on the aromatic ring. The influence of these groups is found to be pharmacokinetic; that is, they influence such factors as absorption of the drug and plasma protein binding. For example, cloxacillin is better absorbed through the gut wall than oxacillin, whereas flucloxacillin is less bound to plasma protein, resulting in higher levels of the free drug in the blood supply.

Having pointed out the advantages of these drugs over methicillin, it is worth putting things into context by pointing out that these three penicillins have inferior activity to the original penicillins when they are used against bacteria without the penicillinase enzyme. They also prove to be inactive against Gram-negative bacteria.

To summarize, acid resistant penicillins would be the first choice of drug against an infection. However, if the bacteria proved resistant because of a penicillinase enzyme, then the therapy would be changed to a penicillinase-resistant penicillin.

Narrow spectrum of activity

One problem has cropped up in everything described so far; most penicillins show a poor activity against Gram-negative bacteria. There are several reasons for this resistance.

Permeability barrier

It is difficult for penicillins to invade a Gram-negative bacterial cell because of the make-up of the cell wall. Gram-negative bacteria have a coating on the outside of their cell wall which consists of a mixture of fats, sugars, and proteins (Fig. 14.34). This coating can act as a barrier in various ways. For example, the outer surface may have an overall negative or positive charge depending on its constituent triglycerides. An excess of phosphatidylglycerol would result in an overall anionic charge, whereas an excess of lysylphosphatidylglycerol would result in an overall cationic charge. Penicillin has a free carboxylic acid which, if ionized, would be repelled by the former type of cell membrane.

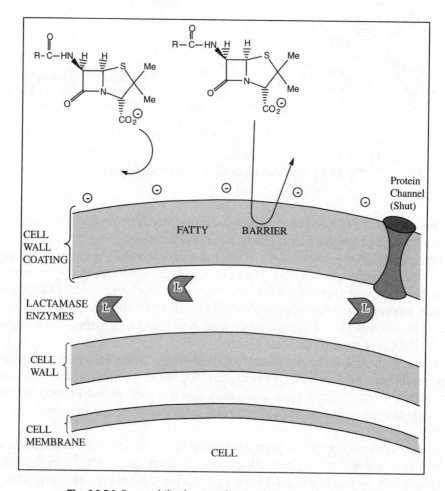

Fig. 14.34 Permeability barrier of a Gram-negative bacterial cell.

Alternatively, the fatty portion of the coating may act as a barrier to the polar, hydrophilic penicillin molecule.

The only way in which penicillin can negotiate such a barrier is through protein channels in the outer coating. Unfortunately, most of these are usually closed.

High levels of transpeptidase enzyme produced

The transpeptidase enzyme is the enzyme attacked by penicillin. In some Gram-negative bacteria, a lot of transpeptidase enzyme is produced, and the penicillin is incapable of inactivating all the enzyme molecules present.

Modification of the transpeptidase enzyme

A mutation may occur which allows the bacterium to produce a transpeptidase enzyme which is not antagonized by penicillin.

Presence of β-lactamase

We have already seen that β-lactamases are enzymes which degrade penicillin. They are situated between the cell wall and its outer coating.

Transfer of the β-lactamase·enzyme

Bacteria can transfer small portions of DNA from one cell to another through structures called plasmids. These are small pieces of circular bacterial DNA. If the transferred DNA contains the code for the β-lactamase enzyme, then the recipient cell acquires immunity.

Tackling the problem of narrow activity spectrum

One, some, or all of these factors might be at work, and therefore it is impossible to devise a sensible strategy to solve the problem completely. The search for broad-spectrum antibiotics has been one of trial and error which involved making a huge variety of analogues. These changes were again confined to variations in the side chain and gave the following results:

• Hydrophobic groups on the side chain (e.g. penicillin G) favour activity against Gram-positive bacteria, but result in poor activity against Gram-negative bacteria.

• If the hydrophobic character is increased, there is little effect on the Gram-positive activity, but what activity there is against Gram-negative bacteria drops even more.

• Hydrophilic groups on the side chain have either little effect on Gram-positive activity (e.g. penicillin T) or cause a reduction of activity (e.g. penicillin N; Fig. 14.35). However, they lead to an increase in activity against Gram-negative bacteria.

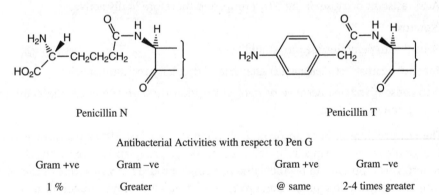

Penicillin N   Penicillin T

Antibacterial Activities with respect to Pen G

Gram +ve	Gram −ve	Gram +ve	Gram −ve
1 %	Greater	@ same	2-4 times greater

Fig. 14.35 Effect of hydrophilic groups on the side chain on antibacterial activity.

• Enhancement of Gram-negative activity is found to be greatest if the hydrophilic group (e.g. NH_2, OH, CO_2H) is attached to the carbon that is *alpha* to the carbonyl group on the side chain.

Those penicillins having useful activity against both Gram-positive and Gram-negative bacteria are known as broad-spectrum antibiotics. There are two classes of broad-spectrum antibiotics. Both have an *alpha* hydrophilic group. However, in one class the hydrophilic group is an amino function, as in ampicillin or amoxycillin (Fig. 14.36), while in the other the hydrophilic group is an acid group, as in carbenicillin (Fig. 14.40).

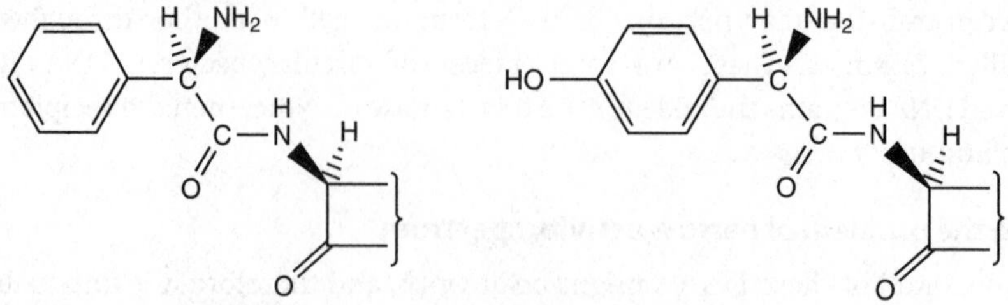

AMPICILLIN (Penbritin)   AMOXYCILLIN (Amoxil)

Fig. 14.36 Class I broad spectrum antibiotics.

Class I broad-spectrum antibiotics – ampicillin and amoxycillin (Beechams 1964)

Ampicillin is the second most used penicillin in medical practice. Amoxycillin differs merely in having a phenolic group. It has similar properties, but is better absorbed through the gut wall.

Properties of ampicillin and amoxycillin are as follows:

- Active versus Gram-positive bacteria and against those Gram-negative bacteria which do not produce penicillinase.

- Acid resistant because of the NH_2 group, and therefore orally active.

- Non-toxic.

- Sensitive to penicillinase (no 'shield').

- Inactive against *Pseudomonas aeruginosa* (a particularly resistant species).

- Can cause diarrhoea because of poor absorption through the gut wall, leading to disruption of gut flora.

The last problem of poor absorption through the gut wall stems from the dipolar nature of the molecule since it has both a free amino group and a free carboxylic acid function. This problem can be alleviated by using a prodrug in which one of the polar groups is masked with a protecting group. This group is removed metabolically once the prodrug has been absorbed through the gut wall. Three examples are shown in Fig. 14.37.

These three compounds are all prodrugs of ampicillin. In all three examples, the esters used to mask the carboxylic acid group seem rather elaborate and one may ask why a simple methyl ester is not used. The answer is that methyl esters of penicillins are not metabolized in man. Perhaps the bulkiness of the penicillin skeleton being so close to the ester functional group prevents the esterases from binding the penicillin.

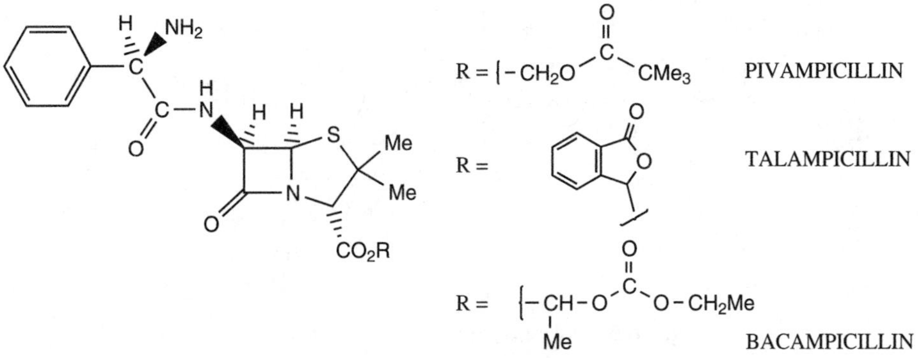

Fig. 14.37 Prodrugs used to aid absorption of antibiotic through gut wall.

Fortunately, it is found that acyloxymethyl esters are susceptible to non-specific esterases. These 'extended' esters contain a second ester group further away from the penicillin nucleus, and which is more exposed to attack. The products formed from hydrolysis are inherently unstable and decompose spontaneously to reveal the free carboxylic acid (Fig. 14.38) and formaldehyde. The release of formaldehyde is not ideal since this is a toxic chemical. However, it is formed naturally in the body through enzymic demethylation of various compounds found in the diet, and the levels produced with drugs such as penicillin cause little problem since they are only taken for a short duration of time.

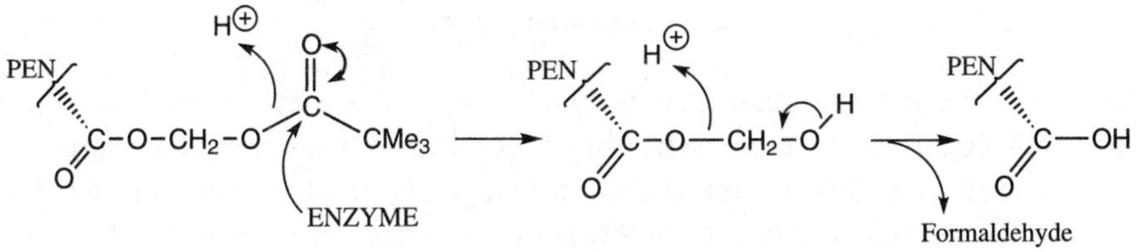

Fig. 14.38 Decomposition of acyloxymethyl esters.

Such extended esters can be used to prepare prodrugs of other penicillins, but one has to be careful that one doesn't go to the other extreme and make the penicillin too lipophilic that it ends up with poor water solubility. For example, the 1-acyloxyalkyl ester of penicillin G is too lipophilic and has poor solubility in water. Fortunately, the problem can easily be avoided by making the extended ester more polar (e.g. by attaching valine) (Fig. 14.39).

Class II broad-spectrum antibiotics – carbenicillin

Carbenicillin (Fig. 14.40) has an activity against a wider range of Gram-negative bacteria than ampicillin. It is resistant to most penicillinases and is also active against the stubborn *Pseudomonas aeruginosa*. This particular organism is known as an

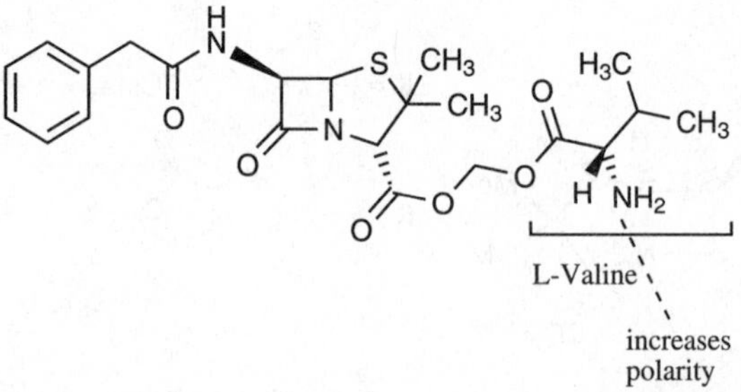

Fig. 14.39 Polar extended ester for penicillin G.

'opportunist' pathogen since it strikes patients when they are in a weakened condition. The organism is usually present in the body, but is kept under control by the body's own defence mechanisms. However, if these defences are weakened for any reason (e.g. shock or chemotherapy), then the organism can strike.

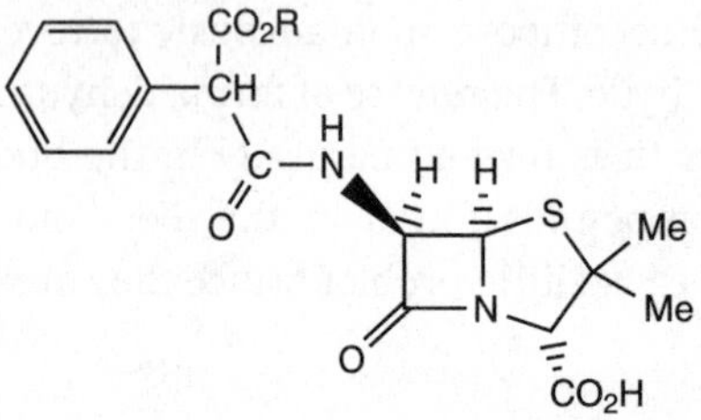

R = H   CARBENICILLIN

R = Ph   CARFECILLIN

Fig. 14.40

This can prove a real problem in hospitals where there are many susceptible patients suffering from cancer or cystic fibrosis. Burn victims are particularly prone to infection and this can lead to septicaemia which can be fatal. The organism is also responsible for serious lung infections. Carbenicillin represents one of the few penicillins which is effective against this organism. However, there are drawbacks to carbenicillin. It shows a marked reduction in activity against Gram-positive bacteria (note the hydrophilic acid group). It is also acid sensitive and has to be injected. In general, carbenicillin is used against penicillin-resistant Gram-negative bacteria. The broad activity against Gram-negative bacteria is due to the hydrophilic acid group (ionized at pH7) on the side chain. It is particularly interesting to note that the stereochemistry of this group is important. The *alpha* carbon is asymmetric and only one of the two enantiomers is active. This implies that the acid group is involved in some sort of binding interaction with the target enzyme.

Carfecillin (Fig. 14.40) is the prodrug for carbenicillin and shows an improved absorption through the gut wall. An aryl ester is used here. Aryl esters are chemically more susceptible to hydrolysis than simple alkyl esters owing to the electron withdrawing inductive effect of the aryl ring. An extended ester is not required since the aryl ester is not shielded by the β-lactam ring.

Synergism of penicillins with other drugs

There are several examples in medicinal chemistry where the presence of one drug enhances the activity of another. In many cases this can be dangerous, leading to an effective overdose of the enhanced drug. In some cases it can be useful. There are two interesting examples whereby the activity of penicillin has been enhanced by the presence of another drug.

One of these is the effect of clavulanic acid, described in Section 14.5.3.

The other is the administration of penicillins with a compound called probenecid (Fig. 14.41). Probenecid is a moderately lipophilic carboxylic acid and, as such, is similar to penicillin. It is found that probenecid can block facilitated transport of penicillin through the kidney tubules. In other words, probenecid slows down the rate at which penicillin is excreted by competing with it in the excretion mechanism. As a result, penicillin levels in the bloodstream are enhanced and the antibacterial activity increases – a useful tactic if faced with a particularly resistant bacterium.

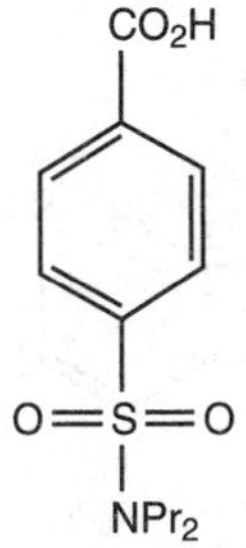

Fig. 14.41 Probenecid.

14.5.2 Cephalosporins

Discovery and structure of cephalosporin C

The second major group of β-lactam antibiotics to be discovered was the cephalosporins. The first cephalosporin was cephalosporin C, isolated in 1948 from a fungus obtained from sewer waters on the island of Sardinia. Although its antibacterial properties were recognized at the time, it was not until 1961 that the structure was established. It is perhaps hard for modern chemists to appreciate how difficult and painstaking structure determination could be, even in the post-war period. The advent of NMR spectroscopy in the sixties and seventies has revolutionized the field so that, if a new fungal metabolite is discovered today, its structure can be worked out in a matter of days rather than a matter of years.

The structure of cephalosporin C (Fig. 14.42) has similarities to that of penicillin in that it has a bicyclic system containing a four-membered β-lactam ring. However, this time the β-lactam ring is fused with a six-membered dihydrothiazine ring. This larger ring relieves the strain in the bicyclic system to some extent, but it is still a reactive system. A study of the cephalosporin skeleton reveals that cephalosporins can be derived from the same biosynthetic precursors as penicillin, i.e. cysteine and valine (Fig. 14.43).

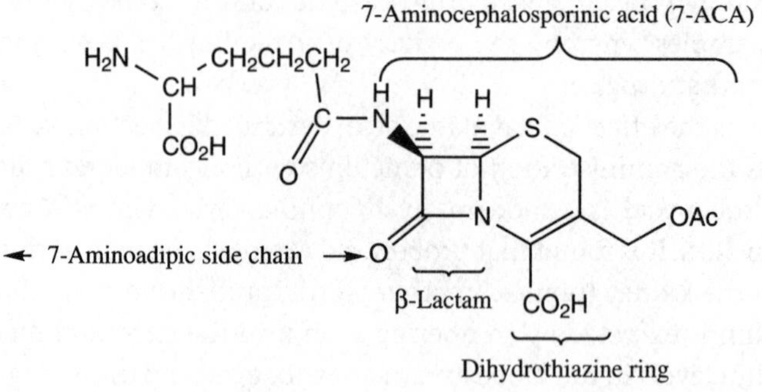

Fig. 14.42 Cephalosporin C.

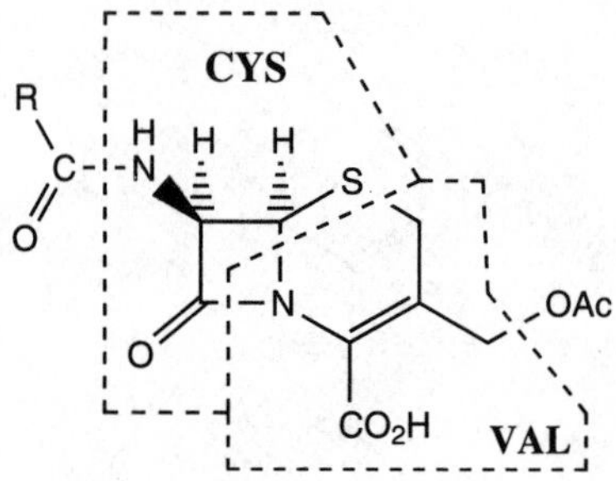

Fig. 14.43 Cephalosporin skeleton.

Properties of cephalosporin C

The properties of cephalosporin C can be summarized as follows:

- Difficult to isolate and purify because of a highly polar side chain
- Low potency (one-thousandth of penicillin G)
- Not absorbed orally
- Non-toxic
- Low risk of allergenic reactions
- Relatively stable to acid hydrolysis compared to penicillin G
- More stable than penicillin G to penicillinase (equivalent to oxacillin)
- Good ratio of activity against Gram-negative and Gram-positive bacteria.

Cephalosporin C has few clinical uses, is not particularly potent and at first sight seems rather uninteresting. However, its importance lies in its potential as a lead compound to something better. This potential resides in the last property mentioned above. Cephalosporin C may have low activity, but the antibacterial activity which it *does* have is more evenly directed against Gram-negative and Gram-positive bacteria than is the case with penicillins. By modifying cephalosporin C, we might be able to increase the potency while retaining the breadth of activity against both Gram-positive and Gram-negative bacteria. Another inbuilt advantage of cephalosporin C over penicillin is that it already has greater resistance to acid hydrolysis and to penicillinase enzymes. Cephalosporin C itself has been used in the treatment of urinary tract infections since it is found to concentrate in the urine and survive the body's hydrolytic enzymes.

Structure–activity relationships of cephalosporin C

Many analogues of cephalosporin C have been made and the structure–activity relationship (SAR) conclusions are as follows.

- The β-lactam ring is essential
- A free carboxyl group is needed at position 4
- The bicyclic system is essential
- The stereochemistry of the side groups and the rings is important.

These results tally closely with those obtained for the penicillins and, once again, there are only a limited number of places in which modifications can be made (Fig. 14.44). Those places are:

- the 7-acylamino side chain
- the 3-acetoxymethyl side chain
- substitution at carbon 7.

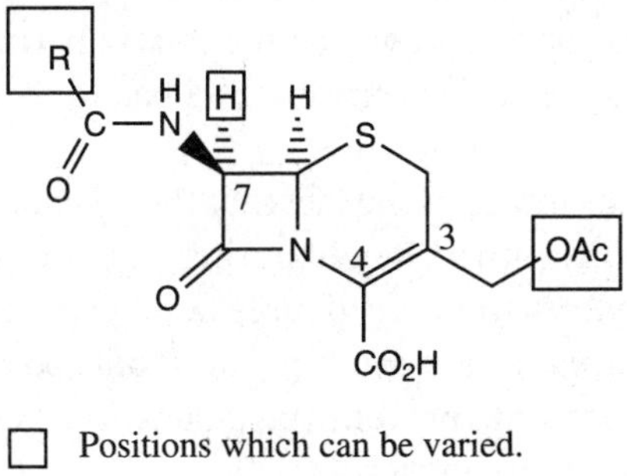

☐ Positions which can be varied.

Fig. 14.44 Positions for possible modification of cephalosporin C.

Analogues of cephalosporin C by variation of the 7-acylamino side chain

Access to analogues with varied side chains at the 7-position initially posed a problem. Unlike penicillins, it proved impossible to obtain cephalosporin analogues by fermentation. Similarly, it was not possible to obtain the 7-ACA (7-amino-cephalosporinic acid) skeleton (Fig. 14.45) either by fermentation or by enzymic hydrolysis of cephalosporin C, thus preventing the semisynthetic approach analogous to the preparation of penicillins from 6-APA.

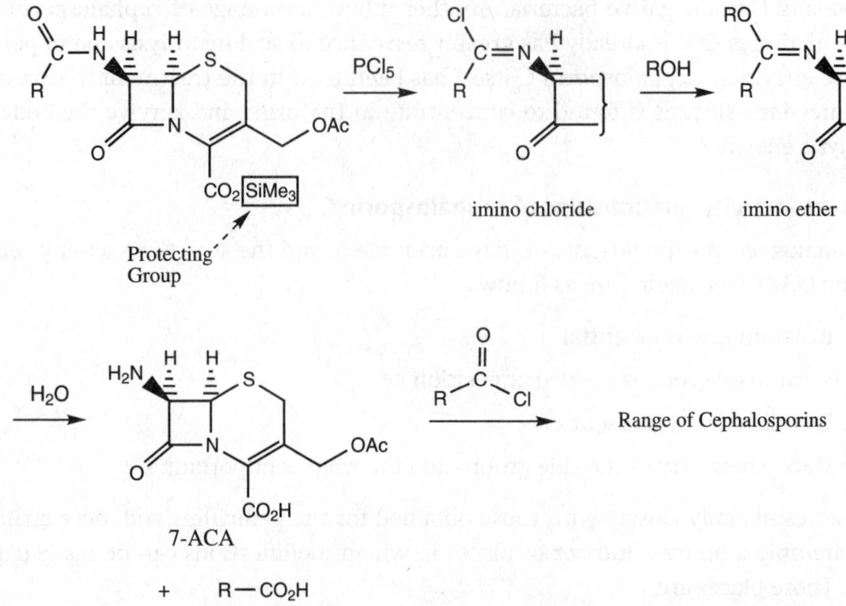

Fig. 14.45 Synthesis of 7-aminocephalosporinic acid and cephalosporin analogues.

Therefore, a way had to be found of obtaining 7-ACA from cephalosporin C by chemical hydrolysis. This is not an easy task. After all, a secondary amide has to be hydrolysed in the presence of a highly reactive β-lactam ring. Normal hydrolytic procedures are not suitable and so a special method had to be worked out, as shown in Fig. 14.45.

The strategy used takes advantage of the fact that the β-lactam nitrogen is unable to share its lone pair of electrons with its neighbouring carbonyl group. The first step of the procedure requires the formation of a double bond between the nitrogen on the side chain and its neighbouring carbonyl group. This is only possible for the secondary amide group since ring constraints prevent the β-lactam nitrogen forming a double bond within the β-lactam ring (see Section 14.5.1).

A chlorine atom is now introduced to form an imino chloride, which can then be reacted with an alcohol to give an imino ether. This product is now more susceptible to hydrolysis than the β-lactam ring and so treatment with aqueous acid successfully gives the desired 7-ACA, which can then be acylated to give a range of analogues. The most commonly used of these cephalosporin analogues is cephalothin (Fig. 14.46).

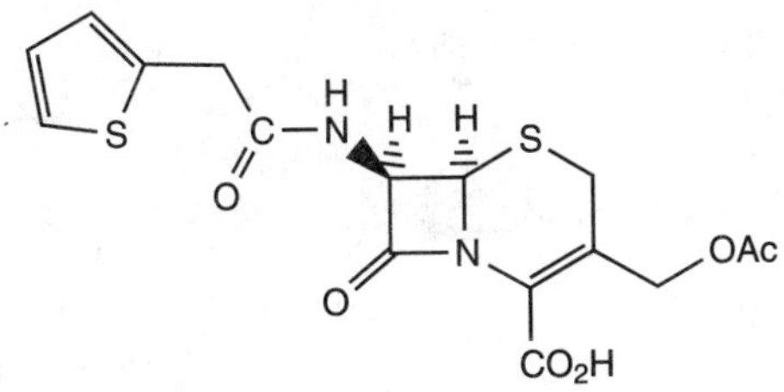

Fig. 14.46 Cephalothin.

The properties of cephalothin are as follows:

- Less active than penicillin G versus cocci and Gram-positive bacilli.
- More active than penicillin G versus some Gram-negative bacilli (*S. aureus* and *E. coli*).
- Resistant to penicillinase from *S. aureus* infections.
- Not active against *Pseudomonas aeruginosa*.
- Poorly absorbed in the gastrointestinal tract and has to be injected.
- Metabolized in man by deacetylation to give a free 3-hydroxymethyl group which has reduced activity.
- Less chance of allergic reactions and can be used for patients with allergies to penicillin.

The study of several analogues has demonstrated the following SAR results relevant to the 7-acylamino side chain:

- Best activity is obtained if the *alpha* carbon is monosubstituted (i.e. RCH$_2$CO-7-ACA). Further substitution leads to a drop in Gram-positive activity.
- Lipophilic substituents on the aromatic or heteroaromatic ring increase the Gram-positive activity and decrease the Gram-negative activity.

Analogues of cephalosporin C by variation of the 3-acetoxymethyl side chain

The first observation which can be made about this area of the molecule is that losing the 3-acetyl group releases the free alcohol group and results in a drop of activity (Fig. 14.47). This hydrolysis occurs metabolically and therefore it would be useful if this process was blocked to prolong the activity of cephalosporins. An example is cephaloridine (Fig. 14.48), which contains a pyridinium group in place of the acetoxy group.

Properties of cephaloridine are as follows:

- Stable to metabolism.
- Soluble in water because of the positive charge.
- Low serum protein binding leads to good levels of free drug in the circulation.

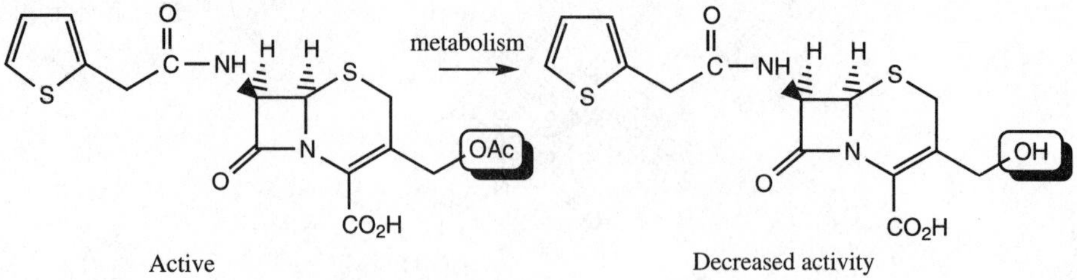

Fig. 14.47 Metabolic hydrolysis of cephalothin.

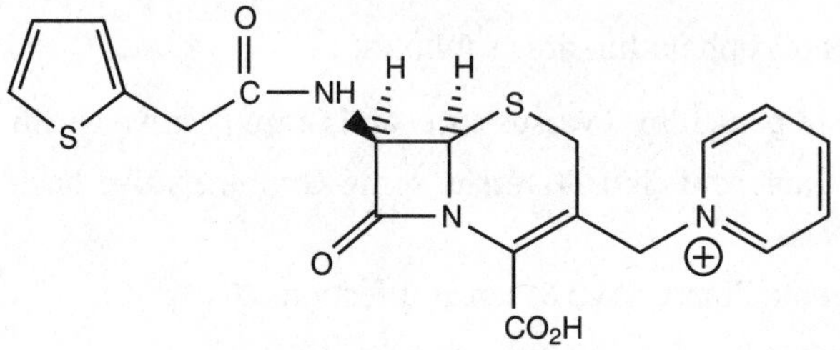

Fig. 14.48 Cephaloridine.

- Excellent activity against Gram-positive bacteria.
- Same activity as cephalothin against Gram-negative bacteria.
- Slightly lower resistance than cephalothin to penicillinase.
- Some kidney toxicity at high doses.
- Poorly absorbed through gut wall and has to be injected.

A second example is cephalexin (Fig. 14.49), which has no substitution at position 3. This is one of the few cephalosporins which is absorbed through the gut wall and can be taken orally. This better absorption appears to be related to the presence of the 3-methyl group. Usually, the presence of such a group lowers the activity of cephalosporins but, if the correct 7-acylamino group is present as in cephalexin, then activity can be retained. The mechanism of the absorption through the gut wall is poorly understood and therefore it is not clear why the 3-methyl group is so advantageous.

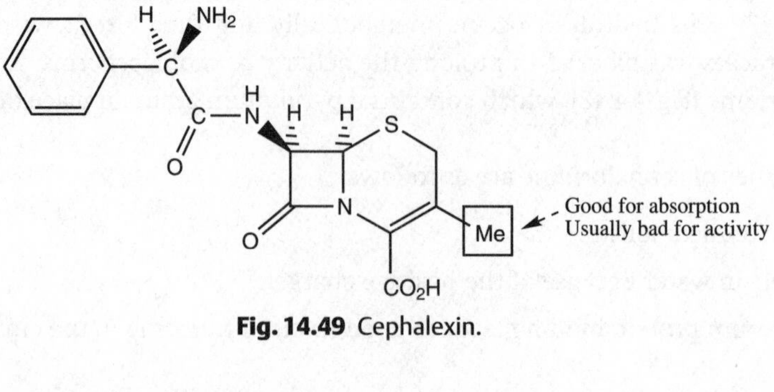

Fig. 14.49 Cephalexin.

The activity of cephalexin against Gram-positive bacteria is lower than injectable cephalosporins, but it is still useful. The activity versus Gram-negative bacteria is similar to the injectable cephalosporins.

Synthesis of 3-methylated cephalosporins

The synthesis of 3-methylated cephalosporins from cephalosporins is very difficult and it is easier to start from the penicillin nucleus, as shown in Fig. 14.50. The synthesis, which was first demonstrated by Eli Lilly Pharmaceuticals, involves a ring expansion in which the five-membered thiazolidine ring in penicillin is converted to the six-membered dihydrothiazine ring in cephalosporin.

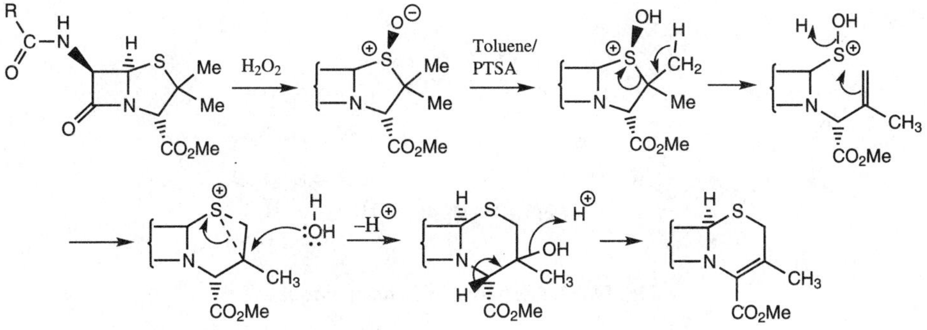

Fig. 14.50 Synthesis of 3-methylated cephalosporins.

Summary of properties of cephalosporins

The following conclusions can be drawn on the analogues studied to this point:

• Injectable cephalosporins of clinical use have a high activity against a large number of Gram-positive and Gram-negative organisms, including the penicillin-resistant staphylococci.

• Most cephalosporins are poorly absorbed through the gut wall.

• In general, cephalosporins have lower activity than comparable penicillins, but a better range. This implies that the enzyme which is attacked by penicillin and cephalosporin has a binding site which fits the penam skeleton better than the cephem skeleton.[1]

• The ease of oral absorption appears to be related to an *alpha* amino group on the 7-acyl substituent, plus an uncharged group at position 3.

The cephalosporins mentioned so far are all useful agents but, as with penicillins, the appearance of resistant organisms has posed a problem. Gram-negative organisms, in particular, appear to have a β-lactamase which can degrade even those

---

[1] Penam and cephem refer to the penicillin and cephalosporin ring systems respectively.

cephalosporins which are resistant to β-lactamase enzymes in Gram-positive species. Attempts to introduce some protection against these lactamases by means of steric shields (compare Section 14.5.1) were successful, but led to inactive compounds. Clearly the introduction of such groups in cephalosporins not only prevents access to the β-lactamase enzyme, but also to the target transpeptidase enzyme.

The next advance came when it was discovered that cephalosporins substituted at the 7-position were active.

Analogues of cephalosporin C by substitution at position 7

The only substitution which has been useful at position 7 has been the introduction of the 7-*alpha* methoxy group to give a class of compounds known as the cephamycins (Fig. 14.51).

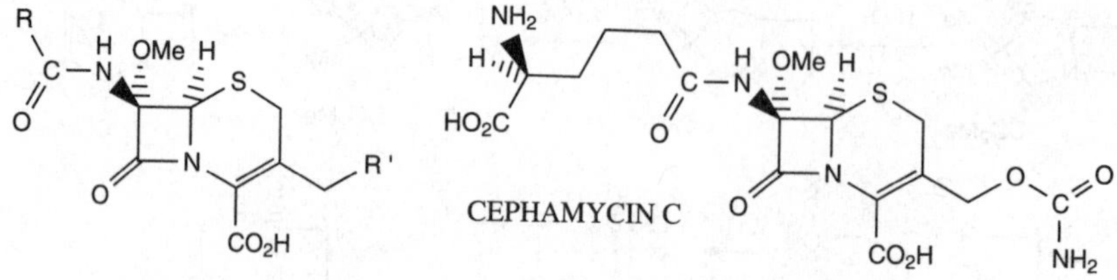

Fig. 14.51 Cephamycin C and analogues.

The parent compound, cephamycin C, was isolated from a culture of *Streptomyces clavuligerus* and was the first β-lactam to be isolated from a bacterial source. Modification of the side chain gave cefoxitin (Fig. 14.52), which showed a broader spectrum of activity than most cephalosporins because of greater resistance to penicillinase enzymes. This increased resistance is thought to be due to the steric hindrance provided by the extra methoxy group. However, it is interesting to note that introduction of the methoxy group at the corresponding 6-*alpha* position of penicillins results in loss of activity. Modifications of the cephamycins are aimed at increasing Gram-positive activity while retaining Gram-negative activity, as in cefoxitin (Fig. 14.52).

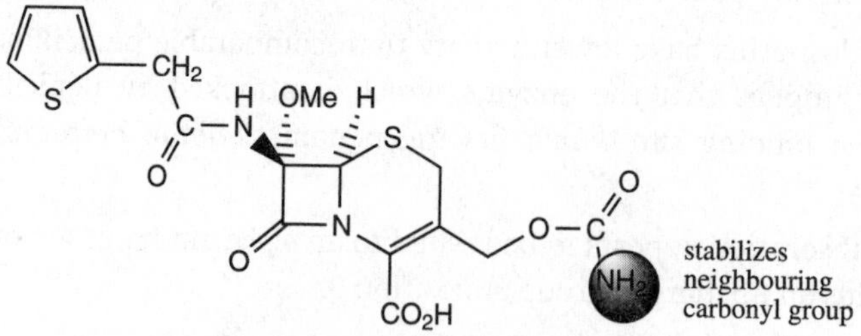

stabilizes
neighbouring
carbonyl group

Fig. 14.52 Cefoxitin.

Properties of cefoxitin are as follows:

- Stable to β-lactamases.

- Stable to mammalian hydrolytic enzymes (owing to NH_2 in place of CH_3; compare Section 15.9.2).

- Broader spectrum of activity than previous cephalosporins.

- Poor absorption through the gut wall and therefore administered by injection.

- Painful at injection site and therefore administered with a local anaesthetic.

- Poor activity against *Pseudomonas aeruginosa*.

Second- and third-generation cephalosporins – oximinocephalosporins

Research is continually being carried out to try and discover cephalosporins with an improved spectrum of activity or which are active against particularly resistant bacteria. One group of cephalosporins which has resulted from this effort has been the oximinocephalosporins. The first useful agent in this class of compounds was cefuroxime (Glaxo; Fig. 14.53) which, like cefoxitin, has good resistance to β-lactamases and mammalian esterases. The drug is very safe, has a wide spectrum of activity, and is useful against organisms which have become resistant to penicillin. However, it is not active against 'difficult' bacteria such as *Pseudomonas aeruginosa*, and it also has to be injected. Various modifications have resulted in another injectable cephalosporin – ceftazidime (Fig. 14.54).

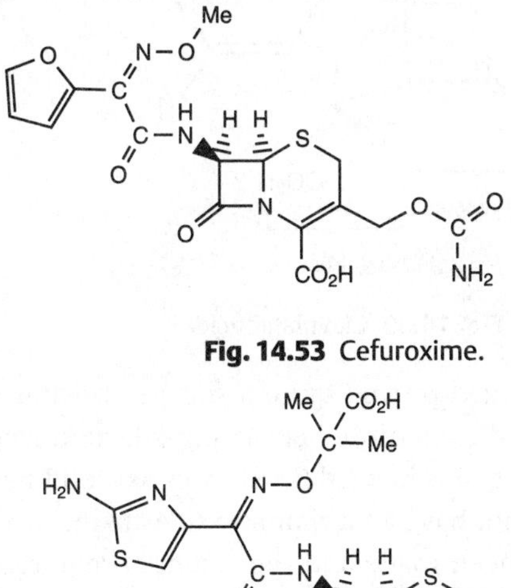

Fig. 14.53 Cefuroxime.

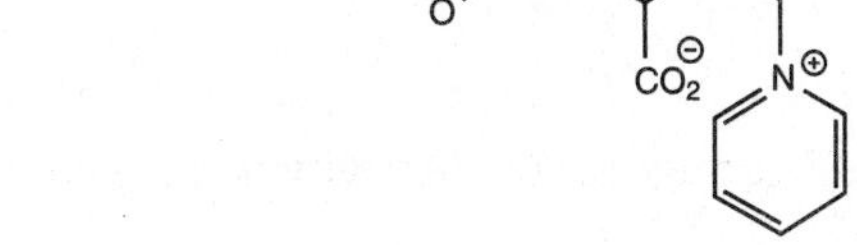

Fig. 14.54 Ceftazidime.

This drug is particularly useful since it is effective against *Pseudomonas aeruginosa*. The new five-membered thiazolidine ring was incorporated, since the literature shows that it is advantageous in other cephalosporin systems. We have already seen how the presence of a pyridinium ring can make cephalosporins more stable to metabolism, and this is also included in the structure.

14.5.3 Novel β-lactam antibiotics

Although penicillins and cephalosporins are the best known and most researched β-lactams, there are other β-lactam structures which are of great interest in the anti-bacterial field.

Clavulanic acid (Beechams 1976)

Clavulanic acid (Fig. 14.55) was isolated from *Streptomyces clavuligerus* by Beechams in 1976. It has weak and unimportant antibiotic activity. However, it is a powerful and irreversible inhibitor of most β-lactamases[1] and, as such, is now used in combination with traditional penicillins such as amoxycillin (Augmentin). This allows the amount of amoxycillin to be reduced and also increases the spectrum of activity.

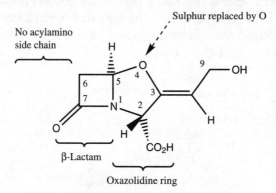

Fig. 14.55 Clavulanic acid.

The structure of clavulanic acid proved quite a surprise once it was determined, since it was the first example of a naturally occurring β-lactam ring which was not fused to a sulfur-containing ring. It is instead fused to an oxazolidine ring structure. It is also unusual in that it does not have an acylamino side chain.

Many analogues have now been made and the essential requirements for β-lactamase activity are:

- The β-lactam ring
- The double bond
- The double bond has the Z configuration. (Activity is reduced but not eliminated if the double bond is *E*)
- No substitution at C6

[1] It must be realized that there are various types of β-lactamases. Clavulanic acid is effective against most but not all.

- (R)-stereochemistry at positions 2 and 5
- The carboxylic acid group.

The variability allowed is therefore strictly limited to the 9-hydroxyl group. Small hydrophilic groups appear to be ideal, suggesting that the original hydroxyl group is involved in a hydrogen bonding interaction with the active site of the β-lactamase.

Clavulanic acid is a mechanism-based irreversible inhibitor and could be classed as a suicide substrate (see Chapter 4). The drug fits the active site of β-lactamase, and the β-lactam ring is opened by a serine residue in the same manner as penicillin. However, the acyl–enzyme intermediate then reacts further with another enzymic nucleophilic group (possibly NH_2) to bind the drug irreversibly to the enzyme (Fig. 14.56). The mechanism requires the loss or gain of protons at various stages and an amino acid such as histidine present in the active site would be capable of acting as a proton donor/acceptor (compare the mechanism of acetylcholinesterase in Chapter 15).

Thienamycin (Merck 1976)

Thienamycin (Fig. 14.57) was isolated from *Streptomyces cattleya*. It is potent, with an extraordinarily broad range of activity against Gram-positive and Gram-negative bacteria (including *P. aeruginosa*). It has low toxicity and shows a high resistance to β-lactamases. This resistance has been ascribed to the presence of the hydroxyethyl side chain. However, it shows poor metabolic and chemical stability, and is not absorbed from the gastrointestinal tract. Therefore, analogues with increased chemical stability and oral activity would be useful.

The big surprise concerning the structure of thienamycin is the missing sulfur atom and acylamino side chain, both of which were thought to be essential to antibacterial activity. Furthermore, the stereochemistry of the side chain at substituent 6 is opposite from the usual stereochemistry in penicillins.

Olivanic acids

The olivanic acids (e.g. MM13902) (Fig. 14.58) were isolated from strains of *Streptomyces olivaceus* and are carbapenam structures like thienamycin. They have very strong β-lactamase activity, in some cases 1000 times more potent than clavulanic acid. They are also effective against the β-lactamases which can break down cephalosporins. These β-lactamases are unaffected by clavulanic acid. Unfortunately, olivanic acids are susceptible to metabolic degradation in the kidney.

Nocardicins

At least seven nocardicins (e.g. nocardicin A; Fig. 14.59) have been isolated from natural sources by the Japanese company Fujisawa. They show moderate activity *in vitro* against a narrow group of Gram-negative bacteria, including *P. aeruginosa*. However, it is surprising that they should show any activity at all since they contain a single β-lactam ring unfused to any other ring system. The presence of a fused second ring has always been thought to be essential to strain the β-lactam ring sufficiently for antibacterial activity.

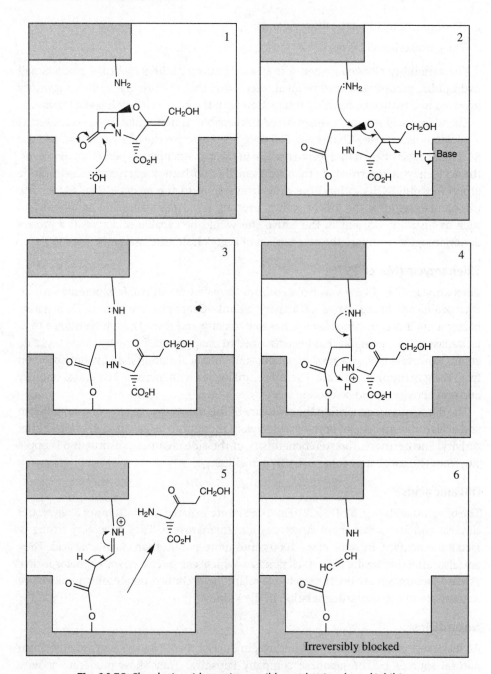

Fig. 14.56 Clavulanic acid as an irreversible mechanism-based inhibitor.

One explanation for the surprising activity of the nocardicins is that they operate via a different mechanism from penicillins and cephalosporins. There is some evidence supporting this in that the nocardicins are inactive against Gram-positive

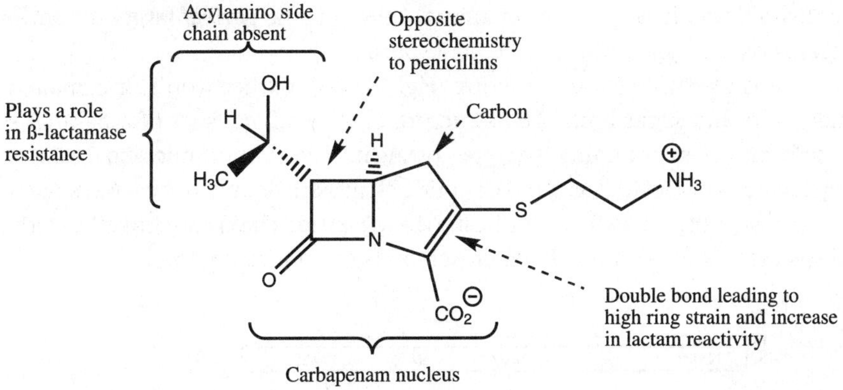

Fig. 14.57 Thienamycin.

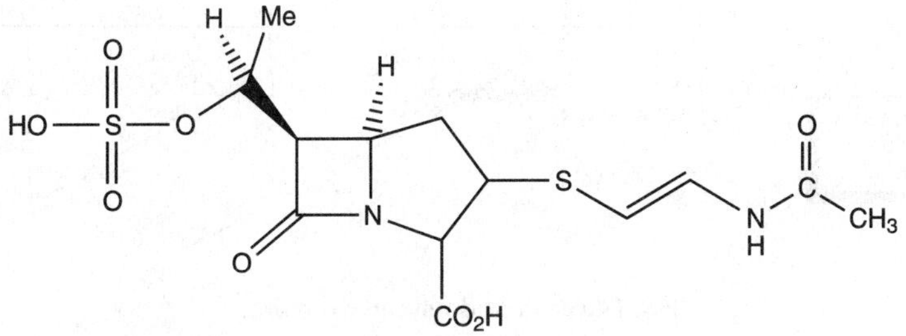

Fig. 14.58 MM13902.

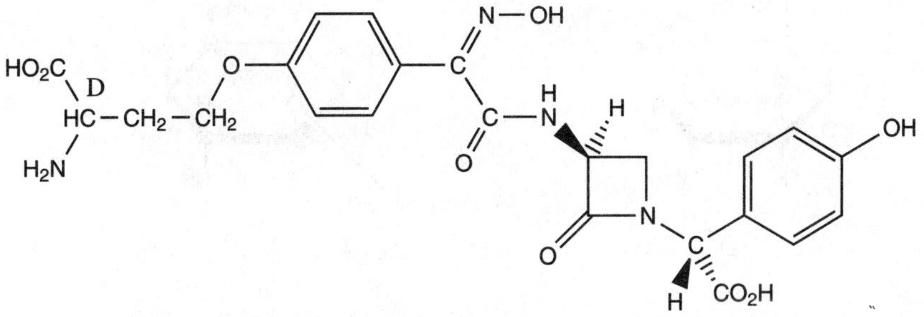

Fig. 14.59 Nocardicin A.

bacteria and generally show a different spectrum of activity from the other β-lactam antibiotics. It is possible that these compounds act on cell wall synthesis by inhibiting a different enzyme. They also show low levels of toxicity.

14.5.4 The mechanism of action of penicillins and cephalosporins

Bacteria have to survive a large range of environmental conditions, such as varying pH, temperature, and osmotic pressure. Therefore, they require a robust cell wall.

Since this cell wall is not present in animal cells, it is the perfect target for antibacterial agents such as penicillins and cephalosporins.

The wall is a peptidoglycan structure (Fig. 14.60). In other words, it is made up of peptide units and sugar units. The structure of the wall consists of a parallel series of sugar backbones containing two types of sugar (N-acetylmuramic acid (NAM) and N-acetylglucosamine (NAG); Fig. 14.61). Peptide chains are bound to the NAM sugars and, in the final step of cell wall biosynthesis, these peptide chains are linked together by the displacement of D-alanine from one chain by glycine in another.

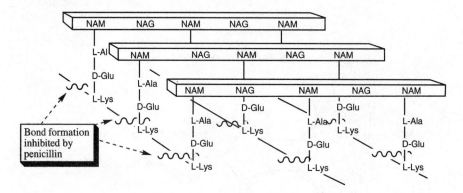

Fig. 14.60 Peptidoglycan structure.

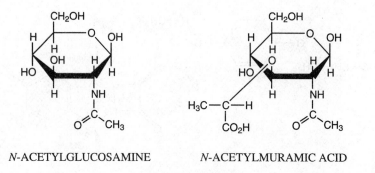

N-ACETYLGLUCOSAMINE   *N*-ACETYLMURAMIC ACID

Fig. 14.61 Sugars contained in cell wall structure of bacteria.

It is this final cross-linking reaction which is inhibited by penicillins and cephalosporins, so that the cell wall framework is not meshed together (Fig. 14.62). As a result, the wall becomes 'leaky'. Since the salt concentrations inside the cell are greater than those outside the cell, water enters the cell, the cell swells, and eventually lyses (bursts). The enzyme responsible for the cross-linking reaction is known as the transpeptidase enzyme.

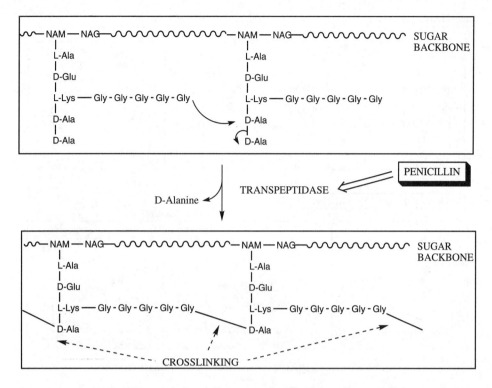

Fig. 14.62 Cross-linking of bacteria cell walls inhibited by penicillin.

It has been proposed that penicillin has a conformation which is similar to the transition-state conformation taken up by D-Ala-D-Ala, the portion of the amino acid chain involved in the cross-linking reaction (Fig. 14.63). Since this is the reaction centre for the transpeptidase enzyme, it is quite an attractive theory to postulate that the enzyme mistakes the penicillin molecule for the D-Ala-D-Ala moiety and accepts the penicillin into its active site. Once penicillin is in the active site, the normal enzymatic reaction would be carried out on the penicillin.

In the normal mechanism (Fig. 14.63), the amide bond between the two alanine units on the peptide chain is split. The terminal alanine departs the active site, leaving the peptide chain bound to the active site. The terminal glycine of the pentaglycyl chain can then enter the active site and form a peptide bond to the alanine group, thus removing it from the active site.

The enzyme can attack the β-lactam ring of penicillin and open it in the same way as it did with the amide bond. However, penicillin is cyclic and, as a result, the molecule is not split in two and nothing leaves the active site. Subsequent hydrolysis of the acyl group does not take place, presumably because glycine is unable to reach the site because of the bulkiness of the penicillin molecule.

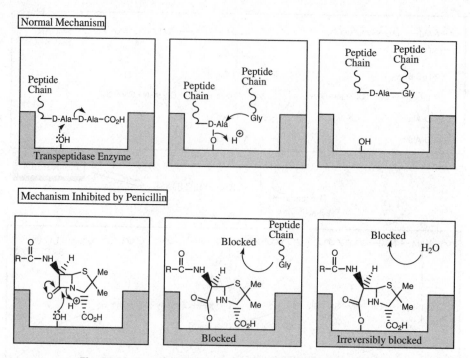

Fig. 14.63 Cross-linking mechanism by transpeptidase enzyme.

However, there is some doubt over this theory since there are one or two anomalies. For example, 6-methylpenicillin (Fig. 14.64) is a closer analogue to D-Ala-D-Ala. It should fit the active site better and have higher activity. On the contrary, it is found to have lower activity.

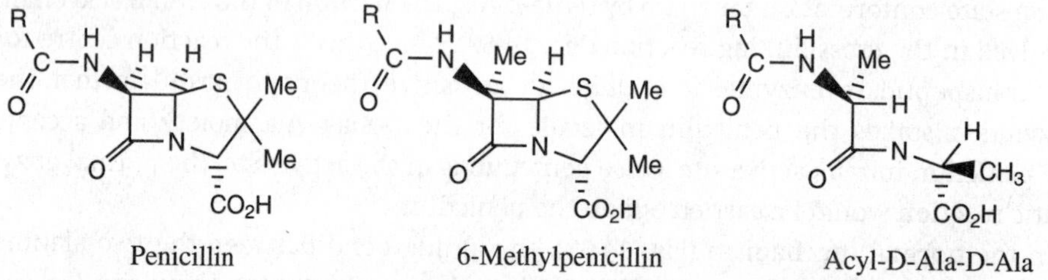

Penicillin   6-Methylpenicillin   Acyl-D-Ala-D-Ala

Fig. 14.64

An alternative proposition is that penicillin does not bind to the active site itself, but binds instead to a site nearby. By doing so, the penicillin structure overlaps the active site and prevents access to the normal reagents – the umbrella effect (see Section 5.7.2). If a nucleophilic group (not necessarily in the active site) attacks the β-lactam ring, the penicillin becomes bound irreversibly, permanently blocking the active site (Fig. 14.65).

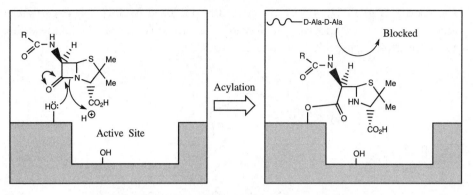

Fig. 14.65 Alternative 'umbrella' mechanism of inhibition.

14.5.5 Other drugs which act on bacterial cell wall biosynthesis

It is important to appreciate that penicillin and cephalosporins are not the only antibacterial agents which inhibit cell wall biosynthesis. The antibacterial agents vancomycin, cycloserine and bacitracin also inhibit biosynthesis, though at different stages. To synthesize the cell wall, NAM (with its pentapeptide substituent) and NAG are linked together in the cytoplasm of the cell to form a disaccharide, and then transported across the cell membrane by a carrier lipid which releases it from the cell membrane to the growing cell wall (Fig. 14.66).

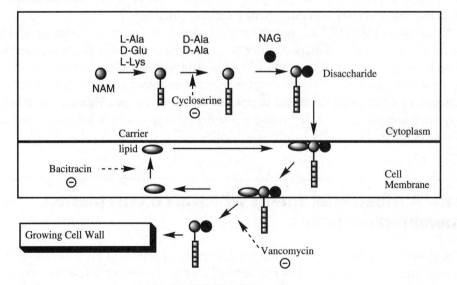

Fig. 14.66 Cell wall biosynthesis.

Cycloserine (Fig. 14.67) is a simple molecule produced by *Streptomyces garyphalus* which acts within the cytoplasm and prevents the addition of the two D-alanine amino acids to the growing peptide chain attached to NAM.

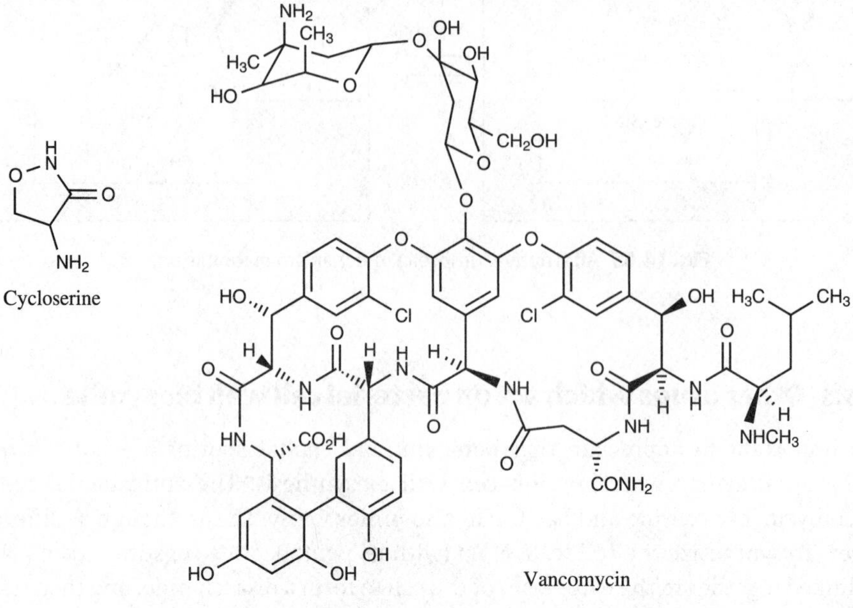

Cycloserine

Vancomycin

Fig. 14.67 Other agents used against cell wall biosynthesis.

Bacitracin is a polypeptide complex produced by *Bacillus subtilis* which interferes with the transport mechanism carrying the disaccharide across the cell membrane, thus starving the cell wall of disaccharide building units.

Vancomycin (Fig. 14.67) is a glycopeptide produced by *Streptomyces orientalis* which prevents the release of the disaccharide from its lipid carrier. It is the main standby drug for methicillin-resistant strains of *S.aureus,* and is usually given by injection since it is not absorbed orally. However, vancomycin is sometimes given orally to treat the appearance of an organism called *Clostridium difficile* in the gut. This organism may appear following the use of broad-spectrum antibiotics, and is harmful since it produces toxins.

14.6 Antibacterial agents which act on the plasma membrane structure

The peptides valinomycin (Fig. 14.68) and gramicidin A (Fig. 14.71) both act as ion conducting antibiotics and allow the uncontrolled movement of ions across the cell membrane. Unfortunately, both these agents show no selective toxicity for bacterial cells over mammalian cells and are therefore useless as therapeutic agents. Their mechanism of action is interesting nevertheless.

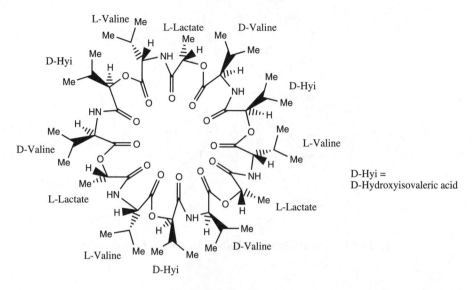

Fig. 14.68 Valinomycin.

Valinomycin is a cyclic structure containing three molecules of L-valine, three molecules of D-valine, three molecules of L-lactic acid, and three molecules of D-hydroxyisovalerate. These four components are linked in an ordered fashion such that there is an alternating sequence of ester and amide linking bonds around the cyclic structure. This is achieved by the presence of a lactic or hydroxyisovaleric acid unit between each of the six valine units. Further ordering can be observed by noting that the L and D portions of valine alternate around the cycle, as do the lactate and hydrox-yisovalerate units.

Valinomycin acts as an ion carrier and in some ways could be looked upon as an inverted detergent. Since it is cyclic, it forms a doughnut-type structure where the polar carbonyl oxygens of the ester and amide groups face inside, while the hydrophobic side chains of the valine and hydroxyisovalerate units point outwards. This is clearly favoured since the hydrophobic side chains can interact via van der Waals interactions with the fatty lipid interior of the cell membrane, while the polar hydrophilic groups are clustered together in the centre of the doughnut to produce a hydrophilic environment.

This hydrophilic centre is large enough to accommodate an ion and it is found that a 'naked' potassium ion (i.e. no surrounding water molecules) fits the space and is complexed by the amide carboxyl groups (Fig. 14.69).

Valinomycin can therefore 'collect' a potassium ion from the inner surface of the membrane, carry it across the membrane and deposit it outside the cell, thus

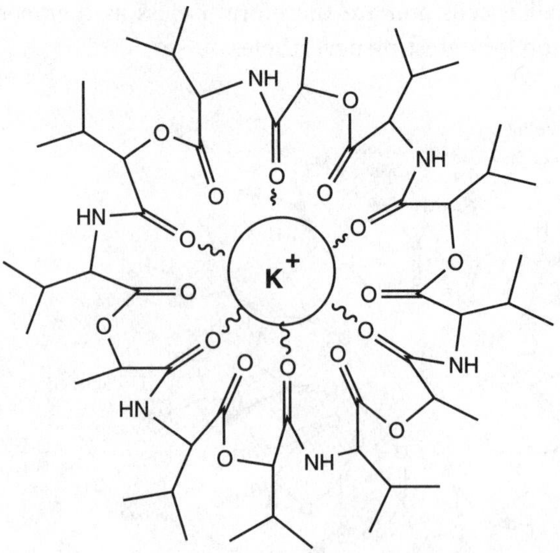

Fig. 14.69 Potassium ion in the hydrophilic centre of valinomycin.

disrupting the ionic equilibrium of the cell (Fig. 14.70). Normally, cells have a high concentration of potassium and a low concentration of sodium. The fatty cell membrane prevents passage of ions between the cell and its environment, and ions can only pass through the cell membrane aided by specialized and controlled ion transport systems. Valinomycin introduces an uncontrolled ion transport system, which proves fatal.

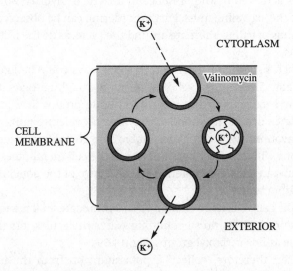

Fig. 14.70 Valinomycin disrupts the ionic equilibrium of a cell.

Valinomycin is specific for potassium ions over sodium ions. One might be tempted to think that sodium ions would be too small to be properly complexed. However, the real reason is that sodium ions do not lose their surrounding water 'coat' very easily and would have to be transported as the hydrated ion. As such, they are too big for the central cavity of valinomycin.

Gramicidin A (Fig. 14.71) is a peptide containing 15 amino acids which is thought to coil into a helix such that the outside of the helix is hydrophobic and interacts with the membrane lipids, while the inside of the helix contains hydrophilic groups, thus allowing the passage of ions. Therefore, gramicidin A could be viewed as an escape tunnel through the cell membrane. In fact, one molecule of gramicidin would not be long enough to traverse the membrane and it has been proposed that two gramicidin helices align themselves end-to-end to achieve the length required (Fig. 14.72).

Val-Gly-Ala-Leu-Ala-Val-Val-Val-Trp-Leu-Trp-Leu-Trp-Leu-Trp-NH-CH$_2$-CH$_2$-OH

Fig. 14.71 Gramicidin A.

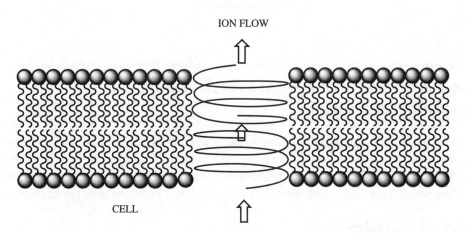

Fig. 14.72 Gramicidin helices aligned end-to-end traversing membrane.

The polypeptide antibiotic polymyxin B (Fig. 14.73) also operates within the cell membrane. It shows selective toxicity for bacterial cells over animal cells, which appears to be related to the ability of the compound to bind selectively to the different plasma membranes. The mechanism of this selectivity is not fully understood.

Polymyxin B acts like valinomycin, but it causes the leakage of small molecules such as nucleosides from the cell. The drug is injected intramuscularly and is useful against *Pseudomonas* strains which are resistant to other antibacterial agents.

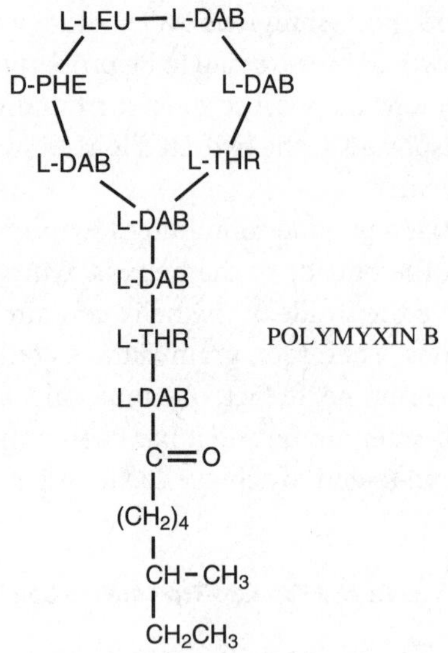

```
              L-LEU —— L-DAB
             /                \
         D-PHE              L-DAB
            \                   \
          L-DAB              L-DAB
              \              /
               \        L-THR
                \      /
                 L-DAB
                   |
                 L-DAB
                   |
                 L-THR        POLYMYXIN B
                   |
                 L-DAB
                   |
                 C═O
                   |
                (CH₂)₄
                   |
                 CH – CH₃
                   |
                 CH₂CH₃
```

Fig. 14.73 Polypeptide antibiotic. (DAB = α, γ-Diaminobutyric acid with peptide link through the α-amino group.)

14.7 Antibacterial agents which impair protein synthesis

Examples of such agents are the rifamycins, which act against RNA, and the aminoglycosides, tetracyclines, and chloramphenicol, which all act against the ribosomes. Selective toxicity is due either to different diffusion rates through the cell barriers of different cell types or to a difference between the target enzymes of different cells.

14.7.1 Rifamycins

Rifampicin (Fig. 14.74) is a semisynthetic rifamycin made from rifamycin B – an antibiotic isolated from *Streptomyces mediterranei*. It inhibits Gram-positive bacteria and works by binding non-covalently to RNA polymerase and inhibiting RNA synthesis. The DNA-dependent RNA polymerases in eukaryotic cells are unaffected, since the drug binds to a peptide chain not present in the mammalian RNA polymerase. It is therefore highly selective.

The drug is mainly used in the treatment of tuberculosis and staphylococci infections that resist penicillin. It is a very useful antibiotic, showing a high degree of selectivity against bacterial cells over mammalian cells. Unfortunately, it is also expensive, which discourages its use against a wider range of infections. The flat naphthalene ring and several of the hydroxyl groups are essential for activity.

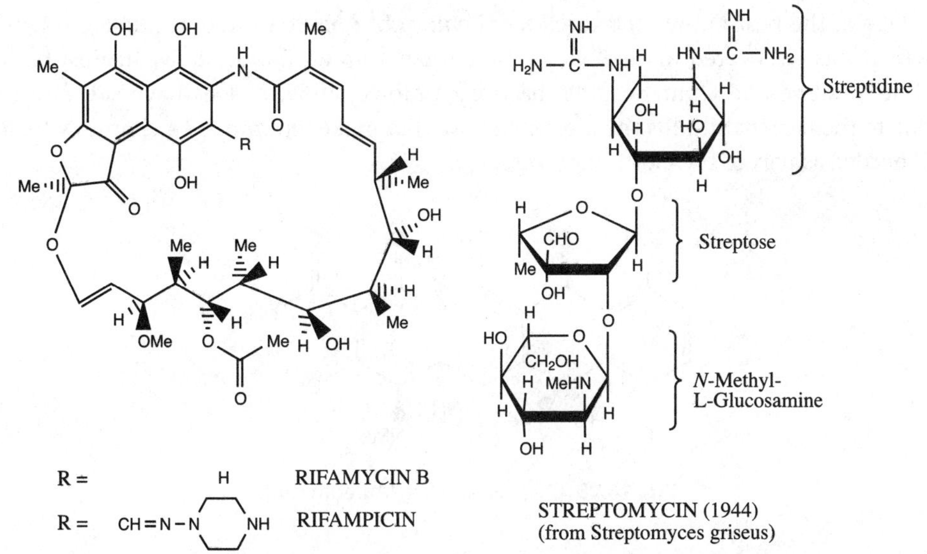

Fig. 14.74 Antibacterial agents which impair protein synthesis.

The selectivity of this antibiotic is interesting since both bacterial cells and mammalian cells contain the enzyme RNA polymerase. However, as we have seen, the enzyme in bacterial cells contains a peptide chain not present in mammalian RNA polymerase. Presumably this chain was lost from the mammalian enzyme during long years of evolution.

14.7.2 Aminoglycosides

Streptomycin (Fig. 14.74) (from *Streptomyces griseus*, 1944) is an example of an important aminoglycoside. Streptomycin was the next most important antibiotic to be discovered after penicillin and proved to be the first antibiotic effective against the lethal disease *tuberculous meningitis*. The drug works by inhibiting protein synthesis. It binds to the 30S ribosomal subunit and prevents the growth of the protein chain as well as preventing the recognition of the triplet code on mRNA.

Aminoglycosides are fast acting, but they can also cause ear and kidney problems if the dose levels are not carefully controlled. The aminoglycoside antibiotics used to be the only compounds effective against the particularly resistant *P. aeruginosa* (see earlier) and it is only recently that alternative treatments have been unveiled (see above).

14.7.3 Tetracyclines

The tetracyclines as a whole have a broad spectrum of activity and are the most widely prescribed form of antibiotic after penicillins. They are also capable of attacking the malarial parasite.

One of the best known tetracyclines is chlortetracycline (Aureomycin) (Fig. 14.75) which was discovered in 1948. It is a broad-spectrum antibiotic, active against both Gram-positive and Gram-negative bacteria. Unfortunately, it does have side-effects due to the fact that it kills the intestinal flora that make vitamin K – a vitamin which is needed as part of the clotting process.

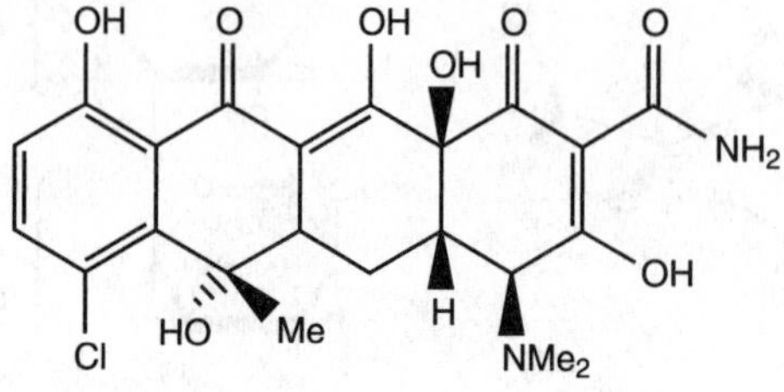

Fig. 14.75 Chlortetracycline (Aureomycin).

Chlortetracycline inhibits protein synthesis by binding to the 30S subunit of ribosomes and prevents aminoacyl-tRNA from binding. This prevents the codon–anticodon interaction from taking place. Protein release is also inhibited.

There is no reason why tetracyclines should not attack protein synthesis in mammalian cells as well as in bacterial cells. In fact, they can. Fortunately, bacterial cells accumulate the drug far more efficiently than mammalian cells and are therefore more susceptible.

14.7.4 Chloramphenicol

Chloramphenicol (Fig. 14.76) was originally isolated from *Streptomyces venezuela*, but is now prepared synthetically. It has two asymmetric centres, but only the *R,R*-isomer is active.

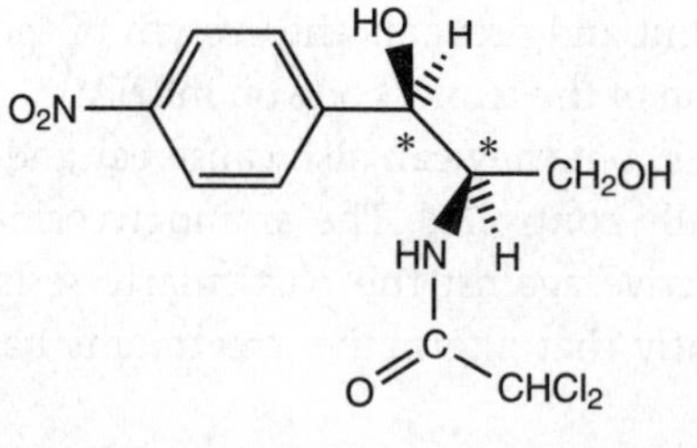

Fig. 14.76 Chloramphenicol (from *Streptomyces venezuela*).

SAR studies demonstrate that there must be a substituent on the aromatic ring which can 'resonate' with it (i.e. NO_2). The *R,R*-propanediol group is essential. The OH groups must be free and presumably are involved in hydrogen bonding. The dichloroacetamide group is important, but can be replaced by other electronegative groups.

Chloramphenicol binds to the 50S subunit of ribosomes and appears to act by inhibiting the movement of ribosomes along mRNA, probably by inhibiting the peptidyl transferase reaction by which the peptide chain is extended.

Chloramphenicol is the drug of choice against typhoid and is also used in severe bacterial infections which are insensitive to other antibacterial agents. It has also found widespread use against eye infections. However, the drug should only be used in these restricted scenarios since it is quite toxic, especially to bone marrow. The NO_2 group is suspected to be responsible for this, although intestinal bacteria are capable of reducing this group to an amino group.

14.7.5 Macrolides

The best known example of this class of compounds is erythromycin - a metabolite produced by the microorganism *Streptomyces erythreus*. The structure (Fig. 14.77) consists of a macrocyclic lactone ring with a sugar and an aminosugar attached. The sugar residues are important for activity.

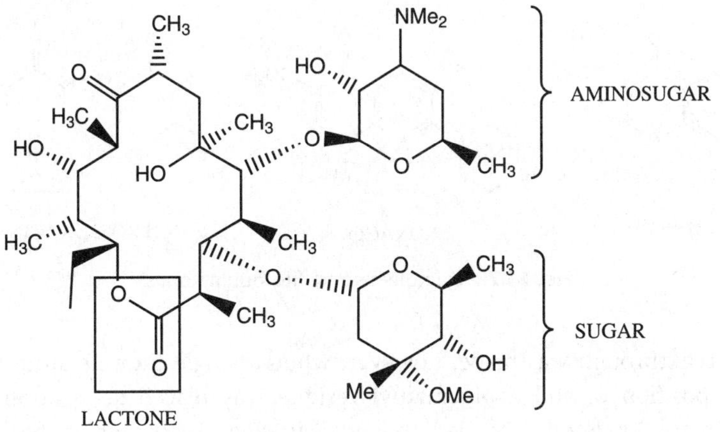

Fig. 14.77 Erythromycin.

Erythromycin acts by binding to the 50S subunit by an unknown mechanism. It works in the same way as chloramphenicol by inhibiting translocation, where the elongated peptide chain attached to tRNA is shifted back from the aminoacyl site to the peptidyl site. Erythromycin was used against penicillin-resistant staphylococci, but newer penicillins are now used for these infections. It is, however, the drug of choice against legionnaire's disease.

14.8 Agents which act on nucleic acid transcription and replication

14.8.1 Quinolones and fluoroquinolones

The quinolone and fluoroquinolone antibacterial agents are relatively late arrivals on the antibacterial scene, but are proving to be very useful therapeutic agents. They are particularly useful in the treatment of urinary tract infections and also for the treatment of infections which prove resistant to the more established antibacterial agents. In the latter case, microorganisms which have gained resistance to penicillin may have done so by mutations affecting cell wall biosynthesis. Since the quinolones and fluoroquinolones act by a different mechanism, such mutations provide no protection against these agents.

Nalidixic acid (Fig. 14.78) was the first therapeutically useful agent in this class of compounds. It is active against Gram-negative bacteria and is useful in the short-term therapy of urinary tract infections. It can be taken orally, but unfortunately bacteria can rapidly develop resistance to it. Various analogues have been synthesized which have similar properties to nalidixic acid, but provide no great advantage.

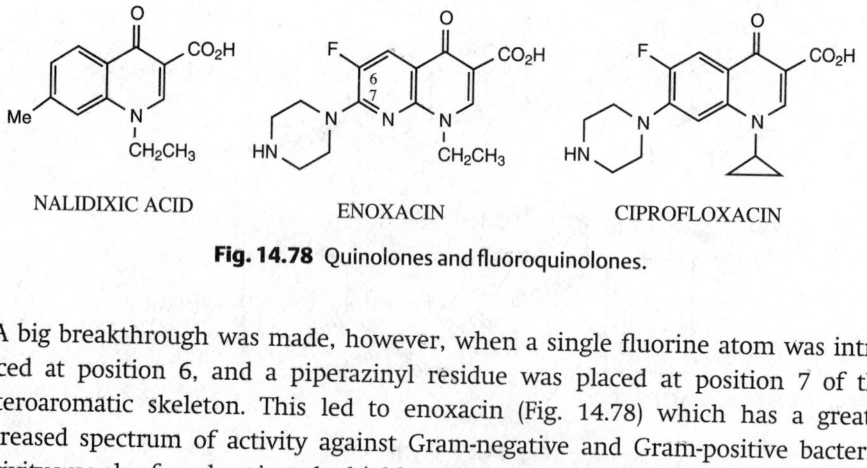

NALIDIXIC ACID   ENOXACIN   CIPROFLOXACIN

Fig. 14.78 Quinolones and fluoroquinolones.

A big breakthrough was made, however, when a single fluorine atom was introduced at position 6, and a piperazinyl residue was placed at position 7 of the heteroaromatic skeleton. This led to enoxacin (Fig. 14.78) which has a greatly increased spectrum of activity against Gram-negative and Gram-positive bacteria. Activity was also found against the highly resistant *P. aeruginosa*.

Further adjustments led to ciprofloxacin (Fig. 14.78), now the agent of choice in treating travellers' diarrhoea. It has been used in the treatment of a large range of infections involving the urinary, respiratory, and gastrointestinal tracts as well as infections of skin, bone, and joints. It has been claimed that ciprofloxacin may be the most active broad-spectrum antibacterial agent on the market. Furthermore, bacteria are slow in acquiring resistance to ciprofloxacin, in contrast to nalidixic acid.

The quinolones and fluoroquinolones are thought to act on the bacterial enzyme deoxyribonucleic acid gyrase (DNA gyrase). This enzyme catalyses the supercoiling of chromosomal DNA into its tertiary structure. A consequence of this is that replication and transcription are inhibited and the bacterial cell's genetic code remains unread. At present, the mechanism by which these agents inhibit DNA gyrase is unclear.

14.8.2 Aminoacridines

Aminoacridines such as proflavine (Fig. 14.2) are topical antibacterial agents which were used in the Second World War for the treatment of surface wounds. Their mechanism of action is described in Chapter 7.

14.9 Drug resistance

With such a wide range of antibacterial agents available in medicine, it may seem surprising that medicinal chemists are still actively seeking new and improved antibacterial agents. The reason for this is due mainly to the worrying ability of bacteria to acquire resistance to current drugs. For example, 60% of *Streptococcus pneumoniae* strains are resistant to β-lactams, while 60% of *Staphylococcus aureus* strains are resistant to methicillin – the so called MRSA bacterium which has been dubbed the 'superbug' by the media. The last resort in treating *S. aureus* infections is vancomycin, but resistance is beginning to appear to that antibiotic as well (the *Visa* bacterium), and there have already been outbreaks in various hospitals. Some strains of *Enterococcus faecalis* appearing in urinary and wound infections are resistant to all known antibiotics and are untreatable. If antibiotic resistance continues to grow, medicine could well be plunged back to the 1930s. Indeed, many of today's advanced surgical procedures would become too risky to carry out due to the risks of infection. Old diseases which were thought to be conquered could make a comeback and are already doing so. For example, a new antibiotic-resistant strain of tuberculosis (MDRTB) appeared in New York and took 4 years and 10 million dollars to bring under control.

Drug resistance can be due to a variety of factors. For example, the bacterial cell may change the structure of its cell membrane and prevent the drug from entering the cell. Alternatively, an enzyme may be produced which destroys the drug. Another possibility is that the cell counteracts the action of the drug. For example, if the drug is targeting a specific enzyme, then the bacterium may synthesize an excess of the enzyme. All these mechanisms require some form of control. In other words, the cell must have the necessary genetic information. This genetic information can be obtained by mutation or by the transfer of genes between cells.

14.9.1 Drug resistance by mutation

Bacteria multiply at such a rapid rate that there is always a chance that a mutation will render a bacterial cell resistant to a particular agent. This feature has been known for a long time and is the reason why patients should fully complete a course of antibacterial treatment even though their symptoms may have disappeared well before the end of the course.

If this rule is adhered to, the vast majority of the invading bacterial cells will be wiped out, leaving the body's own defence system to mop up any isolated survivors or resistant cells. If, however, the treatment is stopped too soon, then the body's defences struggle to cope with the survivors. Any isolated resistant cell is then given the chance to multiply, resulting in a new infection which will, of course, be completely resistant to the original drug.

These mutations occur naturally and randomly and do not require the presence of the drug. Indeed, it is likely that a drug-resistant cell is present in a bacterial population even before the drug is encountered. This was demonstrated with the identification of streptomycin-resistant cells from old cultures of a bacterium called *E. coli* which had been freeze-dried to prevent multiplication before the introduction of streptomycin into medicine.

14.9.2 Drug resistance by genetic transfer

A second way in which bacterial cells can acquire drug resistance is by gaining that resistance from another bacterial cell. This occurs because it is possible for genetic information to be passed on directly from one bacterial cell to another. There are two main methods by which this can take place: transduction and conjugation.

In transduction, small segments of genetic information known as plasmids are transferred by means of bacterial viruses (bacteriophages), leaving the resistant cell and infecting a non-resistant cell. If the plasmid brought to the infected cell contains the gene required for drug resistance, then the recipient cell will be able to use that information and gain resistance. For example, the genetic information required to synthesize β-lactamases can be passed on in this way, rendering bacteria resistant to penicillins. The problem is particularly prevalent in hospitals where currently over 90% of staphylococcal infections are resistant to antibiotics such as penicillin, erythromycin, and tetracycline. It may seem odd that hospitals should be a source of drug-resistant strains of bacteria. In fact, they are the perfect breeding ground. Drugs commonly used in hospitals are present in the air in trace amounts. It has been shown that breathing in these trace amounts kills sensitive bacteria in the nose and allows the nostrils to act as a breeding ground for resistant strains.

In conjugation, bacterial cells pass genetic material directly to each other. This is a method used mainly by Gram-negative, rod-shaped bacteria in the colon, and involves two cells building a connecting bridge of sex pili through which the genetic information can pass.

14.9.3 Other factors affecting drug resistance

The more useful a drug is, the more it will be used and the greater the possibilities of resistant bacterial strains emerging. The original penicillins were used widely in human medicine, but were also commonly used in veterinary medicine. Antibacterial agents have also been used in animal feeding to increase animal weight and this, more than anything else, has resulted in drug-resistant bacterial strains. It is sobering to think that many of the original bacterial strains which were treated so dramatically with penicillin V or penicillin G are now resistant to those early penicillins. In contrast, these two drugs are still highly effective antibacterial agents in poorer, developing nations in Africa, where the use (and abuse) of the drug has been far less widespread.

The ease with which different bacteria acquire resistance varies. For example, *S. aureus* is notorious for its ability to acquire drug resistance due to the ease with which it can undergo transduction. On the other hand, the microorganism responsible for syphilis seems incapable of acquiring resistance and is still susceptible to the original drugs used against it.

14.9.4 The way ahead

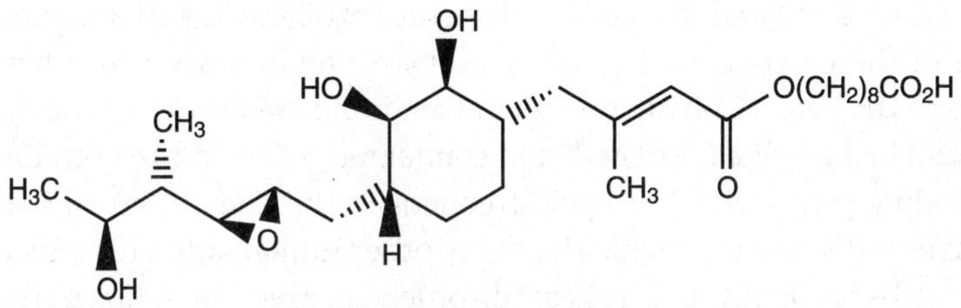

Fig. 14.79 Mupirocin

The ability of bacteria to gain resistance to drugs is an ever present challenge to the medicinal chemist and it is important to continue designing new antibacterial agents. Identifying potential new targets is essential in this never ending battle, and genomic projects offer hope for the future in that respect. One example of a possible set of new targets are the enzymes known as aminoacyl tRNA synthetases. These enzymes are an ancient group of enzymes responsible for attaching amino acids to their tRNAs (see Chapter 7). Since they are ancient, there is a considerable sequence divergence between the bacterial and human enzymes. Therefore, selective inhibition is possible. Isoleucyl tRNA synthetase is one such enzyme which is known to be inhibited by mupirocin (Bactroban) (Fig. 14.79), a clinically useful antibiotic isolated from *Pseudomonas fluorescens* that has activity against MRSA. Research is also being carried out to find novel inhibitors for *S.aureus* tyrosine tRNA synthetase.

15 Cholinergics, anticholinergics, and anticholinesterases

In Chapter 14, we discussed the medicinal chemistry of antibacterial agents and noted the success of these agents in combating many of the diseases which have afflicted mankind over the years. This success was aided in no small way by the fact that the 'enemy' could be identified, isolated, and conquered – first in the petri dish, then in the many hiding places which it could frequent in the body. After this success, the medicinal chemist set out to tackle the many other human ailments which were not infection-based – problems such as heart disorders, depression, schizophrenia, ulcers, autoimmune disease, and cancer. In all these ailments, the body itself has ceased to function properly in some way or other. There is no 'enemy' as such.

So what can medicinal chemistry do if there is no enemy to fight, save for the human body's inefficiency? The first logical step is to understand what exactly has gone wrong.

However, the mechanisms and reaction systems of the human body can be extremely complex. A vast array of human functions proceed each day with the greatest efficiency and with the minimum of outside interference. Breathing, digestion, temperature control, excretion, posture – these are all day-to-day operations which we take for granted – until they go wrong, of course! Considering the complexity of the human body, it is perhaps surprising that its workings don't go wrong more often than they do.

Even if the problem is identified, what can a mere chemical do amidst a body filled with complex enzymes and interrelated chemical reactions? If it is even possible for a single chemical to have a beneficial effect, which of the infinite number of organic compounds would we use?

The problem might be equated with finding the computer virus which has invaded your home computer software, or perhaps trying to trace where a missing letter went, or finding the reason for the country's balance of payments deficit.

However, all is not doom and gloom. The ancient herbal remedies of the past helped to open the curtain on some of the body's jealously guarded secrets. Even the toxins of snakes, spiders, and plants gave important clues to the workings of the body and provided lead compounds to possible cures. Over the past 100 years or so, many biologically active compounds have been extracted from their natural sources, then purified and identified. Chemists have subsequently rung the changes on these lead compounds until an effective drug was identified. The process depended on trial and effort, chance and serendipity, but with this effort came a better understanding of how the body works and how drugs interact with the body. In more recent years, rapid advances in the biological sciences and molecular modelling have resulted in medicinal chemistry moving from being a game of chance to being a science, where the design of new drugs is based on logical theories.

In this chapter, we are going to concentrate on one particular field of medicinal chemistry – cholinergic and anticholinergic drugs. These are drugs which act on the peripheral and central nervous systems. We shall concentrate on the former to start with.

15.1 The peripheral nervous system

The peripheral nervous system (Fig. 15.1) is, as the name indicates, that part of the nervous system which is outside the central nervous system (CNS, the brain and spinal column).

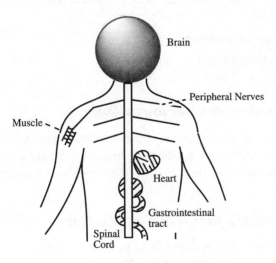

Fig. 15.1 The peripheral nervous system.

There are many divisions and subdivisions of the peripheral system, which can lead to confusion. The first distinction we can make is between the following:

• sensory nerves (nerves which take messages from the body to the CNS)

- motor nerves (nerves which carry messages from the CNS to the rest of the body).

We need only concern ourselves with the latter – the motor nerves.

15.2 Motor nerves of the peripheral nervous system

These nerves take messages from the CNS to various parts of the body, such as skeletal muscle, smooth muscle, cardiac muscle, and glands. The messages can be considered as 'electrical pulses'. However, the analogy with electricity should not be taken too far since the pulse is a result of ion flow across the membranes of nerves and not a flow of electrons (see Appendix 2).

It should be evident that the workings of the human body depend crucially on an effective motor nervous system. Without it, we would not be able to operate our muscles and we would end up as flabby blobs, unable to move or breathe. We would not be able to eat, digest, or excrete our food since the smooth muscle of the gastro-intestinal (GI) tract and the urinary tract are innervated by motor nerves. We would not be able to control body temperature since the smooth muscle controlling the diameter of our peripheral blood vessels would cease to function. Finally, our heart would resemble a wobbly jelly rather than a powerful pump. In short, if the motor nerves failed to function, we would be in a mess! Let us now look at the motor nerves in more detail.

The motor nerves of the peripheral nervous system have been divided into three subsystems (Fig. 15.2):

- the somatic motor nervous system
- the autonomic motor nervous system
- the enteric nervous system.

15.2.1 The somatic motor nervous system

These are nerves which carry messages from the CNS to the skeletal muscles. There are no synapses (junctions) en route and the neurotransmitter at the neuromuscular junction is acetylcholine. The final result of such messages is contraction of skeletal muscle.

15.2.2 The autonomic motor nervous system

These nerves carry messages from the CNS to smooth muscle, cardiac muscle, and the adrenal medulla. This system can be divided into two subgroups.

Parasympathetic nerves

These leave the CNS, travel some distance, then synapse with a second nerve which then proceeds to the final synapse with smooth muscle. The neurotransmitter at both synapses is acetylcholine.

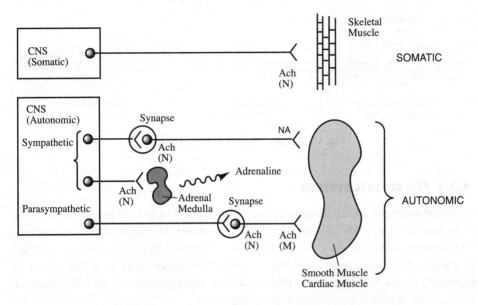

Fig. 15.2 Motor nerves of the peripheral nervous system. N, Nicotinic receptor; M, muscarinic receptor.

Sympathetic nerves

These leave the CNS, but almost immediately synapse with a second nerve (the neurotransmitter here is acetylcholine) which then proceeds to the same target organs as the parasympathetic nerves. However, they synapse with different receptors on the target organs and use a different neurotransmitter – noradrenaline (for their actions, see Section 15.4 and Chapter 16).

The only exception to this are the nerves which go directly to the adrenal medulla. The neurotransmitter released here is noradrenaline and this stimulates the adrenal medulla to release the hormone adrenaline. This hormone then circulates in the blood system and interacts with noradrenaline receptors as well as other adrenaline receptors not directly 'fed' with nerves.

Note that the nerve messages are not sent along continuous 'telephone lines'. Gaps (synapses) occur between different nerves and also between nerves and their target organs (Fig. 15.3). If a nerve wishes to communicate its message to another nerve or a target organ, it can only do so by releasing a chemical. This chemical has to cross the synaptic gap and bind to receptors on the target cell to pass on the message. This interaction between neurotransmitter and receptor can then stimulate other processes which, in the case of a second nerve, leads to the message being continued. Since these chemicals effectively carry the message from one nerve to another, they have become known as chemical messengers, or neurotransmitters. The very fact that they are chemicals and that they carry out a crucial role in nerve transmission allows the medicinal chemist to design and synthesize organic compounds which can mimic (agonists) or block (antagonists) the natural neurotransmitters.

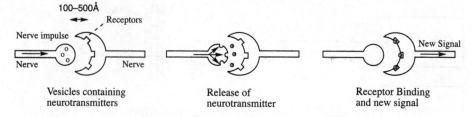

Fig. 15.3 Signal transmission at a synapse.

15.2.3 The enteric system

The third subgroup of the peripheral nervous system is called the enteric system and is located in the walls of the intestine. It receives messages from sympathetic and parasympathetic nerves, but it also responds to local effects to provide local reflex pathways which are important in the control of GI function. A large variety of neurotransmitters are involved, including serotonin, neuropeptides, and ATP. Nitrogen oxide is also involved as a chemical messenger.

15.3 The neurotransmitters

There are a large variety of neurotransmitters in the CNS and the enteric system, but as far as the majority of the peripheral nervous system is concerned we need only consider two – acetylcholine and noradrenaline (Fig. 15.4).

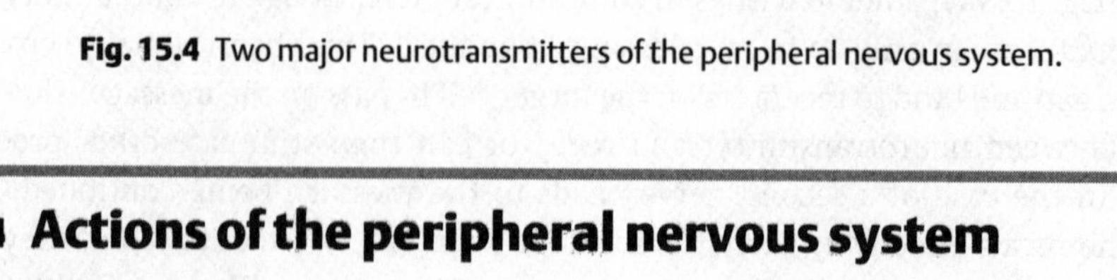

Acetylcholine

R = H   Noradrenaline
R = Me   Adrenaline

Fig. 15.4 Two major neurotransmitters of the peripheral nervous system.

15.4 Actions of the peripheral nervous system

The actions of the peripheral nervous system can be divided into two systems – somatic and autonomic. The autonomic system can be further classified as sympathetic or parasympathetic.

Somatic

Stimulation of the somatic peripheral system leads to the contraction of skeletal muscle.

Autonomic

Sympathetic

Noradrenaline is released at target organs and leads to the contraction of cardiac muscle and an increase in heart rate. It relaxes smooth muscle and reduces the contractions of the GI and urinary tracts. It also reduces salivation and reduces dilatation of the peripheral blood vessels.

In general, the sympathetic nervous system promotes the 'fight or flight' response by shutting down the body's housekeeping roles (digestion, defecation, urination, etc.), and stimulating the heart. The stimulation of the adrenal medulla releases the hormone adrenaline, which reinforces the action of noradrenaline.

Parasympathetic

The stimulation of the parasympathetic system leads to the opposite effects from those of the sympathetic system. Acetylcholine is released at the target organs and reacts with receptors specific to it and not to noradrenaline.

Note that the sympathetic and parasympathetic nervous systems oppose each other in their actions and could be looked upon as a brake and an accelerator. The analogy is not quite apt since both systems are always operating and the overall result depends on which effect is the stronger.

Failure in either of these systems would clearly lead to a large variety of ailments involving heart, skeletal muscle, digestion, etc. Such failure might be the result of either a deficit or an excess of neurotransmitter. Therefore, treatment would involve the administration of drugs which could act as agonists or antagonists, depending on the problem.

However, there is a difficulty with this approach. Usually, the problem that we wish to tackle occurs at a certain location where there might, for example, be a lack of neurotransmitter. Application of an agonist to make up for low levels of neurotransmitter at the heart might solve the problem there, but would lead to problems elsewhere in the body (e.g. the digestive system). At these other locations, the levels of neurotransmitter would be at normal levels, and applying an agonist would then lead to an 'overdose' and cause unwanted side-effects. Therefore, drugs showing selectivity to certain parts of the body over others are clearly preferred.

This selectivity has been achieved to a great extent with both the cholinergic agonists/antagonists and the noradrenaline agonists/antagonists. We will concentrate on the former in this chapter.

15.5 The cholinergic system

15.5.1 The cholinergic signalling system

Let us look first at what happens at synapses involving acetylcholine as the neuro-transmitter. Fig. 15.5 shows the synapse between two nerves and the events involved when a message is transmitted from one nerve cell to another. The same general process takes place when a message is passed from a nerve cell to a muscle cell.

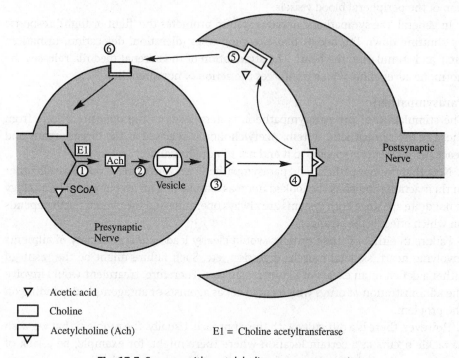

▽   Acetic acid

☐   Choline

⬡   Acetylcholine (Ach)   E1 = Choline acetyltransferase

Fig. 15.5 Synapse with acetylcholine as neurotransmitter.

1. Stage 1 involves the biosynthesis of acetylcholine (Fig. 15.6). Acetylcholine is synthesized in the nerve ending of the presynaptic nerve from choline and acetyl coenzyme A. The reaction is catalysed by the enzyme choline acetyltransferase.

2. Acetylcholine is incorporated into membrane-bound vesicles by means of a specific carrier protein.

3. The arrival of a nerve signal leads to an opening of calcium ion channels and an increase in intracellular calcium concentration. This induces the vesicles to fuse with the cell membrane and in doing so release the transmitter into the synaptic gap.

4. Acetylcholine crosses the synaptic gap and binds to the cholinergic receptor, leading to stimulation of the second nerve.

5. Acetylcholine moves to an enzyme called acetylcholinesterase which is situated on the postsynaptic nerve and which catalyses the hydrolysis of acetylcholine to produce choline and ethanoic acid.

6. Choline binds to the choline receptor on the presynaptic nerve and is taken up into the cell by a carrier protein to continue the cycle.

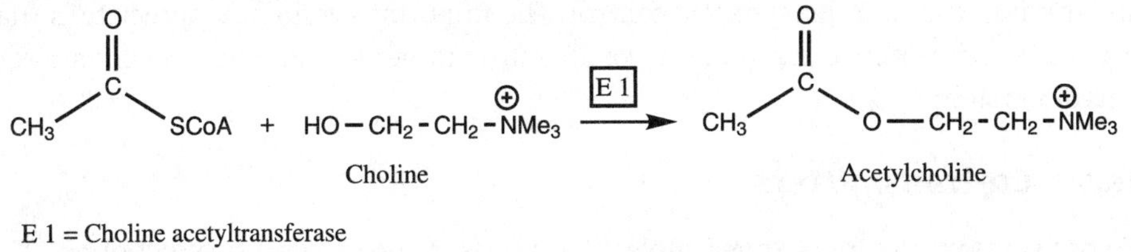

$$CH_3 \overset{O}{\underset{}{\overset{\|}{C}}} SCoA \ + \ HO-CH_2-CH_2-\overset{\oplus}{N}Me_3 \ \xrightarrow{\boxed{E\,1}} \ CH_3 \overset{O}{\underset{}{\overset{\|}{C}}} O-CH_2-CH_2-\overset{\oplus}{N}Me_3$$

Choline   Acetylcholine

E 1 = Choline acetyltransferase

Fig. 15.6 Biosynthesis of acetylcholine.

The most important thing to note about this process is that there are several stages where it is possible to use drugs to either promote or inhibit the overall process. The greatest success so far has been with drugs targeted at stages 4 and 5 (i.e. the cholinergic receptor and the acetylcholinesterase enzyme).

We will look at these in more detail in subsequent sections.

15.5.2 Presynaptic control systems

Cholinergic receptors (called autoreceptors) are also present at the terminal of the presynaptic nerve (Fig. 15.7). The purpose of these receptors is to provide a means of local control over nerve transmission. When acetylcholine is released from the nerve, some of it will find its way to these autoreceptors and switch them on. This has the effect of inhibiting further release of acetylcholine.

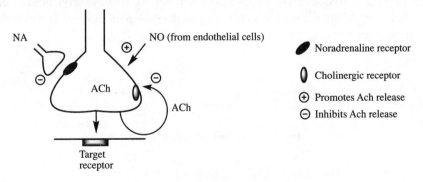

Fig. 15.7 Presynaptic control systems.

The presynaptic nerve also contains receptors for noradrenaline, which act as another control system for acetylcholine release. Branches from the sympathetic nervous system lead to the cholinergic synapses and, when the sympathetic nervous system is active, noradrenaline is released and binds to these receptors. Once again, the effect is to inhibit acetylcholine release. This indirectly enhances the activity of noradrenaline at its target organs by lowering cholinergic activity.

The chemical messenger nitric oxide can also influence acetylcholine release, but in this case it promotes release.

A large variety of other chemical messengers, including cotransmitters (see below) are also implicated in presynaptic control. The important thing to appreciate is that presynaptic receptors offer another possible drug target to influence the cholinergic nervous system.

15.5.3 Co-transmitters

Co-transmitters are messenger molecules released along with acetylcholine. The particular co-transmitter released depends on the location and target cell of the nerves. Each co-transmitter interacts with its own receptor on the postsynaptic cell. Co-transmitters have a variety of structures and include peptides such as vasoactive intestinal peptide, gonadotrophin-releasing hormone (GnRH) and substance P. The roles of these agents appear to be as follows:

• They are longer lasting and reach more distant targets than acetylcholine, leading to longer lasting effects.

• The balance of co-transmitters released varies under different circumstances (e.g. presynaptic control) and so can produce different effects.

15.6 Agonists at the cholinergic receptor

One point might have occurred to the reader. If there is a lack of acetylcholine acting at a certain part of the body, why do we not just give the patient more acetylcholine? After all, it is easy enough to make in the laboratory (Fig. 15.8).

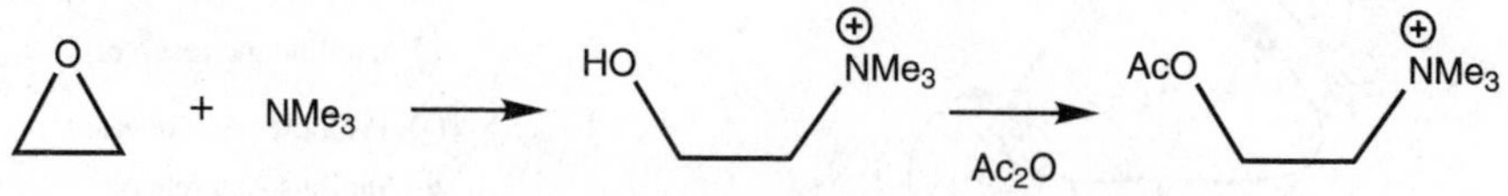

Fig. 15.8 Synthesis of acetylcholine.

There are three reasons why this is not feasible:

• Acetylcholine is easily hydrolysed in the stomach by acid catalysis and cannot be given orally.

• Acetylcholine is easily hydrolysed in the blood, both chemically and by enzymes (esterases).

• There is no selectivity of action. Acetylcholine will switch on all acetylcholine receptors in the body.

Therefore, we need analogues of acetylcholine which are more stable to hydrolysis and which are more selective with respect to where they act in the body. We shall look at selectivity first.

There are two ways in which selectivity can be achieved. First, some drugs might be distributed more efficiently to one part of the body than another. Second, cholinergic receptors in various parts of the body might be slightly different. This difference would have to be quite subtle – not enough to affect the interaction with the natural neurotransmitter acetylcholine but enough to distinguish between two different synthetic analogues.

We could, for example, imagine that the binding site for the cholinergic receptor is a hollow into which the acetylcholine molecule could fit (Fig. 15.9). We might then imagine that some cholinergic receptors in the body have a 'wall' bordering this hollow, while other cholinergic receptors do not. Thus, a synthetic analogue of acetylcholine which is slightly bigger than acetylcholine itself would bind to the latter receptor, but would be unable to bind to the former receptor because of the wall.

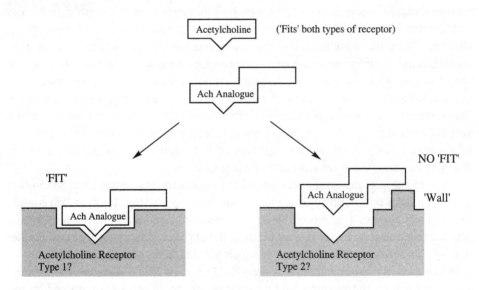

Fig. 15.9 Binding sites for two cholinergic receptors.

This theory might appear to be wishful thinking, but it is now established that cholinergic receptors in different parts of the body are indeed subtly different.

This is not just a peculiarity of acetylcholine receptors. Subtle differences have been observed for other types of receptors, such as those for dopamine, noradrenaline, and serotonin, and there are many types and subtypes of receptor for each chemical messenger (see Chapter 6).

To return to the acetylcholine receptor, how do we know if there are different types? As is often the case, the first clues came from the action of natural compounds. It was discovered that the compounds nicotine (present in tobacco) and muscarine (the active principle of a poisonous mushroom) (Fig. 15.10) were both acetylcholine agonists, but that they had different physiological effects.

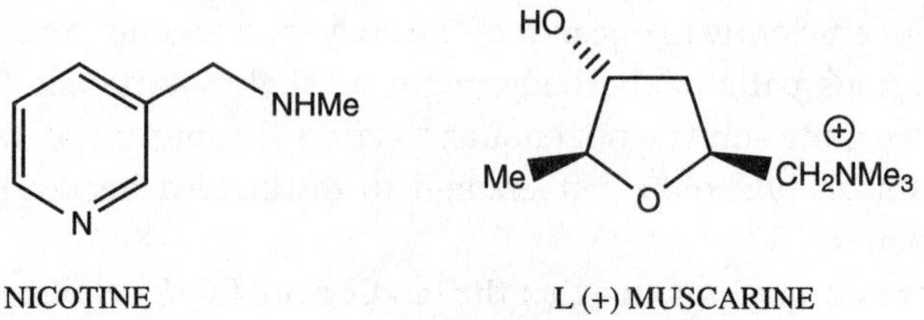

NICOTINE   L (+) MUSCARINE

Fig. 15.10

Nicotine was found to be active at the synapses between two different nerves and also the synapses between nerves and skeletal muscle, but had poor activity elsewhere. Muscarine was active at the synapses of nerves with smooth muscle and cardiac muscle, but showed poor activity at the sites where nicotine was active. From these results, it was concluded that there was one type of acetylcholine receptor on skeletal muscles and at nerve synapses (the nicotinic receptor), and a different sort of acetylcholine receptor on smooth muscle and cardiac muscle (the muscarinic receptor). Therefore, muscarine and nicotine were the first compounds to indicate that receptor selectivity was possible. Unfortunately, these two compounds are not suitable as medicines since they have undesirable side-effects[1]. However, the principle of selectivity was proven and the race was on to design novel drugs which had the selectivity of nicotine or muscarine, but not the side-effects[2].

The first stage in any drug development is to study the lead compound and to find out which parts of the molecule are important to activity so that they can be retained in future analogues (i.e. structure–activity relationships; SAR). These results also provide information about what the binding site of the cholinergic receptor looks like and help in deciding what changes are worth making in new analogues.

In this case, the lead compound is acetylcholine itself. The results described below are valid for both the nicotinic and muscarinic receptors and were obtained by the synthesis of a large range of analogues.

---

[1] This is because of interactions with other receptors, such as the receptors for dopamine or noradrenaline. In the search for a good drug, it is important to gain two types of selectivity – selectivity for one class of receptor over another (e.g. the acetylcholine receptor in preference to a noradrenaline receptor), and selectivity for receptor types (e.g. the muscarinic receptor in preference to a nicotinic receptor).

[2] The search for increasingly selective drugs has led to the discovery that there are subtypes of receptors. In other words, not every muscarinic receptor is the same throughout the body. At present, five subtypes of the muscarinic receptor have been discovered by cloning and have been labelled m_1–m_5. Four of these have been identified in tissues; M_1–M_4.

15.7 Acetylcholine – structure, SAR, and receptor binding

- The positively charged nitrogen atom is essential to activity. Replacing it with a neutral carbon atom eliminates activity.
- The distance from the nitrogen to the ester group is important.
- The ester functional group is important.
- The overall size of the molecule cannot be altered greatly. Bigger molecules have poorer activity.
- The ethylene bridge between the ester and the nitrogen atom cannot be extended (Fig. 15.11).
- There must be two methyl groups on the nitrogen. A larger, third alkyl group is tolerated, but more than one large alkyl group leads to loss of activity.
- Bigger ester groups lead to a loss of activity.

Conclusions: clearly, there is a tight fit between acetylcholine and its binding site which leaves little scope for variation. The above findings tally with a receptor binding site as shown in Fig. 15.12.

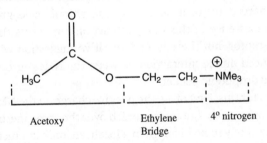

Fig. 15.11 Acetylcholine.

It is proposed that important hydrogen bonding interactions exist between the ester group of the acetylcholine molecule and an asparagine residue. It is also thought that a small hydrophobic pocket exists which can accommodate the methyl group of the ester, but nothing larger. This interaction is thought to be more important in the muscarinic receptor than the nicotinic receptor.

Now let us look at the NMe_3^+ group. The evidence suggests that this group is placed in a hydrophobic pocket lined with three aromatic amino acids. It is also thought that the pocket contains two smaller hydrophobic pockets which are large enough to accommodate two of the three methyl substituents on the NMe_3^+ group. The third methyl substituent on the nitrogen is positioned in an open region of the binding site

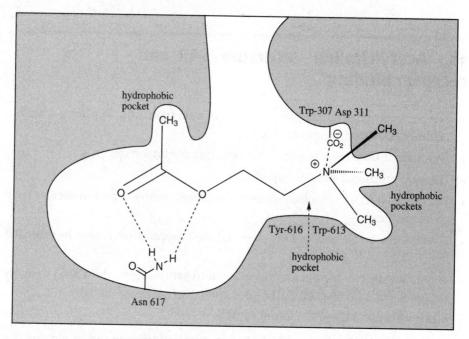

Fig. 15.12 Muscarinic receptor binding site.

and so it is possible to replace it with other groups. A strong ionic interaction has been proposed between the charged nitrogen atom and the anionic side group of an aspartic acid residue. The existence of this ionic interaction represents the classical view of the cholinergic receptor, but there is an alternative suggestion which states that there may be an induced dipole interaction between the NMe_3^+ group and the aromatic residues in the hydrophobic pocket.

There are several reasons for this. First of all, the positive charge on the NMe_3^+ group is not localized on the nitrogen atom, but is also spread over the three methyl groups. Such a diffuse charge is less likely to be involved in a localized ionic interaction and it has been shown by model studies that NMe_3^+ groups can be stabilized by binding to aromatic rings. It might seem strange that a hydrophobic group like an aromatic ring should be capable of stabilizing a positively charged group. However, one must remember that aromatic rings are electron rich, as shown by the fact that they can undergo reaction with electrophiles. It is thought that the diffuse positive charge on the NMe_3^+ group is capable of distorting the π electron cloud of aromatic rings to induce a dipole moment (Fig. 4.8). Dipole interactions between the NMe_3^+ group and an aromatic residue such as tyrosine would then account for the binding. The fact that three aromatic amino acids are present in the pocket adds weight to the argument.

Of course, it is possible that both types of binding interactions are taking place which will please both parties!

A large amount of effort has been expended trying to find out the active conformation of acetylcholine, i.e. the shape adopted by the neurotransmitter when it binds to the cholinergic receptor. This has been no easy task since acetylcholine is a highly flexible molecule, in which bond rotation along the length of its chain can lead to at least nine possible stable conformations (or shapes) (Fig. 15.13).

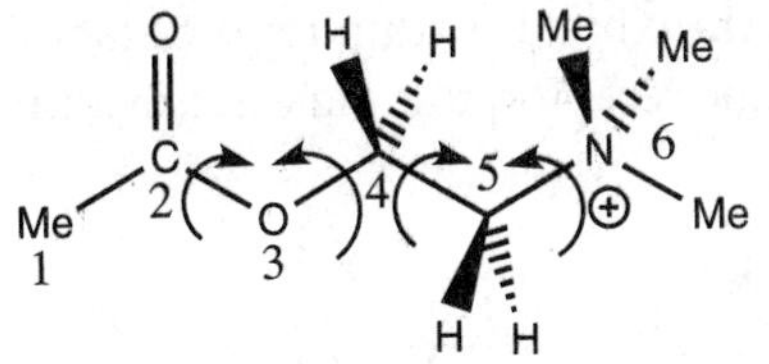

Fig. 15.13 Bond rotations in acetylcholine leading to different conformations.

In the past, it was assumed that a flexible neurotransmitter such as acetylcholine would interact with its receptor in its most stable conformation. In the case of acetylcholine, that would be the conformation represented by the sawhorse and Newman projections shown in Fig. 15.14.

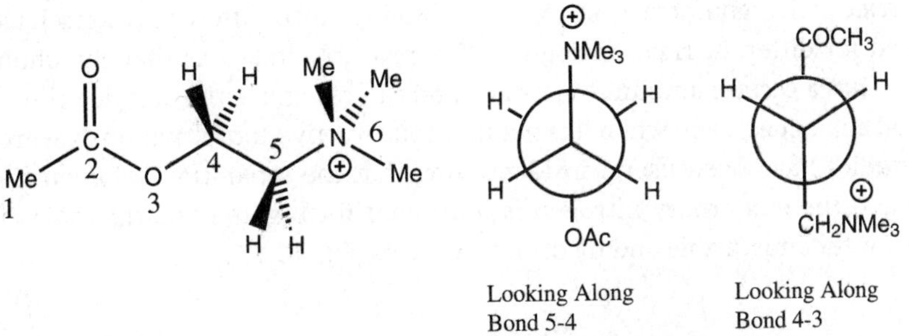

Looking Along
Bond 5-4

Looking Along
Bond 4-3

Fig. 15.14 The sawhorse and Newman projections of acetylcholine.

This assumption is invalid since there is not a great energy difference between alternative conformations (e.g. the gauche conformation shown in Fig. 15.15). The energy gained from the neurotransmitter–receptor binding interaction would be more than sufficient to compensate for any such difference.

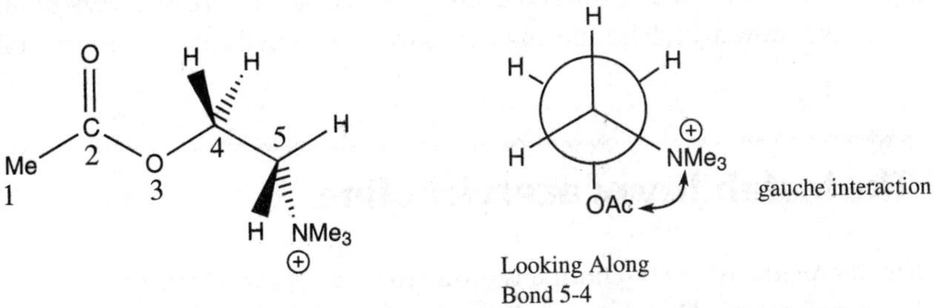

gauche interaction

Looking Along
Bond 5-4

Fig. 15.15 A gauche conformation for acetylcholine.

To try and establish the 'active' conformation of acetylcholine, rigid cyclic molecules have been studied which contain the skeleton of acetylcholine within their structure (e.g. muscarine and the analogues shown in Fig. 15.16). In these structures, the portion of the acetylcholine skeleton which is included in a ring is locked into a particular conformation, since bonds within rings cannot freely rotate. If such molecules bind to the cholinergic receptor, this indicates that this particular conformation is 'allowed' for activity.

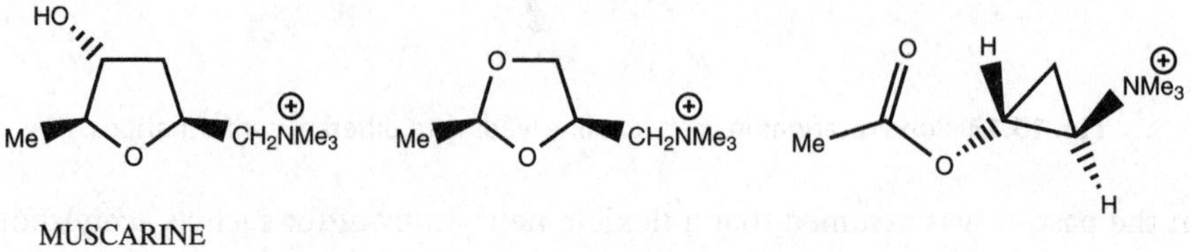

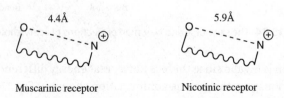

MUSCARINE

Fig. 15.16 Rigid molecules incorporating the acetylcholine skeleton (C-C-O-C-C-N).

Many such structures have been prepared, but it has not been possible to identify one *specific* active conformation for acetylcholine since different results have been obtained for different rigid analogues. This probably indicates that the cholinergic receptor has a certain amount of latitude and can recognize the acetylcholine within the rigid analogues, even when it is not in the ideal active conformation. Nevertheless, such studies have been useful in identifying that the separation between the ester group and the quaternary nitrogen is important for binding and that this distance differs for the muscarinic and nicotinic receptors (Fig. 15.17).

4.4Å   5.9Å

Muscarinic receptor   Nicotinic receptor

Fig. 15.17 Pharmacophore of acetylcholine.

Having identified the binding interactions and pharmacophore of acetylcholine, we shall now look at how acetylcholine analogues were designed with improved stability.

15.8 The instability of acetylcholine

As described previously, acetylcholine is prone to hydrolysis. Why is this and how can the stability be improved? The reason for acetylcholine's instability can be explained by considering one of the conformations that the molecule can adopt (Fig. 15.18).

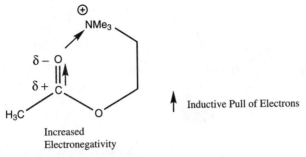

Fig. 15.18 Neighbouring group participation.

In this conformation, the positively charged nitrogen interacts with the carbonyl oxygen and has an electron withdrawing effect. To compensate for this, the oxygen atom pulls electrons towards it from the neighbouring carbon atom and, as a result, makes that carbon atom electron deficient and more prone to nucleophilic attack. Water is a poor nucleophile, but since the carbonyl group is now more electrophilic, hydrolysis takes place relatively easily. This influence of the nitrogen ion is known as neighbouring group participation or anchimeric assistance.

We shall now look at how the problem of hydrolysis was overcome, but it should be appreciated that we are doing so with the benefit of hindsight. At the time the problem was tackled, the SAR studies were incomplete and the format of the cholinergic receptor binding site was unknown. In fact, it was the very analogues which were made to try and solve the problem of hydrolysis that led to a better understanding of the receptor binding site.

15.9 Design of acetylcholine analogues

To tackle the inherent instability of acetylcholine, two approaches are possible:

- steric hindrance
- electronic stabilization.

15.9.1 Steric hindrance

The principle involved here can be demonstrated with methacholine (Fig. 15.19).

This analogue of acetylcholine contains an extra methyl group on the ethylene bridge. The reasons for putting it there are twofold. First, it is to try and build in a shield for the carbonyl group. The bulky methyl group should hinder the approach of any potential nucleophile and slow down the rate of hydrolysis. It should also hinder binding to the esterase enzymes, thus slowing down enzymatic hydrolysis. The results were encouraging, with methacholine proving three times more stable to hydrolysis than acetylcholine.

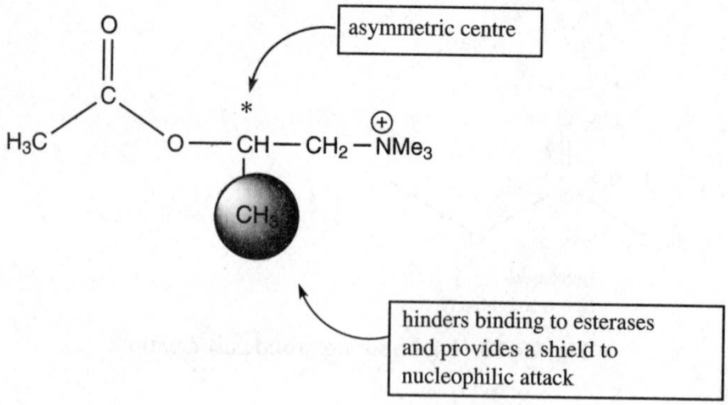

Fig. 15.19 Methacholine.

The obvious question to ask now is, why not put on a bigger alkyl group like an ethyl group or a propyl group? Alternatively, why not put a bulky group on the acyl half of the molecule, since this would be closer to the carbonyl centre and have a greater shielding effect?

In fact, these approaches were tried, but failed. We should already know why – the fit between acetylcholine and its receptor is so tight that there is little scope for enlarging the molecule. The extra methyl group is as much as we can get away with. Larger substituents certainly cut down the chemical and enzymatic hydrolysis, but they also prevent the molecule binding to the cholinergic receptor.

In conclusion, attempts to increase the steric shield beyond the methyl group certainly increase the stability of the molecule, but decrease its activity since it cannot fit the cholinergic receptor.

One other very useful result was obtained from methacholine. It was discovered that the introduction of the methyl group led to significant muscarinic activity and very little nicotinic activity. Therefore, methacholine showed a good selective action for the muscarinic receptor. This result is perhaps more important than the gain in stability.

The good binding to the muscarinic receptor can be explained if we compare the active conformation of methacholine with muscarine (Fig. 15.20). The methyl group of methacholine can occupy the same position as a methylene group in muscarine.

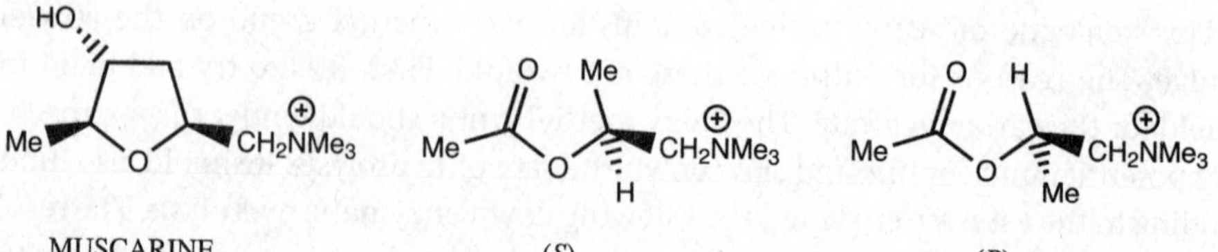

Fig. 15.20 Comparison of muscarine and the R- and S-enantiomers of methacholine.

Note, however, that methacholine can exist as two enantiomers (R and S) and only the S-enantiomer matches the structure of muscarine. The two enantiomers of methacholine have been isolated and the S-enantiomer is the more active enantiomer, as expected. It is not used therapeutically, however.

15.9.2 Electronic effects

The best example of this approach is provided by carbachol (Fig. 15.21), a long acting cholinergic agent which is resistant to hydrolysis. In carbachol, the acyl methyl group has been replaced by an NH_2 group which is of comparable size and can therefore fit the receptor.

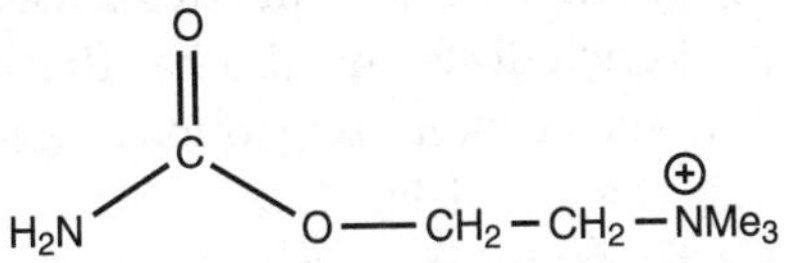

Fig. 15.21 Carbachol.

The resistance to hydrolysis is due to the electronic effect of the carbamate group. The resonance structures shown in Fig. 15.22 demonstrate how the lone pair from the nitrogen atom is fed into the carbonyl group such that the group's electrophilic character is eliminated. As a result, the carbonyl is no longer susceptible to nucleophilic attack.

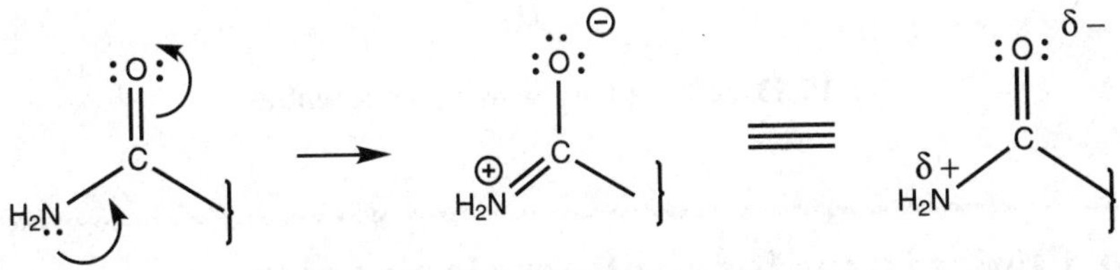

Fig. 15.22 Resonance structures of carbachol.

Carbachol is certainly stable to hydrolysis and is the right size to fit the cholinergic receptor, but it was by no means a foregone conclusion that it would be active. After all, a hydrophobic methyl group has been replaced with a polar NH_2 group and this implies that a polar group has to fit into a hydrophobic pocket in the receptor.

Fortunately, carbachol does fit and is active. Since the methyl group of acetylcholine has been replaced with an amino group without affecting the biological activity, we can call the amino group a 'bioisostere' of the methyl group.

It is worth emphasizing that a bioisostere is a group which can replace another group without affecting the pharmacological activity of interest. Thus, the amino group is a bioisostere of the methyl group as far as the cholinergic receptor is concerned, but not as far as the esterase enzymes are concerned.

Therefore, the inclusion of an electron donating group such as the amino group has greatly increased the chemical and enzymatic stability of our cholinergic agonist. Unfortunately, it was found that carbachol showed very little selectivity between the muscarinic and nicotinic receptors. However, it was used clinically for the treatment of glaucoma. Here, it was applied locally, thus avoiding the problems of receptor selectivity. Glaucoma arises when the aqueous contents of the eye cannot be drained. This raises the pressure on the eye and can lead to blindness. Agonists cause eye muscles to contract, thus relieving the blockage and allowing drainage.

15.9.3 Combining steric and electronic effects

We have already seen that a β-methyl group slightly increases the stability of acetylcholine analogues through steric effects and also has the advantage of introducing some selectivity. Clearly, it would be interesting to add a β-methyl group to carbachol. The compound obtained is bethanechol (Fig. 15.23) which, as expected, is both stable to hydrolysis and selective in its action. It is occasionally used therapeutically in stimulating the GI tract and urinary bladder after surgery. (Both these organs are 'shut down' with drugs during surgery.)

Fig. 15.23 Bethanechol (* = asymmetric centre).

15.10 Clinical uses for cholinergic agonists

Muscarinic agonists:

- Treatment of glaucoma.
- 'Switching on' the GI and urinary tracts after surgery.
- Treatment of certain heart defects by decreasing heart muscle activity and heart rate.

Nicotinic agonists:

• Treatment of myasthenia gravis, an autoimmune disease where the body has produced antibodies against its own cholinergic receptors. This leads to a reduction in the number of available receptors and so fewer messages reach the muscle cells. This in turn leads to severe muscle weakness and fatigue. Administering an agonist increases the chance of activating what few receptors remain. An example of a selective nicotinic agonist is shown in Fig. 15.24. However, this particular compound is not used clinically and the use of anticholinesterases (see Section 15.16) is preferred for the treatment of this disease.

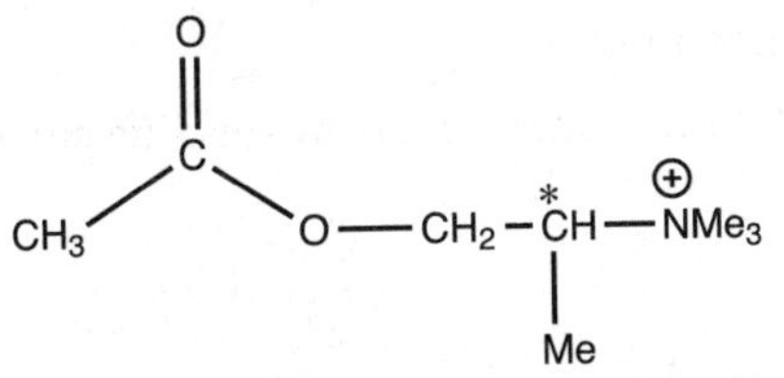

$$CH_3 - \overset{\overset{O}{\parallel}}{C} - O - CH_2 - \overset{*}{CH} - \overset{\oplus}{NMe_3} \qquad * \text{ asymmetric centre}$$
$$|$$
$$Me$$

Fig. 15.24 A selective nicotinic agonist.

15.11 Antagonists of the muscarinic cholinergic receptor

15.11.1 Actions and uses of muscarinic antagonists

Antagonists of the cholinergic receptor are drugs which bind to the receptor but do not 'switch it on'. By binding to the receptor, an antagonist acts like a plug at the receptor site and prevents the normal neurotransmitter (i.e. acetylcholine) from binding (Fig. 15.25). Since acetylcholine cannot 'switch on' its receptor, the overall effect on the body is the same as if there was a lack of acetylcholine. Therefore, antagonists have the opposite clinical effect from agonists.

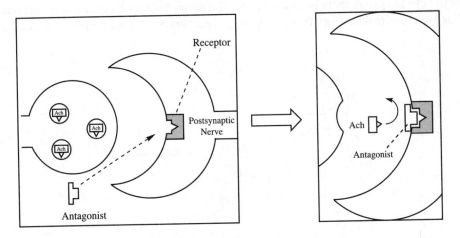

Fig. 15.25 Action of an antagonist.

The antagonists described in this section act only at the muscarinic receptor and therefore affect nerve transmissions to the smooth muscle of the GI tract, urinary tract, glands and the CNS. The clinical effects and uses of these antagonists reflect this fact.

Clinical effects:

- Reduction of saliva and gastric secretions
- Reduction of the motility of the GI and urinary tracts by relaxing smooth muscle
- Dilatation of eye pupils
- CNS effects.

Clinical uses:

- Shutting down the GI and urinary tracts during surgery
- Ophthalmic examinations
- Relief of peptic ulcers
- Treatment of Parkinson's disease
- Anticholinesterase poisoning
- Motion sickness.

15.11.2 Muscarinic antagonists

The first antagonists were natural products and in particular alkaloids (nitrogen-containing compounds derived from plants).

Atropine

Atropine (Fig. 15.26) is obtained from the roots of belladonna (deadly nightshade) and is included in a root extract which was once used by Italian women to dilate the pupils of the eye to appear more beautiful (hence the name belladonna). Clinically, atropine has been used to decrease GIT motility and to counteract anticholinesterase poisoning.

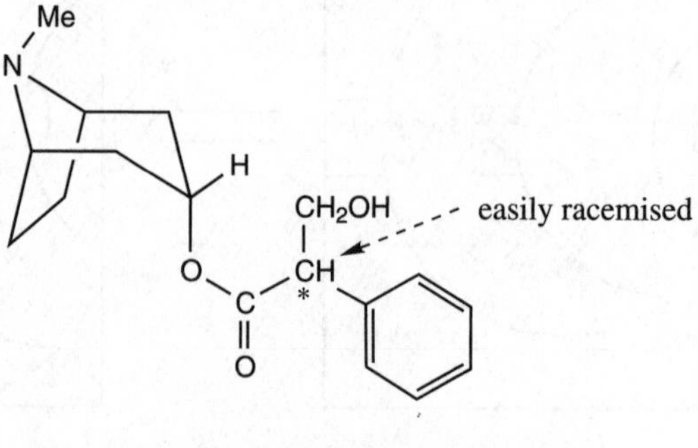

Fig. 15.26 Atropine.

Atropine has an asymmetric centre (*) and therefore two enantiomers are possible. Usually, natural products exist exclusively as one enantiomer. This is also true for atropine, which is present in the plant species *Solanaceae* as a single enantiomer called hyoscyamine. However, as soon as the natural product is extracted into solution, the asymmetric centre racemizes such that atropine is obtained as a racemic mixture and not as a single enantiomer. The asymmetric centre in atropine is easily racemized since it is next to a carbonyl group. The proton attached to the asymmetric centre is acidic and as a result is easily replaced.

Hyoscine (1879–84)

Hyoscine (or scopolamine) (Fig. 15.27) is obtained from the thorn apple and is very similar in structure to atropine. Hyoscine has been used in treating motion sickness.

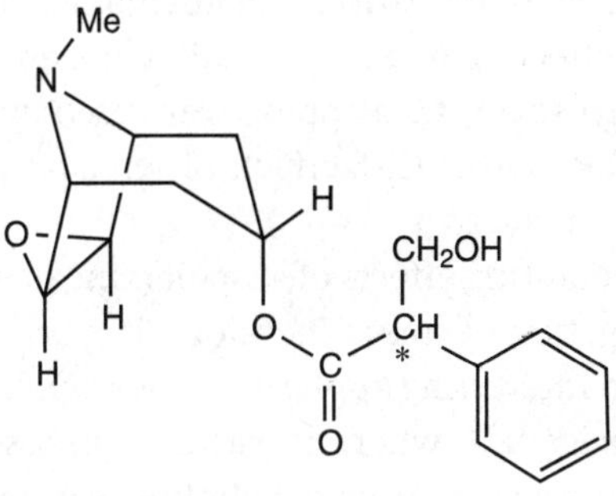

Fig. 15.27 Hyoscine (scopolamine).

These two compounds can bind to and block the cholinergic receptor, but why should they? At first sight, they do not look anything like acetylcholine. If we look more closely, though, we can see that a basic nitrogen and an ester group are present, and if we superimpose the acetylcholine skeleton on to the atropine skeleton, the distance between the ester and the nitrogen groups are similar in both molecules (Fig. 15.28).

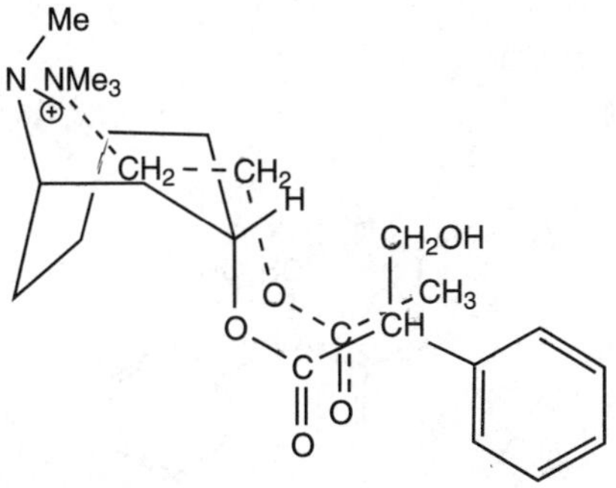

Fig. 15.28 Acetylcholine skeleton superimposed on to the atropine skeleton.

There is, of course, the problem that the nitrogen in atropine is uncharged, whereas the nitrogen in acetylcholine is quaternary and has a full positive charge. This implies that the nitrogen atom in atropine is protonated when it binds to the cholinergic receptor.

Therefore, atropine can be seen to have the two important binding features of acetylcholine – a charged nitrogen when protonated, and an ester group. It is, therefore, able to bind to the receptor, but is unable to 'switch it on'. Since atropine is a larger molecule than acetylcholine, it is capable of binding to other binding groups outside the acetylcholine binding site. As a result, it interacts differently with the receptor, and does not induce the same conformational changes as acetylcholine.

Structural analogues based on atropine

Since both atropine and hyoscine are tertiary amines rather than quaternary salts, they are able to cross the blood–brain barrier and antagonize muscarinic receptors in the brain. This leads to CNS effects. For example, hallucinogenic activity is brought on with high doses, and both hyoscine and atropine were used by witches of the middle ages to produce that very effect. Other CNS effects observed in atropine poisoning are restlessness, agitation, and hyperactivity.

In recent times, the disorientating effects of scopolamine have seen it being used as a truth drug for the interrogation of spies. Therefore, it is no surprise to find it cropping up in various novels. An interesting application for scopolamine was described in Jack Higgin's novel *Day of Judgement*, where it was used in association with succinyl choline to torture one hapless victim. Succinyl choline was applied to the conscious victim to create initial convulsive muscle spasms, followed by paralysis, inability to breathe, agonizing pain and a living impression of death. Scopolamine was then used to erase the memory of this horror, such that the impact would be just as bad when the process was repeated!

To reduce CNS side-effects, quaternary salts of atropine are often used clinically. For example, ipratropium (Fig. 15.29) is used as a bronchodilator and atropine methonitrate (Fig. 15.30) is used to lower motility in the GI tract.

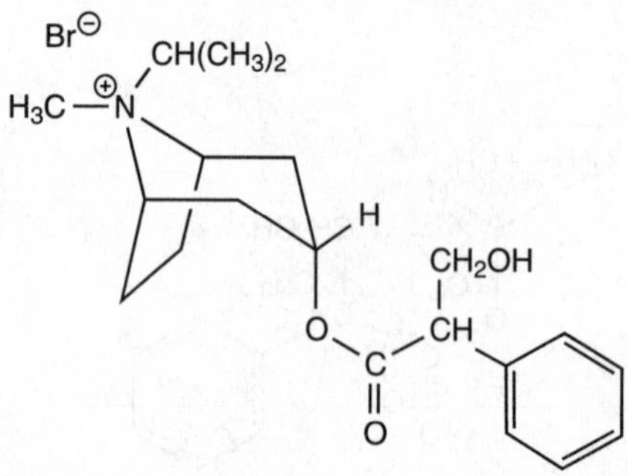

Fig. 15.29 Ipratropium.

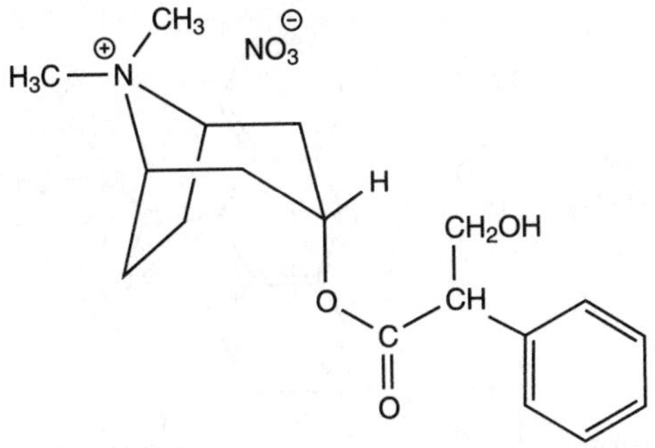

Fig. 15.30 Atropine methonitrate.

A large number of different analogues of atropine were synthesized to investigate its structure–activity relationships. The following groups were important:

- The basic nitrogen is an important binding group and interacts in the ionic form
- The aromatic ring is important
- The ester is important.

It was further discovered that the complex ring system was not necessary for antagonist activity and that simplification could be carried out. For example, amprotropine (Fig. 15.31) has an ester group separated from an amine by three carbon atoms. Chain contraction to two carbon atoms can be carried out without loss of activity, and a large variety of active antagonists have been prepared having the general formula shown in Fig. 15.33 (e.g. tridihexethyl bromide and propantheline chloride; Fig. 15.32).

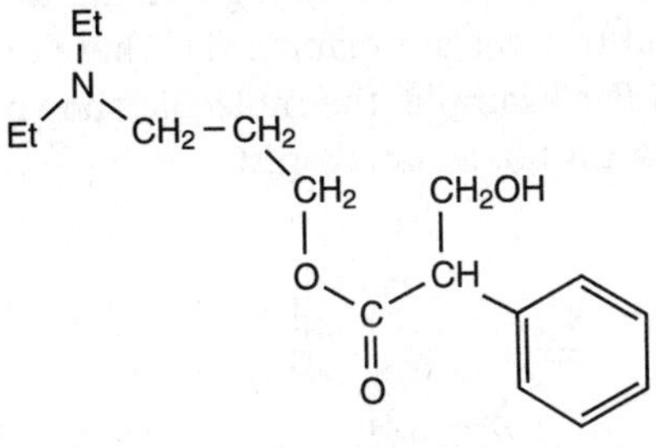

Fig. 15.31 Amprotropine.

These studies came up with the following generalizations:

- The alkyl groups (R) on nitrogen can be larger than methyl (in contrast to agonists).
- The nitrogen can be tertiary or quaternary, whereas agonists must have a quaternary nitrogen (note, however, that the tertiary nitrogen is probably charged when it interacts with the receptor).

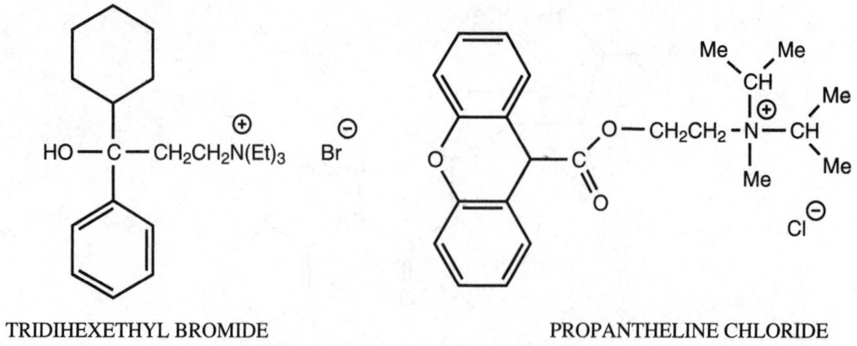

TRIDIHEXETHYL BROMIDE   PROPANTHELINE CHLORIDE

Fig. 15.32 Two analogues of atropine.

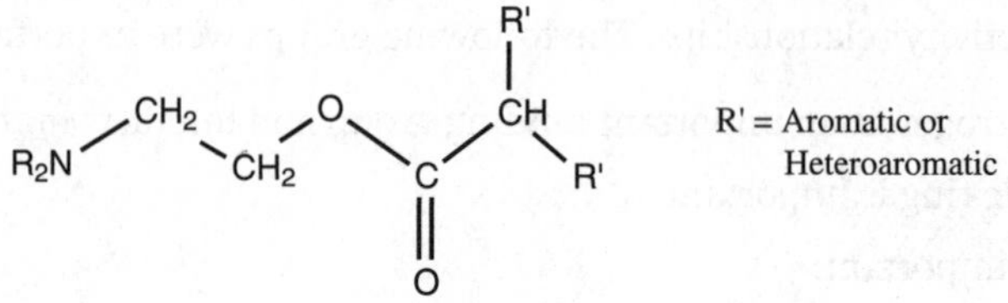

R' = Aromatic or Heteroaromatic

Fig. 15.33 General structure of muscarinic antagonists.

• Very large acyl groups are allowed (R' = aromatic or heteroaromatic rings). This is in contrast with agonists where only the acetyl group is permitted.

It is the last point which appears to be the most crucial in determining whether a compound will act as an antagonist or not. The acyl group has to be bulky, but it also has to have that bulk arranged in a certain manner (i.e. there must be some sort of branching in the acyl group). For example, the molecule shown in Fig. 15.34 has a large unbranched acyl group but is not an antagonist.

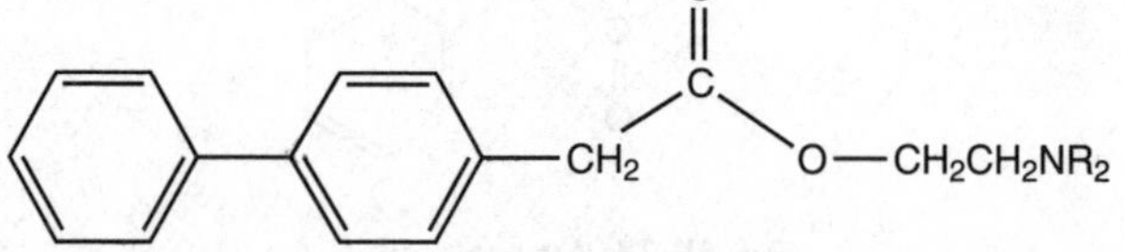

Fig. 15.34 Analogue with no branching in the acyl group.

The conclusion which can be drawn from these results is that there must be other binding regions on the receptor surface next to the normal acetylcholine binding site. These regions must be hydrophobic since most antagonists have aromatic rings. The overall shape of the acetylcholine binding site plus the extra binding regions would have to be T- or Y-shaped to explain the importance of branching in antagonists (Fig. 15.35).

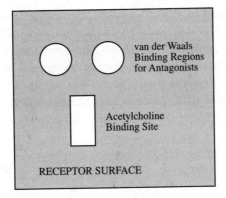

Fig. 15.35 Binding sites on the receptor surface.

A structure such as propantheline, which contains the complete acetylcholine skeleton as well as the hydrophobic acyl side chain, not surprisingly binds more strongly to the receptor than acetylcholine itself (Fig. 15.36).

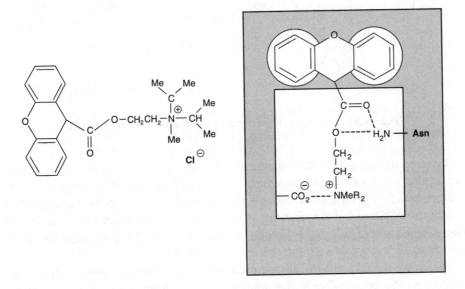

Fig. 15.36 Propantheline which binds strongly to the receptor.

The extra bonding interaction means that the conformational changes induced in the receptor (if any are induced at all) will be different from those induced by acetylcholine and will fail to induce the secondary biological response. As long as the antagonist is bound, acetylcholine is unable to bind and pass on its message.

A large variety of antagonists have proved to be useful medicines, with many showing selectivity for specific organs. For example, some act at the intestine to relieve spasm (atropine methonitrate), some are useful as anti-asthmatics (ipratr-opium), some are used in eye drops to dilate pupils for ophthalmic examination (tropicamide and cyclopentolate; Fig. 15.37), some are used centrally to counteract movement disorders due to Parkinson's disease (benzhexol, benztropine; Fig. 15.37), some act selectively to decrease gastric secretions, while others are useful in ulcer therapy. This selectivity of action owes more to the distribution properties of the drug than to receptor selectivity (i.e. the compounds can reach one part of the body more easily than another).

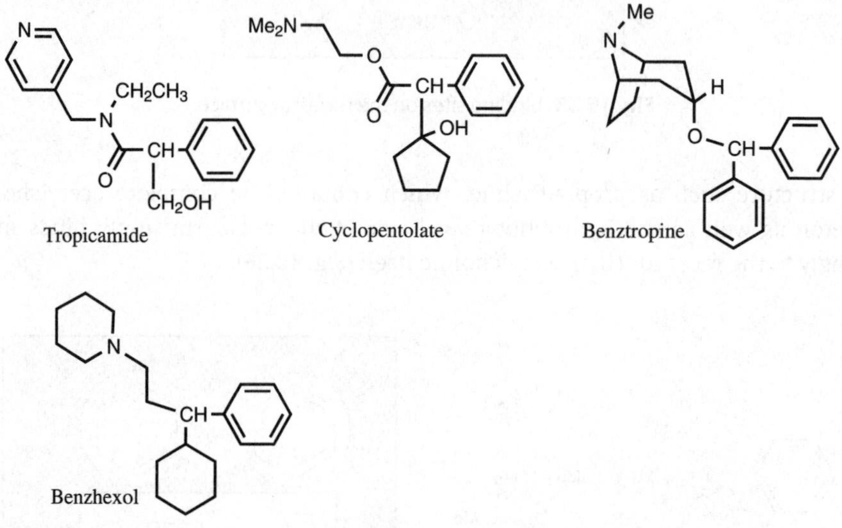

Fig. 15.37 Some examples of clinically useful cholinergic antagonists.

However, the antagonist pirenzepine (Fig. 15.38), which is used in the treatment of peptic ulcers, is a selective M_1 antagonist with no activity against M_2 receptors.[3]

Since antagonists bind more strongly than agonists, they are better compounds to use for the labelling and identification of receptors on tissue preparations. An antagonist, labelled with a radioactive isotope of H or C, binds strongly to the receptor and the radioactivity reveals where the receptor is located.

Ideally, we would want the antagonist to bind irreversibly in this situation. Such binding would be possible if the antagonist could form a covalent bond to the receptor. One useful tactic is to take an established antagonist and to incorporate a reactive chemical centre into the molecule. This reactive centre is usually elec-trophilic so that it will react with any suitably placed nucleophile close to the binding site (for example, the OH of a serine residue or the SH of a cysteine residue). In theory,

---

[3] The search for increasingly selective drugs has led to the discovery that there are receptor subtypes of the muscarinic receptor. In other words, not every muscarinic receptor is the same throughout the body. At present, four subtypes of the muscarinic receptor have been discovered and have been labelled M_1–M_4. More may still be discovered.

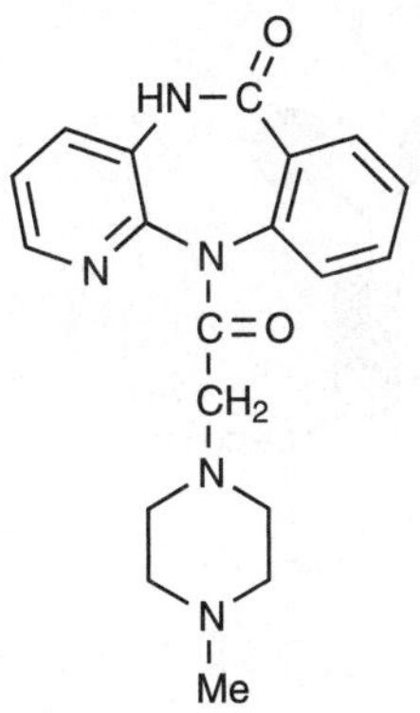

Fig. 15.38 Pirenzepine.

the antagonist should bind to the receptor in the usual way and the electrophilic group will react with any nucleophilic amino acid within range. The resulting alkylation irreversibly binds the antagonist to the receptor through a covalent bond.

In practice, the procedure is not always as simple as this, since the highly reactive electrophilic centre might react with another nucleophilic group before it reaches the receptor binding site. One way to avoid this problem is to include a latent reactive centre which can only be activated once the antagonist has bound to the receptor binding site. One favourite method is photoaffinity labelling, where the reactive centre is activated by light. Chemical groups such as diazoketones or azides can be converted to highly reactive carbenes and nitrenes, respectively, when irradiated (Fig. 15.39).

15.12 Antagonists of the nicotinic cholinergic receptor

15.12.1 Applications of nicotinic antagonists

Nicotinic receptors are present in nerve synapses at ganglia, as well as at the neuromuscular synapse. However, drugs are able to show a level of selectivity between these two sites, mainly because of the distinctive routes which have to be taken to reach them. Antagonists of ganglionic nicotinic receptor sites are not therapeutically useful since they cannot distinguish between the ganglia of the sympathetic nervous system and the ganglia of the parasympathetic nervous system (both use nicotinic receptors). Consequently, they have many side-effects. However, antagonists of the neuromuscular junction are therapeutically useful and are known as neuromuscular blocking agents.

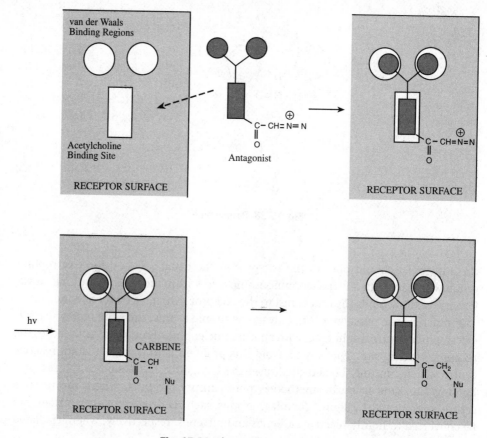

Fig. 15.39 Photoaffinity labelling.

15.12.2 Nicotinic antagonists

Curare (1516) and tubocurarine

Curare was first identified when Spanish soldiers in South America found themselves the unwilling victims of poisoned arrows. It was discovered that the Indians were putting a poison on to the tips of their arrows. This poison was a crude, dried extract from a plant called *Chondrodendron tomentosum* and caused paralysis as well as stopping the heart. We now know that curare is a mixture of compounds. The active principle, however, is an antagonist of acetylcholine which blocks nerve transmissions from nerve to muscle.

It might seem strange to consider such a compound for medicinal use, but at the right dose levels and under proper control, there are very useful applications for this sort of action. The main application is in the relaxation of abdominal muscles in preparation for surgery. This allows the surgeon to use lower levels of general anaesthetic than would otherwise be required and therefore increase the safety margin for operations.

Curare, as mentioned above, is actually a mixture of compounds, and it was not until 1935 that the active principle (tubocurarine) was isolated. The determination of the structure took even longer and was not established until 1970 (Fig. 15.40). Tubocurarine was used clinically as a neuromuscular blocker. However, it had unwanted side-effects and better agents are now available, as we will see. These side-effects arose since tubocurarine also acted as an antagonist at the nicotinic receptors of the autonomic nervous system (Fig. 15.2).

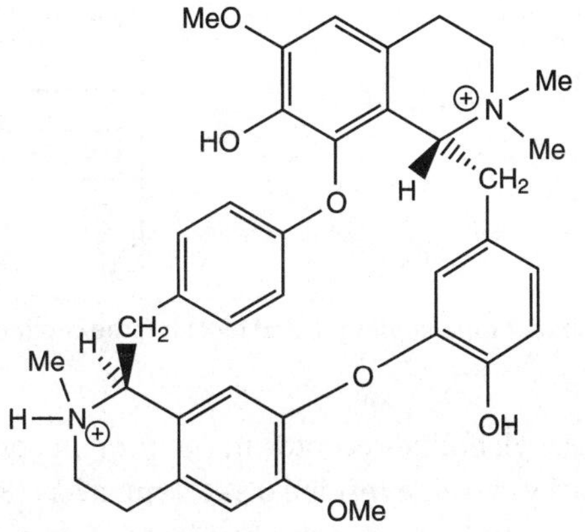

Fig. 15.40 Tubocurarine.

The structure of tubocurarine presents a problem to our theory of receptor binding, since, although it has a couple of charged nitrogen centres, there is no ester present to interact with the acetyl binding region. Studies on the compounds discussed so far show that the positively charged nitrogen on its own is not sufficient for good binding, so why should tubocurarine bind to and block the cholinergic nicotinic receptor?

The answer lies in the fact that the molecule has two positively charged nitrogen atoms (one tertiary which is protonated, and one quaternary). Originally, it was believed that the distance between the two centres (1.15 nm) might be equivalent to the distance between two separate cholinergic receptors and that the large tubocurarine molecule could act as a bridge between the two receptor sites, thus spreading a blanket over the two receptors and blocking access to acetylcholine. However pleasing that theory may be, the dimensions of the nicotinic receptor make this unlikely. The receptor, as we shall see in Section 15.14, is a protein dimer made up of two identical protein complexes separated by 9–10 nm; this is far too large for the tubocurarine molecule to bridge (Fig. 15.41a).

Another possibility is that the tubocurarine molecule bridges two acetylcholine binding sites within the one protein complex. Since there are two such sites within the complex, this appears an attractive alternative theory. However, the two sites are further apart than 1.15 nm and so this too seems unlikely. It has now been proposed that one of the positively charged nitrogens on tubocurarine binds to the anionic

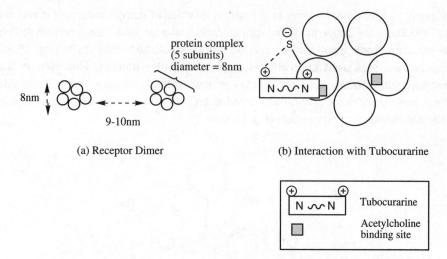

(a) Receptor Dimer   (b) Interaction with Tubocurarine

Fig. 15.41 Tubocurarine binding to and blocking the cholinergic receptor.

binding region of the acetylcholine receptor in the protein complex, while the other nitrogen binds to a nearby cysteine residue 0.9–1.2 nm away (Fig. 15.41b).

Despite the uncertainty surrounding the bonding interactions of tubocurarine, it seems highly probable that two ionic bonding regions are involved. Such an interaction is extremely strong and would more than make up for the lack of the ester binding interaction.

It is also clear that the distance between the two positively charged nitrogen atoms is crucial to activity. Therefore, analogues which retain this distance should also be good antagonists. Strong evidence that this is so comes from the fact that the simple molecule decamethonium is a good antagonist.

Decamethonium and suxamethonium

Decamethonium (Fig. 15.42) is as simple an analogue of tubocurarine as one could imagine. It is a straight-chain molecule and, as such, is capable of a large number of conformations. The fully extended conformation places the nitrogen atoms 1.4 nm apart, but bond rotations can result in another conformation which positions the nitrogen centres 1.14 nm apart, which compares well with the equivalent distance in tubocurarine (1.15 nm) (see also Section 13.11).

$$\overset{\oplus}{Me_3N}(CH_2)_{10}\overset{\oplus}{NMe_3}$$

Fig. 15.42 Decamethonium.

The drug binds strongly to cholinergic receptors and has proved a useful clinical agent. However, it suffers from several disadvantages. For example, when it binds initially to the nicotinic receptor, it acts as an agonist rather than an antagonist. In other words, it switches on the receptor and this leads to a brief contraction of the muscle. Once this effect has passed, the drug remains bound to the receptor – blocking access to acetylcholine – and thus acts as an antagonist. (A theory on how such an effect might take place is described in Chapter 5.) Unfortunately, it binds too strongly and, as a result, patients take a long time to recover from its effects. It is also not completely selective for the neuromuscular junction and has an effect on acetylcholine receptors at the heart. This leads to an increased heart rate and a fall in blood pressure.

The problem we now face in designing a better drug is the opposite problem from the one we faced when trying to design acetylcholine agonists. Instead of stabilizing a molecule, we now want to introduce some sort of instability – a sort of timer control whereby the molecule can be switched off quickly and become inactive. Success was first achieved by introducing ester groups into the chain while retaining the distance between the two charged nitrogens, to give suxamethonium (Fig. 15.43).

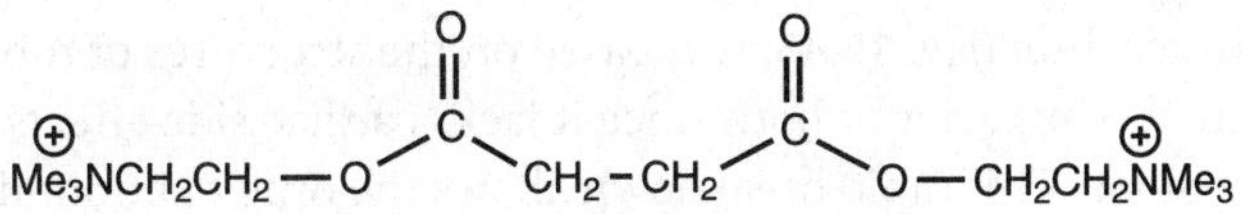

$$Me_3NCH_2CH_2 - O - \overset{\overset{\displaystyle O}{\|}}{C} - CH_2-CH_2 - \overset{\overset{\displaystyle O}{\|}}{C} - O - CH_2CH_2NMe_3$$

Fig. 15.43 Suxamethonium.

The ester groups are susceptible to chemical and enzymatic hydrolysis. Once hydrolysis occurs, the molecule can no longer bridge the two receptor sites and becomes inactive. Suxamethonium has a fast onset and short duration of action of 5–10 minutes, but suffers from various side-effects.[4] Furthermore, about one person in every 2000 lacks the enzyme which hydrolyses suxamethonium. However, it is still used clinically in short surgical procedures such as the insertion of tracheal tubes.

Pancuronium and vecuronium

The design of pancuronium and vecuronium (Fig. 15.44) was based on tubocurarine, but involved a steroid nucleus to act as the 'spacer' between the two nitrogen groups. The distance between the quaternary nitrogens is 1.09 nm, as compared to 1.15 nm in tubocurarine. Acyl groups were also added to introduce two acetylcholine skeletons into the molecule to improve affinity for the receptor sites. These compounds have a faster onset of action than tubocurarine and do not affect blood pressure. They are not as rapid in onset as suxamethonium and have a longer duration of action (45 minutes). However, their main advantage is that they have fewer side-effects and they are widely used clinically.

---

[4] Both decamethonium and suxamethonium have effects on the autonomic ganglia, and this explains some of their side-effects.

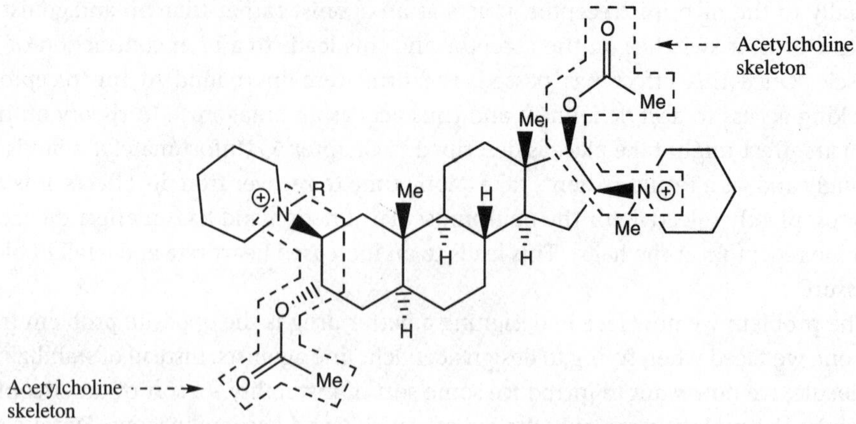

Fig. 15.44 Pancuronium (R=Me) and vecuronium (R=H).

Atracurium

The design of atracurium (Fig. 15.45) was based on the structures of tubocurarine and suxamethonium. It is superior to both since it lacks cardiac side-effects and is rapidly broken down in blood. This rapid breakdown allows the drug to be administered as an intravenous drip.

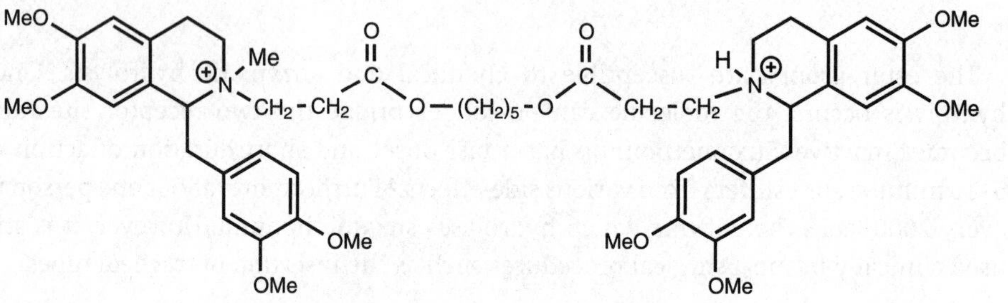

Fig. 15.45 Atracurium.

The rapid breakdown was designed into the molecule by incorporating a self-destruct mechanism. At blood pH (slightly alkaline at 7.4), the molecule can undergo a Hofmann elimination (Fig. 15.46). Once this happens, the compound is inactivated since the positive charge on the nitrogen is lost. It is a particularly clever example of drug design in that the very element responsible for the molecule's biological activity promotes its deactivation.

The important features of atracurium are as follows.

The spacer

This is the 13-atom connecting chain which connects the two quaternary centres and separates the two centres.

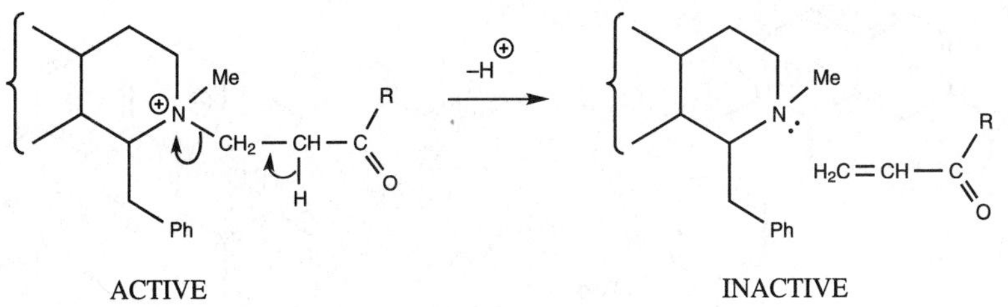

Fig. 15.46 Hofmann elimination of atracurium.

The blocking units

The cyclic structures at either end of the molecule block the receptor site from acetylcholine.

The quaternary centres

These are essential for receptor binding. If one is lost through Hofmann elimination, the binding interaction is too weak and the antagonist leaves the binding site.

The Hofmann elimination

The ester groups within the spacer chain are crucial to the rapid deactivation process. Hofmann eliminations normally require strong alkaline conditions and high temperatures – hardly normal physiological conditions. However, if a good electron withdrawing group is present on the carbon *beta* to the quaternary nitrogen centre, it allows the reaction to proceed under much milder conditions. The electron withdrawing group increases the acidity of the hydrogens on the *beta* carbon such that they are easily lost. The Hofmann elimination does not occur at acid pH, and so the drug is stable in solution at a pH of 3–4 and can be stored safely in a refrigerator.

Since the drug only acts very briefly (approximately 30 minutes), it has to be added intravenously for as long as it is needed. As soon as surgery is over, the intravenous drip is stopped and antagonism ceases almost instantaneously.

Another major advantage of a drug which is deactivated by a chemical mechanism rather than by an enzymatic mechanism is that deactivation occurs at a constant rate between patients. With previous neuromuscular blockers, deactivation depended on metabolic mechanisms involving enzymic deactivation and/or excretion. The efficiency of these processes varies from patient to patient and is particularly poor for patients with kidney failure or with low levels of plasma esterases.

Mivacurium (Fig. 15.47) is a newer drug which is chemically very similar to atracurium and which is rapidly inactivated by plasma enzymes as well as by the Hofmann elimination. It has a faster onset (around 2 minutes) and shorter duration (around 15 minutes), although it has a longer duration if the patients have liver disease or enzyme deficiencies.

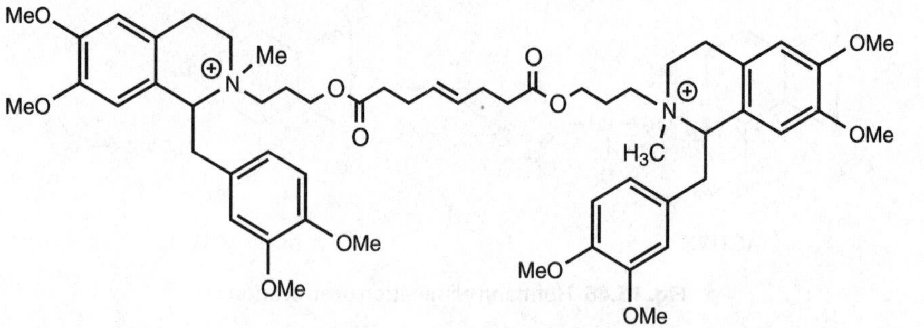

Fig. 15.47 Mivacurium.

15.13 Other cholinergic antagonists

Local anaesthetics and barbiturates appear to prevent the changes in ion permeability which would normally result from the interaction of acetylcholine with the nicotinic receptor. They do not, however, bind to the acetylcholine binding site. It is believed that they bind instead to the part of the receptor which is on the inside of the cell membrane, perhaps binding to the ion channel itself and blocking it.

Certain snake toxins have been found to bind irreversibly to the nicotinic receptor, thus blocking cholinergic transmissions. These include toxins such as *alpha* bungarotoxin from the Indian cobra. The toxin is a polypeptide containing 70 amino acids which cross-links the *alpha* and *beta* subunits of the cholinergic receptor (see Section 15.14.).

15.14 The nicotinic receptor – structure

The nicotinic receptor has been successfully isolated from the electric ray (*Torpedo marmorata*), found in the Atlantic Ocean and Mediterranean Sea, allowing the receptor to be carefully studied. As a result, a great deal is known about its structure and operation.

It is a protein complex made up of five subunits, two of which are the same. The five subunits (two *alpha*, one *beta*, *gamma*, and *delta*) form a cylindrical or barrel shape which traverses the cell membrane as shown in Fig. 15.48 (see also Section 6.2.1).

The centre of the cylinder can therefore act as an ion channel for sodium. A gating or lock system is controlled by the interaction of the receptor with acetylcholine. When acetylcholine is unbound, the gate is shut. When acetylcholine binds, the gate is opened.

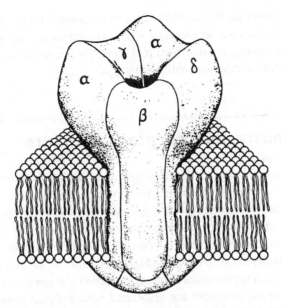

Fig. 15.48 Schematic diagram of the nicotinic receptor. Taken from C. M. Smith and A. M. Reynard, *Textbook of Pharmacology*, WB Saunders and Co. (1992).

The amino acid sequence for each subunit has been established and it is known that there is extensive secondary structure. The binding site for acetylcholine is situated on the *alpha* subunit and therefore there are two binding sites per receptor protein.

It is usually found that the nicotinic receptors occur in pairs linked together by a disulphide bridge between the delta subunits (Fig. 15.49).

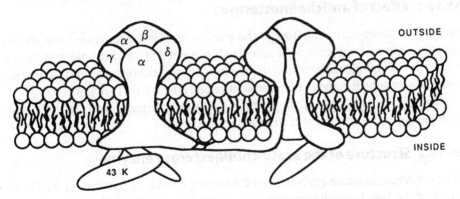

Fig. 15.49 Nicotinic receptor pair. (Taken from T. Nogrady, *Medicinal Chemistry, a biochemical approach*, 2nd edn, Oxford University Press (1988).)

This is the make-up of the nicotinic receptor at neuromuscular junctions. The nicotinic receptors at ganglia and in the CNS are more diverse in nature, involving different α- and β-subunits. This allows drugs to act selectively on neuromuscular versus neuronal receptors. For example, decamethonium is only a weak antagonist at autonomic ganglia, whereas epibatidine (Fig. 8.7) (extracted from a South American frog) is a selective agonist for neuronal receptors. The snake toxin α-bungarotoxin is specific for receptors at neuromuscular junctions.

15.15 The muscarinic receptor – structure

Muscarinic receptors belong to the superfamily of 7-TM receptors (see Chapter 6), which operate by activation of a signal transduction process. Five subtypes of muscarinic receptors have been discovered by gene cloning and are labelled m_1–m_5. Four of these have been identified in the body and are labelled M_1–M_4 (corresponding to m_1–m_4). These subtypes tend to be concentrated in specific tissues. For example, M_2 receptors occur mainly in the heart while M_4 receptors are found mainly in the CNS. M_2 receptors are also used for the autoreceptors on cholinergic nerves (see Section 15.5.2).

The m_1, m_3, and m_5 receptors are associated with a signal transduction process involving the secondary messenger IP_3, while the m_2 and m_4 receptors involve a process which inhibits the production of the secondary messenger cyclic AMP. Lack of M_1 activity is thought to be associated with dementia.

15.16 Anticholinesterases and acetylcholinesterase

15.16.1 Effect of anticholinesterases

Anticholinesterases are inhibitors of the enzyme acetylcholinesterase – the enzyme which hydrolyses acetylcholine. If acetylcholine is not destroyed, it can return to reactivate the cholinergic receptor and so the effect of an anticholinesterase is to increase levels of acetylcholine and to increase cholinergic effects (Fig. 15.50).

Therefore, an inhibitor of the acetylcholinesterase enzyme will have the same biological effect as an agonist at the cholinergic receptor.

15.16.2 Structure of the acetylcholinesterase enzyme[5]

The acetylcholinesterase enzyme has a fascinating tree-like structure (Fig. 15.51). The trunk of the tree is a collagen molecule which is anchored to the cell membrane. There are three branches (disulphide bridges) leading from the trunk, each of which holds the acetylcholinesterase enzyme above the surface of the membrane. The

---

[5] A soluble cholinesterase enzyme called butyrylcholinesterase is also present in various tissues and plasma. This enzyme has broader substrate specificity than acetylcholinesterase and can hydrolyse a variety of esters.

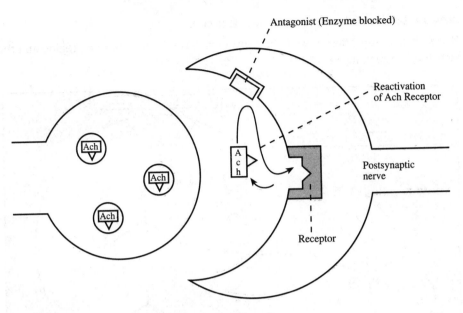

Fig. 15.50 Effect of anticholinesterases.

enzyme itself is made up of four protein subunits, each of which has an active site. Therefore, each enzyme tree has twelve active sites. The trees are rooted immediately next to the acetylcholine receptors so that they will efficiently capture acetylcholine molecules as they depart the receptor. In fact, the acetylcholinesterase enzyme is one of the most efficient enzymes known.

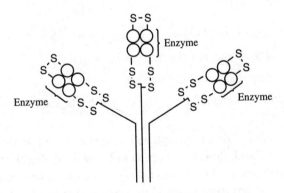

Fig. 15.51 The acetylcholinesterase enzyme.

15.16.3 The active site of acetylcholinesterase

The design of anticholinesterases depends on the shape of the enzyme active site, the binding interactions involved with acetylcholine, and the mechanism of hydrolysis.

15.16.3.1 Binding interactions at the active site

There are two important areas to be considered – the anionic binding region and the ester binding region (Fig. 15.52).

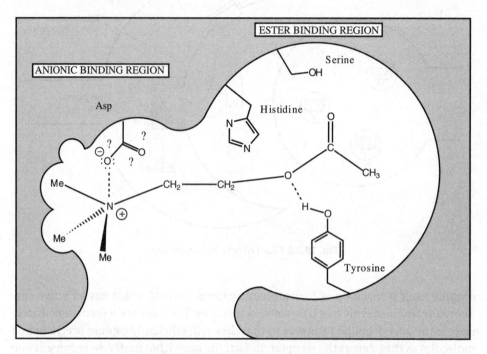

Fig. 15.52 Binding interactions at the active site.

Note that:

- Acetylcholine binds to the cholinesterase enzyme by:
  (a) ionic bonding to an Asp residue (but see below),
  (b) hydrogen bonding to a tyrosine residue.
- The histidine and serine residues at the catalytic site are involved in the mechanism of hydrolysis.
- The anionic binding region in acetylcholinesterase is very similar to the anionic binding region in the cholinergic receptor and may be identical. There are thought to be two hydrophobic pockets large enough to accommodate methyl residues but nothing larger. The positively charged nitrogen is thought to be bound to a negatively charged aspartate residue, and may also interact with aromatic amino acids by induced dipole interactions (see Section 15.7).

15.16.3.2 Mechanism of hydrolysis

The histidine residue acts as an acid/base catalyst throughout the mechanism, while serine plays the part of a nucleophile. This is not a particularly good role for serine since an aliphatic alcohol is a poor nucleophile. In fact, serine by itself is unable to hydrolyse an ester. However, the fact that histidine is close by to provide acid/base catalysis overcomes that disadvantage. There are several stages to the mechanism (Fig. 15.53).

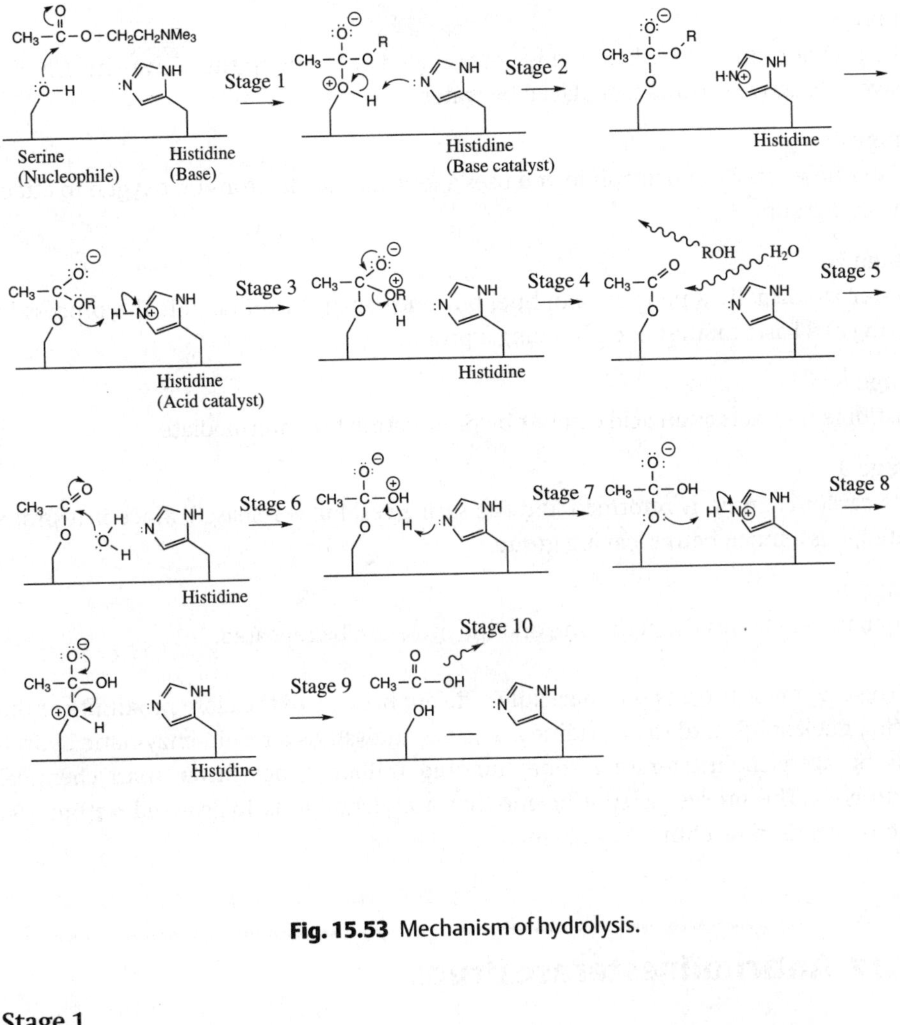

Fig. 15.53 Mechanism of hydrolysis.

Stage 1

Acetylcholine approaches and binds to the acetylcholinesterase enzyme. Serine acts as a nucleophile and uses a lone pair of electrons to form a bond to the ester of acetylcholine. Nucleophilic addition to the ester takes place and opens up the carbonyl group.

Stage 2

The histidine residue catalyses this reaction by acting as a base and removing a proton, thus making serine more nucleophilic.

Stage 3

The histidine now acts as an acid catalyst and protonates the 'OR' portion of the intermediate, turning it into a much better leaving group.

Stage 4

The carbonyl group reforms and expels the alcohol portion of the ester (i.e. choline).

Stage 5

The acyl portion of acetylcholine is now covalently bound to the active site. Choline leaves the active site and is replaced by water.

Stage 6

Water now acts as a nucleophile and uses a lone pair of electrons on oxygen to attack the acyl group.

Stage 7

Water is normally a poor nucleophile, but once again histidine aids the process by acting as a basic catalyst and removing a proton.

Stage 8

Histidine now acts as an acid catalyst by protonating the intermediate.

Stage 9

The carbonyl group is reformed and the serine residue is released. Since it is protonated, it is a much better leaving group.

Stage 10

Ethanoic acid leaves the active site and the cycle can be repeated.

The enzymatic process is remarkably efficient because of the close proximity of the serine nucleophile and the histidine acid/base catalyst. As a result, enzymatic hydrolysis by acetylcholinesterase is one hundred million times faster than chemical hydrolysis. The process is so efficient that acetylcholine is hydrolysed within 100 microseconds of reaching the enzyme.

15.17 Anticholinesterase drugs

Obviously, by the very nature of their being, anticholinesterase drugs must be inhibitors; that is, they stop the enzyme from hydrolysing acetylcholine. This inhibition can be either reversible or irreversible depending on how the drug reacts with the active site. There are two main groups of acetylcholinesterases which we shall consider – carbamates and organophosphorus agents.

15.17.1 The carbamates

15.17.1.1 Physostigmine

As in so many fields of medicinal chemistry, it was a natural product which provided the lead to this group of compounds. The natural product was physostigmine (also called eserine) which was discovered in 1864 as a product of the poisonous calabar beans (the ordeal bean) from West Africa. Extracts of these beans were fed to criminals to assess whether they were guilty or innocent. Death indicated a guilty verdict. The structure was established in 1925 (Fig. 15.54) and physostigmine is still used clinically to treat glaucoma.

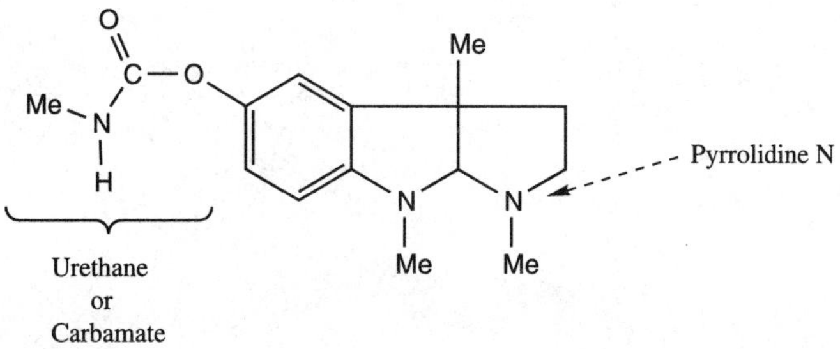

Fig. 15.54 Physostigmine.

Structure–activity relationships:

- The carbamate group is essential to activity
- The benzene ring is important
- The pyrrolidine nitrogen (which is ionized at blood pH) is important.

Working backwards, the positively charged pyrrolidine nitrogen is clearly important since it must bind to the anionic binding region of the enzyme.

The benzene ring may be involved in some extra hydrophobic bonding with the active site. Alternatively, it may be important in the mechanism of inhibition since it provides a good leaving group.

The carbamate group is the crucial group responsible for physostigmine's inhibitory properties and, to understand why, we have to look again at the mechanism of hydrolysis at the active site (Fig. 15.55). This time we shall see what happens when physostigmine and not acetylcholine is the substrate for the reaction.

The first four stages proceed as normal, with histidine catalysing the nucleophilic attack of the serine residue on physostigmine (stages 1 and 2). The alcohol portion (this time a phenol) is expelled with the aid of acid catalysis from histidine (stages 3 and 4), and the phenol leaves the active site to be replaced by a water molecule.

However, the next stage turns out to be extremely slow. Despite the fact that histidine can still act as a basic catalyst, water finds it difficult to attack the carbamoyl

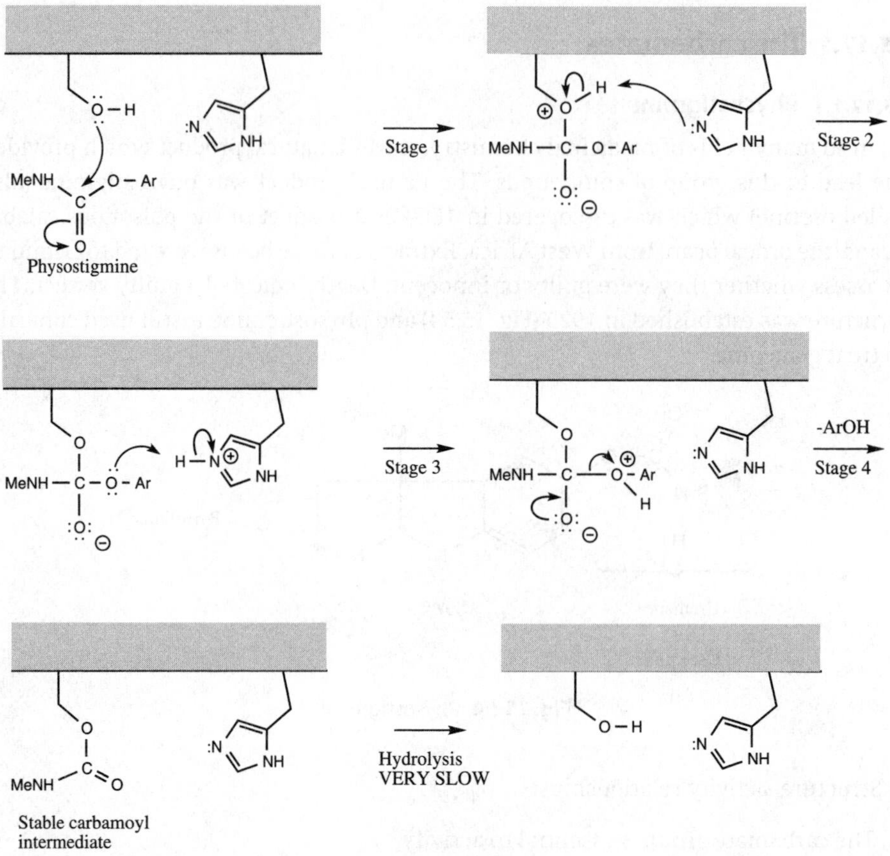

Fig. 15.55 Mechanism of inhibition.

intermediate. This step becomes the rate determining step for the whole process and the overall rate of hydrolysis of physostigmine compared to acetylcholine is forty million times slower. As a result, the cholinesterase active site becomes 'bunged up' and is unable to react with acetylcholine.

Why is this final stage so slow?

The carbamoyl–enzyme intermediate is stabilized because the nitrogen can feed a lone pair of electrons into the carbonyl group. This drastically reduces the electrophilic character and reactivity of the carbonyl group (Fig. 15.56). This is the same electronic influence which stabilizes carbachol and makes it resistant towards hydrolysis (see Section 15.9.2).

15.17.1.2 Analogues of physostigmine

Physostigmine has limited medicinal use since it has serious side-effects, and as a result it has only been used in the treatment of glaucoma or as an antidote for atropine poisoning. However, simpler analogues have been made and have been used in the

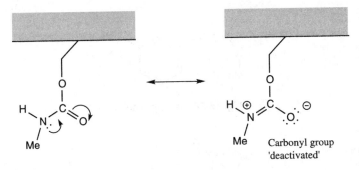

Fig. 15.56 Stabilization of the carbamoyl–enzyme intermediate.

treatment of myasthenia gravis and as an antidote to curare. These analogues retain the important features mentioned above.

Miotine (Fig. 15.57) still has the necessary carbamate, aromatic, and tertiary aliphatic nitrogen groups. It is active as an antagonist but suffers from the following disadvantages:

- It is susceptible to chemical hydrolysis.
- It can cross the blood–brain barrier as the free base. This results in side-effects because of its action in the CNS.

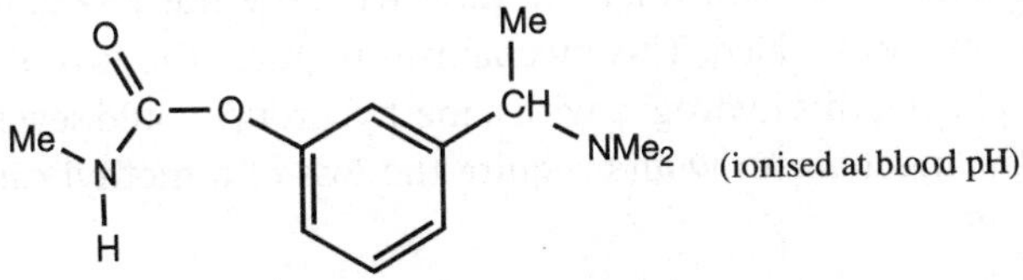

Fig. 15.57 Miotine.

Neostigmine (Fig. 15.58) was designed to deal with both the problems described above. First of all, a quaternary nitrogen atom is present so that there is no chance of the free base being formed. Since the molecule is permanently charged, it cannot cross the blood–brain barrier and cause CNS side-effects[5].

Increased stability to hydrolysis is achieved by using a dimethylcarbamate group rather than a methylcarbamate group. There are two possible explanations for this, based on two possible hydrolysis mechanisms.

Mechanism 1 (Fig. 15.59) involves nucleophilic substitution by a water molecule. The rate of the reaction depends on the electrophilic character of the carbonyl group and, if this is reduced, the rate of hydrolysis is reduced.

[5] The blood–brain barrier is a series of lipophilic membranes which coat the blood vessels feeding the brain and which prevent polar molecules from entering the CNS. The fact that it exists can be useful in a case like this, since polar molecules can be designed which are unable to cross it. However, its presence can be disadvantageous when trying to design drugs to act in the CNS itself.

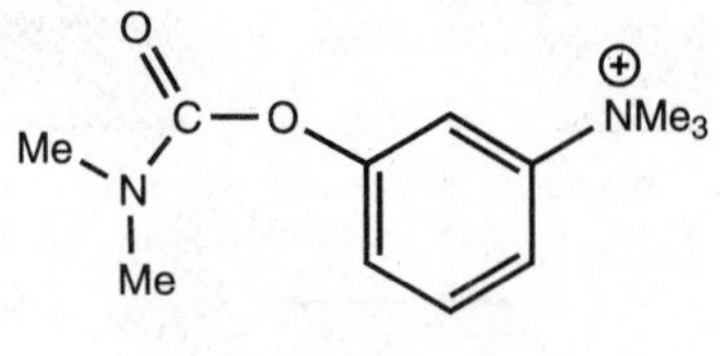

Fig. 15.58 Neostigmine.

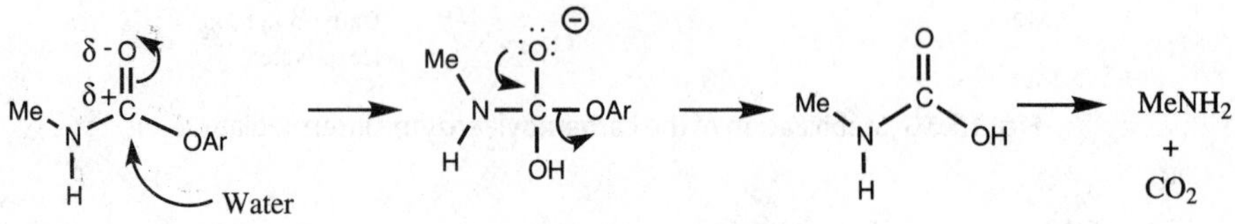

Fig. 15.59 Mechanism 1.

We have already seen how the lone pair of the neighbouring nitrogen can reduce the electrophilic character of the carbonyl group. The presence of a second methyl group on the nitrogen has an inductive 'pushing' effect which increases electron density on the nitrogen and further encourages the nitrogen lone pair to interact with the carbonyl group.

Mechanism 2 (Fig. 15.60) is a fragmentation whereby the phenolic group is lost before the nucleophile is added. This mechanism requires the loss of a proton from the nitrogen. Replacing this hydrogen with a methyl group would severely inhibit the reaction since the mechanism would require the loss of a methyl cation – a highly disfavoured process.

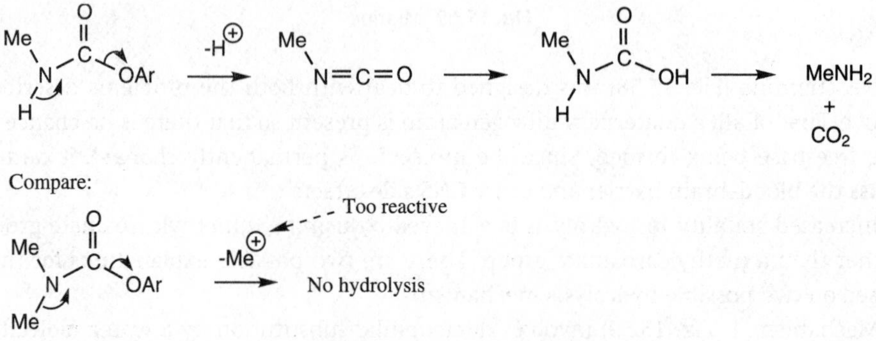

Compare:

Fig. 15.60 Mechanism 2.

Whichever mechanism is involved, the presence of the second methyl group acts to discourage the process. Two further points to note about neostigmine are the following:

- The quaternary nitrogen is 4.7Å away from the ester group.

- The direct bonding of the quaternary centre to the aromatic ring reduces the number of conformations that the molecule can take up. This is an advantage (assuming that the active conformation is still retained), since the molecule is more likely to be in the active conformation when it approaches the active site.

Neostigmine has proved a useful agent and is still in use today. It is given intravenously to reverse the actions of neuromuscular blockers and is used orally in the treatment of myasthenia gravis. The latter is an autoimmune disease where the body destroys the nicotinic receptors on muscle, leading to muscle weakness and fatigue. Inhibiting the hydrolysis of acetylcholine allows a greater activation of the remaining receptors.

15.17.2 Organophosphorus compounds

Organophosphorus agents were designed as nerve gases during the Second World War, but were fortunately never used. In peace time, organophosphate agents have been used as insecticides and medicines. We shall deal with the nerve gases first.

15.17.2.1 Nerve gases

The nerve gases dyflos and sarin (Fig. 15.61) were discovered and perfected long before their mode of action was known. Dyflos, which has an ID_{50} of 0.01 mg/kg, was developed as a nerve gas in the Second World War. It inhibits acetylcholinesterase by irreversibly phosphorylating the serine residue at the active site (Fig. 15.62).

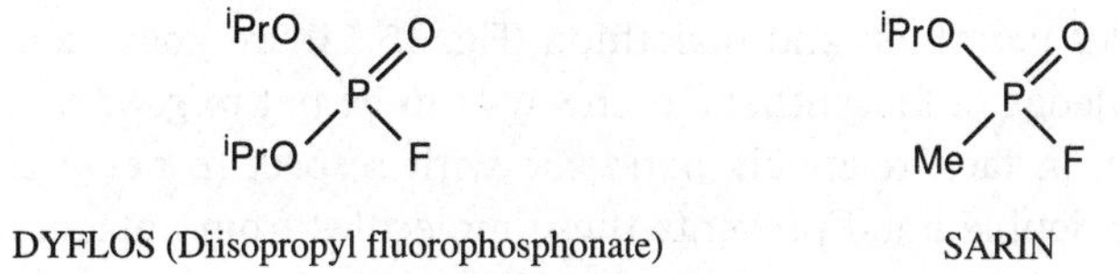

DYFLOS (Diisopropyl fluorophosphonate)   SARIN

Fig. 15.61 Nerve gases.

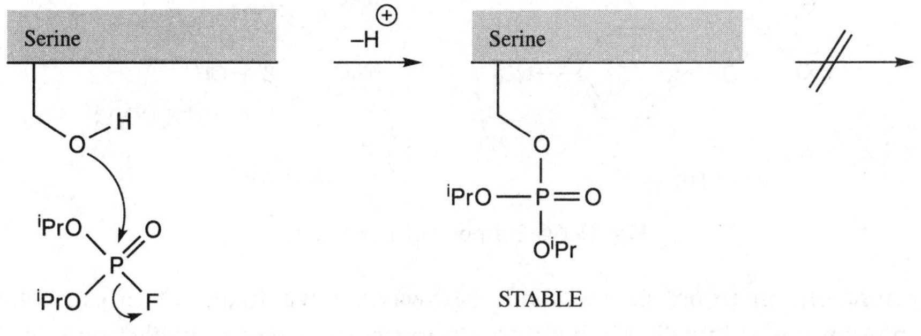

Fig. 15.62 Action of dyflos.

The mechanism is the same as before, but the phosphorylated adduct which is formed after the first three stages is extremely resistant to hydrolysis. Consequently, the enzyme is permanently inactivated. Acetylcholine cannot be hydrolysed and, as a result, the cholinergic system is continually stimulated. This results in permanent contraction of skeletal muscle, leading to death.

15.17.2.2 Medicines

As mentioned earlier, the nerve agents were discovered before their mechanism of action was known. However, once it was known that they acted on the acetyl-cholinesterase enzyme, compounds such as ecothiopate (Fig. 15.63) were designed to fit the active site more accurately by including a quaternary amine to bind with the anionic region. This meant that lower doses would be more effective.

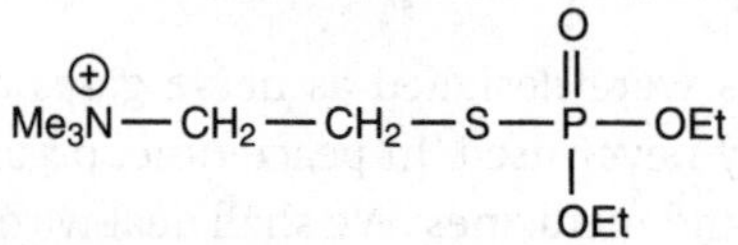

$$Me_3\overset{\oplus}{N}-CH_2-CH_2-S-\underset{\underset{OEt}{|}}{\overset{\overset{O}{\|}}{P}}-OEt$$

Fig. 15.63 Ecothiopate.

Ecothiopate is used medicinally in the form of eye drops for the treatment of glaucoma and has advantages over dyflos, which has also been used in this way. Unlike dyflos, ecothiopate slowly hydrolyses from the enzyme over a matter of days.

15.17.2.3 Insecticides

The insecticides parathion and malathion (Fig. 15.64) are good examples of how a detailed knowledge of biosynthetic pathways can be put to good use. Parathion and malathion are, in fact, relatively non-toxic with respect to nerve gases. The phosphorus–sulfur double bond prevents these molecules from antagonizing the active site on the acetylcholinesterase enzyme. The equivalent compounds containing a phosphorus–oxygen double bond are, on the other hand, lethal compounds.

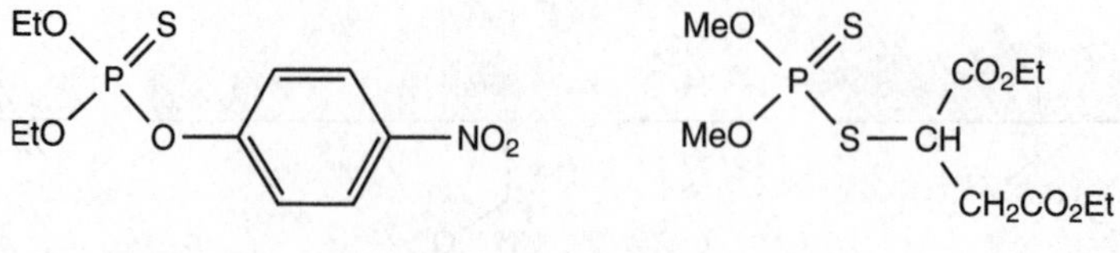

PARATHION   MALATHION

Fig. 15.64 Examples of insecticides.

Fortunately, there are no metabolic pathways in mammals which can convert the phosphorus–sulfur double bond to a phosphorus–oxygen double bond. Such a pathway does, however, exist in insects. In the latter species, parathion and malathion act as prodrugs. They are metabolized by oxidative desulfurization to give the active anticholinesterases which irreversibly bind to the insects' acetylcholinesterase enzymes and lead to death. In mammals, the same compounds are metabolized in a different way to give inactive compounds which are then excreted (Fig. 15.65). Nevertheless, despite their relative safety, organophosphate insecticides are not safe chemicals and cause serious side-effects if farmers do not handle them with care. Parathion has high lipid solubility and is easily absorbed through mucous membranes and can even be absorbed through skin.

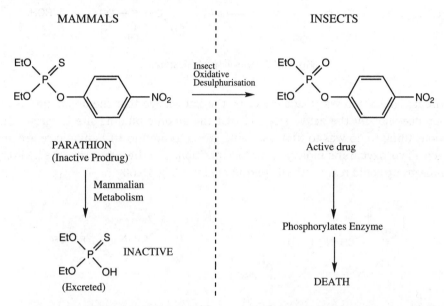

Fig. 15.65 Metabolization of insecticides in mammals and insects.

15.18 Pralidoxime – an organophosphate antidote

Pralidoxime (Fig. 15.66) represents one of the early examples of rational drug design. It is an antidote to organophosphate poisoning if given quickly enough and was designed as such. The problem faced in designing an antidote to organophosphate poisoning is to find a drug which will displace the organophosphate molecule from serine. This requires hydrolysis of the phosphate–serine bond, but this is a strong bond and not easily broken. Therefore, a stronger nucleophile than water is required.

The literature revealed that phosphates can be hydrolysed with hydroxylamine (Fig. 15.67). This proved too toxic a compound to be used on humans, so the next stage was to design an equally reactive nucleophilic group which would specifically target the acetylcholinesterase enzyme. If such a compound could be designed, then there was less chance of the antidote taking part in toxic side reactions.

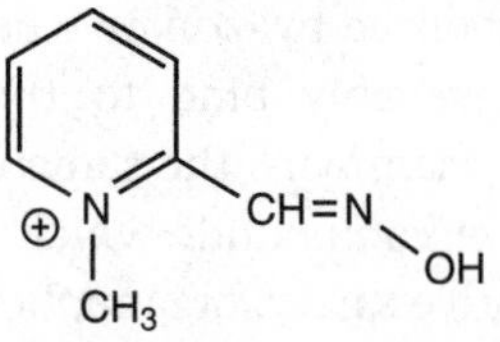

Fig. 15.66 Pralidoxime.

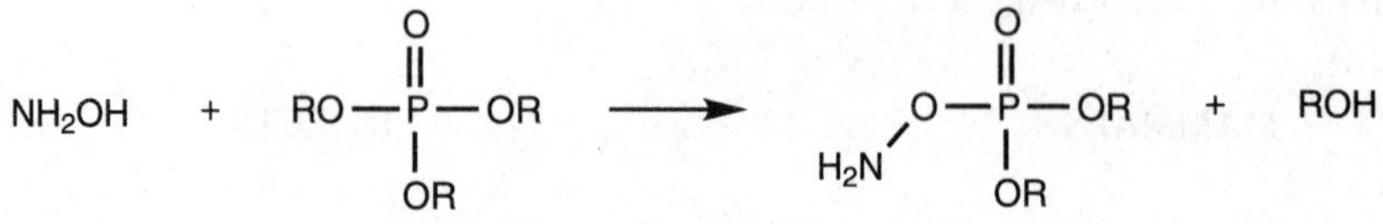

Fig. 15.67 Hydrolysis of phosphates.

The designers' job was made easier by the knowledge that the organophosphate group does not fill the active site and that the anionic binding site is vacant. The obvious thing to do was to find a suitable group to bind to this anionic centre and attach a hydroxylamine moiety to it. Once positioned in the active site, the hydroxylamine group could react with the phosphate ester (Fig. 15.68).

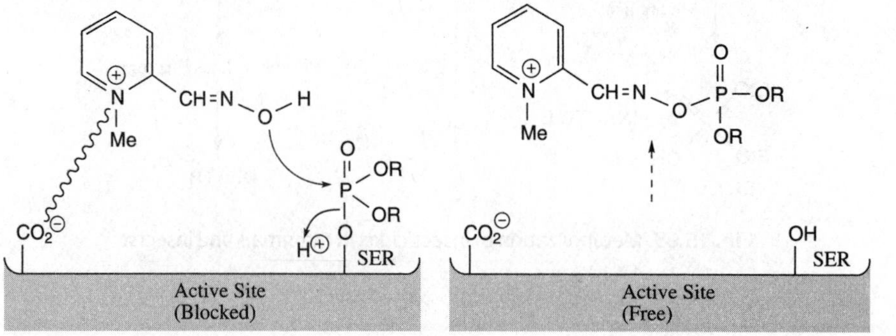

Fig. 15.68 Hydroxylamine group reaction with the phosphate ester.

Pralidoxime was the result. The positive charge is provided by a methylated pyridine ring and the nucleophilic side group is attached to the *ortho* position, since it was calculated that this would place the nucleophilic hydroxyl group in exactly the correct position to react with the phosphate ester. The results were spectacular, with pralidoxime showing a potency as an antidote one million times greater than hydroxylamine.

Since pralidoxime has a quaternary nitrogen, it is fully charged and cannot pass through the blood–brain barrier into the CNS. Pro-2-PAM (Fig. 15.69) is a prodrug of pralidoxime which avoids this problem. Since Pro-2-PAM is a tertiary amine it can pass through the blood–brain barrier and, once it has entered the CNS, it is oxidized to pralidoxime.

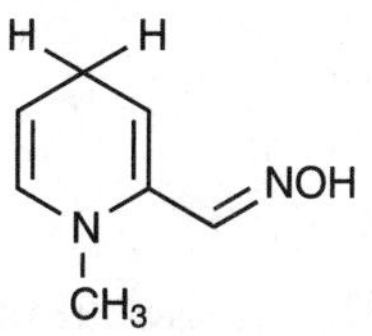

Fig. 15.69 ProPAM.

15.19 Anticholinesterases as 'smart drugs'

Acetylcholine is an important neurotransmitter in the CNS as well as in the peripheral nervous system. In recent years, it has been proposed that the memory loss, intellectual deterioration, and personality changes associated with Alzheimer's disease may in part be due to loss of cholinergic nerves in the brain. Although Alzheimer's disease is primarily a disease of the elderly, it can strike victims as young as 30 years. The disease destroys neurons in the brain (including cholinergic receptors) and is associated with the appearance of plaques and tangles of nerve fibres.

Research has been carried out into the use of anticholinesterases for the treatment of Alzheimer's disease – the so called 'smart drugs'. There is no evidence that such compounds can assist general memory improvement and so students studying for exams may not find such compounds of much use! The treatment does not offer a cure for Alzheimer's disease either, but it can alleviate the symptoms by allowing the brain to make more use of the cholinergic receptors still surviving. Unlike anticholinesterases acting in the periphery, 'smart drugs' have to cross the blood–brain barrier and so structures containing quaternary nitrogen atoms are not suitable. The first drug to be used for the treatment of Alzheimer's was tacrine (Cognex) (Fig. 15.70). However, this is an extremely toxic drug. Other agents which have been introduced since include donepezil, metrifonate (an organophosphate), galanthamine (obtained from daffodils or snowdrop bulbs), anabaseine (from ants and marine worms), physostigmine and rivastigmine (Exelon). Rivastigmine (an analogue of physostigmine) is the first drug to be approved in all EU countries. It shows selectivity for the brain and has beneficial effects on cognition, memory, concentration, and functional abilities (day-to-day tasks or hobbies). The drug has a short half-life, reducing the risk of accumulation or drug–drug interactions.

A club moss (*Huperzia serrata*) extract used for centuries in Chinese herbal medicine to treat symptoms varying from confusion to schizophrenia contains a novel alkaloid called Huperzine A (HupA), which acts as an anticholinesterase. Binding is very specific and so the drug can be used in small doses, thus minimizing the risk of side-effects. HupA has been undergoing clinical trials in China and has been shown to have memory enhancing effects.

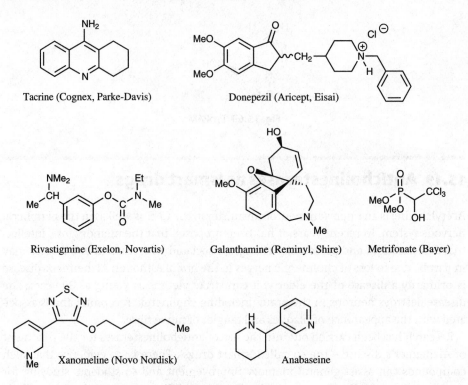

Tacrine (Cognex, Parke-Davis)

Donepezil (Aricept, Eisai)

Rivastigmine (Exelon, Novartis)

Galanthamine (Reminyl, Shire)

Metrifonate (Bayer)

Xanomeline (Novo Nordisk)

Anabaseine

Fig. 15.70 'Smart drugs'.

16 The adrenergic nervous system

16.1 The adrenergic system

16.1.1 The peripheral nervous system

In Chapter 15, we studied the cholinergic system and the important role it plays in the peripheral nervous system. Acetylcholine is the crucial neurotransmitter in the cholinergic system and has specific actions at various synapses and tissues. The other important player in the peripheral nervous system (Fig. 15.2) is the adrenergic system, which makes use of the chemical messengers adrenaline and noradrenaline. Noradrenaline (norepinephrine) is the neurotransmitter released by the sympathetic nerves which feed smooth muscle and cardiac muscle, whereas adrenaline (epinephrine) is a hormone released along with noradrenaline from the adrenal medulla, and which circulates in the blood supply to reach adrenergic receptors.

The action of noradrenaline at various tissues is the opposite to that of acetylcholine, which means that tissues are under a dual control. For example, if noradrenaline has a stimulant activity at a specific tissue, acetylcholine has an inhibitory activity at that same tissue. Both the cholinergic and adrenergic systems have a 'background' activity which would be analogous to driving a car with one foot on the brake and one foot on the accelerator at the same time. The overall effect on the tissue depends on which effect is predominant.

The adrenergic nervous system has a component which the cholinergic system does not have, and that is the facility to release the hormone adrenaline during times of stress. Adrenaline is responsible for the 'fight or flight' response. Whenever we are faced with danger or a challenge, the body automatically responds by releasing adrenaline. Adrenaline activates adrenergic receptors around the body, preparing it for immediate physical action, whether that be to fight the perceived danger or to flee from it. This means that the organs required for physical activity are activated, whilst those which are not important are suppressed. For example, adrenaline stimulates the heart and dilates the blood vessels to muscles such that the latter are supplied with sufficient blood for physical activity. At the same time, smooth muscle activity in the gastrointestinal tract is suppressed since digestion is not an immediate priority. This 'fight or flight' response is clearly an evolutionary advantage and stood early man in good stead when faced with an unexpected encounter with a grumpy old bear. Nowadays, it is unlikely that you will meet a grizzly bear on your way to the super- market, but the 'fight ot flight' response is still functional when faced with the modern dangers of crazy drivers. It also functions in any situation of stress, such as an imminent exam, important football game or public performance. In general, the effects of noradrenaline are the same as adrenaline, although noradrenaline constricts blood vessels to skeletal muscle rather than dilates them.

16.1.2 The central nervous system

There are also adrenergic receptors in the central nervous system (CNS), and nora- drenaline is important in many CNS functions, including sleep, emotion, temperature regulation, and appetite. However, the emphasis in this chapter will be on the periph- eral role of adrenergic agents.

16.2 Adrenergic receptors

16.2.1 Types of adrenergic receptor

In Chapter 15, we saw that there were two types of cholinergic receptor, with subtypes of each. The same holds true for adrenergic receptors. The two main types of adrenergic receptor are called the α- and β-adrenoreceptors. Both are G-protein-coupled receptors (see Chapter 6), but differ in the type of G-protein with which they couple (G_o for the α-adrenoreceptor; G_s for the β-adrenoreceptors).

For each type of receptor, there are various receptor subtypes which have slightly different structures. The α-adrenoreceptor consists of α_1- and α_2-subtypes, which differ in structure and also differ in the type of secondary message which is produced. α_1-Receptors produce inositol triphosphate and diacylglycerol as secondary messengers (see Chapter 6), while α_2-receptors inhibit the production of the secondary messenger cyclic AMP.

The β-adrenoreceptor consists of β_1-, β_2-, and β_3-subtypes, all of which activate the formation of cyclic AMP.

To complicate matters slightly further, both the α_1- and α_2-adrenoreceptors have further subclassifications (α_{1A}, α_{1B}, α_{1C}, α_{2A}, α_{2B}, α_{2C}).

All of these adrenergic receptor types and subtypes are 'switched on' by adrenaline and noradrenaline, but the fact that they have slightly different structures means that it should be possible to design drugs which will be selective and 'switch on' only a few or even just one of them. This is crucial in designing drugs with minimal side-effects which will act at specific organs in the body for, as we shall see, the various adreno- receptors are not evenly distributed around the body.

16.2.2 Distribution of receptors

As discussed above, the various adrenoreceptor types and subtypes are not uniformly distributed around the body and some tissues contain more of one type of adrenoreceptor than others. Table 16.1 describes various tissues, the types of adrenoreceptor which predominate at these tissues and the effect of activating these receptors.

One or two points are worth highlighting here:

- Activation of α-receptors generally contracts smooth muscle (except in the gut), whereas activation of β-receptors generally relaxes smooth muscle. This latter effect

Table 16.1

Organ or tissue	Predominant adrenoreceptors	Effect of activation	Physiological effect
Heart muscle	β_1	Muscle contraction	Increased heart rate and force
Bronchial smooth muscle	α_1	Smooth muscle contraction	Closes airways
	β_2	Smooth muscle relaxation	Dilates and opens airways
Arteriole smooth muscle (not supplying muscles)	α	Smooth muscle contraction	Constricts arterioles and increases blood pressure (hypertension)
Arteriole smooth muscle (supplying muscle)	β_2	Smooth muscle relaxation	Dilates arterioles and increases blood supply to muscles
Veins	α	Smooth muscle contraction	Constricts veins and increases blood pressure (hypertension)
	β_2	Smooth muscle relaxation	Dilates veins and decreases blood pressure (hypotension)
Liver	α_1 and β_2	Activates enzymes which metabolize glycogen and deactivates enzymes which synthesize glycogen	Breakdown of glycogen to produce glucose
GI tract smooth muscle	α_1, α_2, and β_2	Relaxation	'Shuts down' digestion
Kidney	β_2	Increases renin secretion	Increases blood pressure
Fat cells	β_3	Activates lipases	Fat breakdown

is mediated through the most common of the β-adrenoreceptors – the β_2-receptor. In the heart, the β_1-adrenoreceptors predominate and activation results in contraction of muscle.

• Different types of adrenoreceptor explain why adrenaline can have different effects at different parts of the body. For example, the blood vessels supplying skeletal muscle have mainly β_2-adrenoreceptors and are dilated by adrenaline, whereas the blood vessels elsewhere have mainly α-adrenoreceptors and are constricted by adrenaline. Since there are more blood vessels constricted than dilated in the system, the overall effect of adrenaline is to increase the blood pressure, yet at the same time provide sufficient blood for the muscles in the fight or flight response.

16.2.3 Clinical effects

The main clinical use for adrenergic agonists is in the treatment of asthma. Activation of β_2-adrenoreceptors causes the smooth muscles of the bronchi to relax, thus widening the airways.

The main uses for adrenergic antagonists are in treating angina and hypertension. Agents which block the α-receptors act on the α-receptors of blood vessels, causing relaxation of smooth muscle, dilatation of the blood vessels, and a drop in blood pressure. Agents which block β_1-receptors act on β_1-receptors in the heart, slowing down the heart rate and reducing the force of contractions. β-Blockers also have a range of other effects in other parts of the body which combine to lower blood pressure.

16.3 Adrenergic transmitters

The adrenergic transmitters adrenaline and noradrenaline are the natural chemical messengers which 'switch on' the receptors in the adrenergic nervous system. They belong to a group of compounds called the catecholamines – so called because they have an alkylamine chain linked to a catechol ring (the 1,2-benzenediol ring; Fig. 16.1).

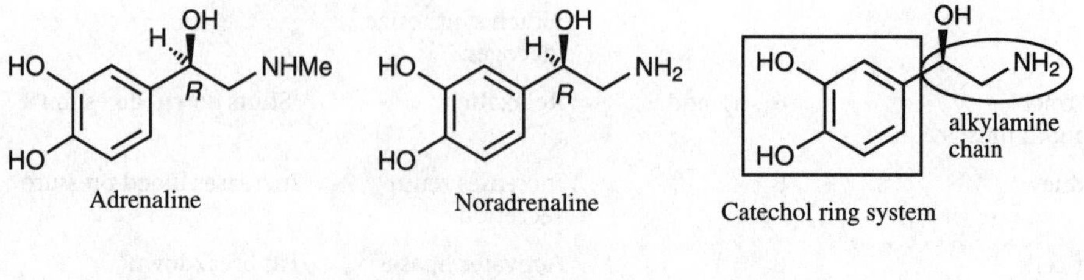

Fig. 16.1 Adrenergic transmitters.

16.4 Biosynthesis of catecholamines

The biosynthesis of noradrenaline and adrenaline starts from the amino acid L-tyrosine (Fig. 16.2). The enzyme tyrosine hydroxylase catalyses the introduction of a second phenol group to form L-dopa which is then decarboxylated to give dopamine – an important neurotransmitter in its own right. Dopamine is then hydroxylated to noradrenaline, which is the end product in adrenergic nerves. However, in the adrenal medulla, noradrenaline is N-methylated to form adrenaline. The biosynthesis of the catecholamines is controlled by regulation of tyrosine hydroxylase – the first enzyme in the pathway. This enzyme is inhibited by the end product of biosynthesis – noradrenaline, thus allowing self-regulation of catecholamine synthesis and control of catecholamine levels.

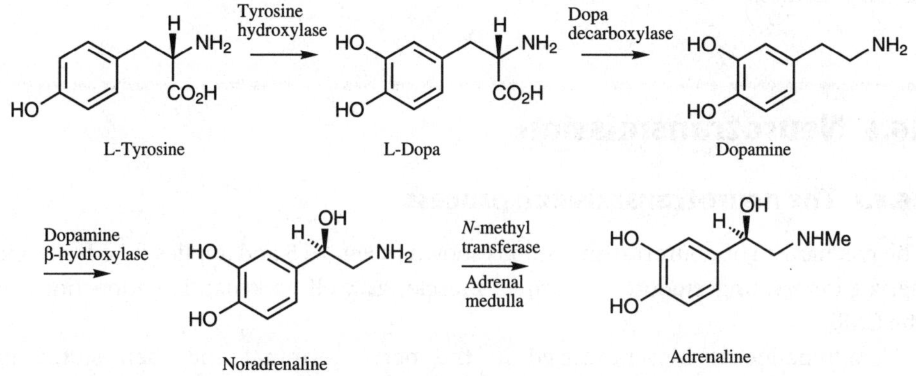

Fig. 16.2 Biosynthesis of noradrenaline and adrenaline.

16.5 Metabolism of catecholamines

Metabolism of catecholamines in the periphery takes place within cells and involves two enzymes – monoamine oxidase (MAO) and catechol O-methyltransferase (COMT).

MAO converts catecholamines to their corresponding aldehydes. These compounds are inactive as adrenergic agents and undergo further metabolism, as shown in Fig. 16.3 for noradrenaline. The final carboxylic acid is polar and is excreted in the urine.

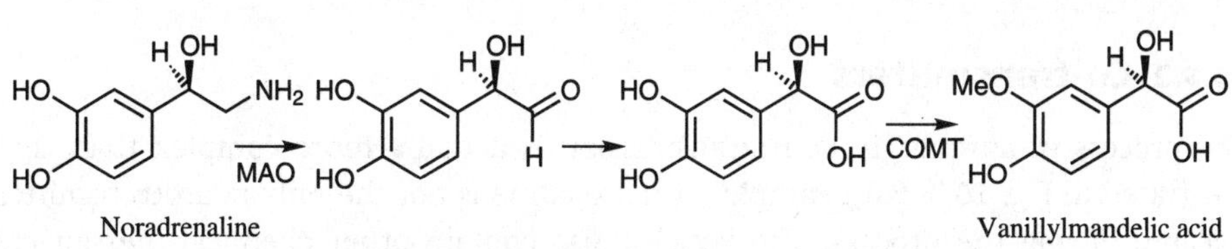

Fig. 16.3 Metabolism with monoamine oxidase.

An alternative metabolic route is possible which results in the same product. This time the enzyme COMT catalyses the methylation of one of the phenolic groups of the catecholamine. The methylated product is oxidized, then converted to the final carboxylic acid and excreted (Fig. 16.4).

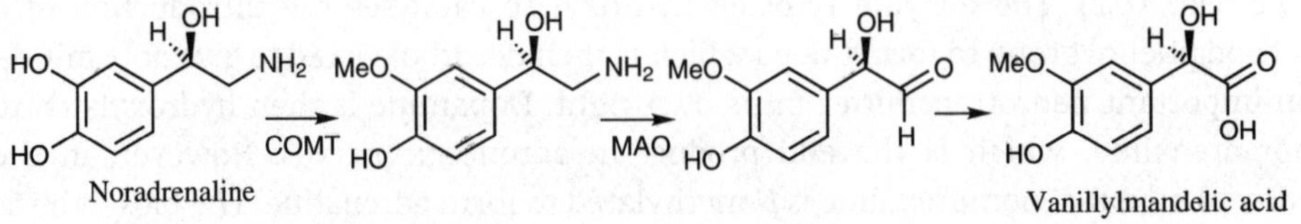

Fig. 16.4 Metabolism with COMT.

Metabolism in the CNS is slightly different, but still involves MAO and COMT as the initial enzymes.

16.6 Neurotransmission

16.6.1 The neurotransmission process

The mechanism of neurotransmission is shown in Fig. 16.5 and applies for adrenergic nerves innervating smooth or cardiac muscle, as well as synaptic connections in the CNS.

Noradrenaline is biosynthesized in the nerve terminal and then stored in membrane-bound vesicles. When a nerve impulse arrives at the nerve terminal it stimulates the opening of calcium ion channels and this promotes the fusion of the vesicles with the cell membrane to release noradrenaline.

The neurotransmitter then diffuses to adrenergic receptors on the target cell, where it binds and activates the receptor, leading to the signalling process which will eventually result in a cellular response.

Once the message has been received, noradrenaline departs and is taken back up into the nerve terminal through active transport by means of a carrier protein which is specific for noradrenaline. Once back in the cell, noradrenaline is repackaged into the vesicles. Some of the noradrenaline is metabolized before it has the chance to be repackaged, but this is balanced out by noradrenaline biosynthesis.

16.6.2 Co-transmitters

The process of adrenergic neurotransmission is actually more complex than that illustrated in Fig. 16.5. For example, noradrenaline is not the only neurotransmitter released during the process. The vesicles also contain other chemical messengers which act as co-transmitters. Adenosine triphosphate (ATP) and a protein called

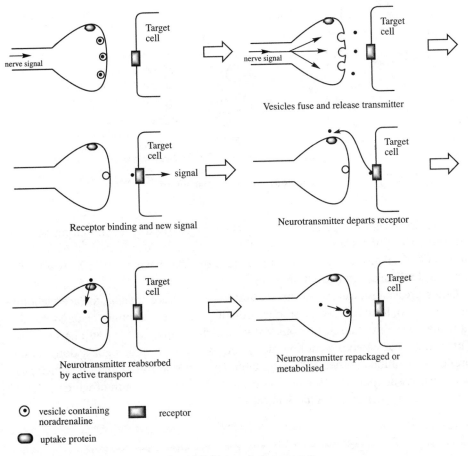

Fig. 16.5 Transmission process.

chromogranin A are released along with noradrenaline. They interact with their own specific receptors on the target cell and allow a certain variation in the speed and type of message which the target cell receives. For example, ATP leads to a fast response in smooth muscle contraction.

16.6.3 Presynaptic receptors and control

A further feature of the neurotransmission process not shown in Fig. 16.5 is the existence of presynaptic receptors which have a controlling effect on noradrenaline release (Fig. 16.6). There are a variety of these receptors, each of which responds to a specific chemical messenger. For example, there is an adrenergic receptor (α_2-adrenoreceptor) which interacts with released noradrenaline and has an inhibitory effect on further release of noradrenaline. Thus, noradrenaline acts to control its own release by a negative feedback system.

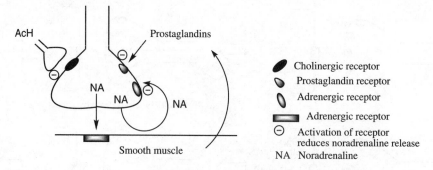

Fig. 16.6 Presynaptic receptors.

There are receptors specific for prostaglandins released from the target cell. For example, the prostaglandin PGE_2 appears to inhibit transmission, whereas $PGF_{2\alpha}$ appears to facilitate it. Thus, the target cell itself can have some influence on the adrenergic signals coming to it.

Other presynaptic receptors (muscarinic receptors) are specific for acetylcholine and serve to inhibit release of noradrenaline. These receptors respond to side branches of the cholinergic nervous system which synapse onto the adrenergic nerve. This means that when the cholinergic system is active, it sends signals along its side branches to inhibit adrenergic transmission. Therefore, as the cholinergic activity to a particular tissue increases, the adrenergic activity decreases, both of which enhance the overall cholinergic effect (cf. Section 15.5.2).

16.7 Drug targets

Having studied the nerve transmission process, it is now possible to identify several potential drug targets which will affect the process (Fig. 16.7):

1. Biosynthetic enzymes

2. Vesicle carriers

3. Exocytosis and release of transmitter

4. Adrenergic receptors

5. Uptake by carrier protein

6. Metabolic enzymes

7. Presynaptic adrenergic receptors.

We shall first concentrate on the adrenergic receptor and agents which interact with it. Later on in the chapter we will consider some of the other possible drug targets.

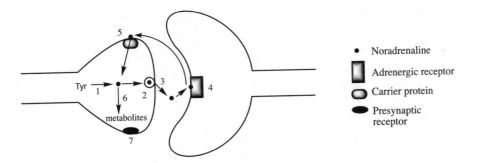

Fig. 16.7 Drug targets affecting nordrenaline transmission.

16.8 The adrenergic binding site

The adrenergic receptors are G-protein-linked receptors which consist of seven transmembrane helices (see Chapter 6). Three of these helices (TM3, 5, and 6) are involved in the binding site, illustrated for the β-adrenoreceptor in Fig. 16.8. Mutagenesis studies have indicated the importance of an aspartic acid residue (Asp-113), a phenylalanine residue (Phe-290) and two serine residues (Ser-207 and Ser-204). These groups can bind to adrenaline or noradrenaline as shown in the figure. The serine residues interact with the phenolic groups of the catecholamine via hydrogen bonding. The aromatic ring of Phe-290 interacts with the catechol ring by van der Waals interactions, while Asp-113 interacts with the protonated nitrogen of the catecholamine by ionic bonding. There is also a hydrogen bonding interaction between a hydrogen bonding group of the receptor and the alcohol function of the catecholamine.

16.9 Structure–activity relationships

16.9.1 Important binding groups on catecholamines

Support for the above binding site interactions is provided by structure–activity (SAR) studies on catecholamines. These emphasize the importance of having the alcohol group, the intact catechol ring system with both phenolic groups unsubstituted, and the ionized amine (Fig. 16.9).

Some of the evidence supporting these conclusions follows.

The alcohol group

The R-enantiomer of noradrenaline is more active than the S-enantiomer, indicating that the secondary alcohol is involved in a hydrogen bonding interaction. Compounds lacking the hydroxyl group (e.g. dopamine) have a greatly reduced interaction.

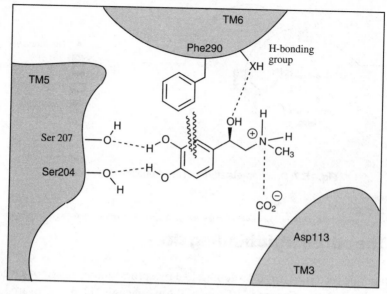

Fig. 16.8 Adrenergic binding site.

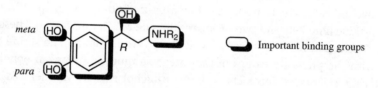

Fig. 16.9 Important binding groups.

However, some activity is retained, indicating that the alcohol group is important but not essential.

The amine

The amine is normally protonated at physiological pH, and the positive charge resulting from this is important since replacing nitrogen with carbon results in a large drop in activity. Activity is also affected by the number of substituents on the nitrogen. Primary and secondary amines have good adrenergic activity, whereas tertiary amines and quaternary ammonium salts do not.

The phenol substituents

The phenol groups are very important groups. For example, tyramine, amphetamine, and ephedrine (Fig. 16.10) have no affinity for adrenoreceptors. Nevertheless, the phenol groups can be replaced with other groups capable of interacting with the binding site by hydrogen bonding. This is particularly true for the *meta* phenol group which can be replaced with a variety of groups (e.g. CH_2OH, CH_2CH_2OH, NH_2, NHMe, NHCOR, NMe_2, $NHSO_2R$).

Alkyl substitution

Alkyl substitution on the side chain linking the aromatic ring to the amine decreases activity at both α- and β-adrenergic receptors. This may be a steric interaction which blocks hydrogen bonding to the alcohol, or which prevents the molecule adopting the active conformation.

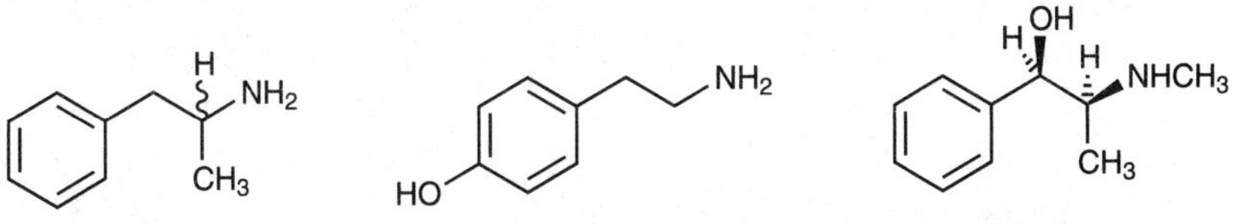

Fig. 16.10 Agents which have no affinity for the adrenergic receptor.

16.9.2 Selectivity for α versus β-adrenoreceptors

SAR's also demonstrate certain features which introduce a level of receptor selectivity between the α- and β-adrenoreceptors.

N-Alkyl substitution

It was discovered early on that adrenaline has the same potency for both types of adrenoreceptor, whereas noradrenaline has a greater potency for α-adrenoreceptors over β-adrenoreceptors. This indicated that an N-alkyl substituent has a role to play in receptor selectivity. Further work demonstrated that increasing the size of the N-alkyl substituent resulted in loss of potency at the α-receptor but an increase in potency at β-receptors. For example, the synthetic analogue isoprenaline (Fig. 16.11) is a powerful β-stimulant devoid of α-agonist activity.

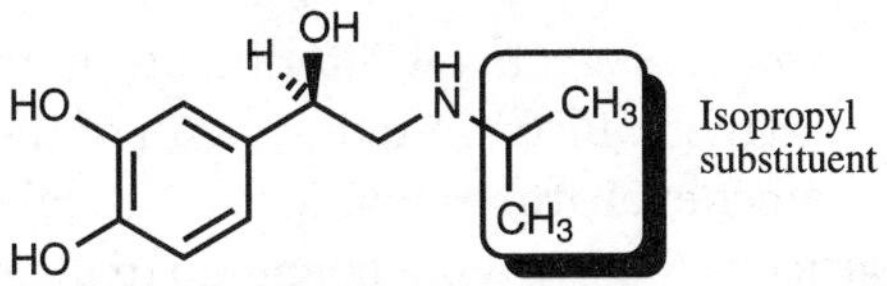

Fig. 16.11 Isoprenaline.

The presence of a bulky N-alkyl group (e.g. isopropyl or *tert*-butyl) is particularly good for β-activity. These results indicate that the β-adrenoreceptor has a hydrophobic pocket into which a bulky alkyl group can fit, whereas the α-adrenoreceptor does not (Fig. 16.12).

Phenol groups

The phenol groups seem particularly important for the β-receptor. If the phenol groups are absent, activity drops more significantly for the β-receptor than for the α-receptor.

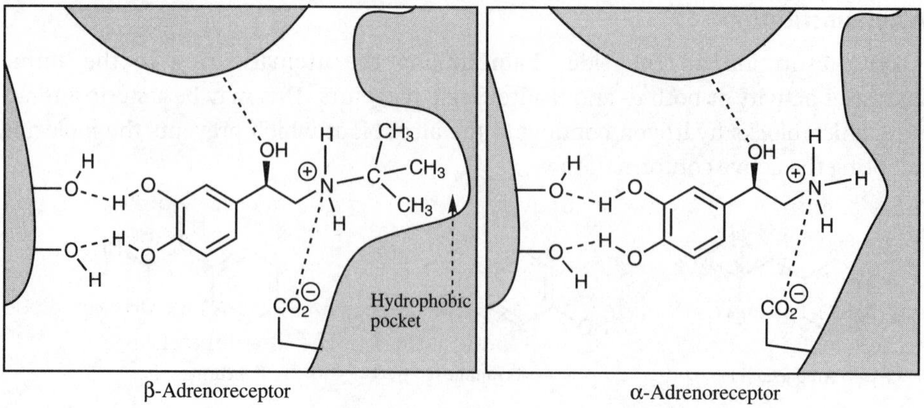

β-Adrenoreceptor   α-Adrenoreceptor

Fig. 16.12

α-Methyl substitution

Addition of an α-methyl group (e.g. α-methylnoradrenaline; Fig. 16.13) increases α_2-receptor selectivity.

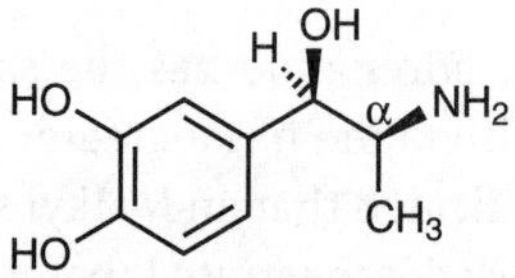

Fig. 16.13 α-Methylnoradrenaline.

Extension

As mentioned above, isopropyl or *tert*-butyl substituents on the amine nitrogen are particularly good for β-selectivity. Increasing the length of the alkyl chain offers no advantage, but if a polar functional group is placed at the end of the alkyl group, the situation changes. In particular, adding a phenol group to the end of a 2C alkyl chain results in a dramatic rise in activity, demonstrating that an extra polar binding region has been accessed which can take part in hydrogen bonding. For example, the structure shown in Fig. 16.14 had an increased activity by a factor of 800.

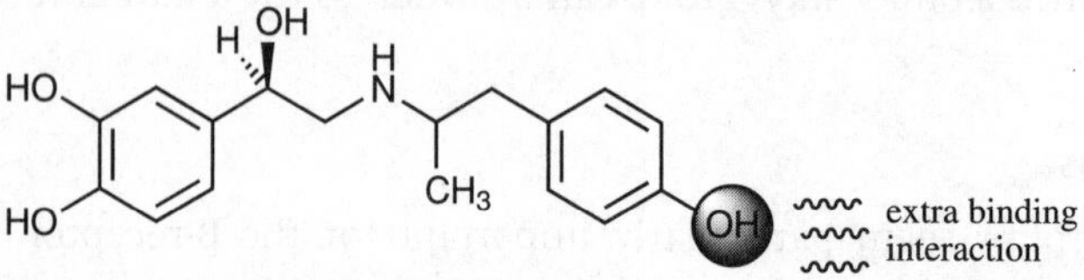

extra binding
interaction

Fig. 16.14 Extension tactics resulting in an additional hydrogen bonding interaction.

16.10 Adrenergic agonists

16.10.1 Adrenaline

Adrenaline itself is an obvious agonist for the overall adrenergic system and it is frequently used in emergency situations such as cardiac arrest or anaphylactic reactions where the patient is hypersensitive to some foreign chemical (e.g. bee stings or penicillin). Adrenaline is also administered with local anaesthetics to constrict blood vessels and to prolong the local anaesthetic activity at the site of injection.

However, apart from these conditions, it is preferable to have agonists which are selective for specific adrenoreceptors.

16.10.2 α_1-, α_2-, β_1-, and β_3-Agonists

In general, there is limited scope for agonists at these receptors, although there is potential for antiobesity drugs which act on the β_3-receptor. The β_1-agonist dobutamine (Fig. 16.15) is used to treat cardiogenic shock. Agonists acting on other adrenoreceptors are less useful since these agents constrict blood vessels, raise blood pressure, and can cause cardiovascular problems.

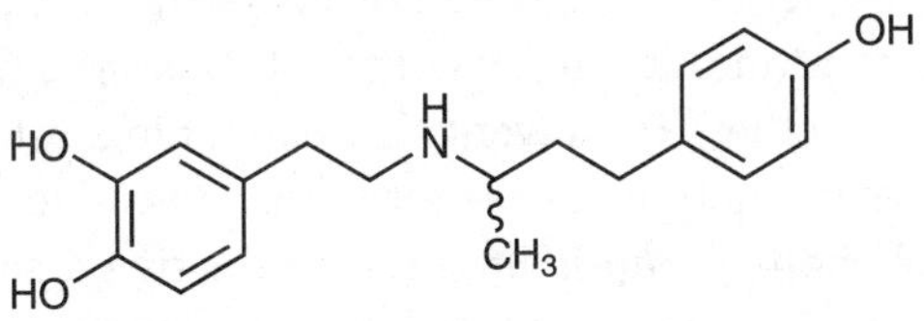

Fig. 16.15 Dobutamine.

16.10.3 β_2-Agonists

The most useful adrenergic agonists in medicine today are the β_2-agonists. These can be used to relax smooth muscle in the uterus to delay premature labour. However, β_2-agonists are more commonly used for the treatment of asthma. Activation of the β_2-adrenoreceptor results in smooth muscle relaxation and, since β_2-receptors predominate in bronchial smooth muscle, this leads to dilatation of the airways.

Adrenaline was one of the early compounds used to dilate the airways and it is still used today in cases of emergency. However, it only has a short duration of action and, since it has no selectivity for β_2-receptors, it 'switches on' all possible adrenergic receptors leading to a whole range of side-effects, and in particular cardiovascular side-effects.

To increase the selectivity of action, adrenaline was replaced with isoprenaline (Fig. 16.11) as an anti-asthmatic agent. Isoprenaline was taken by inhalation and was selective for β-receptors over α-receptors because of its bulky N-alkyl substituent.

Unfortunately, isoprenaline showed no selectivity between the different subtypes of β-receptor. Therefore, it not only activated the β_2-receptors in the airways, it also activated the β_1-receptors of the heart, again leading to unwanted cardiovascular effects. Therefore, the search was on to find an agonist selective to β_2-receptors which could be inhaled and which would also have a long duration of action.

Further research demonstrated that selectivity between different types of β-receptor could be obtained by introducing alkyl substituents onto the side chain linking the aromatic ring and the amine, and/or varying the alkyl substituents on the nitrogen. For example, isoetharine (Fig. 16.16) was shown to be selective for β_2-receptors. Unfortunately, like adrenaline, isoetharine was short lasting.

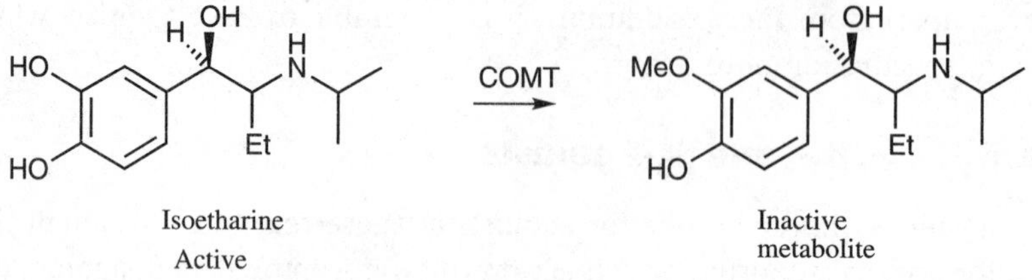

Isoetharine
Active

Inactive
metabolite

Fig. 16.16 Metabolism of isoetharine.

The short duration of action for isoetharine and adrenaline is due to the fact that they are taken up by tissues and inactivated by enzymes such as catechol-*O*-methyltransferase (COMT), resulting in methylation of the *meta* hydroxy group and formation of an inactive ether. To prevent this happening, attempts were made to modify the *meta* phenol group to make it more resistant to metabolism. This was no easy task since the phenolic group is important to activity, so it was necessary to replace it with a group which could still bind to the receptor and retain biological activity, but which would not be recognized by the metabolic enzyme. Various factors had to be taken into account in finding a suitable group to replace the phenol, for at the time this work was carried out, it was not clear exactly why this moiety was important. Was it taking part in hydrogen bonding, or was it ionized and taking part in ionic bonding? Did the phenol have an important electronic influence on the aromatic ring which affected binding, either through inductive or resonance effects? Was the size of the phenol group important? To test these various possibilities, the *meta* phenol group was replaced with a variety of different substituents.

A carboxylic acid group (A; Fig. 16.17) was tried first, but this compound had no activity at all. An ester and an amide were then tried (B and C). However, these compounds proved to be β-antagonists rather than agonists.

The introduction of a sulphonamide group ($MeSO_2NH$) was more successful, resulting in a long lasting selective β_2-agonist called soterenol (Fig. 16.18). However, this compound was never used clinically because a better compound was obtained in salbutamol. Here, the *meta* phenol group of the catecholamine skeleton was replaced with a hydroxymethylene group. Salbutamol has the same potency as isoprenaline, but is 2000 times less active on the heart. It has a duration of 4 hours, is not taken up

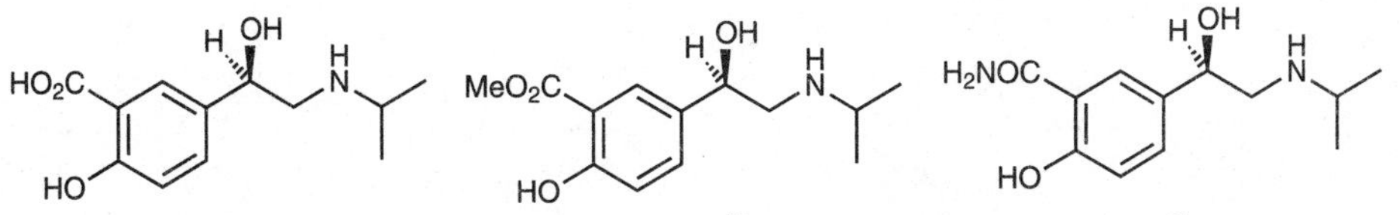

Fig. 16.17 Variation of the *meta* substituent.

by carrier proteins, and is not metabolized by COMT. Instead, it is more slowly metabolized to a phenolic sulphate. Thus, this modification proved successful in making the compound unrecognizable to COMT, whilst still being recognized by the adrenergic receptor (see also Section 10.5.6) The *R*-enantiomer is 68 times more active than the *S*-enantiomer, and salbutamol soon became a market leader in 26 countries for the treatment of asthma.

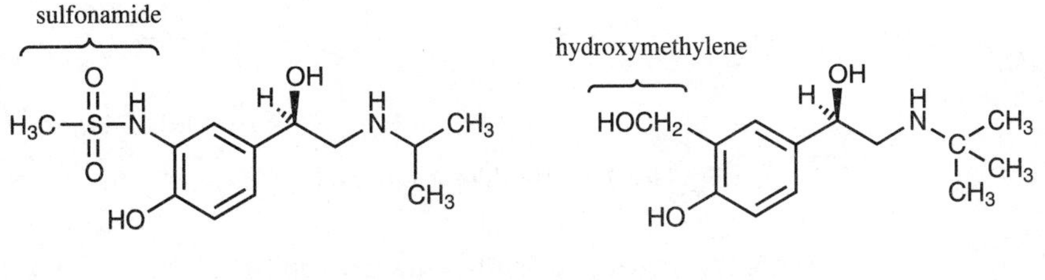

Fig. 16.18 Selective β_2 agonists.

Salbutamol can be synthesized from aspirin, as shown in Fig. 16.19. Fries rearrangement of aspirin produces a ketoacid which is then esterified. A bromoketone is then prepared which allows the introduction of the amino group by nucleophilic substitution. The methyl ester and ketone are then reduced, and finally the *N*-benzyl protecting group is removed by hydrogenolysis.

Several analogues of salbutamol were synthesized using variations or extensions of this synthesis.

For example, analogues were prepared to test whether the *meta* CH_2OH group could be modified further. These and previous results demonstrated the following requirements for the *meta* substituent:

- The *meta* substituent had to be capable of taking part in hydrogen bonding (e.g. $MeSO_2NHCH_2$; $HCONHCH_2$; $H_2NCONHCH_2$).
- *meta* substituents which had an electron withdrawing effect on the ring had poor activity (e.g. CO_2H).
- *meta* substituents which were bulky were bad for activity since they could prevent the substituent adopting the necessary conformation for hydrogen bonding.
- The CH_2OH group could be extended to CH_2CH_2OH but no further.

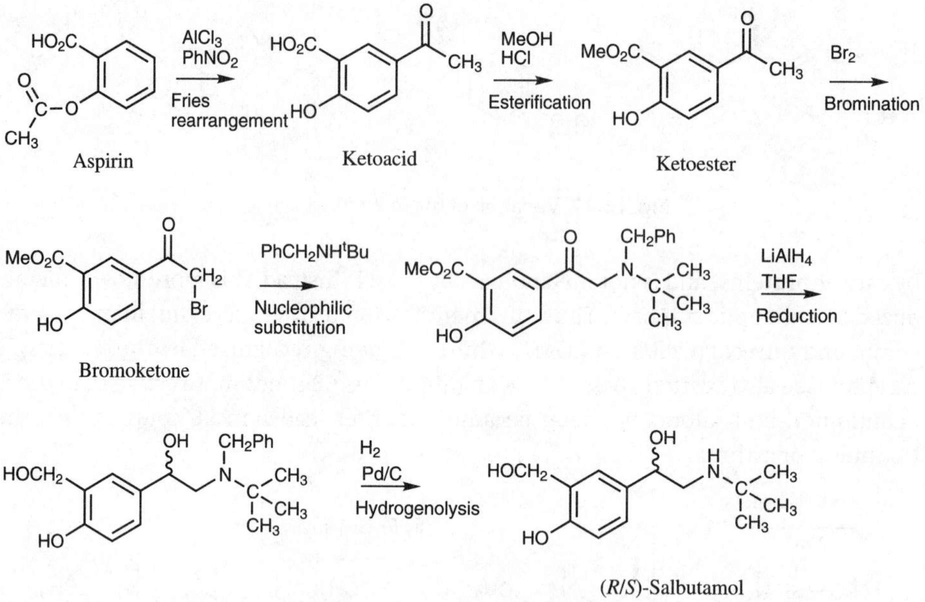

Fig. 16.19 Synthesis of salbutamol.

Having identified the advantages of a hydroxymethyl group at the *meta* position, attention now turned to the *N*-alkyl substituents. Salbutamol itself has a bulky tertiary butyl group. *N*-Arylalkyl substituents were now added which would be capable of reaching the polar region of the binding site described earlier (extension) (see Section 16.9.2). For example, salmefamol (Fig. 16.20) is 1.5 times more active than salbutamol and has a longer duration of action (6 hours). The drug is given by inhalation, but in severe attacks it may be given intravenously.

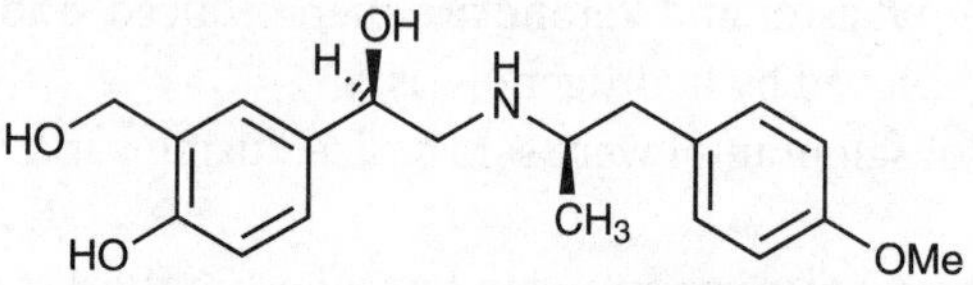

Fig. 16.20 Salmefamol.

Further developments were carried out to find a longer lasting agent to cope with nocturnal asthma, a condition which occurs at about 4 am (commonly called the morning dip). It was decided to increase the lipophilicity of the drug on the assumption that the new drug would bind more strongly to the tissue in the vicinity of the adrenoreceptor and thus be available to act for longer. Increased lipophilicity was introduced by using a long hydrocarbon chain as one of the *N*-substituents. This led to salmeterol (Fig. 16.21), which has twice the potency of salbutamol and an extended action of 12 hours.

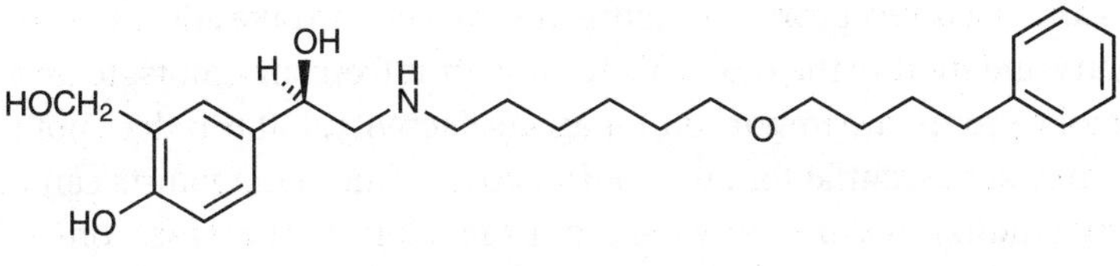

Fig. 16.21 Salmeterol.

16.11 Adrenergic receptor antagonists

16.11.1 α-Blockers

In general, the uses of α-antagonists are limited, being restricted to selective α_1-antagonists which have been used to treat hypertension or to control urinary output.

Prazosin (Fig. 16.22) was the first α_1-selective antagonist to be used for the treatment of hypertension, but it is short acting. Longer lasting drugs such as doxazosin and terazosin are better since they are given as once daily doses. These agents relieve hypertension by blocking the actions of noradrenaline or adrenaline at the α_1-receptors of smooth muscle in blood vessels. This results in relaxation of the smooth muscle and dilatation of the blood vessels, leading to a lowering in blood pressure.

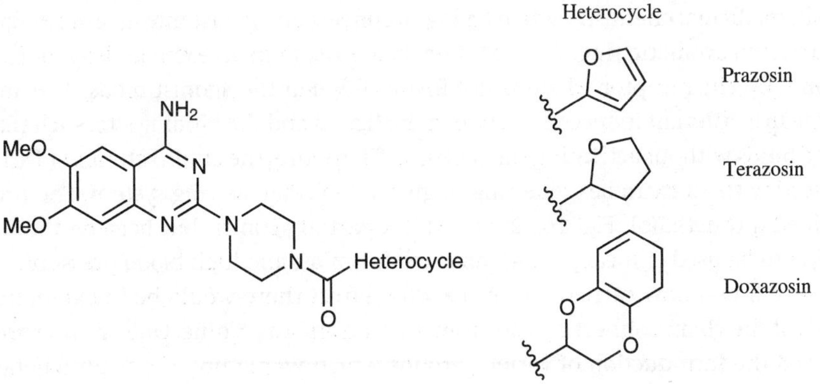

Fig. 16.22 α_1-Selective antagonists.

16.11.2 β-Blockers

16.11.2.1 First generation β-blockers

The most useful adrenergic antagonists used in medicine today are the β-blockers, which were originally designed to act as antagonists at the β_1-receptors of the heart.

The first goal in the development of these agents was to achieve selectivity for β-receptors with respect to α-receptors. Isoprenaline (Fig. 16.11) was chosen as the lead compound. Although this is an agonist and not an antagonist, it was active at β-receptors and not α-receptors. Therefore, the goal was to take advantage of the inherent specificity and modify the molecule to convert it from an agonist to an antagonist.

The phenolic groups are important for agonist activity, but this does not necessarily mean that they are essential for antagonist activity since antagonists can often block receptors by binding in different ways from the agonist. Therefore, one of the early experiments was to replace the phenol groups with other substituents. For example, replacing the phenolic groups of isoprenaline with chloro substituents produced dichloroisoprenaline (DCI) (Fig. 16.23). This compound was a partial agonist. In other words, it has some agonist activity, but it was weaker than a pure agonist. Nevertheless, DCI could block the natural chemical messengers from binding and could therefore be viewed as an antagonist since it lowered adrenergic activity.

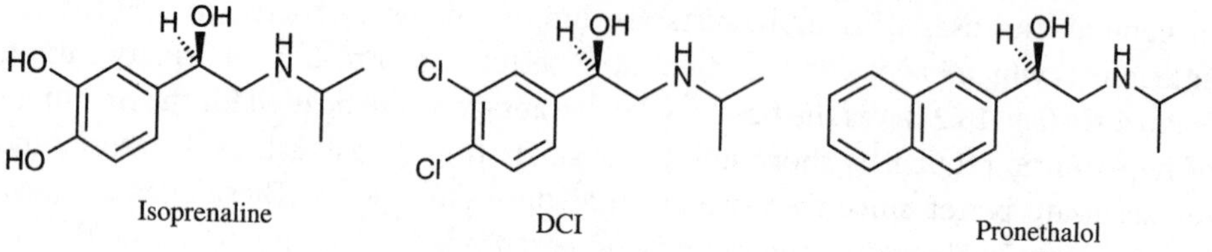

Isoprenaline   DCI   Pronethalol

Fig. 16.23 Partial β-agonists.

The next stage was to try and remove the partial agonist activity. A common tactic used in medicinal chemistry when trying to convert an agonist into an antagonist is to add an extra aromatic ring. This can sometimes result in an extra hydrophobic interaction with the receptor which is not involved when the agonist binds. This in turn can mean a different induced fit between the ligand and the binding site, such that the ligand binds without activating the receptor. Therefore, the chloro groups of DCI were replaced with an extra benzene ring to give a naphthalene ring system. The product obtained (pronethalol; Fig. 16.22) was still a partial agonist, but became the first β-blocker to be used clinically for angina, arryhthmias, and high blood pressure.

Research was now carried out to see what effect there would be in extending the length of the chain connecting the aromatic ring and the amine. One of these projects involved the introduction of various groups which would link the naphthalene ring and the ethanolamine portion of the molecule (Fig. 16.24). At this stage an element of good fortune came into play. It was intended to synthesize the target structure shown in Fig. 16.25 starting from β-naphthol. However, β-naphthol was not in stock at the time and so the synthesis was carried out using α-naphthol, which *was* available.

This led to propranolol (Fig. 16.25), which was found to be a pure antagonist and which was 10–20 times more active than pronethalol. Propranolol was introduced into the clinic for the treatment of angina and has since become the benchmark against which all β-blockers are rated. The *S*-enantiomer is the active enantiomer[1],

---

[1] Although the *S*-enantiomer of aryloxypropanolamines and the *R*-enantiomer of arylethanolamines are the active forms, the absolute configuration at the asymmetric centre is the same for both classes of compound.

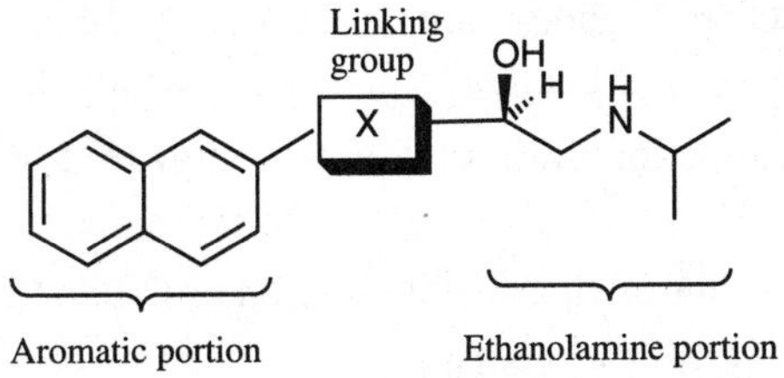

Fig. 16.24 Chain extension tactics.

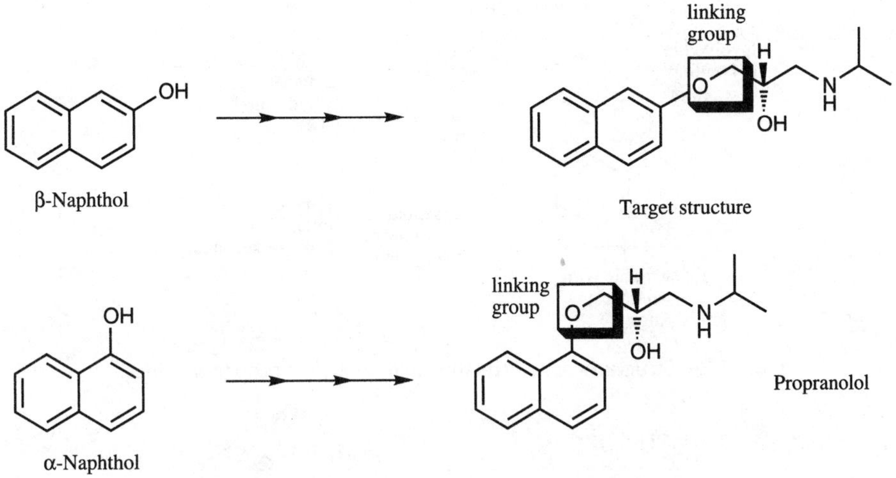

β-Naphthol

Target structure

α-Naphthol

Propranolol

Fig. 16.25 Development of propranolol.

although propranolol is used clinically as a racemate. Later, when the original target structure was synthesized, it was found to be less active.

16.11.2.2 Structure–activity relationships of aryloxypropanolamines (Fig. 16.26)

Propranolol is an example of an aryloxypropanolamine structure. A large number of aryloxypropanolamines have been synthesized using the route shown in Fig. 16.27 and from these results it was found that:

• Branched bulky N-alkyl groups (e.g. ^{i}Pr and ^{t}Bu) are good for β-antagonist activity, suggesting an interaction with a hydrophobic pocket in the binding site (compare β-agonists).

• Variation of the aromatic ring system is possible and heteroaromatic rings can be introduced (e.g. timolol and pindolol; Fig. 16.28).

• Substitution on the side chain methylene group increases metabolic stability but lowers activity.

• The alcohol group on the side chain is essential for activity.

- Replacing the ether O on the side chain with S, CH_2, or NMe is detrimental although a tissue selective β-blocker has been obtained using NH for O.

- Longer alkyl substituents than isopropyl or *tert*-butyl are less effective (but see next point).

- Adding an arylethyl group such as $CHMe_2$-CH_2Ph or $CHMe$-CH_2 Ph is beneficial.

- The amine nitrogen must be secondary.

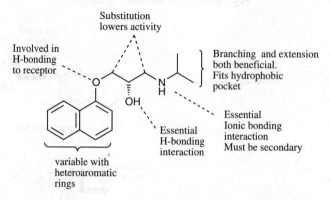

Fig. 16.26 Structure–activity relationships of aryloxypropanolamines.

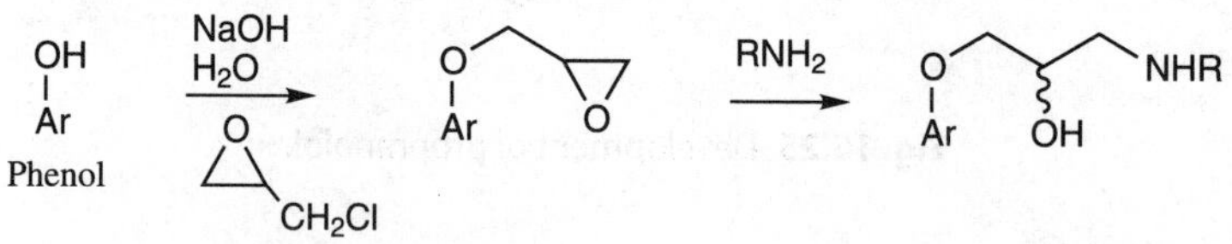

Fig. 16.27 Synthesis of aryloxypropanolamines.

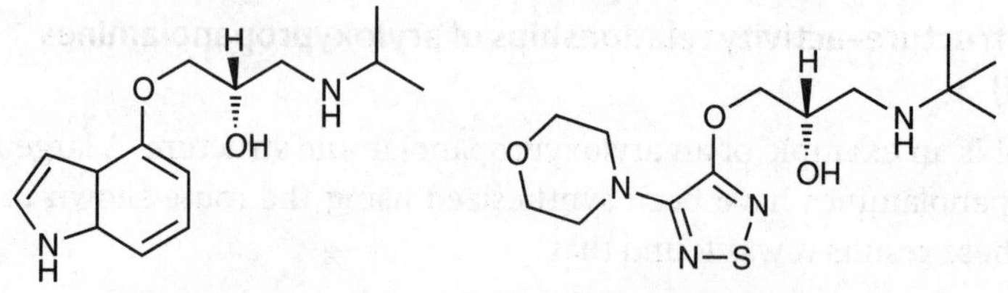

Fig. 16.28 β₁-Antagonists containing heteroaromatic ring systems.

Pindolol   Timolol

16.11.2.3 Clinical effects of first generation β-blockers

The effects of propranolol and other first generation β-blockers depend on how active the patient is. At rest, propranolol causes little change in heart rate, output, or pressure. However, if the patient exercises or becomes excited, propranolol reduces the resulting effects of circulating adrenaline. The β-blockers were originally intended for use in angina since they were targeted on the heart. However, an unexpected bonus was the fact that they also had antihypertensive activity (lowering of blood pressure). Indeed, the β-blockers are now more commonly used as antihypertensives rather than for the treatment of angina. The antihypertensive effect arises from a number of factors resulting from the effect of β-blockers at various parts of the body[2]:

- Action at the heart to reduce cardiac output.
- Action at the kidneys to reduce renin release. (Renin catalyses formation of angiotensin I which is quickly converted to angiotensin II which is a potent vaso-constrictor.)
- Action in the CNS to lower the overall activity of the sympathetic nervous system.

The side-effects of first generation β-blockers include the following:

- Bronchoconstriction in asthmatics, which means that these patients need β$_2$-receptor stimulation during treatment with β-blockers.
- Tiredness of limbs because of reduced cardiac output.
- CNS effects (dizziness, dreams, and sedation), especially with lipophilic β-blockers such as propranolol, pindolol, and oxprenol which can cross the blood–brain barrier.
- Heart failure for patients on the verge of a heart attack. The β-blockers produce a fall in the resting heart rate and this may push some patients over the threshold.
- Inhibition of noradrenaline release at synapses.

16.11.2.4 Selective β$_1$-blockers (second generation β-blockers)

Propranolol is a non-selective β-antagonist which acts as an antagonist at β$_2$-receptors as well as β$_1$-receptors. Normally this is not a problem, but it does pose serious problems if the patient is asthmatic since the use of propranolol could initiate an asthmatic attack by antagonizing the β$_2$-receptors in bronchial smooth muscle. This would lead to contraction of bronchial smooth muscle and closure of the airways.

Practolol (Fig. 16.29) is not as potent as propranolol, but it is a selective cardiac β$_1$-antagonist which does not block vascular or bronchial β$_2$-receptors. It was much safer for asthmatic patients and since it was more polar than propranolol it had fewer CNS effects.

Practolol was marketed as the first cardioselective β$_1$-blocker for the treatment of angina and hypertension, but had to be withdrawn after a few years because of unexpected but serious side-effects in a very small number of patients (e.g. skin rashes, eye problems, and peritonitis).

---

[2] These effects overide the effect that the β-blockers have on blood vessels where blocking the β-receptors would normally cause a constriction of the blood vessels and lead to a rise in blood pressure.

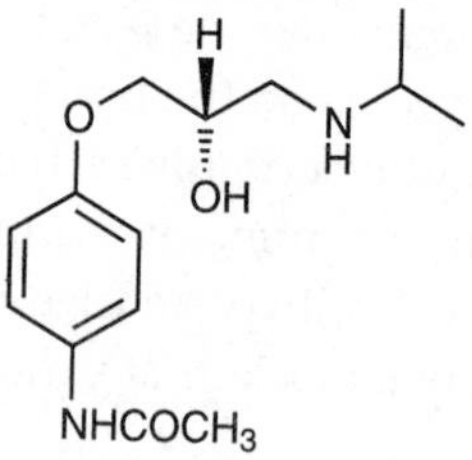

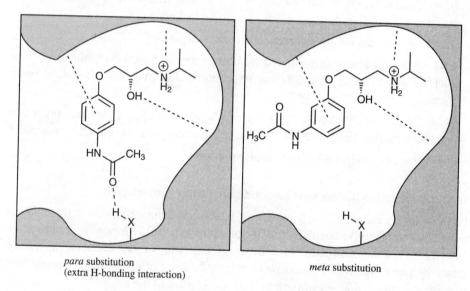

Fig. 16.29 Practolol.

Further investigations were carried out and it was demonstrated that the amido group had to be in the *para* position of the aromatic ring rather than the *ortho* or *meta* positions if the structure was to retain selectivity for the cardiac β_1-receptors. This implied that there was an extra hydrogen bonding interaction in the β_1-receptors (Fig. 16.30) but not the β_2-receptors.

para substitution
(extra H-bonding interaction)

meta substitution

Fig. 16.30 Bonding interactions with β_1-receptors.

Replacement of the acetamido group with other groups capable of hydrogen bonding led to a series of cardioselective β_1-blockers which reached the market (e.g. acebutolol, atenolol, metoprolol, and betaxolol; Fig. 16.31).

16.11.2.5 Third generation β-blockers

The *N*-alkyl groups in first and second generation β-blockers are normally isopropyl or *tert*-butyl groups. Extension tactics involving the addition of arylalkyl groups to the nitrogen atom resulted in a series of third generation β-blockers which bind to the β_1-receptor using an additional hydrogen bonding interaction, e.g. epanolol and

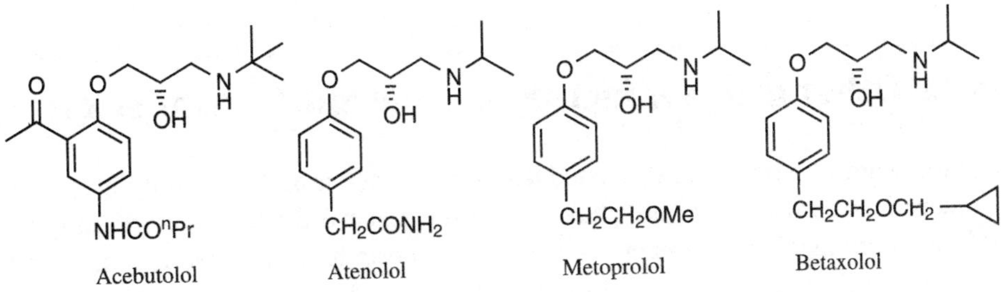

Fig. 16.31 Second generation β-blockers.

primidolol (Fig. 16.32). Related to these structures is xamoterol which is a very selective β_1-partial agonist used in the treatment of mild heart failure. As an agonist it provides cardiac stimulation when the patient is at rest, but it acts as a β-blocker during strenuous exercise, when greater amounts of adrenaline and noradrenaline are being generated.

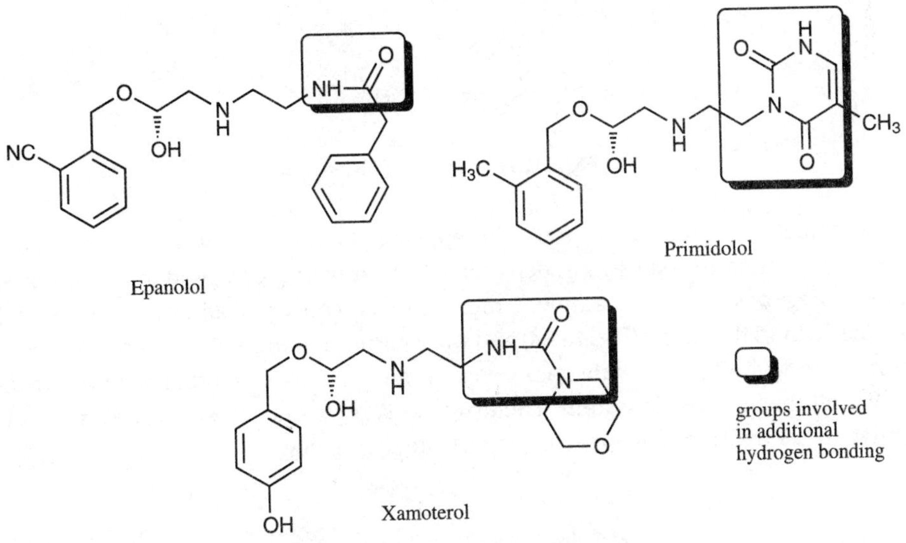

Fig. 16.32 Third generation β-blockers.

16.11.2.6 Other clinical uses for β-blockers

β-Blockers have a range of other clinical uses apart from cardiovascular medicine. They are used to counteract overproduction of catecholamines resulting from an enlarged thyroid gland or tumours of the adrenal gland. They can also be used to alleviate the trauma of alcohol and drug withdrawal as well as relieving the stress associated with situations such as exams, public speaking, and public performances. Timolol and betaxolol are used in the treatment of glaucoma, although their mechanism of action is not clear, while propranolol is used to treat anxiety and migraine.

16.12 Other drugs affecting adrenergic transmission

In the previous sections, we discussed drugs which act as agonists or antagonists at adrenergic receptors. However, there are various other drug targets involved in the adrenergic transmission process which are important in controlling adrenergic activity. In this section we shall briefly cover some of the most important aspects of these.

16.12.1 Drugs which affect the biosynthesis of adrenergics

In Section 16.4, we identified tyrosine hydroxylase as the regulatory enzyme for catecholamine biosynthesis. As such, it is a potential drug target. For example, α-methyltyrosine (Fig. 16.33) inhibits tyrosine hydroxylase and is sometimes used clinically to treat tumour cells which overproduce catecholamines.

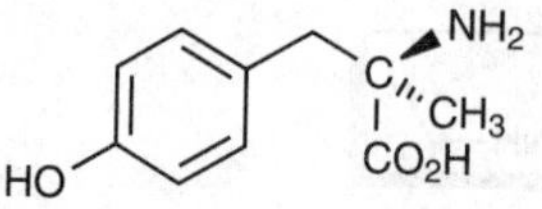

Fig. 16.33 α-Methyltyrosine.

It is sometimes possible to 'fool' the enzymes of the biosynthetic process into accepting an unnatural substrate such that a 'false' transmitter is produced and stored in the storage vesicles. For example, α-methyldopa is converted and stored in vesicles as α-methylnoradrenaline (Fig. 16.34) and so displaces noradrenaline. Such false transmitters are less effective than noradrenaline and so this is another way of down-regulating the adrenergic system. The drug has serious side-effects, however, and is limited to the treatment of hypertension in late pregnancy.

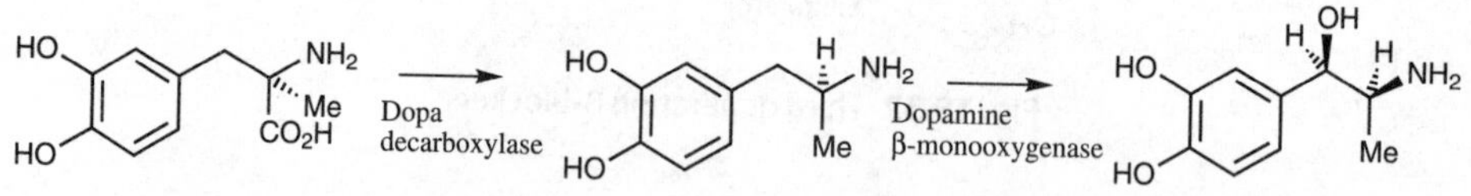

Fig. 16.34 False transmitter – α-methylnoradrenaline.

A similar example is the use of α-methyl-*m*-tyrosine in the treatment of shock. This unnatural amino acid is accepted by the enzymes of the biosynthetic pathway and converted to metaraminol (Fig. 16.35).

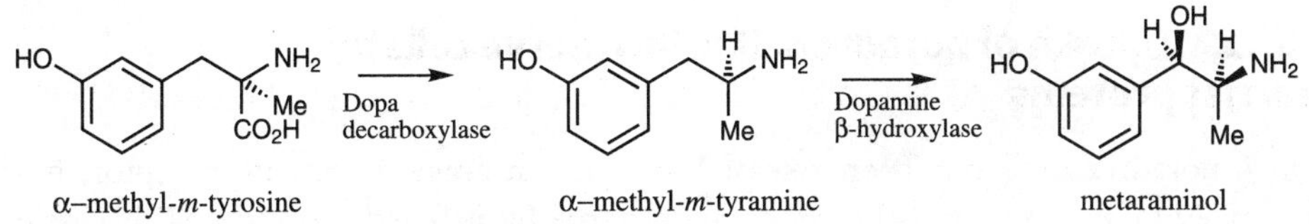

Fig. 16.35 False transmitter – metaraminol.

16.12.2 Uptake of noradrenaline into storage vesicles

The uptake of noradrenaline into storage vesicles can be inhibited by drugs. The natural product reserpine (Fig. 16.36) binds to the transport protein responsible for carrying noradrenaline into the vesicles and so noradrenaline accumulates in the cytoplasm where it is metabolized by MAO. As noradrenaline levels drop, adrenergic activity drops. Reserpine was once used as an antihypertensive agent but has serious side-effects such as depression, and so it is no longer used.

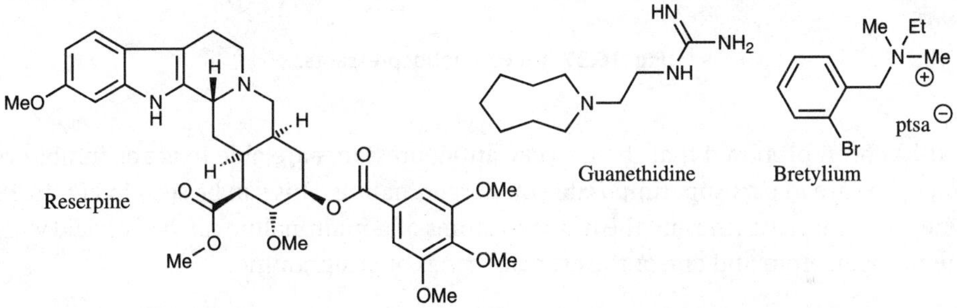

Fig. 16.36 Agents which affect adrenergic activity (ptsa, *para* toluenesulphonate).

16.12.3 Release of noradrenaline from storage vesicles

The storage vesicles are also the site of action for the drugs guanethidine and bretylium. These agents inhibit the release of noradrenaline from these vesicles at the nerve terminals.

Guanethidine is taken up into the nerves and storage vesicles by the same carrier proteins as noradrenaline, and displaces noradrenaline in the same way as reserpine. However, the drug also prevents exocytosis of the vesicle and so prevents release of the vesicle's contents into the synaptic gap. Guanethidine is an effective antihypertensive agent, but is no longer used clinically because of the side-effects resulting from such a non-specific inhibition of adrenergic nerve transmission.

Bretylium works in the same way as guanethidine and is sometimes used clinically to treat irregular heart rythyms.

16.12.4 Uptake of noradrenaline into nerve cells by carrier proteins

Once noradrenaline has been released and has interacted with its receptor, it is normally taken back into the presynaptic nerve by its carrier protein. This carrier protein is an important target for various drugs which inhibit noradrenaline uptake and thus prolong adrenergic activity.

The tricyclic antidepressants (e.g. desipramine, imipramine, and amitriptyline; Fig. 16.37) work by inhibiting noradrenaline reuptake in the CNS and stimulating adrenergic activity.

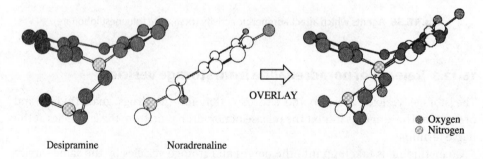

Desipramine   Imipramine   Amitriptyline

Fig. 16.37 Tricyclic antidepressants.

It has been proposed that the tricyclic antidepressants are able to act as inhibitors since they are in part superimposable on noradrenaline. This can be seen in Fig. 16.38 where the aromatic ring and the nitrogen atoms of noradrenaline can be overlaid with the nitrogen atom and one of the aromatic rings of desipramine.

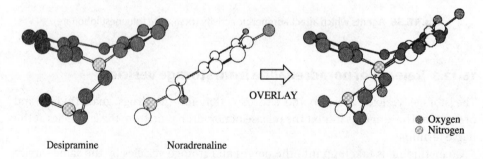

OVERLAY

● Oxygen
◎ Nitrogen

Desipramine   Noradrenaline

Fig. 16.38 Overlay of desipramine and noradrenaline.

Note that the tricyclic system of desipramine is V-shaped. This means that the second aromatic ring is held above the plane of the noradrenaline structure following overlay. Planar tricyclic structures would be expected to be less active as inhibitors since the second aromatic ring would then occupy the space required for the amine nitrogen.

Cocaine also inhibits noradrenaline uptake when it is chewed from coca leaves, but this time the inhibition is in the peripheral nervous system rather than the CNS. Chewing coca leaves was well known to the Inca nation as a means of increasing endurance and suppressing hunger, and the Incas would chew the leaves whenever they were faced with situations requiring long periods of physical effort or stamina. By chewing the coca leaves, cocaine is absorbed into the systemic blood supply and predominantly acts on peripheral adrenergic receptors to increase adrenergic activity. Nowadays, cocaine abusers prefer to smoke cocaine, which results in the drug entering the CNS more efficiently. There, it inhibits the uptake of dopamine rather than noradrenaline, resulting in its CNS effects.

Some amines, such as tyramine, amphetamine, and ephedrine (Fig.16.10), closely resemble noradrenaline in structure and are transported into the nerve cell by noradrenaline's carrier proteins. Once in the cell, they are taken up into the vesicles. Since these amines are competing with noradrenaline for the carrier proteins, noradrenaline is more slowly reabsorbed into the nerve cells. Moreover, as the foreign amines are transported into the nerve cell, noradrenaline is transported out by those same carrier proteins. Both of these facts mean that more noradrenaline is available to interact with its receptors. Therefore, amphetamines and similar amines have an indirect agonist effect on the adrenergic system.

16.12.5 Metabolism

Inhibition of the enzymes responsible for the metabolism of noradrenaline should prolong noradrenaline activity. We have seen how amines such as tyramine, amphetamine, and ephedrine inhibit the reuptake of noradrenaline into the presynaptic nerve. These amines also inhibit one of the important enzymes involved in the metabolism of noradrenaline, MAO. This in turn leads to a buildup in noradrenaline levels and an increase in adrenergic activity.

Monoamine oxidase inhibitors (MAOI) such as phenelzine, iproniazid, and tranylcypromine (Fig. 16.39) have been used clinically as antidepressants, but other classes of compound such as the tricyclic antidepressants are now favoured since they have fewer side-effects. It is important to realize that the MAOI's not only affect noradrenaline levels but also the levels of other neurotransmitters which are normally metabolized by these enzymes. In particular, dopamine and serotonin levels are both increased by MOAI's. Because of these widespread effects, it is difficult to be sure what the mechanism of the antidepressant activity actually is.

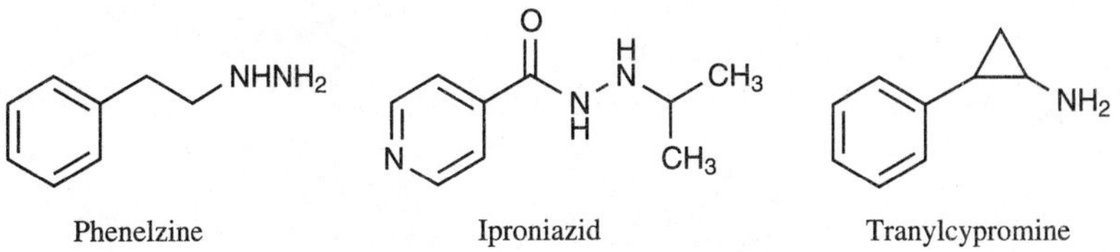

Phenelzine   Iproniazid   Tranylcypromine

Fig. 16.39 Monoamine oxidase inhibitors.

There are serious problems associated with the clinical use of MAOI's because of interactions with other drugs and food. A well known example of this is the 'cheese reaction'. Ripe cheese contains tyramine, which is normally metabolized by MAO's in the gut wall and liver and so never enters the systemic circulation. This changes once MOA's are inhibited. Tyramine can then circulate round the body, enhancing the adrenergic system and leading to acute hypertension and severe headaches.

17 The opium analgesics

17.1 Introduction

17.1.1 History of opium

We are now going to look in detail at one of the oldest fields in medicinal chemistry, yet one where true success has proved elusive – the search for a safe, orally active, and non-addictive analgesic based on the opiate structure.

It is important to appreciate that the opiates are not the only compounds which are of use in the relief of pain and that there are several other classes of compounds, including aspirin, which combat pain. These compounds, however, operate by different mechanisms from those employed by the opiates, and therefore relieve a different, 'sharper' kind of pain. The opiates have proved ideal for the treatment of 'deep' chronic pain and work in the central nervous system (CNS).

The term 'opium alkaloids' has been used rather loosely to cover all narcotic analgesics, whether they be synthetic compounds, partially synthetic, or extracted from plant material. To be precise, we should really only use the term for those natural compounds which have been extracted from opium – the sticky exudate obtained from the poppy (*Papaver somniferum*). The term alkaloid refers to a natural product which contains a nitrogen atom and is therefore basic in character. There are, in fact, several thousand alkaloids which have been extracted and identified from various plant sources, and examples of some of the better known alkaloids are shown in Fig. 17.1. These compounds provide a vast 'library' of biologically active compounds which can be used as lead compounds in many possible fields of medicinal chemistry. However, we are only interested at present in the alkaloids derived from opium.

The opiates are perhaps the oldest drugs known to man. The use of opium was recorded in China over 2000 years ago and was known in Mesopotamia before that. Because of opium's properties, the Greeks dedicated the opium poppy to Thanatos (the God of death), Hypnos (the God of sleep) and Morpheus (the God of dreams). Later physicians prescribed opium for a whole range of afflictions, including chronic headache, vertigo, epilepsy, asthma, colic, fevers, dropsies, leprosies, melancholy, and 'troubles to which women are subject'.

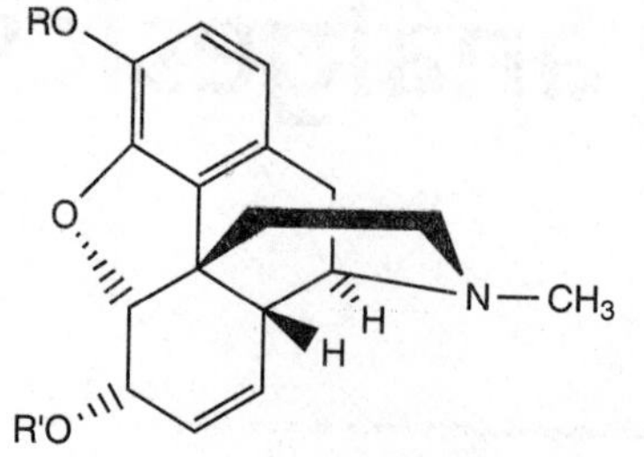

R = R' = H   Morphine
R = Me   R' = H   Codeine
R = R' = Ac   Diamorphine (Heroin)

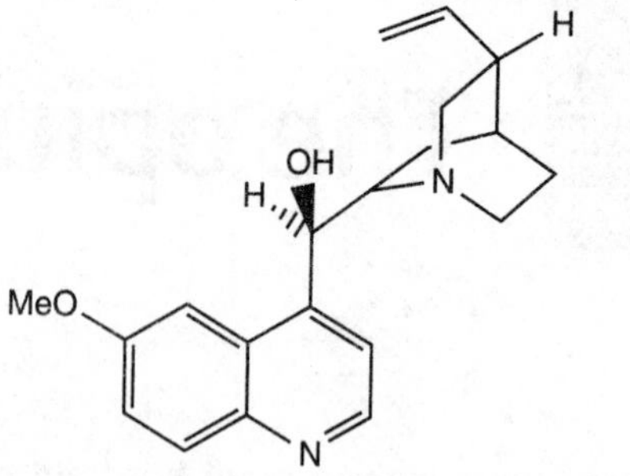

Quinine

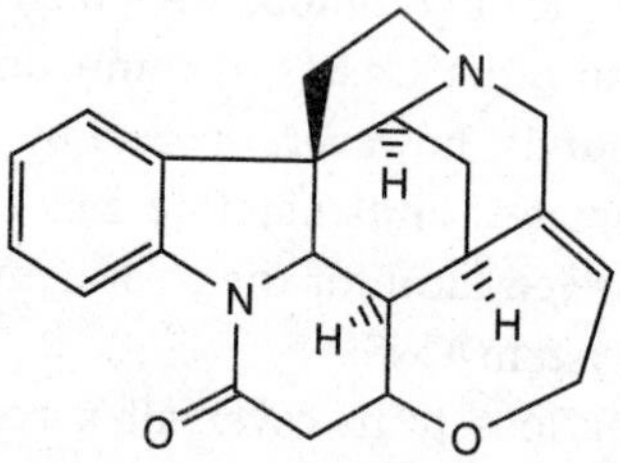

Strychnine

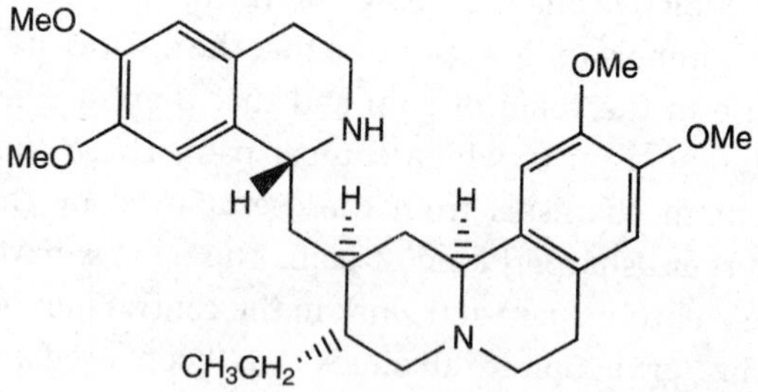

Emetine

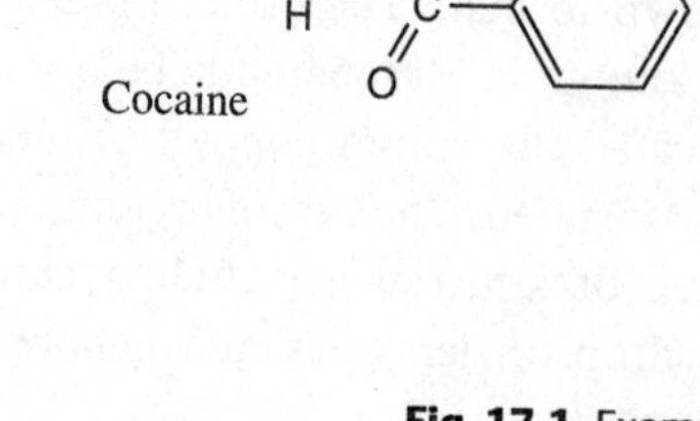

Cocaine

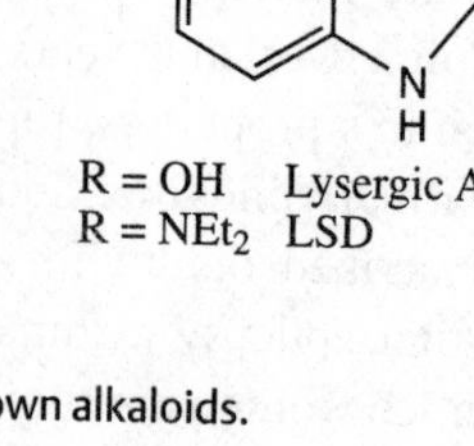

R = OH   Lysergic Acid
R = NEt$_2$ LSD

Fig. 17.1 Examples of well known alkaloids.

As a result of this, its fame spread and, by AD 632, knowledge of opium had reached Spain in the west and Persia and India in the east. It was cultivated in Persia, India, Malaysia, and China and was used primarily as a sleeping draught (sedative) and in the treatment of diarrhoea. It's use in medicine is quoted in a 12th century prescription: 'Take opium, mandragora and henbane in equal parts and mix with water. When you want to saw or cut a man, dip a rag in this and put it to his nostrils. He will sleep so deep that you may do what you wish'.

Its use as an analgesic came much later in the 16th century, when Paracelsus introduced preparations of opium known as laudanum. However, problems were also reported with its use. Doctors stated that stopping the drug after long-term use led to 'great and intolerable distresses, anxieties and depression of the spirit . . .'. These were the first reports of addiction and withdrawal symptoms.

Opium was first marketed in Britain by Thomas Dover, a one time pirate who had taken up medicine[1]. Dover prepared a powder containing opium, liquorice, saltpetre, and ipecacuanha. The last named compound is an emetic and had the advantage of making the consumers sick should they take too much of the concoction.

Another popular remedy of the day was 'Godfrey's cordial' which contained opium, molasses, and sassafras. This was used as a teething aid, for rheumatic pains, and for diarrhoea. These preparations were freely available in grocery shops without prescription or restriction, despite the fact that many people became addicted to them.

It has to be appreciated that in these days opium and the opium trade were considered to be as legitimate as tobacco or tea, and that this view continued right up to the twentieth century. Indeed, during the nineteenth century the opium trade led directly to a war between the United Kingdom and China.

During the 19th century, China was ruled by an elite class who considered all foreigners as nothing better than barbarians and wanted nothing to do with them. As a result, trade barriers were set up against all foreign imports and China strove to be totally self-sufficient. Many nations felt aggrieved over this, including the British who were buying tea from Canton and were not allowed to trade in return. Eventually, Britain thought it had the answer in opium.

Up until the early 17th century, China had grown its own opium for use as an ingredient in cakes and as a medicine but, strangely enough, it was the introduction of tobacco which changed all this. Tobacco was discovered in the 15th century and sailors introduced the habit into the far east. In China, smoking became so widespread that the Emperor Tsung Chen forbade the use of tobacco in 1644. Deprived of tobacco, the population started smoking opium instead! By the end of the century about a quarter of the population was using the drug and the local crops were insufficient to keep up with demand. Since India was a major producer of opium, the British East India company saw this as an opportunity to import the drug into China and, by the 1830s, the company was supplying one million pounds worth of opium per year.

---

[1] Dover had another claim to fame. During his seafaring days, he rescued the marooned Alexander Selkirk from an uninhabited island. This was the inspiration for Defoe's Robinson Crusoe.

However, because of China's embargo most of this had to be smuggled in via the port of Canton by British and American merchants.

Eventually, the Chinese authorities decided to act. They seized and burnt a shipload of opium, then closed the port of Canton to the British. The British traders were outraged and appealed to Lord Palmerston, the British Foreign secretary. Relations between the two countries steadily deteriorated and led to the Opium Wars of 1839–42. China was quickly defeated and was forced to lease Hong Kong to Britain as a trading port. They were also forced to accept the principles of free trade and to pay reparations of 21 million pounds. It may seem odd now, but at the time the British saw the Opium War as a just war aimed at defending the principles of free trade. Furthermore, China was seen as a tyrannical regime where justice was harsh and penalties were even harsher (e.g. death by a thousand cuts). However, the Opium War was first and foremost a trade war, and it wasn't long before the other European nations and the USA picked their own fight with China and imposed their own trade settlements and trading ports.

In the mid 19th century, opium was smoked in much the same way as cigarettes are today, and opium dens were as much a part of London society as coffee shops. These dens were used by many of the romantic authors of the day, including Thomas de Quincy, Edgar Allan Poe, and Samuel Taylor Coleridge. De Quincy even wrote a book recording his opium experiences (*Confessions of an English Opium Eater*) and was consuming around 4 pints of laudanum a week when he wrote *The Rime of the Ancient Mariner*. A later poem called *Dejection* may have been inspired by his experience of withdrawal symptoms.

Towards the end of the 19th century, doubts were beginning to grow about the long-term effects of opium and its addictive properties. However, there were many who leapt to its defence, and economic arguments were as powerful a weapon then as they are today. Indeed, in 1882 a parliamentary report stated, 'If Indian opium was stopped at once it would be a very frightful calamity indeed. I should say that one third of the adult population of China would die for want of opium'. Nevertheless, doubts persisted and a motion was put forward in parliament in 1893 stating that the 'opium trade was morally indefensible'. However, the motion was heavily defeated.

It wasn't until Chinese immigrants introduced opium on a large scale into the USA, Australia, and South America that governments really cracked down on the trade. In 1909, the International Opium Commission was set up and, by 1914, 34 nations had agreed to curb opium production and trade. By 1924, 62 countries had signed up and the League of Nations took over the role of control, requiring countries to limit the use of narcotic drugs to medicine alone. Unfortunately, many farmers in India, Pakistan, Afghanistan, Turkey, Iran, and the Golden Triangle (the borders of Burma, Thailand, and Laos) depended on the opium trade for survival, and as a result the trade went underground and has continued to this day.

17.1.2 Isolation of morphine

Opium contains a complex mixture of almost 25 alkaloids. The principle alkaloid in the mixture, and the one responsible for analgesic activity, is morphine, named after the ancient god of sleep – Morpheus. Although pure morphine was isolated in 1803, it was not until 1833 that chemists at Macfarlane & Co. (now Macfarlane–Smith) in Edinburgh were able to isolate and purify it on a commercial scale. However, since morphine was poorly absorbed orally, it was little used in medicine until the hypo-dermic syringe was invented in 1853, allowing doctors to inject morphine directly into the blood supply.

Morphine was then found to be a particularly good analgesic and sedative, and was far more effective than crude opium. But there was also a price to be paid. Morphine was used during the American Civil war (1861–65) and the Franco–Prussian war. However, there was poor understanding about safe dose levels, the effects of long-term use, and the increased risks of addiction, tolerance, and respiratory depression. As a result, many casualties were either killed by overdoses or became addicted to the drug.

At this stage, it is worth pointing out that all drugs have side-effects of one sort or another. This is usually due to the drug not being specific enough in its action, and interacting with receptors other than the one of interest. One reason for drug devel-opment is to try and eliminate the side-effects without losing the useful activity. Therefore, the medicinal chemist has to try and modify the structure of the original drug molecule to make it more specific for the target receptor. Admittedly, this has often been a case of trial and error in the past, but there are various strategies which can be employed (see Chapter 9). The development of narcotic analgesics is a good example of the traditional approach to medicinal chemistry and provides good exam-ples of the various strategies which can be employed in drug development. We can identify several stages:

Stage 1
Recognition that a natural plant or herb (opium from the poppy) has a pharmacolog-ical action.

Stage 2
Extraction and identification of the active principle (morphine).

Stage 3
Synthetic studies (full and partial synthesis).

Stage 4
Structure–activity relationships – the synthesis of analogues to see which parts of the molecule are important to biological activity.

Stage 5
Drug development – the synthesis of analogues to try and improve activity or reduce side-effects.

Stage 6

Theories on the analgesic receptors. Synthesis of analogues to test theories.

Stages 5 and 6 are the most challenging and rewarding parts of the procedure as far as the medicinal chemist is concerned, since the possibility exists of improving on what nature has provided. In this way, the chemist hopes to gain a better understanding of the biological process involved, which in turn suggests further possibilities for new drugs.

17.2 Morphine

17.2.1 Structure and properties

By 19th century standards, morphine was an extremely complex molecule and provided a huge challenge to chemists. By 1881, the functional groups on morphine had been identified, but it took many more years to establish the full structure. In those days the only way to find the structure of a complicated molecule was to break it down into simpler fragments which were already known and could be identified. So, for example, the degradation of morphine with strong basic solutions to produce methylamine gas established that there was an $N–CH_3$ fragment in the molecule.

From these fragments, chemists would propose a structure. This would be like trying to work out the structure of a bombed cathedral from the rubble. Once a structure had been proposed, chemists would then attempt to synthesize the structure. If the properties of the synthesized compound were the same as the natural compound, then the structure would be established. This was a long drawn out affair, made all the more difficult since there were fewer of the synthetic reagents or procedures which are available today. As a result, it was not until 1925 that Sir Robert Robinson proposed the correct structure. A full synthesis of morphine was achieved in 1952 and the structure proposed by Robinson was finally established when it was studied by X-ray crystallography in 1968 (164 years after the original isolation).

Morphine (Fig. 17.2) is the active principle of opium and is still one of the most effective painkillers available to medicine. It is especially good for treating dull, constant pain rather than sharp, periodic pain. It acts in the brain and appears to work by elevating the pain threshold, thus decreasing the brain's awareness of pain. Unfortunately, it has a large number of side-effects which include the following:

- depression of the respiratory centre
- constipation
- excitation
- euphoria
- nausea

- pupil constriction
- tolerance
- dependence.

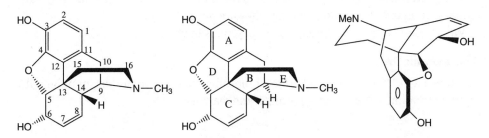

Fig. 17.2 Structure of morphine.

Some side-effects are not particularly serious. Some, in fact, can be advantageous. Euphoria, for example, is a useful side-effect when treating pain in terminally ill patients. Other side-effects, such as constipation, are uncomfortable but can give clues to other possible uses for opiate-like structures. For example, opiate structures are widely used in cough medicines and the treatment of diarrhoea.

The dangerous side-effects of morphine are those of tolerance and dependence, allied with the effects morphine can have on breathing. In fact, the most common cause of death from a morphine overdose is by suffocation. Tolerance and dependence in the one drug are particularly dangerous and lead to severe withdrawal symptoms when the drug is no longer taken.

Withdrawal symptoms associated with morphine include anorexia, weight loss, pupil dilatation, chills, excessive sweating, abdominal cramps, muscle spasms, hyper-irritability, lacrimation, tremor, increased heart rate, and increased blood pressure. No wonder addicts find it hard to kick the habit!

The isolation and structural identification of morphine mark the first two stages of our story and have already been described. The molecule contains five rings, labelled A–E, and has a pronounced T shape. It is basic because of the tertiary amino group, but it also contains a phenolic group, an alcohol group, an aromatic ring, an ether bridge, and a double bond. The next stage in the procedure is to find out which of these functional groups is essential to the analgesic activity.

17.2.2 Structure–activity relationships

The story of how morphine's secrets were uncovered is presented here in a logical step-by-step fashion. However, in reality this was not how the problem was tackled at the time. Different compounds were made in a random fashion depending on the ease of synthesis, and the logical pattern followed on from the results obtained. By presenting the development of morphine in the following manner, we are distorting history but we do get a better idea of the general strategies and the logical approach to drug development as a whole.

The first and easiest morphine analogues which can be made are those involving peripheral modifications of the molecule (that is, changes which do not affect the basic skeleton of the molecule). In this approach, we are looking at the different functional groups and discovering whether they are needed or not.

We now look at each of these functional groups in turn.

The phenolic OH

Codeine (Fig. 17.3) is the methyl ether of morphine and is also present in opium. It is used for treating moderate pain, coughs, and diarrhoea.

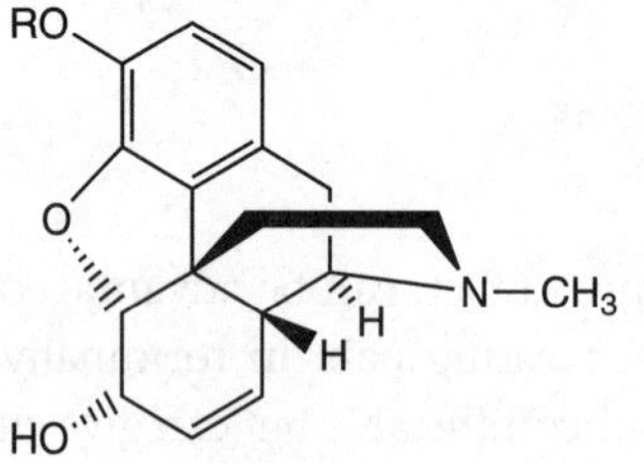

R = Me   CODEINE
R = Et   3-ETHYLMORPHINE   } Analgesic
R = Acetyl 3-ACETYLMORPHINE   Activity

Fig. 17.3

By methylating the phenolic OH, the analgesic activity drops drastically and codeine is only 0.1% as active as morphine. This drop in activity is observed in other analogues containing a masked phenolic group. Clearly, a free phenolic group is crucial for analgesic activity.

However, the above result refers to isolated receptors in laboratory experiments. If codeine is administered to patients, its analgesic effect is 20% that of morphine – much better than expected. Why is this so?

The answer lies in the fact that codeine can be metabolized in the liver to give morphine. The methyl ether is removed to give the free phenolic group. Thus, codeine can be viewed as a prodrug for morphine. Further evidence supporting this is provided by the fact that codeine has no analgesic effect at all if it is injected directly into the brain. By doing this, codeine is injected directly into the CNS and does not pass through the liver. As a result, demethylation does not take place.

This example shows the problems that the medicinal chemist can face in testing drugs. The manner in which the drugs are tested can be just as important as making the drug in the first place.

In all the following examples, the test procedures were carried out on animals or humans and so it must be remembered that there are several possible ways in which a change of activity could have resulted.

The 6-alcohol

The results in Fig. 17.4 show that masking or the complete loss of the alcohol group does not decrease analgesic activity and, in fact, often has the opposite effect. Again, it has to be emphasized that the testing of analgesics has generally been done *in vivo* and that there are many ways in which improved activity can be achieved.

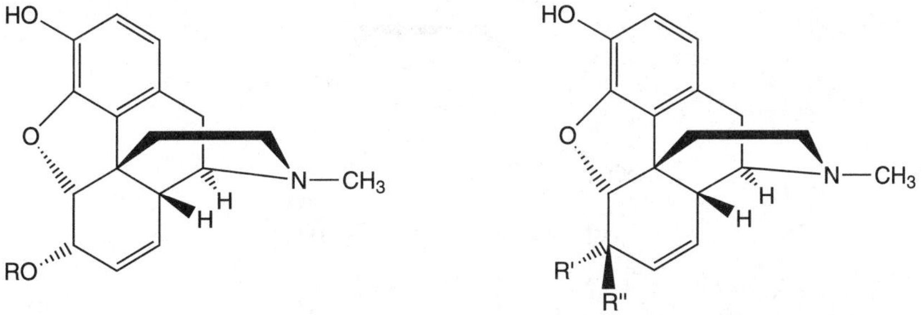

R		Analgesia wrt morphine
Me	Heterocodeine	5x
Et	6-Ethylmorphine	greater
Acetyl	6-Acetylmorphine	4x

R'	R"	Analgesia wrt morphine
H	OH	Increased
H	H	or
Ketone	Ketone	similar

Fig. 17.4 Effect of loss of alcohol group on analgesic activity.

In these examples, the improvement in activity is due to the pharmacokinetic properties of these drugs rather than their affinity for the analgesic receptor. In other words, it reflects how much of the drug can reach the receptor rather than how well it binds to it.

There are a number of factors which can be responsible for affecting how much of a drug reaches its target. For example, the active compound might be metabolized to an inactive compound before it reaches the receptor. Alternatively, it might be distributed more efficiently to one part of the body than another.

In this case, the morphine analogues shown are able to reach the analgesic receptor far more efficiently than morphine itself. This is because the analgesic receptors are located in the brain and, to reach the brain, the drugs have to cross a barrier called the blood–brain barrier. The capillaries which supply the brain are lined by a series of fatty membranes which overlap more closely than in any other part of the body. To enter the brain, drugs have to negotiate this barrier. Since the barrier is fatty, highly polar compounds are prevented from crossing. Thus, the more polar groups a molecule has, the more difficulty it has in reaching the brain. Morphine has three polar groups (phenol, alcohol, and an amine), whereas the analogues above have either lost the polar alcohol group or have it masked by an alkyl or acyl group. They therefore enter the brain more easily and accumulate at the receptor sites in greater concentrations; hence, the better analgesic activity.

It is interesting to compare the activities of morphine, 6-acetylmorphine (Fig. 17.4), and diamorphine (heroin; Fig. 17.5). The most active (and the most dangerous) compound of the three is 6-acetylmorphine, which is four times more active than morphine. Heroin is also more active than morphine by a factor of two, but is less active than 6-acetylmorphine. How do we explain this?

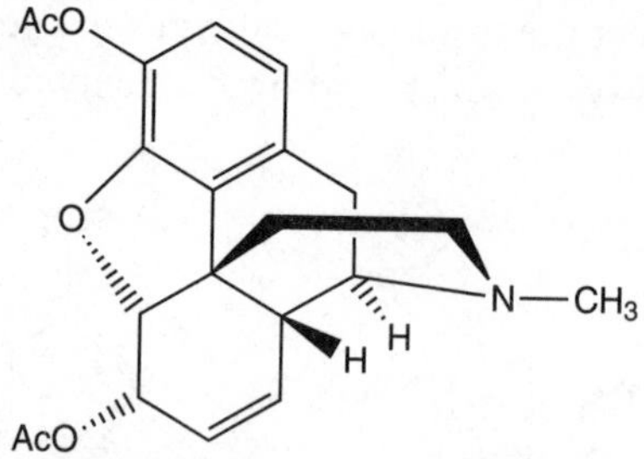

Fig. 17.5 Diamorphine (heroin).

6-Acetylmorphine, as we have seen already, is less polar than morphine and will enter the brain more quickly and in greater concentrations. The phenolic group is free and therefore it will interact immediately with the analgesic receptors.

Heroin has two polar groups which are masked and is therefore the most efficient compound of the three to cross the blood–brain barrier. However, before it can act at the receptor, the acetyl group on the phenolic group has to be removed by esterases in the brain. Therefore, it is more powerful than morphine because it enters the brain more easily, but it is less powerful than 6-acetylmorphine because the 3-acetyl group has to be removed before it can act.

Heroin and 6-acetylmorphine are both more potent analgesics than morphine. Unfortunately, they also have greater side-effects and have severe tolerance and dependence characteristics. Heroin is still used to treat terminally ill patients, such as those dying of cancer, but 6-acetylmorphine is so dangerous that its synthesis is banned in many countries.

To conclude, the 6-hydroxyl group is not required for analgesic activity and its removal can be beneficial to analgesic activity.

The double bond at 7–8

Several analogues, including dihydromorphine (Fig. 17.6) have shown that the double bond is not necessary for analgesic activity.

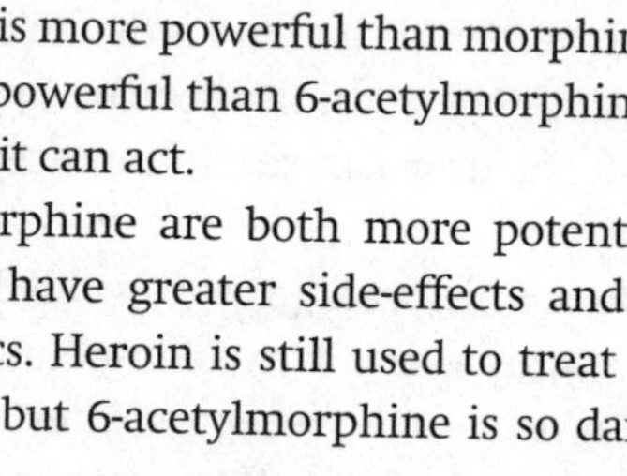

Fig. 17.6 Dihydromorphine.

The *N*-methyl group

The *N*-oxide and *N*-methyl quaternary salts of morphine are both inactive, which might suggest that the introduction of charge destroys analgesic activity (Fig. 17.7). However, we have to remember that these experiments were performed on animals and it is hardly surprising that no analgesia is observed, since a charged molecule has very little chance of crossing the blood–brain barrier. If these same compounds are injected directly into the brain, a totally different result is obtained and both these compounds are found to have similar analgesic activity to morphine. This fact, allied with the fact that neither compound can lose its charge, shows that the nitrogen atom of morphine is ionized when it binds to the receptor.

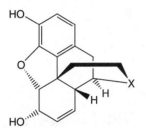

X		Analgesic Activity wrt morphine
NH	Normorphine	25%
N-Oxide		0%
Quaternary salt		0%

Fig. 17.7 Effect of introduction of charge on analgesic activity.

The replacement of the NMe group with NH reduces activity but does not eliminate it. The secondary NH group is more polar than the tertiary NMe group and therefore finds it more difficult to cross the blood–brain barrier, leading to a drop in activity. The fact that significant activity *is* retained shows that the methyl substituent is not essential to activity.

However, the nitrogen itself is crucial. If it is removed completely, all analgesic activity is lost. To conclude, the nitrogen atom is essential to analgesic activity and interacts with the analgesic receptor in the ionized form.

The aromatic ring

The aromatic ring is essential. Compounds lacking it show no analgesic activity.

The ether bridge

As we shall see later, the ether bridge is not required for analgesic activity.

Stereochemistry

At this stage, it is worth making some observations on stereochemistry. Morphine is an asymmetric molecule containing several asymmetric centres, and exists naturally as a single enantiomer. When morphine was first synthesized, it was made as a racemic mixture of the naturally occurring enantiomer plus its mirror image. These were separated and the unnatural mirror image was tested for analgesic activity. It turned out to have no activity whatsoever.

This is not particularly surprising if we consider the interactions which must take place between morphine and its receptor. We have identified that there are at least three important interactions involving the phenol, the aromatic ring, and the amine on morphine. Let us consider a diagrammatic representation of morphine as a T-shaped block with the three groups marked as shown in Fig. 17.8. The receptor has complementary binding groups placed in such a way that they can interact with all three groups. If we now consider the mirror image of morphine, then we can see that it can interact with only one binding region at any one time.

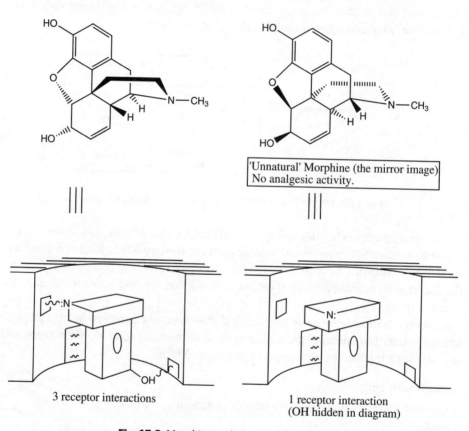

'Unnatural' Morphine (the mirror image) No analgesic activity.

3 receptor interactions

1 receptor interaction (OH hidden in diagram)

Fig. 17.8 Morphine and 'unnatural' morphine.

Epimerization of a single asymmetric centre such as the 14-position (Fig. 17.9) is not beneficial either, since changing the stereochemistry at even one asymmetric centre can result in a drastic change of shape, making it difficult for the molecule to bind to the analgesic receptors.

To summarize, the important functional groups for analgesic activity in morphine are shown in Fig. 17.10.

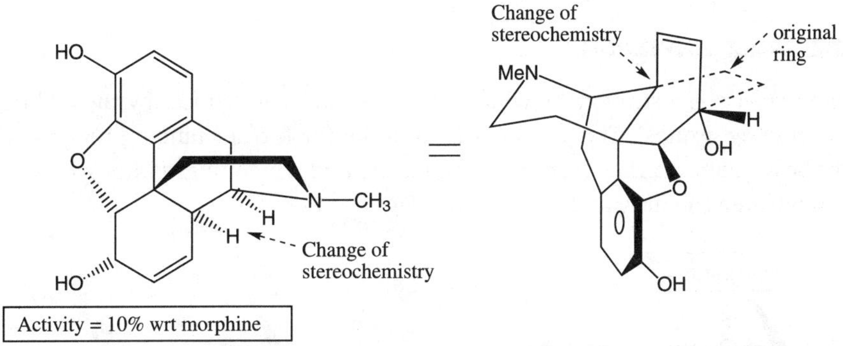

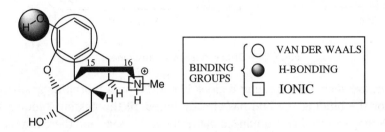

Fig. 17.9 Epimerization of a single asymmetric centre.

	BINDING GROUPS	⎧ ○ VAN DER WAALS

Fig. 17.10 Important functional groups for analgesic activity in morphine.

17.3 Development of morphine analogues

We now move on to consider the development of morphine analogues. As mentioned in Chapter 9, there are several strategies used in drug development. The following have been particularly useful in the development of morphine analogues:

- variation of substituents
- drug extension
- simplification
- rigidification.

17.3.1 Variation of substituents

A series of alkyl chains on the phenolic group give compounds which are inactive or poorly active. We have already identified that the phenol group must be free for analgesic activity.

The removal of the N-methyl group to give normorphine allows a series of alkyl chains to be added to the basic centre. These results are discussed under drug extension since the results obtained are more relevant under that heading.

17.3.2 Drug extension

Drug extension is a strategy by which the molecule is 'extended' by the addition of extra 'binding groups'. The aim here is to probe for further binding regions which might be available in the receptor's binding site and which might improve the interaction between the drug and the receptor (Fig. 17.11).

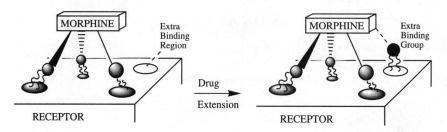

Fig. 17.11 Drug extension of morphine.

However, are such extra binding regions likely? Perhaps morphine is already an ideal fit for the binding site and has already found all the possible binding interactions? Perhaps morphine is produced naturally in the body and there is a receptor specifically for it?

All of these scenarios are certainly possible, but rather unlikely. Alkaloids such as morphine are secondary metabolites which are not essential to the growth of the plant and are only produced once the plant is mature. They could be looked upon as 'luxury items' which are not widespread in the chemistry of life and which have evolved in individual species. It seems highly unlikely then, that the ability to synthesize morphine has evolved separately in poppies and humans. It also seems highly unlikely that morphine evolved in the poppy to act as an analgesic, and so we have to conclude that it is a happy coincidence that morphine is able to interact with a painkilling receptor in the body. For this reason, it is quite possible that a search for further binding regions would be productive. For example, it is perfectly possible that there are four important binding regions in the binding site and that morphine only uses three of them (Fig. 17.11). Therefore, why not add binding groups to the morphine skeleton to search for that fourth binding interaction?

Many analogues of morphine have been made with extra functional groups attached. These have rarely shown any improvement. However, there are two exceptions. The introduction of a hydroxyl group at position 14 has been particularly useful (Fig. 17.12). This might be taken to suggest that there is a possible hydrogen bond interaction taking place between the 14-OH group and a suitable amino acid residue on the receptor. However, an alternative explanation is provided in Section 17.5.

The easiest position at which to add substituents (and the most advantageous) has been the nitrogen atom. The synthesis is easily achieved by removing the *N*-methyl group from morphine to give normorphine, then alkylating the amino group with an

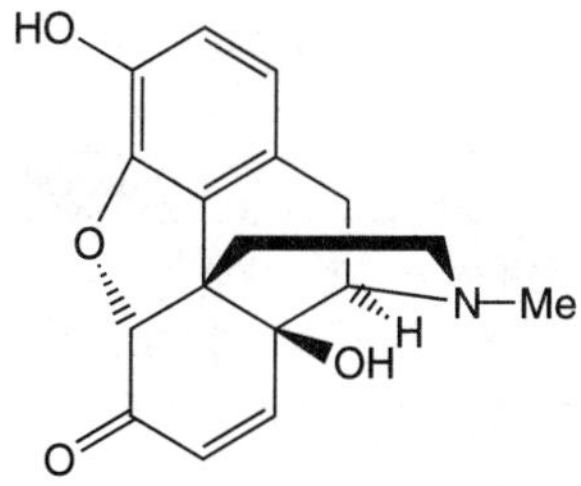

Fig. 17.12 Oxymorphine (2.5× activity of morphine).

alkyl halide. Removal of the *N*-methyl group was originally achieved by a von Braun degradation with cyanogen bromide, but is now more conveniently carried out using a chloroformate reagent such as vinyloxycarbonyl chloride (Fig. 17.13). The final alkylation step can sometimes be profitably replaced by a two-step process involving an acylation to give an amide, followed by reduction.

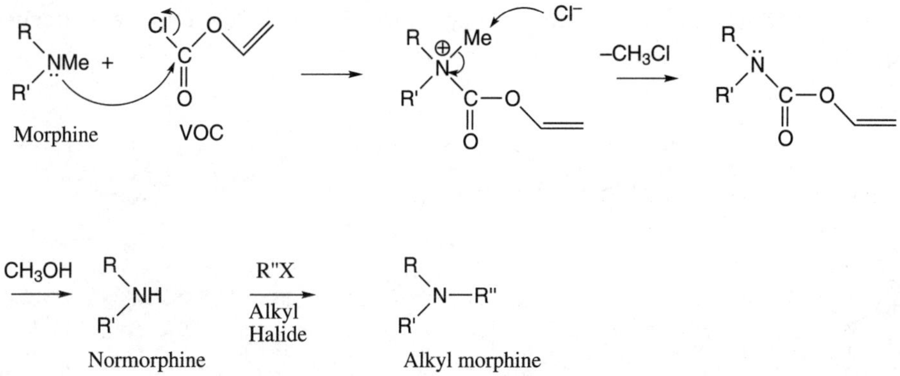

Fig. 17.13 Demethylation and alkylation of the basic centre.

The results obtained from the alkylation studies are quite dramatic. As the alkyl group is increased in size from a methyl to a butyl group, the activity drops to zero (Fig. 17.14). However, with a larger group such as a pentyl or a hexyl group, activity recovers slightly. None of this is particularly exciting, but when a phenethyl group is attached, the activity increases 14-fold – a strong indication that a hydrophobic binding region has been located which interacts favourably with the new aromatic ring (Fig. 17.15).

R =	Me	Et	Pr	Bu	Pentyl, Hexyl	CH_2CH_2Ph
	Agonism decreases Antagonism increases			Zero Activity	Agonists	14 x Activity wrt morphine

Fig. 17.14 Change in activity with respect to alkyl group size.

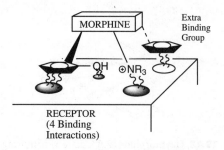

Fig. 17.15 Indication of fourth binding region.

To conclude, the size and nature of the group on the nitrogen is important to the activity spectrum. Drug extension can lead to better binding by making use of additional binding interactions.

Before leaving this subject, it is worth describing another series of important results arising from varying substituents on the nitrogen atom. Spectacular results were obtained when an allyl group or a cyclopropylmethylene group were attached (Fig. 17.16).

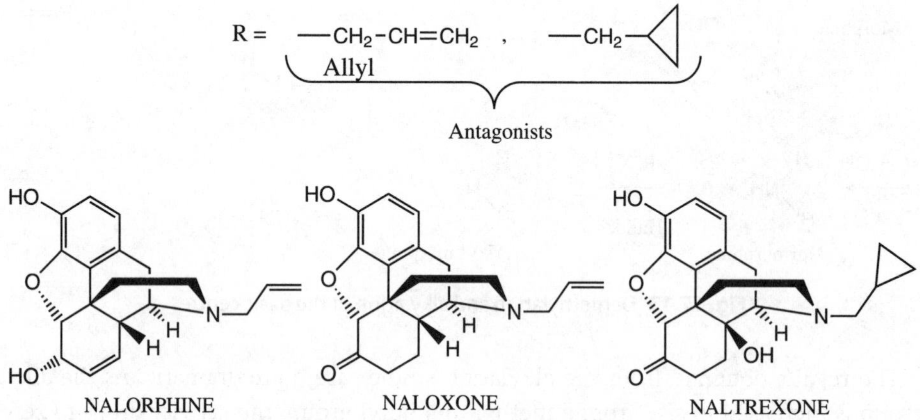

Fig. 17.16 Antagonists to morphine.

No increase in analgesic activity was observed and, in fact, the results were quite the opposite. Naloxone and naltrexone, for example, have no analgesic activity at all, whilst nalorphine retains only weak analgesic activity. However, the important feature about these molecules is that they act as antagonists to morphine. They do this by binding to the analgesic receptors without 'switching them on'. Once they have bound to the receptors, they block morphine from binding. As a result, morphine can no longer act as an analgesic. One might be hard pushed to see an advantage in this and with good reason. If we are just considering analgesia, there is none. However, the fact that morphine is blocked from all its receptors means that none of its side-effects are produced either, and it is the blocking of these effects which make antagonists extremely useful.

In particular, accident victims have sometimes been given an overdose of morphine. If this is not treated, then the casualty may die of suffocation. By administering nalorphine, the antagonist displaces morphine from the receptor and binds more strongly, thus preventing morphine from continuing its action.

Naltrexone is eight times more active than naloxone as an antagonist and is given to drug addicts who have been weaned off morphine or heroin. Since naltrexone blocks the opiate receptors, it blocks the effects which the addicts might seek by restarting their habit and makes it less likely that they will do so.

There is, however, another interesting observation arising from the biological results of these antagonists. For many years, chemists had been trying to find a morphine analogue with analgesic properties, but without the depressant effects on breathing, or the withdrawal symptoms. There had been so little success that many workers believed that the two properties were directly related, perhaps through the same receptor. The fact that the antagonist naloxone blocked both the analgesia and side-effects of morphine did nothing to change that view.

However, the properties of nalorphine offered a glimmer of hope. Nalorphine is a strong antagonist and blocks morphine from its receptors. Therefore, no analgesic activity should be observed. However, a very weak analgesic activity is observed and, what is more, this analgesia appears to be free of the undesired side-effects. This was the first sign that a non-addictive, safe analgesic might be possible.

But how can this be? How can a compound be an antagonist of morphine but also act as an agonist and produce analgesia. If it is acting as an agonist, why is the activity so weak and why is it free of the side-effects?

As we shall see later, there is not one single type of analgesic receptor, but several. Multiple receptors are common (see Chapter 6). We have already seen in Chapter 15 that there are two types of cholinergic receptor – the nicotinic and muscarinic. We also saw in Chapter 16 that there are two main types of adrenergic receptor (α and β).

In the same way, there are at least three types of analgesic receptor. The differences between them are slight, such that morphine cannot distinguish between them and activates them all, but in theory it should be possible to find compounds which would be selective for one type of analgesic receptor over another. However, this is not the way that nalorphine works.

Nalorphine binds to all three types of analgesic receptor and therefore blocks morphine from all three. Nalorphine itself is unable to switch on two of the receptors and is therefore a true antagonist at these receptors. However, at the third type of receptor, nalorphine acts as a weak or partial agonist (see Section 5.8.). In other words, it has activated the receptor, but only weakly. We could imagine how this might occur if the third receptor is controlling something like an ion channel (Fig. 17.17).

Morphine is a strong agonist and interacts strongly with this receptor, leading to a change in receptor conformation which fully opens the ion channel. Ions flow in or out of the cell, resulting in the activation or deactivation of enzymes. Naloxone is a

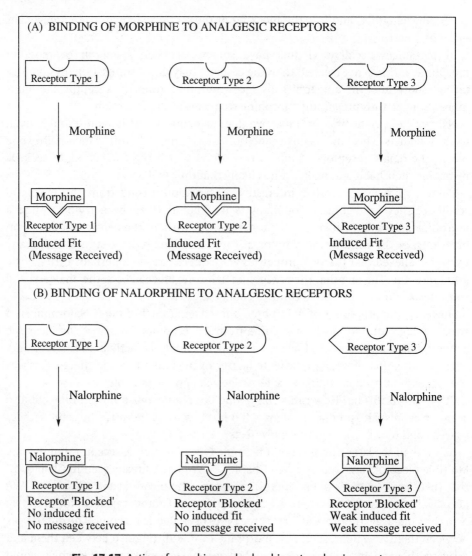

Fig. 17.17 Action of morphine and nalorphine at analgesic receptors.

pure antagonist. It binds strongly, but does not produce the correct change in the receptor conformation. Therefore, the ion channel remains closed. Nalorphine binds to the third receptor and changes the tertiary structure of the receptor very slightly, leading to a slight opening of the ion channel. It is therefore a weak agonist at this receptor, but it is also an antagonist since it blocks morphine from fully 'switching on' the receptor.

The results observed with nalorphine show that activation of this third type of analgesic receptor leads to analgesia without the undesirable side-effects associated with the other two analgesic receptors.

Unfortunately, nalorphine has hallucinogenic side-effects resulting from the activation of a non-analgesic receptor, and is therefore unsuitable as an analgesic, but for the first time a certain amount of analgesia had been obtained without the side-effects of respiratory depression and tolerance.

17.3.3 Simplification or drug dissection

We turn now to more drastic alterations of the morphine structure and ask whether the complete carbon skeleton is really necessary. After all, if we could simplify the molecule, it would be easier to make it in the laboratory. This in turn would allow the chemist to make analogues more easily, more efficiently and more cheaply.

There are five rings present in the structure of morphine (Fig. 17.18) and analogues were made to see which rings could be removed.

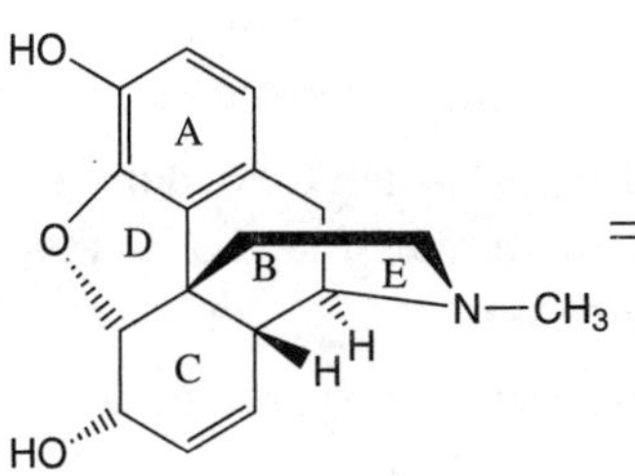
HO
A
O
D
B
E
N—CH3
C
H
H
HO
HO
A
O
D
B
C
H
HO

Fig. 17.18 Removing ring E from morphine.

Removing ring E

Removing ring E leads to a complete loss of activity. This result emphasizes the importance of the basic nitrogen to analgesic activity.

Removing ring D

Removing the oxygen bridge gives a series of compounds called the morphinans which have useful analgesic activity. This demonstrates that the oxygen bridge is not essential. Examples are shown in Fig. 17.19.

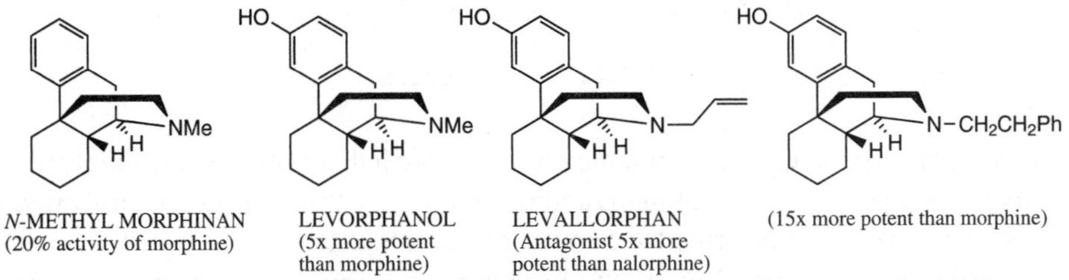
NMe
H H
HO
NMe
H H
HO
N
H H
HO
N-CH2CH2Ph
H H
N-METHYL MORPHINAN
(20% activity of morphine)
LEVORPHANOL
(5x more potent
than morphine)
LEVALLORPHAN
(Antagonist 5x more
potent than nalorphine)
(15x more potent than morphine)

Fig. 17.19 Examples of morphinans.

N-Methyl morphinan was the first such compound tested and is only 20% as active as morphine but, since the phenolic group is missing, this is not surprising. The more relevant levorphanol structure is five times more active than morphine and, although side-effects are also increased, levorphanol has a massive advantage over morphine in that it can be taken orally and lasts much longer in the body. This is because levorphanol is not metabolized in the liver to the same extent as morphine. As might be expected, the mirror image of levorphanol (dextrorphan) has insignificant analgesic activity.

The same strategy of drug extension already described for the morphine structures was also tried on the morphinans with similar results. For example, adding an allyl substituent on the nitrogen gives antagonists. Adding a phenethyl group to the nitrogen greatly increases potency. Adding a 14-OH group also increases activity.

To conclude:

• Morphinans are more potent and longer acting than their morphine counterparts, but they also have higher toxicity and comparable dependence characteristics.

• The modifications carried out on morphine, when carried out on the morphinans, lead to the same biological results. This implies that both types of molecule are binding to the same receptors in the same way.

• The morphinans are easier to synthesize since they are simpler molecules.

Removing rings C and D

Opening both rings C and D gives an interesting group of compounds called the benzomorphans (Fig. 17.20) which are found to retain analgesic activity. One of the simplest of these structures is metazocine, which has the same analgesic activity as morphine. Notice that the two methyl groups in metazocine are *cis* with respect to each other and represent the 'stumps' of the C ring.

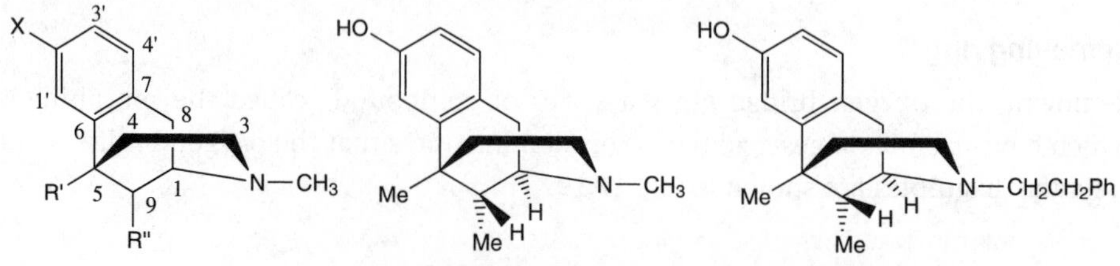

METAZOCINE
(Same potency as morphine)

PHENAZOCINE
(4x more potent than morphine)

Fig. 17.20 Benzomorphans.

If the same type of chemical modifications are carried out on the benzomorphans as were described for the morphinans and morphine, then the same biological effects are observed. This suggests that the benzomorphans interact similarly with the analgesic receptors as the morphinans and morphine analogues. For example, replacing the N-methyl group of metazocine with a phenethyl group gives phenazocine, which is four times more active than morphine and the first compound to have a useful level of analgesia without dependence properties.

Further developments led to pentazocine (Fig. 17.21), which has proved to be a useful long-term analgesic with a very low risk of addiction. A newer compound (bremazocine) has a longer duration, has 200 times the activity of morphine, appears to have no addictive properties, and does not depress breathing.

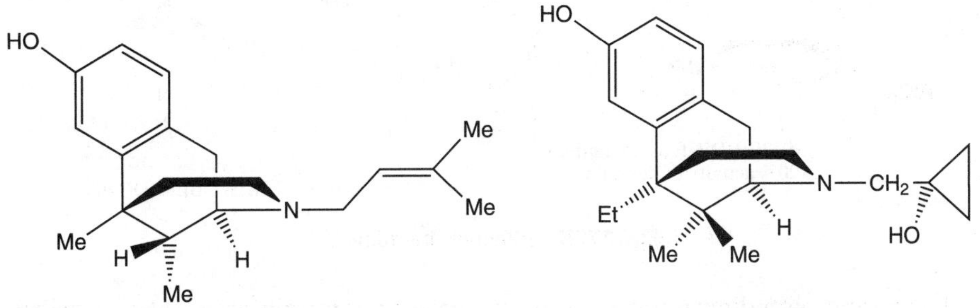

PENTAZOCINE
(33% activity of morphine,
short duration, low
addiction liability)

BREMAZOCINE

Fig. 17.21 Benzomorphans with low rates of dependency.

These compounds appear to be similar in their action to nalorphine in that they act as antagonists at two of the three types of analgesic receptors, but act as an agonist at the third. The big difference between nalorphine and compounds like pentazocine is that the latter are far stronger agonists, resulting in a more useful level of analgesia.

Unfortunately, many of these compounds have hallucinogenic side-effects because of interactions with a non-analgesic receptor.

We shall come back to the interaction of benzomorphans with analgesic receptors later. For the moment, we can make the following conclusions about benzomorphans.

- Rings C and D are not essential to analgesic activity.

- Analgesia and addiction are not necessarily co-existent.

- 6,7-Benzomorphans are clinically useful compounds with reasonable analgesic activity, less addictive liability, and less tolerance.

- Benzomorphans are simpler to synthesize.

Removing rings B, C, and D

Removing rings B, C, and D gives a series of compounds known as 4-phenyl-piperidines. The analgesic activity of these compounds was discovered by chance in the 1940's when chemists were studying analogues of cocaine for antispasmodic properties. Their structural relationship to morphine was only identified when they were found to be analgesics, and is evident if the structure is drawn as shown in Fig. 17.22.

Activity can be increased sixfold by introducing the phenolic group and altering the ester to a ketone to give ketobemidone.

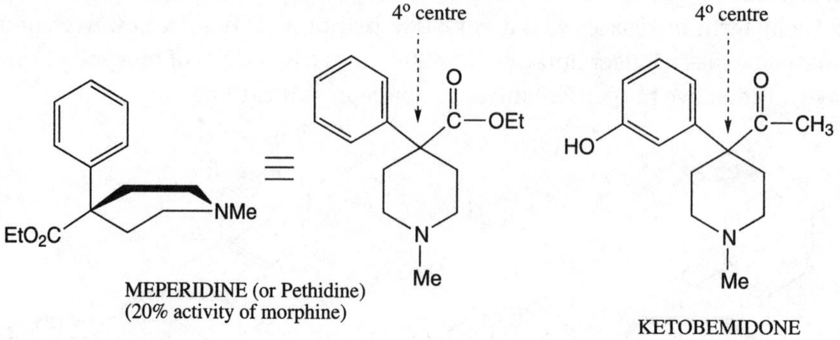

Fig. 17.22 4-Phenyl piperidines.

Meperidine (pethidine) is not as strong an analgesic as morphine and also shares the same undesirable side-effects. However, it has a rapid onset and a shorter duration, and as a result has been used as an analgesic for difficult childbirths. The rapid onset and short duration of action mean that there is less chance of the drug depressing the baby's breathing.

The piperidines are more easily synthesized than any of the previous groups and a large number of analogues have been studied. There is some doubt as to whether they act in the same way as morphine at analgesic receptors, since some of the chemical adaptations we have already described do not lead to comparable biological results. For example, adding allyl or cyclopropyl groups does not give antagonists. The replacement of the methyl group of meperidine with a cinnamic acid residue increases the activity by 30 times, whereas putting the same group on morphine eliminates activity (Fig. 17.23).

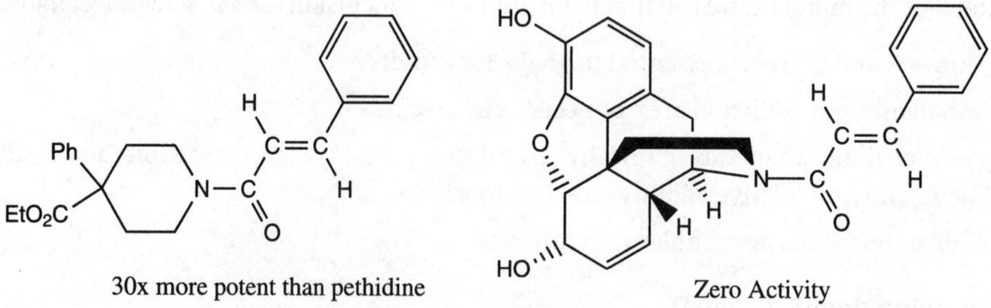

Fig. 17.23 Effect of addition of a cinnamic acid residue on meperidine and morphine.

These results might have something to do with the fact that the piperidines are far more flexible molecules than the previous structures and are thus more likely to bind with receptors in different ways.

One of the most successful piperidine derivatives is fentanyl (Fig. 17.24), which is up to 100 times more active than morphine. The drug lacks a phenolic group, but is very lipophilic. As a result, it can cross the blood–brain barrier efficiently.

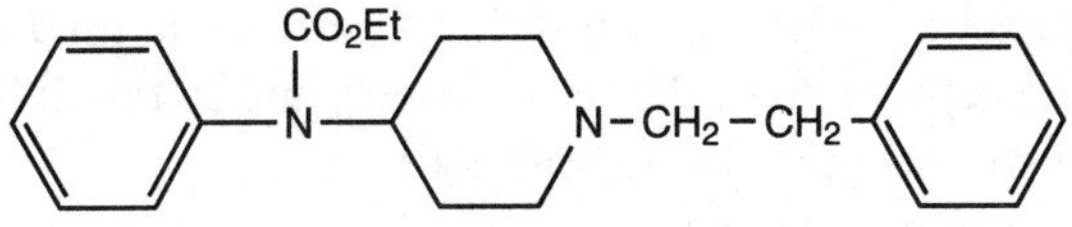

Fig. 17.24 Fentanyl.

To conclude:

- Rings C, D, and E are not essential for analgesic activity.
- Piperidines retain side-effects such as addiction and depression of the respiratory centre.
- Piperidine analgesics are faster acting and have shorter duration.
- The quaternary centre present in piperidines is usually necessary (fentanyl is an exception).
- The aromatic ring and basic nitrogen are essential to activity, but the phenol group is not.
- Piperidine analgesics appear to bind with analgesic receptors in a different manner to previous groups.

Removing rings B, C, D, and E

The analgesic methadone (Fig. 17.25) was discovered in Germany during the Second World War and has proved to be a useful agent, comparable in activity to morphine. Unfortunately, methadone retains morphine-like side-effects. However, it is orally active and has less severe emetic and constipation effects. Side-effects such as sedation, euphoria, and withdrawal are also less severe and therefore the compound has been given to drug addicts as a substitute for morphine or heroin in order to wean them off these drugs. This is not a complete cure since it merely swaps an addiction to heroin/morphine for an addiction to methadone. However, this is considered less dangerous.

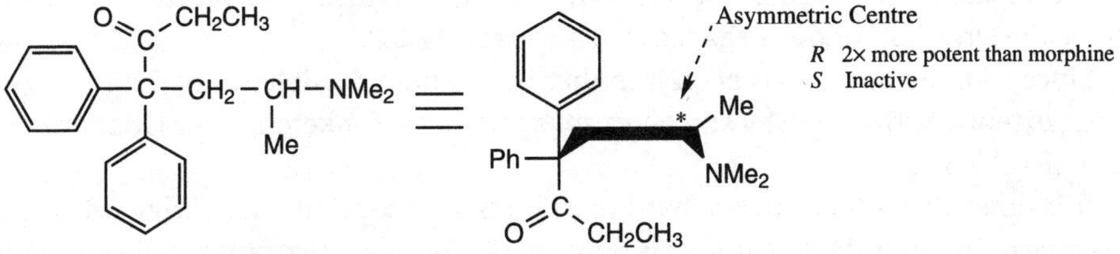

Fig. 17.25 Methadone.

The molecule has a single asymmetric centre and, when the molecule is drawn in the same manner as morphine, we would expect the R-enantiomer to be more active. This proves to be the case, with the R-enantiomer being twice as powerful as morphine, whereas the S-enantiomer is inactive. This is quite a dramatic difference. Since the R- and S-enantiomers have identical physical properties and lipid solubility, they should both reach the receptor site to the same extent, and so the difference in activity is most probably due to receptor–ligand interactions.

Many analogues of methadone have been synthesized, but with little improvement over the parent drug.

17.3.4 Rigidification

Up until now, we have considered minor adjustments of functional groups on the periphery of the morphine skeleton or drastic simplification of the morphine skeleton.

A completely different strategy is to make the molecule more complicated or more rigid. This strategy is usually employed in an attempt to remove the side-effects of a drug or to increase activity.

It is usually assumed that the side-effects of a drug are due to interactions with additional receptors other than the one in which we are interested. These interactions are probably because of the molecule taking up different conformations or shapes. If we make the molecule more rigid so that it takes up fewer conformations, we might eliminate the conformations which are recognized by undesirable receptors, and thus restrict the molecule to the specific conformation which fits the desired receptor. In this way, we would hope to eliminate such side-effects as dependence and respiratory depression. We might also expect increased activity since the molecule is more likely to be in the correct conformation to interact with the receptor.

The best example of this tactic in the analgesic field is provided by a group of compounds known as the oripavines. These structures often show remarkably high activity.

The oripavines are made from an alkaloid which we have not described so far – thebaine (Fig. 17.26). Thebaine can be extracted from opium along with codeine and morphine, and is very similar in structure to both these compounds. However, unlike morphine and codeine, thebaine has no analgesic activity. There is a diene group present in ring C and, when thebaine reacts with methyl vinyl ketone, a Diels Alder reaction takes place to give an extra ring and increased rigidity to the structure (Fig. 17.26).

A comparison with morphine shows that the extra ring sticks out from what used to be the 'crossbar' of the T-shaped structure (Fig. 17.27).

Since a ketone group has been introduced, it is now possible to try the strategy of drug extension, this time by adding various groups to the ketone via a Grignard reaction (Fig. 17.28).

It is noteworthy that the Grignard reaction is stereospecific. The Grignard reagent complexes to both the 6-methoxy group and the ketone, and is then delivered to the less hindered face of the ketone to give an asymmetric centre (Fig. 17.29).

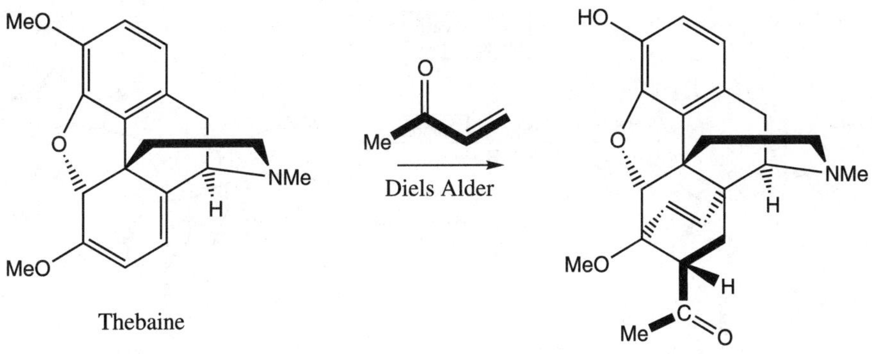

Fig. 17.26 Formation of oripavines.

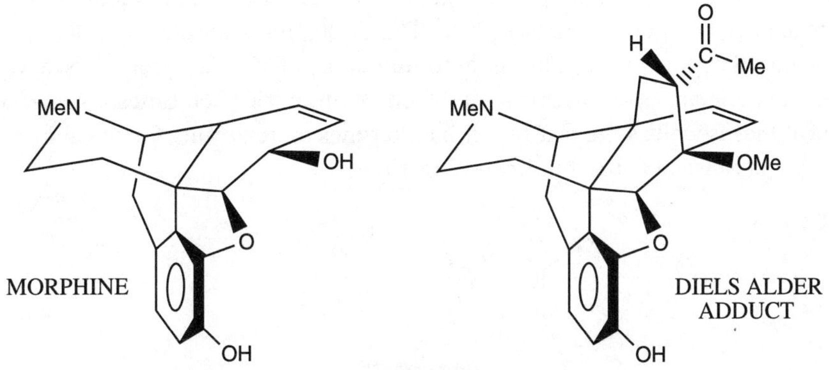

Fig. 17.27 Comparison of morphine and oripavine.

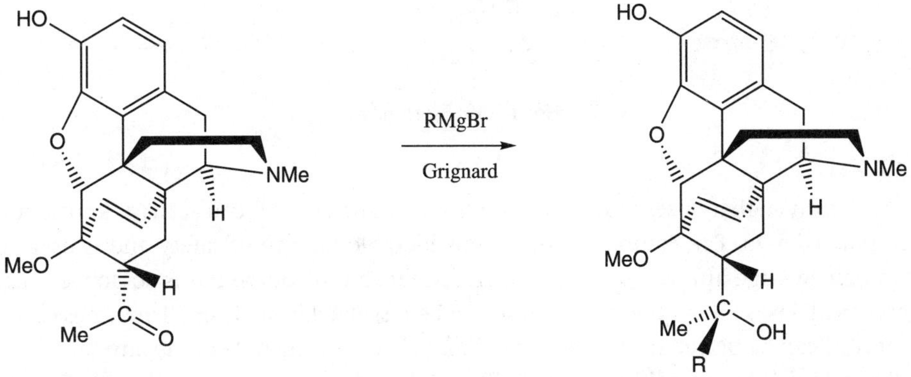

Fig. 17.28 Drug extension.

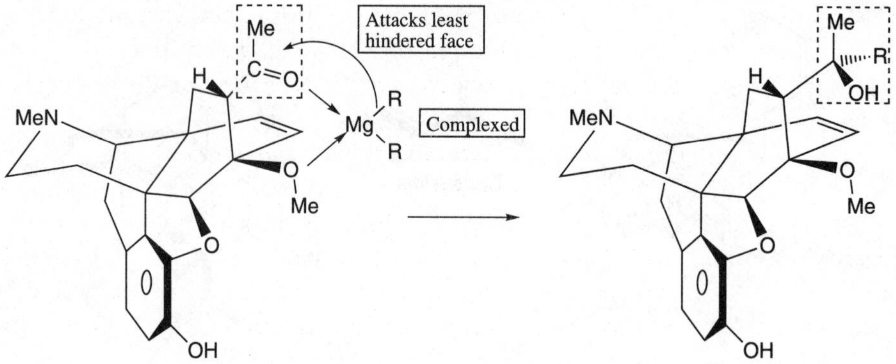

Fig. 17.29 Grignard reaction leads to an asymmetric centre.

By varying the groups added by the Grignard reaction, some remarkably powerful compounds have been obtained. Etorphine (Fig. 17.30), for example, is 10 000 times more potent than morphine. This is a combination of the fact that it is a very hydrophobic molecule and can cross the blood–brain barrier 300 times more easily than morphine, as well as the fact that it has 20 times more affinity for the analgesic receptor site because of better binding interactions.

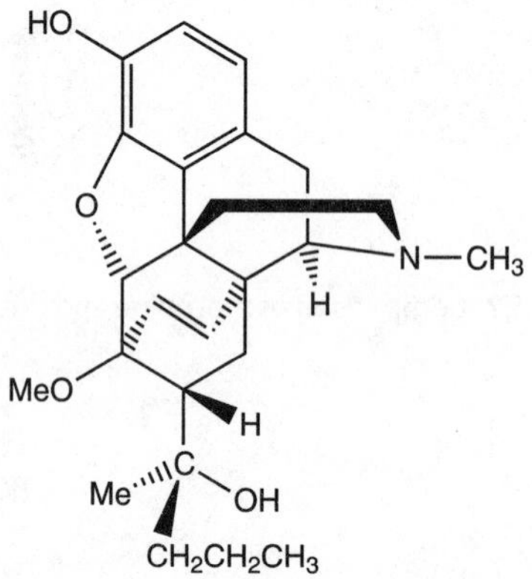

Fig. 17.30 Etorphine.

At slightly higher doses than those required for analgesia, it can act as a 'knock-out' drug or sedative. The compound has a considerable margin of safety and is used to immobilize large animals such as elephants. Since the compound is so active, only very small doses are required and these can be dissolved in such small volumes (1 ml) that they can be placed in crossbow darts and fired into the hide of the animal.

The addition of lipophilic groups (R; Fig. 17.29) is found to improve activity dramatically, indicating the presence of a hydrophobic binding region close by on the receptor[2]. The group best able to interact with this region is a phenethyl substituent, and the product containing this group is even more active than etorphine.

As one might imagine, these highly active compounds have to be handled very carefully in the laboratory.

Because of their rigid structures, these compounds are highly selective agents for the analgesic receptors. Unfortunately, the increased analgesic activity is also accompanied by unacceptable side-effects. It was therefore decided to see whether N-substituents such as an allyl or cyclopropyl group, would give antagonists as found in the morphine, morphinan, and benzomorphan series of compounds. If so, it might be possible to obtain an oripavine equivalent of a pentazocine or a nalorphine – an antagonist with some agonist activity and with reduced side-effects.

Adding a cyclopropyl group gives a very powerful antagonist called diprenorphine (Fig. 17.31), which is 100 times more potent than nalorphine and can be used to reverse the immobilizing effects of etorphine (see above). Diprenorphine has no analgesic activity.

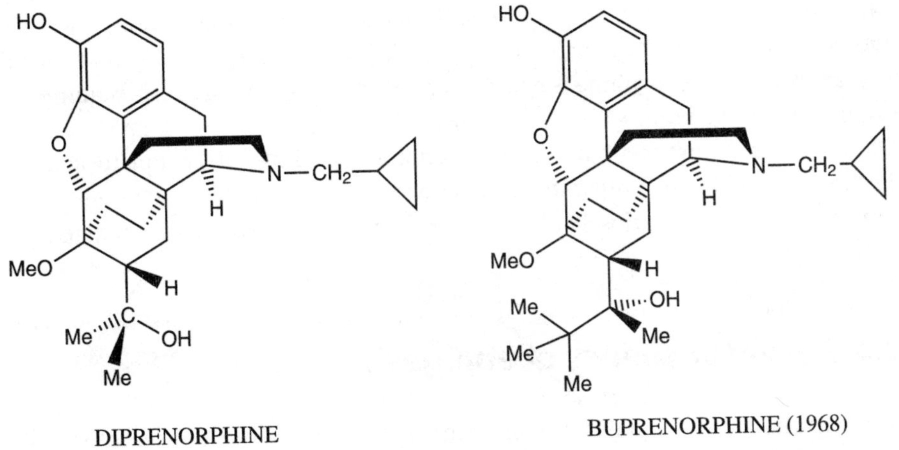

DIPRENORPHINE   BUPRENORPHINE (1968)

Fig. 17.31

Replacing the methyl group derived from the Grignard reagent with a *tert*-butyl group gives buprenorphine (Fig. 17.31), which has similar properties to drugs like nalorphine and pentazocine in that it has analgesic activity with a very low risk of addiction. This feature appears to be related to the slow onset and removal of buprenorphine from the analgesic receptors. Since these effects are so gradual, the receptor system is not subjected to sudden changes in transmitter levels.

Buprenorphine is the most lipophilic compound in the oripavine series of compounds and therefore enters the brain very easily. Usually, such a drug would react quickly with its receptor. The fact that it does not is therefore a feature of its

---

[2] It is believed that the phenylalanine aromatic ring on enkephalins (see later) interacts with this same binding region.

interaction with the receptor rather than the ease with which it can reach the receptor. It is 100 times more active than morphine as an agonist and four times more active than nalorphine as an antagonist. It is a particularly safe drug since it has very little effect on respiration and what little effect it does have actually decreases at high doses. Therefore, the risks of suffocation from a drug overdose are much smaller than with morphine. Buprenorphine has been used in hospitals to treat patients suffering from cancer and also following surgery. Its drawbacks include side-effects such as nausea and vomiting, as well as the fact that it cannot be taken orally. A further use for buprenorphine is as an alternative means to methadone for weaning addicts off heroin.

Buprenorphine binds slowly to analgesic receptors but, once it does bind, it binds very strongly. As a result, less buprenorphine is required to interact with a certain percentage of analgesic receptors than morphine. On the other hand, buprenorphine is only a partial agonist. In other words, it is not very efficient at switching the analgesic receptor on. This means that it is unable to reach the maximum level of analgesia which can be acquired by morphine. Overall, buprenorphine's stronger affinity for analgesic receptors outweighs its relatively weak action, such that buprenorphine can produce analgesia at lower doses than morphine. However, if pain levels are high, buprenorphine is unable to reach the levels of analgesia required and morphine has to be used.

Nevertheless, buprenorphine provides another example of an opiate analogue where analgesia has been separated from dangerous side-effects.

It is time to look more closely at the receptor theories relevant to the analgesics.

17.4 Receptor theory of analgesics

Although it has been assumed for many years that there are analgesic receptors, information about them has only been gained relatively recently (1973). There are at least four different receptors with which morphine can interact, three of which are analgesic receptors. The initial theory on receptor binding (the Beckett–Casy hypothesis) assumed a single analgesic receptor, but this does not invalidate many of the proposals which were made.

17.4.1 Beckett–Casy hypothesis

In this theory, it was assumed that there was a rigid binding site and that morphine and its analogues fitted into the site in a classic lock and key analogy. Based on the results already described, the following features were proposed as being essential if an analgesic was to interact with its receptor (Fig. 17.32).

• There must be a basic centre (nitrogen) which can be ionized at physiological pH to form a positively charged group. This group then forms an ionic bond with a comparable anionic group in the receptor. As a consequence of this, analgesics have to have a pK_a of 7.8–8.9 such that there is an approximately equal chance of the amine being ionized or un-ionized at physiological pH. This is necessary since the analgesic has to cross the blood–brain barrier as the free base, but once across has to be ionized to interact with the receptor. The pK_a values of useful analgesics all match this prediction.

• The aromatic ring in morphine has to be properly orientated with respect to the nitrogen atom to allow a van der Waals interaction with a suitable hydrophobic location on the receptor. The nature of this interaction suggests that there has to be a close spatial relationship between the aromatic ring and the surface of the receptor.

• The phenol group is probably hydrogen bonded to a suitable residue at the receptor site.

• There might be a 'hollow' just large enough for the ethylene bridge of carbons 15 and 16 to fit. Such a fit would help to align the molecule and enhance the overall fit.

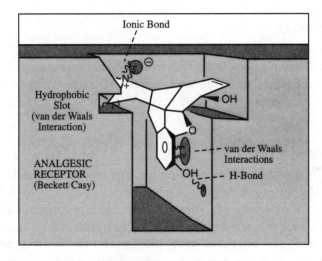

Fig. 17.32 Beckett–Casy hypothesis.

This first theory fitted in well with the majority of results. There can be no doubt that the aromatic ring, phenol, and the nitrogen groups are all important, but there is some doubt as to whether the ethylene bridge is important, since there are several analgesics which lack it (e.g. fentanyl).

The theory also fails to include the extra binding region which was discovered by drug extension. The theory can easily be adapted to allow for this, but other anomalies exist which have already been discussed (e.g. the different results obtained for meperidine compared to morphine when a substituent such as the allyl group is attached to nitrogen). These results strongly suggested that a simple one-receptor theory was not applicable.

17.4.2 Multiple analgesic receptors

The Beckett–Casy theory tried to explain analgesic results based on a single analgesic receptor. It is now known that there are three different analgesic receptors which are associated with different types of side-effects. It is also known that several analgesics show preference for some of these receptors over others. This helps to explain some of the aforementioned anomalies. However, it is important to appreciate that the main points of the original theory still apply for each of the analgesic receptors now to be described. The important binding groups for each receptor are the phenol, the aromatic ring, and the ionized nitrogen centre. Beyond that, there are subtle differences between each receptor which can distinguish between the finer details of different analgesic molecules. As a result, some analgesics show preference for one analgesic receptor over another or interact in different ways.

There are three analgesic receptors which are activated by morphine, and which have been labelled with Greek letters.

The mu receptor (μ)

Morphine binds strongly to this receptor and produces analgesia. Receptor binding also leads to the undesired side-effects of respiratory depression, euphoria, and addiction. We can now see why it is so difficult to remove the side-effects of morphine and its analogues, since the receptor with which they bind most strongly is also inherently involved with these side-effects.

The kappa receptor (κ)

Morphine binds less strongly to this receptor. The biological response is analgesia with sedation and none of the hazardous side-effects. It is this receptor which provides the best hope for the ultimate safe analgesic. The earlier results obtained from nalorphine, pentazocine, and buprenorphine can now be explained.

Nalorphine acts as an antagonist at the μ receptor, thus blocking morphine from acting there. However, it acts as a weak agonist at the κ receptor (as does morphine) and so the slight analgesia observed with nalorphine is due to the partial activation of the κ receptor. Unfortunately, nalorphine has hallucinogenic side-effects. This is caused by nalorphine also binding to a completely different, non-analgesic receptor in the brain called the sigma receptor (σ) (see Section 17.7.4.) where it acts as an agonist.

Pentazocine interacts with the μ and κ receptors in the same way, but is able to 'switch on' the κ receptor more strongly. It too suffers the drawback that it 'switches on' the σ receptor. Buprenorphine is slightly different. It binds strongly to all three analgesic receptors and acts as an antagonist at the δ (see below) and κ receptors, but acts as a partial agonist at the μ receptor to produce its analgesic effect. This might suggest that buprenorphine should suffer the same side-effects as morphine. The fact that it does not is related in some way to the rate at which buprenorphine interacts with the receptor. It is slow to bind but, once it has bound, it is slow to leave.

The delta receptor (δ)

The δ receptor is where the brain's natural painkillers (the enkephalins; see Section 17.6) interact. Morphine can also bind quite strongly to this receptor.

Table 17.1 shows the relative activities of morphine, nalorphine, pentazocine, enkephalins, pethidine, and naloxone. A plus sign indicates that the compound is acting as an agonist. A minus sign means that it acts as an antagonist. A zero sign means that there is no activity or minor activity.

Table 17.1 Relative activities of analgesics.

		Morphine	Nalorphine	Pentazocine	Enkephalins	Pethidine	Naloxone
Mu	Analgesia respiratory depression euphoria addiction	+++	–	–	+	+++	– – –
Kappa	Analgesia sedation	+	+	+	+	+	–
Sigma	Psychotomimetic	0	+	+	0	0	0
Delta	Analgesia	++	–	–	+++	+	–

+, Componds acting as agonists; –, as antagonists. 0, No activity or minor activity.

There is now a search going on for orally active opiate structures which can act as antagonists at the μ receptor, agonists at the κ receptor, and have no activity at the σ receptor. Some success has been obtained, especially with the compounds shown in Fig. 17.33, but even these compounds still suffer from side-effects, or lack the desired oral activity.

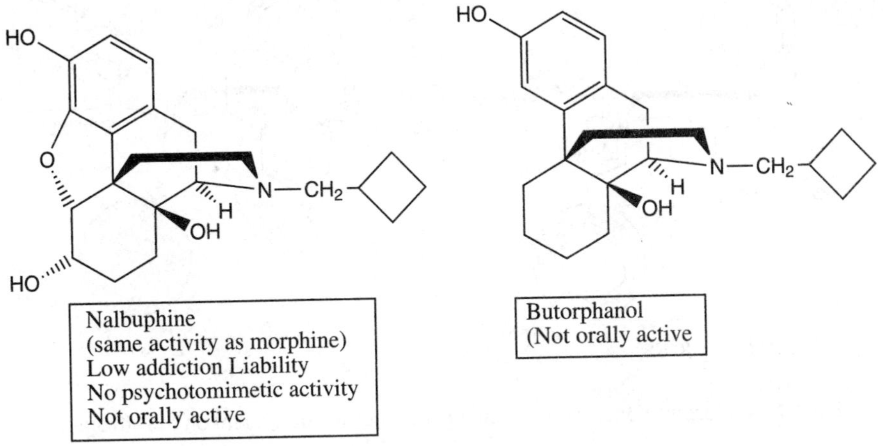

Nalbuphine
(same activity as morphine)
Low addiction Liability
No psychotomimetic activity
Not orally active

Butorphanol
(Not orally active

Fig. 17.33 New analgesic structures.

17.5 Agonists and antagonists

We return now to look at a particularly interesting problem regarding the agonist/antagonist properties of morphine analogues. Why should such a small change as replacing an *N*-methyl group with an allyl group result in such a dramatic change in biological activity such that an agonist becomes an antagonist? Why should a molecule such as nalorphine act as an agonist at one analgesic receptor and an antagonist at another? How can different receptors distinguish between such subtle changes in a molecule?

We shall consider one theory which attempts to explain how these distinctions might take place, but it is important to realize that there are alternative theories. In this particular theory, it is suggested that there are two accessory hydrophobic binding regions present in an analgesic receptor. It is then proposed that a structure will act as an agonist or as an antagonist depending on which of these extra binding regions is used. In other words, one of the hydrophobic binding regions is an agonist binding region, whereas the other is an antagonist binding region.

The model was proposed by Snyder and co-workers[3] and is shown in Figs. 17.34–17.36). In the model, the agonist binding region is further away from the nitrogen and is positioned axially with respect to it. The antagonist region is closer and positioned equatorially.

Let us now consider the morphine analogue containing a phenethyl substituent on the nitrogen (Fig. 17.34). It is proposed that this structure binds as already described, such that the phenol, aromatic ring, and basic centre are interacting with their respective binding regions. If the phenethyl group is in the axial position, the aromatic ring is in the correct position to interact with the agonist binding region. However, if the

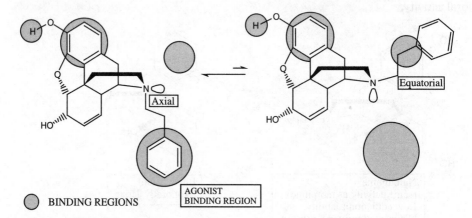

Fig. 17.34 Morphine analogue containing a phenethyl substituent on the nitrogen.

[3] Feinberg, A.P., Creese, I., and Snyder, S.H. (1976). *Proc. Natl. Acad. Sci. USA*, 73, 4215.

phenethyl group is in the equatorial position, the aromatic ring is placed beyond the antagonist binding region and cannot bind. The overall result is increased activity as an agonist.

Let us now consider what happens if the phenethyl group is replaced with an allyl group (Fig. 17.35). In the equatorial position, the allyl group is able to bind strongly to the antagonist binding region, whereas in the axial position it barely reaches the agonist binding region, resulting in a weak interaction.

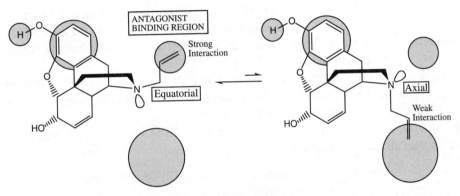

Fig. 17.35 Morphine analogue containing an allyl substituent.

In this theory, it is proposed that a molecule such as phenazocine (with a phenethyl group) acts as an agonist since it can only bind to the agonist binding region. A molecule such as nalorphine (with an allyl group) can bind to both agonist and antagonist regions and therefore acts as an agonist at one receptor and an antagonist at another. The ratio of these effects would depend on the relative equilibrium ratio of the axial and equatorial substituted isomers.

A compound which is a pure antagonist would be forced to have a suitable substituent in the equatorial position. It is believed that the presence of a 14-OH group sterically hinders the isomer with the axial substituent, and forces the substituent to remain equatorial (Fig. 17.36).

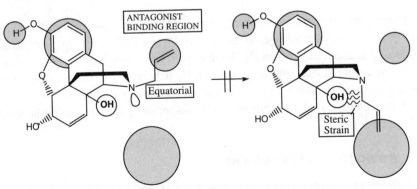

Fig. 17.36 Influence of 14-OH on binding interactions.

17.6 Enkephalins and endorphins

17.6.1 Naturally occurring enkaphalins and endorphins

Morphine, as we have already discussed, is an alkaloid which relieves pain and acts in the CNS. There are two conclusions which can be drawn from this. The first is that there must be analgesic receptors in the CNS. The second conclusion is that there must be chemicals produced in the body which interact with these receptors. Morphine itself is not produced by humans and therefore the body must be using a different chemical as its natural painkiller.

The search for this natural analgesic took many years, but ultimately led to the discovery of the enkephalins and the endorphins. The term enkephalin is derived from the Greek, meaning 'in the head', and that is exactly where the enkephalins are produced. The first enkephalins to be discovered were the pentapeptides Met-enkephalin and Leu-enkephalin.

H-Tyr-Gly-Gly-Phe-Met-OH   H-Tyr-Gly-Gly-Phe-Leu-OH

At least 15 endogenous peptides have now been discovered, varying in length from 5 to 33 amino acids (the enkephalins and the endorphins). These compounds are thought to be neurotransmitters or neurohormones in the brain, and operate as the body's natural painkillers as well as having a number of other roles. They are derived from three inactive precursor proteins – proenkephalin, prodynorphin, or pro-opiomelanocortin (Fig. 17.37).

Proenkephalin
Prodynorphin   Endorphins + Enkephalins
Pro-opiomelanocortin

Fig. 17.37 Production of the body's natural painkillers.

All 15 compounds are found to have either the Met- or the Leu-enkephalin skeleton at their N-terminus, which emphasizes the importance of this pentapeptide structure towards analgesic activity. It has also been shown conclusively that the tyrosine part of these molecules is essential to activity and much has been made of the fact that there is a tyrosine skeleton in the morphine skeleton (Fig. 17.38). It is interesting to note that enkephalins are thought to be responsible for the analgesia resulting from acupuncture.

17.6.2 Analogues of enkephalins

SAR studies on the enkephalins have shown the importance of the tyrosine phenol ring and the tyrosine amino group. Without either, activity is lost. If tyrosine is

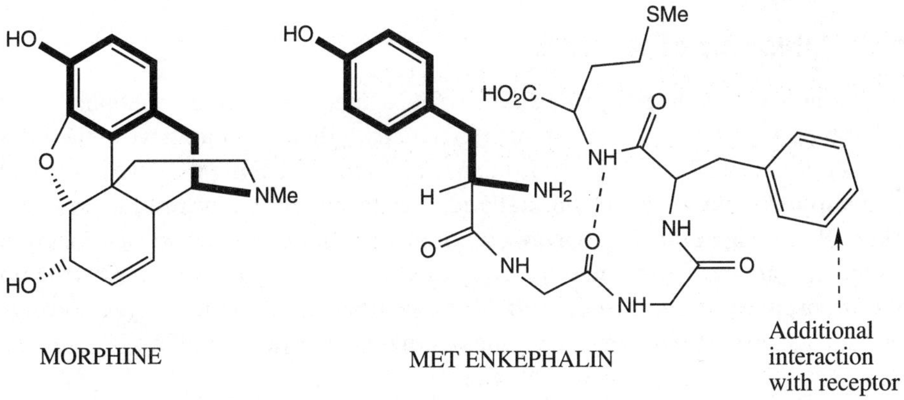

Fig. 17.38 The tyrosine section is essential to activity.

replaced with another amino acid, then activity is also lost (the only exception being D-serine). It has also been found that the enkephalins are easily inactivated by peptidase enzymes *in vivo*. The most labile peptide bond in the enkephalins is that between the tyrosine and glycine residues.

Much work has been done, therefore, to try and stabilize this bond towards hydrolysis. It is possible to replace the amino acid glycine with an unnatural D-amino acid such as D-alanine. Since D-amino acids do not occur naturally, peptidases do not recognize the structure and the peptide bond is not attacked. The alternative tactic of replacing L-tyrosine with D-tyrosine is not possible, since this completely alters the relative orientation of the tyrosine aromatic ring with respect to the rest of the molecule. As a result, the analogue is unable to bind to the analgesic receptor and is inactive.

Putting a methyl group on to the amide nitrogen can also block hydrolysis by peptidases. Another tactic is to use unusual amino acids which are either not recognized by peptidases or which prevent the molecule from fitting the peptidase active site. Examples of these tactics at work are demonstrated in Fig. 17.39. Unfortunately, the enkephalins also have some activity at the μ receptor and so the search for selective agents continues.

H—L-Tyr—Gly—Gly——L-Phe—L-Met–OH	Delta Agonist + a little mu activity
H—L-Tyr—D-AA—Gly——NMe-L-Phe—L-Met–OH	Resistant to peptidase. Orally active.
N,N-Diallyl-L-Tyr—aib——aib——L-Phe—L-Leu–OH	Antagonist to delta receptor. (aib = alpha-aminobutyric acid)
Longer enkaphalins/endorphins	Increase in kappa activity Slight increase in mu activity

Fig. 17.39 Tactics to stabilize the bond between the tyrosine and glycine residues.

17.6.3 Inhibitors of peptidases

An alternative approach is to enhance the activity of natural enkephalins by inhibiting the peptidase enzyme which metabolizes them. Studies have shown that the enzyme responsible for metabolism has a zinc ion present in the active site as well as a hydrophobic pocket which normally accepts the phenylalanine residue present in enkephalins. A dipeptide (Phe–Gly) was chosen as a lead compound and a thiol group was incorporated to act as a binding group for the zinc ion. The result was a structure called thiorphan (Fig. 17.40), which was shown to produce analgesic activity. However, it remains to be seen whether agents such as these will prove useful in the clinic.

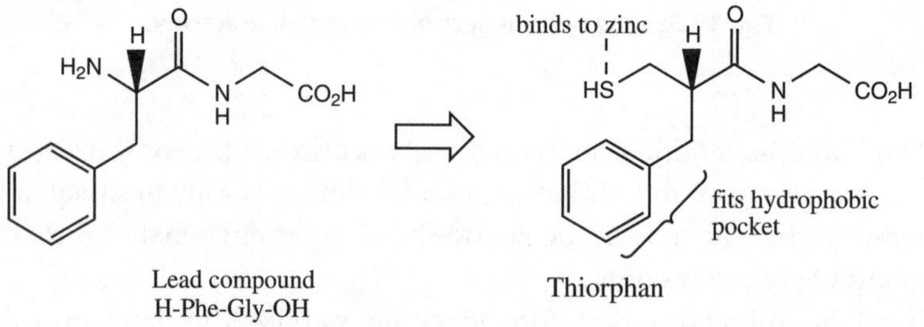

Lead compound
H-Phe-Gly-OH

Thiorphan

Fig. 17.40 Thiorphan.

17.7 Receptor mechanisms

Up until now we have discussed receptors very much as 'black boxes'. The substrate comes along, binds to the receptor, and switches it on. There is a biological response, be it analgesia, sedation, or whatever, but we have given no indication of how this response takes place. Why should morphine cause analgesia just by attaching itself to a receptor protein?

In general, all receptors in the body are situated on the surface of cells and act as communication centres for the various messages being sent from one part of the body to another. The message may be sent through nerves or via hormones, but ultimately the message has to be delivered from one cell to another by a chemical messenger. This chemical messenger has to 'dock' with the receptor which is waiting for it. When it does so, it forces the receptor to change shape. This change in shape of the receptor molecule may force a change in the shape of some neighbouring protein or perhaps an ion channel, resulting in an alteration of ion flows in and out of the cell. Such effects will ultimately have a biological effect, dependent on the cells affected.

We shall now look at the analgesic receptors in a little more detail.

17.7.1 The mu receptor (μ)

As the diagram in Fig. 17.41 demonstrates, morphine binds to the μ receptor and induces a change in shape. This change in conformation opens up an ion channel in the cell membrane and, as a result, potassium ions can flow out of the cell. This flow hyperpolarizes the membrane potential and makes it more difficult for an active potential to be reached (see Appendix 2). Therefore, the frequency of action potential firing is decreased, which results in a decrease in neurone excitability.

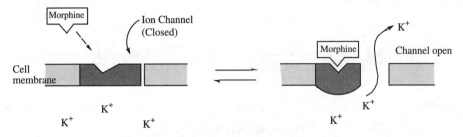

Fig. 17.41 Morphine binding to μ receptor.

This increase in potassium permeability has another indirect effect, since it also decreases the influx of calcium ions into the nerve terminal and this in turn reduces neurotransmitter release.

Both effects, therefore, 'shut down' the nerve and block the pain messages.

Unfortunately, this receptor is also associated with the hazardous side-effects of narcotic analgesics. There is still a search to see if there are possibly two slightly different μ receptors, one which is solely due to analgesia and one responsible for the side-effects.

17.7.2 The kappa receptor (κ)

The κ receptor is directly associated with a calcium channel (Fig. 17.42). When an agonist binds to the κ receptor, the receptor changes conformation and the calcium channel (normally open when the nerve is firing and passing on pain messages) is closed. Calcium is required for the production of the nerve's neurotransmitters and therefore the nerve is shut down and cannot pass on pain messages.

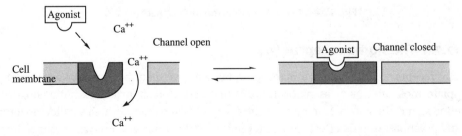

Fig. 17.42 The association of the κ receptor and calcium channels.

The nerves affected by the κ mechanism are those related to pain induced by non-thermal stimuli. This is not the case with the μ receptor, where all pain messages are inhibited. This suggests a different distribution of κ receptors from μ receptors.

17.7.3 The delta receptor (δ)

Like the μ receptor, the nerves containing the δ receptor do not discriminate between pain from different sources.

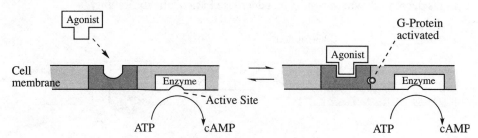

Fig. 17.43 The delta receptor.

The δ receptor is a G-protein-linked receptor (Fig. 17.43) (see Chapter 6). When an agonist binds, the receptor changes shape and triggers a messenger protein (the G-protein) to carry a message to a neighbouring enzyme which catalyses the formation of cyclic adenosine monophosphate (cAMP; Fig. 17.44). The G-protein inactivates the enzyme, preventing the synthesis of cyclic AMP. The transmission of the pain signal requires cyclic AMP to act as a secondary messenger, and so inactivating the enzyme stops the pain transmission.

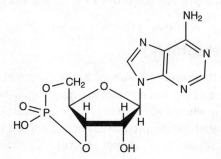

Fig. 17.44 Cyclic adenosine monophosphate (cAMP).

17.7.4 The sigma receptor (σ)

This receptor is not an analgesic receptor, but we have seen that it can be activated by opiate molecules such as nalorphine. When activated, it produces hallucinogenic effects, and the σ receptor may be the receptor associated with the hallucinogenic and psychotomimetic effects of phencyclidine (PCP), otherwise known as 'angel dust' (Fig. 17.45).

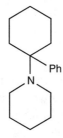

Fig. 17.45 Phencyclidine (PCP) or 'angel dust'.

17.8 The future

There is still a need for analgesic drugs with reduced side-effects. Four approaches are feasible in the field of opiates.

κ Agonists

Such compounds should have much reduced side-effects. However, a completely specific κ agonist has not yet been found. Examples of selective agonists are shown in Fig. 17.46.

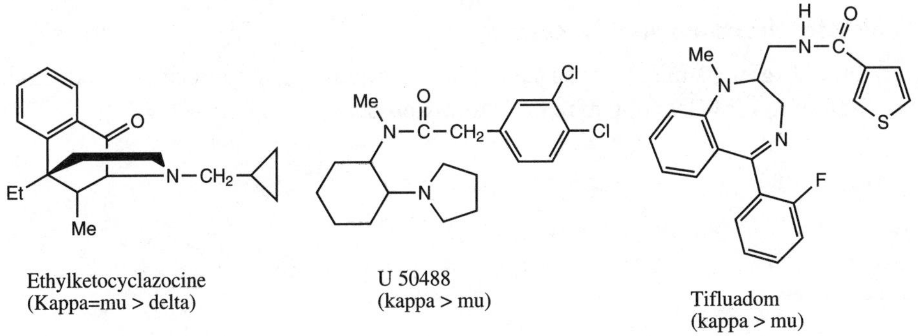

Ethylketocyclazocine
(Kappa=mu > delta)

U 50488
(kappa > mu)

Tifluadom
(kappa > mu)

Fig. 17.46 Examples of selective agonists.

Selectivity between μ receptor subtypes

There might be two slightly different μ receptors, one of which is purely responsible for analgesia (μ_1) and the other solely responsible for unwanted side-effects such as respiratory depression (μ_2). An agent showing selectivity would prove such a theory and be a very useful analgesic.

Peripheral opiate receptors

Peripheral opiate receptors have been identified in the ileum and are responsible for the antidiarrhoeal activity of opiates. If peripheral sensory nerves also possess opiate receptors, drugs might be designed versus these sites and as a result would not need to cross the blood–brain barrier.

Blocking postsynaptic receptors

The actions of the analgesic receptors modify the transmission of pain in sensory nerves, but these are not the receptors involved in the transmission of the pain signal itself. Different receptors are involved which interact with specific chemical messengers totally unrelated to the opiates. Blocking the postsynaptic receptors which are responsible for the transmission of pain with selective antagonists may well be the best approach to treating pain and the best way of eliminating side-effects. One promising lead is provided by the neurotransmitter γ-aminobutanoic acid (GABA; Fig. 17.47), which appears to have a role in the regulation of enkephalinergic neurons and, as such, affects pain pathways. Another chemical messenger involved in the role of pain mediation is an undecapeptide structure called substance P which is an excitatory neurotransmitter.

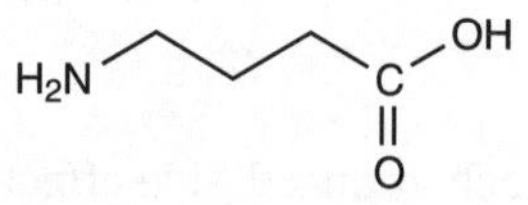

Fig. 17.47 γ-Aminobutanoic acid (GABA).

Agonists for the cannabinoid receptor

Cannabinoid agonists may have a role to play in enhancing the effects of opiate analgesics and may allow less opiate to be administered.

18 Cimetidine – a rational approach to drug design

18.1 Introduction

Many of the past successes in medicinal chemistry have involved the fortuitous discovery of useful pharmaceutical agents from natural sources such as plants or microorganisms. Analogues of these structures were then made in an effort to improve activity and/or to reduce side-effects, but often these variations were carried out on a trial-and-error basis. While this approach yielded a large range of medicinal compounds, it was wasteful with respect to the time and effort involved.

In the last twenty to thirty years, greater emphasis has been placed on rational drug design whereby drugs are designed to interact with a known biological system. For example, when looking for an enzyme inhibitor, a rational approach is to purify the enzyme and to study its tertiary structure by X-ray crystallography. If the enzyme can be crystallized along with a bound inhibitor, then the researcher can identify and study the binding site of the enzyme. The X-ray data can be read into a computer and the binding site studied to see whether new inhibitors can be designed to fit more strongly.

However, it is not often possible to isolate and purify enzymes, and when it comes to membrane-bound receptors, the difficulties becomes even greater. Nevertheless, rational drug design is still possible, even when the receptor cannot be studied directly. One of the early examples of the rational approach to drug design was the development of the anti-ulcer drug cimetidine (Tagamet) (Fig. 18.1), carried out by scientists at Smith, Kline, & French (SK&F).

The remarkable aspect of the cimetidine story lies in the fact that at the onset of the project there were no lead compounds and it was not even known if the necessary receptor protein even existed!

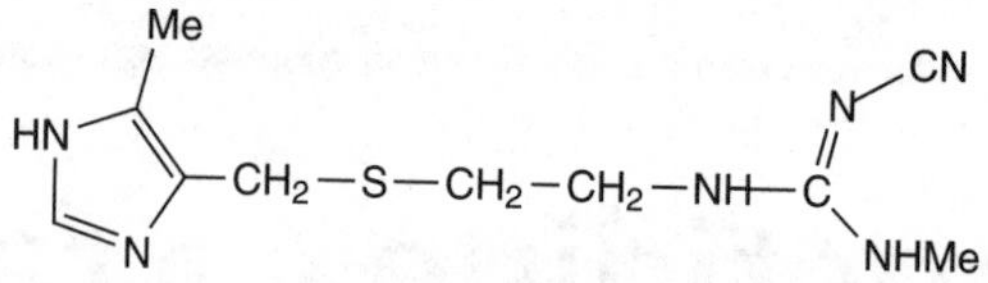

Fig. 18.1 Cimetidine.

18.2 In the beginning – ulcer therapy in 1964

When the cimetidine programme started in 1964, the methods available for treating peptic ulcers were few and generally unsatisfactory.

Ulcers are localized erosions of the mucous membranes of the stomach or duodenum. It is not known how these ulcers arise, but the presence of gastric acid aggravates the problem and delays recovery. Ulcer sufferers often suffered intense pain for many years, and if left untreated, the ulcer could result in severe bleeding and even death. For example, the film star Rudolph Valentino died in 1926 at the age of 31 from a perforated ulcer.

In the early 1960s, the conventional treatment was to try and neutralize gastric acid in the stomach by administering bases such as sodium bicarbonate or calcium carbonate. However, the dose levels required for neutralization were large and caused unpleasant side-effects. Relief was also only temporary and patients were often advised to stick to rigid diets such as strained porridge and steamed fish. Ultimately, the only answer was surgery to remove part of the stomach. It was reasoned that a better approach would be to inhibit the release of gastric acid at source.

Gastric acid (HCl) is released by cells known as parietal cells in the stomach (Fig. 18.2). These parietal cells are innervated with nerves (not shown on the diagram) from the autonomic nervous system (see Chapter 15). When the autonomic nervous system is stimulated, a signal is sent to the parietal cells culminating in the release of the neurotransmitter acetylcholine at the nerve termini. Acetylcholine crosses the gap between nerve and parietal cell and activates the cholinergic receptors of the parietal cells leading to the release of gastric acid into the stomach. The trigger for this process is provided by the sight, smell, or even the thought, of food. Thus, gastric acid is released before food has even entered the stomach.

Nerve signals also stimulate a region of the stomach known as the antrum which contains hormone-producing cells known as G cells. The hormone released is a peptide called gastrin (Fig. 18.3) which is also released when food is present in the stomach. The gastrin moves into the blood supply and travels to the parietal cells further stimulating the release of gastric acid. Release of gastric acid should therefore be inhibited by antagonists blocking either the acetylcholine receptor or the receptor for gastrin.

Agents which block the acetylcholine receptor are known as anticholinergic drugs (see Chapter 15). These agents certainly block the cholinergic receptor in parietal cells

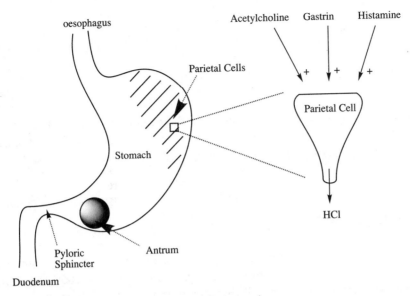

Fig. 18.2 The stomach.

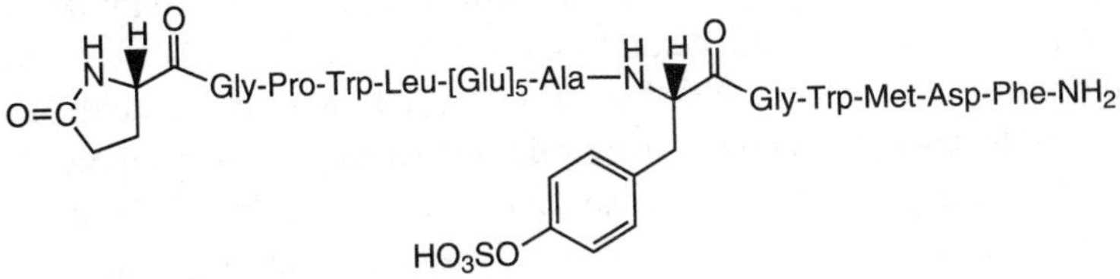

Fig. 18.3 Gastrin.

and inhibit release of gastric acid. Unfortunately, they also inhibit acetylcholine receptors at other parts of the body and cause unwanted side-effects.

Therefore, in 1964, the best hope of achieving an antiulcer agent appeared to be in finding a drug which would block the hormone gastrin. Several research teams were active in this field, but the research team at SK&F decided to follow a different tack altogether.

It was known that histamine (Fig. 18.4) could also stimulate gastric acid release, and it was proposed by the SK&F team that an antihistamine agent might also be effective in treating ulcers. At the time, this was a highly speculative proposal. Although histamine had been shown experimentally to stimulate gastric acid release, it was by no means certain that it played any significant role *in vivo*. Many workers at the time discounted the importance of histamine, especially when it was found that conventional antihistamines failed to inhibit gastric acid release. This result appeared to suggest the absence of histamine receptors in the parietal cells. The fact that histamine did have a stimulatory effect could be explained away by suggesting that histamine coincidentally switched on the gastrin or acetylcholine receptors.

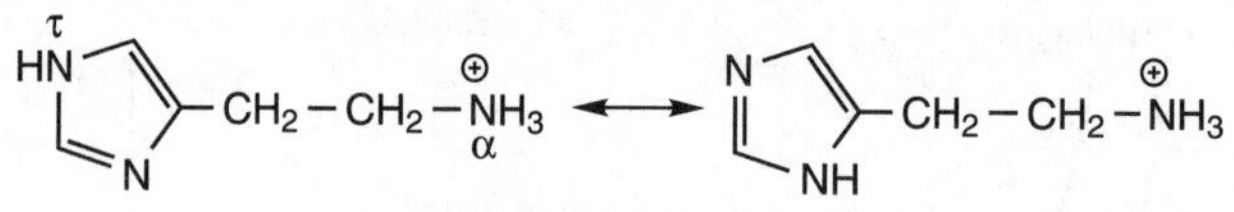

Fig. 18.4 Histamine.

Why then did the SK&F team persevere in their search for an effective antihistamine? What was their reasoning? Before answering that, let us look at histamine itself and the antihistamines available at that time.

18.3 Histamine

Histamine is made up of an imidazole ring which can exist in two tautomeric forms as shown in Fig. 18.4. Attached to the imidazole ring is a two-carbon chain with a terminal α-amino group. The pK_a of this amino group is 9.80, which means that at a plasma pH of 7.4, the side-chain of histamine is 99.6 per cent ionized.

The pK_a of the imidazole ring is 5.74 and so the ring is mostly un-ionized at pH 7.4.

Whenever cell damage occurs, histamine is released and stimulates the dilation and increased permeability of small blood vessels. The advantage of this to the body is that defensive cells (e.g. white blood cells) are released from the blood supply into an area of tissue damage and are able to combat any potential infection. Unfortunately, the release of histamine can also be a problem. For example, when an allergic reaction or irritation is experienced, histamine is released and produces the same effects when they are not really needed.

The early antihistamine drugs were therefore designed to treat conditions such as hay fever, rashes, insect bites, or asthma.

Two examples of these early antihistamines are mepyramine (Fig. 18.5) and diphenhydramine ('Benadryl') (Fig. 18.6).

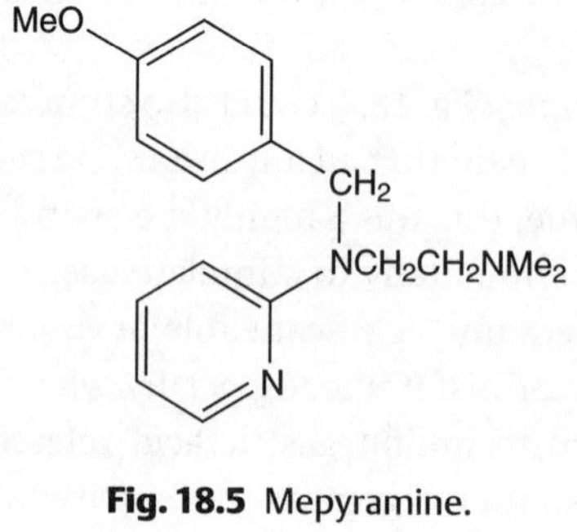

Fig. 18.5 Mepyramine.

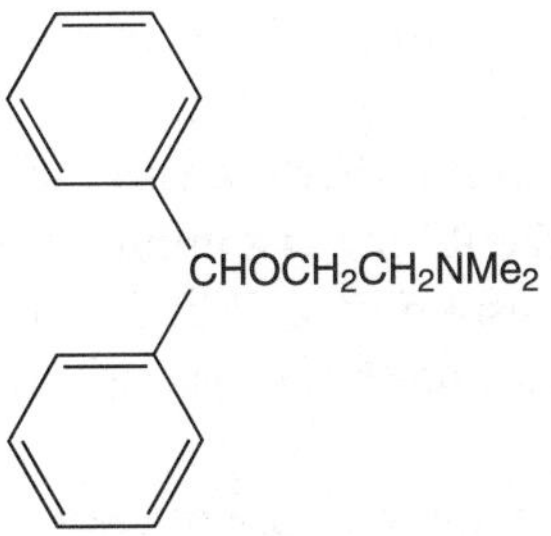

Fig. 18.6 Diphenhydramine.

18.4 The theory – two histamine receptors?

We are now able to return to the question we asked at the end of Section 18.2. Bearing in mind the failure of the known antihistamines to inhibit gastric acid release, why did the SK&F team persevere with the antihistamine approach?

As mentioned above, conventional antihistamines failed to have any effect on gastric acid release. However, they also failed to inhibit other actions of histamine. For example, they failed to fully inhibit the dilation of blood vessels induced by histamine. The SK&F scientists therefore proposed that there might be two different types of histamine receptor, analogous to the two types of acetylcholine receptor mentioned in Chapter 15. Histamine – the natural messenger – would switch both on equally effectively and would not distinguish between them. However, suitably designed antagonists should in theory be capable of making that distinction. By implication, this means that the conventional antihistamines known in the early sixties were already selective in that they were able to inhibit the histamine receptors involved in the inflammation process (classified as H1-receptors), and were unable to inhibit the proposed histamine receptors responsible for gastric acid secretion (classified as H2-receptors).

It was an interesting theory, but the fact remained that there was no known antagonist for the proposed H2-receptors. Until such a compound was found, it could not be certain that the H2-receptors even existed, and yet without a receptor to study, how could one design an antagonist to act with it?

18.5 Searching for a lead – histamine

The SK&F team obviously had a problem. They had a theory but no lead compound. How could they make a start?

Their answer was to start from histamine itself. If histamine was stimulating the release of gastric acid by binding to a hypothetical H2-receptor, then clearly histamine was being 'recognized' by the receptor. The task then was to vary the structure of hist-amine in such a way that it would still be recognized by the receptor, but bind in such a way that it acted as an antagonist rather than an agonist.

It was necessary then to find out how histamine itself was binding to its receptors. Structure–activity studies on histamine and histamine analogues revealed that the binding requirements for histamine to the H1- and the proposed H2-receptors were slightly different.

At the H1-receptor, the essential requirements were as follows:

• The side-chain had to have a positively charged nitrogen atom with at least one attached proton. Quaternary ammonium salts which lacked such a proton were extremely weak in activity.

• There had to be a flexible chain between the above cation and a heteroaromatic ring.

• The heteroaromatic ring did not have to be imidazole, but it did have to contain a nitrogen atom with a lone pair of electrons, *ortho* to the side-chain.

For the proposed H2-receptor, structure–activity studies were carried out to determine whether histamine analogues could bring about the physiological effects proposed for this receptor (e.g. stimulating gastric acid release).

The essential structure–activity requirements (SAR) were the same as for the H1-receptor except that the heteroaromatic ring had to contain an amidine unit (HN-CH-N:). These results are summarized in Fig. 18.7.

From these results, it appeared that the terminal α-amino group was involved in a binding interaction with both types of receptor via ionic or hydrogen bonding, while the nitrogen atom(s) in the heteroaromatic ring bound via hydrogen bonding, as shown in Fig. 18.8.

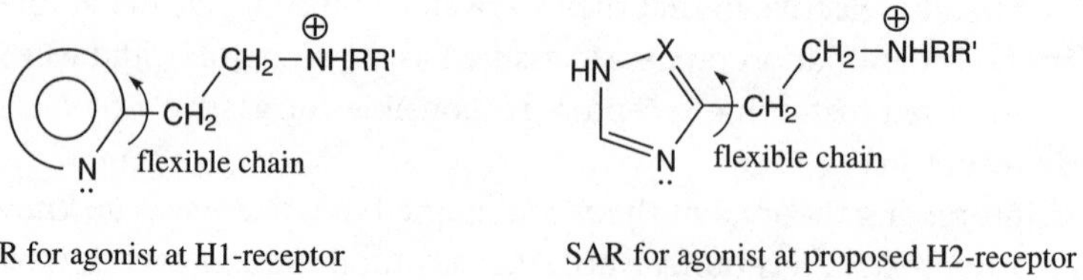

SAR for agonist at H1-receptor   SAR for agonist at proposed H2-receptor

Fig. 18.7 Summary of SAR results.

18.6 Searching for a lead – N^{α}-guanylhistamine

Having gained a knowledge of the structure-activity relationships for histamine, the task was now to design a molecule which would be recognized by the H2-receptor, but which would not activate it. In other words, an agonist had to be converted to an antagonist. In order to do that, it would be necessary to alter the way in which the molecule was bound to the receptor.

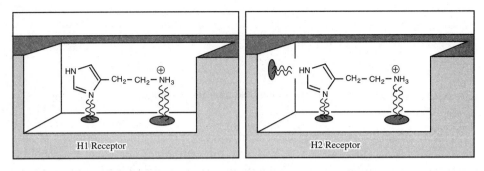

Fig. 18.8 Binding interactions at H1- and H2- receptors.

Pictorially, one can imagine histamine fitting into its receptor site and inducing a change in shape which 'switches the receptor on' (Fig. 18.9). An antagonist might be found by adding a functional group which would bind to another binding region on the receptor and prevent the change in shape required for activation.

This was one of several strategies tried out by the SK&F workers. To begin with, a study of known agonists and antagonists in other fields of medicinal chemistry was carried out. The structural differences between agonists and antagonists for a particular receptor were identified and then similar alterations were tried on histamine.

For example, fusing an aromatic ring on to noradrenaline had been a successful tactic used in the design of antagonists for the noradrenaline receptor (see Section 16.11.2). This same tactic was tried with histamine to give analogues such as the one shown in Fig. 18.10, but none of the compounds synthesized proved to be an antagonist.

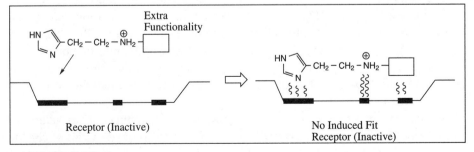

Fig. 18.9 Possible receptor interactions of histamine and an antagonist.

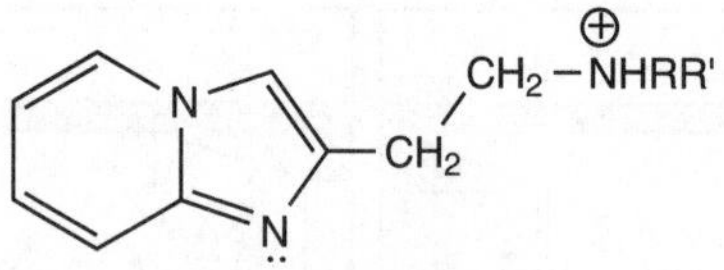

Fig. 18.10 Histamine analogue–not an antagonist.

Another approach which had been used successfully in the development of anti-cholinergic agents (Section 15.11.2) had been the addition of non-polar, hydrophobic substituents. This approach was tried with histamine by attaching various alkyl and arylalkyl groups to different locations on the histamine skeleton. Unfortunately, none of these analogues proved to be antagonists.

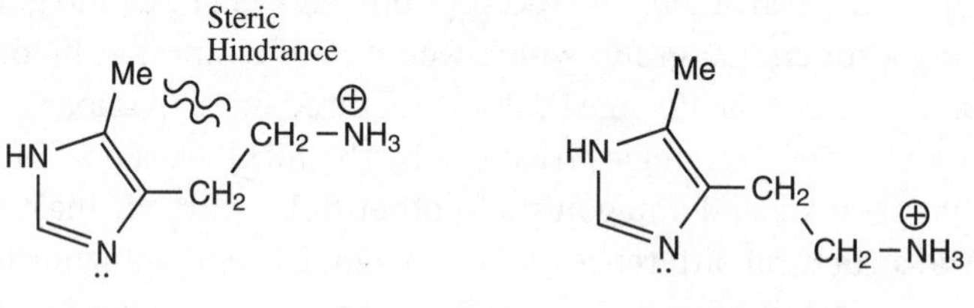

Fig. 18.11 4-Methylhistamine.

However, one interesting result was obtained which was to be relevant to later studies. It was discovered that 4-methylhistamine (Fig. 18.11) was a highly selective H2-agonist, showing far greater activity for the H2-receptor than for the H1-receptor. Why should such a simple alteration produce this selectivity?

4-Methylhistamine (like histamine) is a highly flexible molecule due to its side-chain, but structural studies show that some of its conformations are less stable than others. In particular, conformation I in Fig. 18.11 is disallowed due to a large steric interaction between the 4-methyl group and the side-chain. The selectivity observed suggests that 4-methylhistamine (and by inference histamine) has to adopt two different conformations in order to fit the H1- or the H2-receptor. Since 4-methylhistamine is more active at the H2-receptor, it implies that the conformation required for the H2-receptor is a stable one for 4-methylhistamine (conformation II), whereas the conformation required for the H1-receptor is an unstable one (conformation I).

Despite this interesting result, the SK&F workers were no closer to an H2-antagonist. Two hundred compounds had been synthesized and not one had shown a hint of being an antagonist.

Research up until this stage had concentrated on searching for an additional hydrophobic binding region on the receptor. Now the focus of research switched to see what would happen if the terminal α-NH_3^+ group was replaced with a variety of different polar functional groups. It was reasoned that different polar groups could bond to the same region on the receptor as the NH_3^+ group, but that the geometry of bonding might be altered sufficiently to produce an antagonist. It was from this study that the first crucial breakthrough was achieved with the discovery that N^α-guanylhistamine (Fig. 18.12) was acting very weakly as an antagonist.

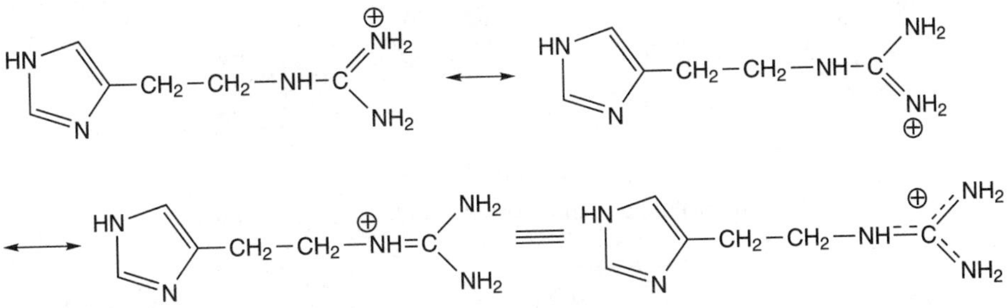

Fig. 18.12 N^α-Guanylhistamine.

This structure had in fact been synthesized early on in the project, but had not been recognized as an antagonist. This is not too surprising since it acts as an agonist! It was not until later pharmacological studies were carried out that it was realized that N^α-guanylhistamine was also acting as an antagonist to histamine. In other words, it was a partial agonist (see Section 5.8.).

N^α-guanylhistamine activates the H2-receptor, but not the same extent as histamine. As a result, the amount of gastric acid released is lower. More importantly, as long as N^α-guanylhistamine is bound to the receptor, it prevents histamine from binding and thus prevents complete receptor activation.

This was the first indication of any sort of antagonism to histamine.

The question now arose as to which part or parts of the N^α-guanylhistamine skeleton were really necessary for this effect. Perhaps the guanidine group itself could act as an antagonist?

Various guanidine structures were synthesized which lacked the imidazole ring, but none had the desired antagonist activity, demonstrating that both the imidazole ring and the guanidine group were required.

The structures of N^α-guanylhistamine and histamine were now compared. Both structures contain an imidazole ring and a positively charged group linked by a two-carbon bridge. The guanidine group is basic and protonated at pH 7.4 so that the analogue has a positive charge similar to histamine. However, the charge on the guanidine group can be spread around a planar arrangement of three nitrogens and can potentially be further away from the imidazole ring (Fig. 18.12). This leads to the possibility that the analogue could be interacting with a binding region on the receptor which is 'out of reach' of histamine. This is demonstrated in Fig. 18.13 and 18.14. Two alternative binding regions might be available for the cationic group – an agonist region where binding leads to activation of the receptor and an antagonist region where binding does not activate the receptor. In Fig. 18.14, histamine is only able to reach the agonist region. However, the analogue with its extended functionality is capable of reaching either region (Fig. 18.13).

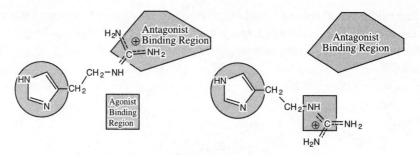

Fig. 18.13 Possible binding modes for N^{α}-Guanylhistamine.

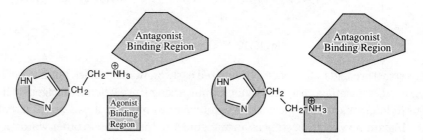

Fig. 18.14 Binding of histamine: agonist mode only.

If most of the analogue molecules bind to the agonist region and the remainder bind to the antagonist region, then this could explain the partial agonist activity. Regardless of the mode of binding, histamine would be prevented from binding and an antagonism would be observed due to the percentage of N^{α}-guanylhistamine bound to the antagonist region.

18.7 Developing the lead – a chelation bonding theory

Variations were now necessary to see if an analogue could be made which would only bind to the antagonist region.

The synthesis of the isothiourea (Fig. 18.15) gave a structure where the nitrogen nearest to the imidazole ring was replaced with a sulfur atom.

The positive charge in this molecule is now restricted to the terminal portion of the chain and should interact more strongly with the proposed antagonist binding region if it is indeed further away.

Fig. 18.15 Isothiourea analogue.

Antagonist activity did increase, but the compound was still a partial agonist, showing that binding was still possible to the agonist region.

Two other analogues were synthesized, where one of the terminal amino groups in the guanidine group was replaced with either a methylthio group or a methyl group (Fig. 18.16). Both the resulting structures were partial agonists, but with poorer antagonist activity.

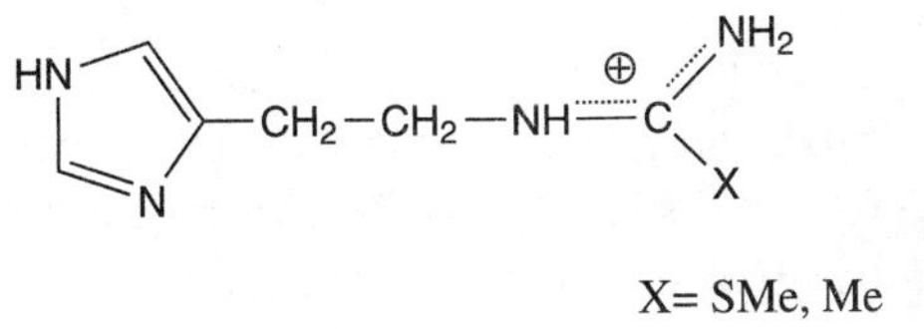

X= SMe, Me

Fig. 18.16 Analogue, where X is a methyl group.

From these results, it was concluded that both terminal amino groups were required for binding to the antagonist binding region. It was proposed that the charged guanidine group was interacting with a charged carboxylate residue on the receptor via two hydrogen bonds (Fig. 18.17). If either of these terminal amino groups were absent, then binding would be weaker, resulting in a lower level of antagonism.

The chain was now extended from a two-carbon unit to a three-carbon unit to see what would happen if the guanidine group was moved further away from the imidazole ring. The antagonist activity increased for the guanidine structure (Fig. 18.18), but

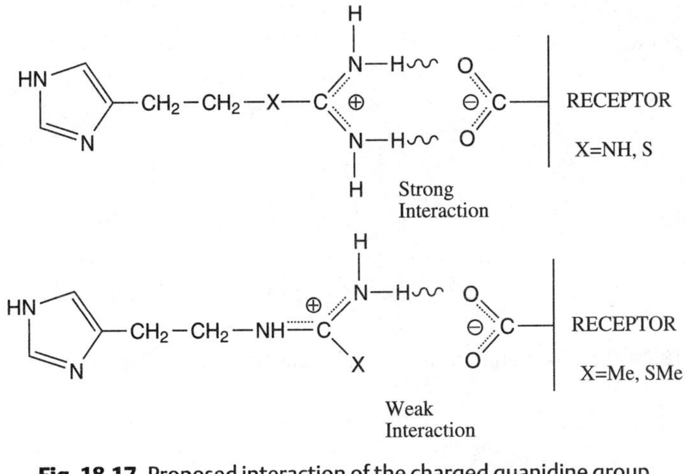

Fig. 18.17 Proposed interaction of the charged guanidine group.

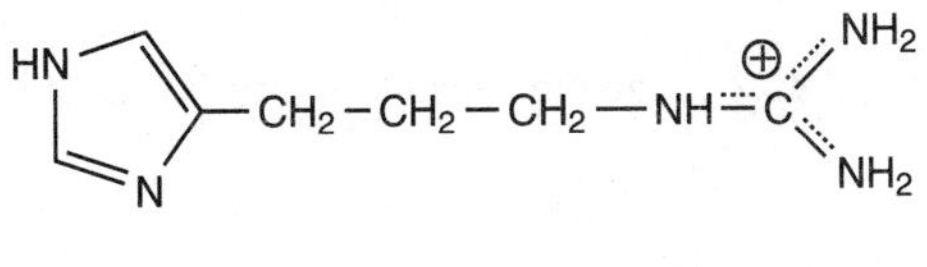

Fig. 18.18 Guanidine structure.

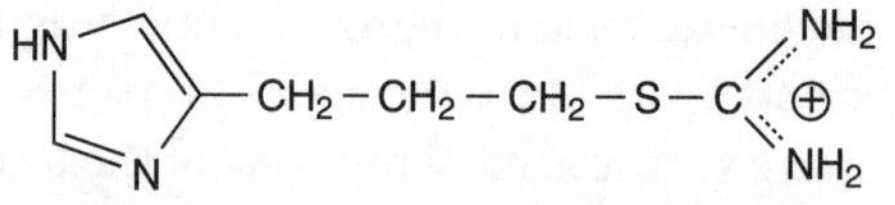

Fig. 18.19 Isothiourea structure.

strangely enough, decreased for the isothiourea structure (Fig. 18.19). It was therefore proposed that with a chain length of two carbon units, hydrogen bonding to the receptor involved the terminal NH_2 groups, but with a chain length of three carbon units, hydrogen bonding involved one terminal NH_2 group along with the NH group within the chain (Fig. 18.20). Support for this theory was provided by the fact that replacing one of the terminal NH_2 groups in the guanidine analogue (Fig. 18.18) with SMe or Me (Fig. 18.21) did not adversely affect the antagonist activity. This was completely different from the results obtained when similar changes were carried out on the two-carbon bridged compound.

These bonding interactions are represented pictorially in Figs 18.22 and 18.23.

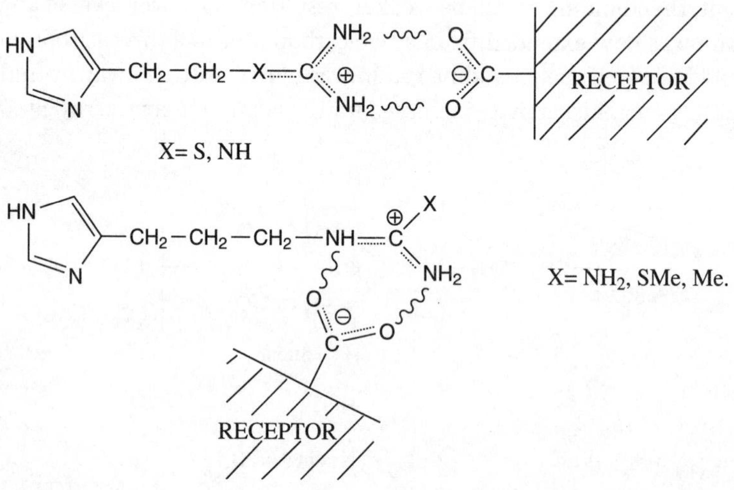

Fig. 18.20 Proposed binding interactions for analogues of different chain length.

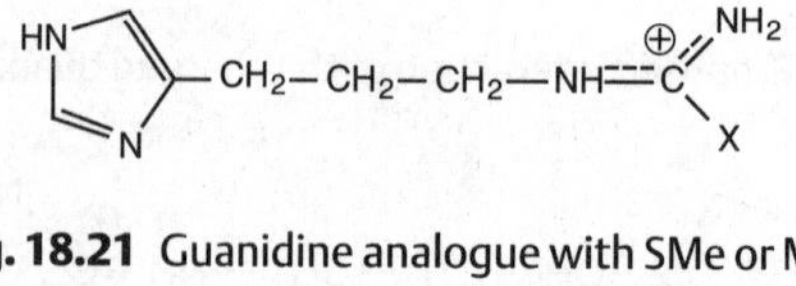

Fig. 18.21 Guanidine analogue with SMe or Me.

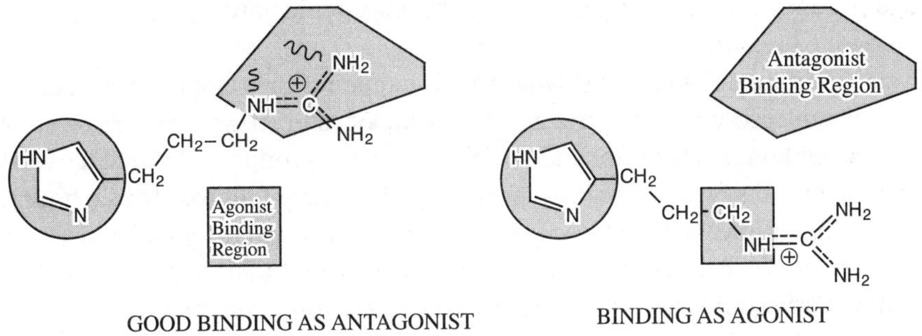

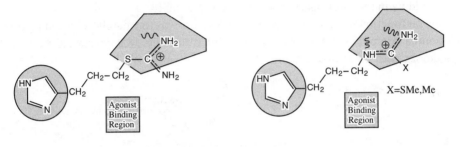

GOOD BINDING AS ANTAGONIST   BINDING AS AGONIST

Fig. 18.22 Binding interactions for 3C bridged analogue.

POOR BINDING AS ANTAGONIST   GOOD BINDING AS ANTAGONIST

Fig. 18.23 Effect of varying the guanidine group on binding to the antagonist region.

18.8 From partial agonist to antagonist – the development of burimamide

The problem now was to completely remove the agonist activity to get compounds with pure antagonist activity. This meant designing a structure which would differentiate between the agonist and antagonist binding regions.

At first sight this looks impossible since both regions appear to involve the same type of bonding. Histamine's activity as an agonist depends on the imidazole ring and the charged amino function, with the two groups taking part in hydrogen and ionic bonding, respectively. However, the antagonist activity of the partial agonists described so far also appears to depend on a hydrogen bonding imidazole ring and an ionic bonding guanidine group.

Fortunately, a distinction can be made between the charged groups.

The structures which show antagonist activity are all capable of forming a chelated bonding structure as previously shown in Fig. 18.20. This interaction involves two hydrogen bonds between two charged species, but it is really necessary for the chelating group to be charged? Could a neutral group also chelate to the antagonist region by hydrogen bonding alone? If so, it might be possible to distinguish between the agonist and antagonist regions, especially since ionic bonding appears necessary for the agonist region.

It was therefore decided to see what would happen if the strongly basic guanidine group was replaced with a neutral group capable of interacting with the receptor by two hydrogen bonds. There are a large variety of such groups, but the SK&F workers limited the options by adhering to a principle which they followed throughout their research programme. Whenever they wished to alter any specific physical or chemical property, they strove to ensure that other properties were changed as little as possible. Only in this way could they rationalize any observed improvement in activity.

Thus, in order to study the effect of replacing the basic guanidine group with a neutral group, it was necessary to ensure that the new group was as similar as possible to guanidine in terms of size, shape, and hydrophobicity.

Several functional groups were tried, but success was ultimately achieved by using a thiourea group. The thiourea derivative SK&F 91581 (Fig. 18.24) proved to be a weak antagonist with no agonist activity.

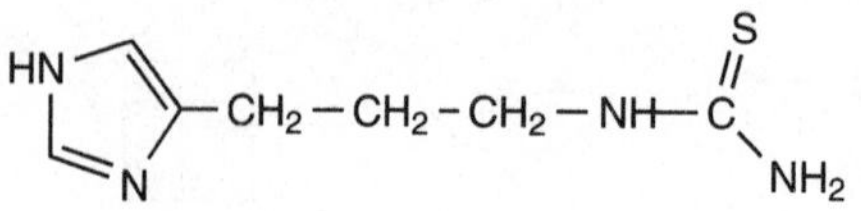

No Agonist Activity
Very Weak Antagonist Activity

Fig. 18.24 SK&F 91581.

Apart from basicity, the properties of the thiourea group are very similar to the guanidine group. Both groups are planar, similar in size, and can take part in hydrogen bonding. Thus, the alteration in biological activity can reasonably be attributed to the differences in basicity between the two groups.

Unlike guanidine, the thiourea group is neutral. This is due to the C=S group, which has an electron withdrawing effect on the neighbouring nitrogens, making them non-basic and more like amide nitrogens.

The fact that a neutral group could bind to the antagonist region and not to the agonist site could be taken to imply that the agonist binding region involves ionic bonding, whereas the antagonist region involves hydrogen bonding.

Further chain extension and the addition of an N-methyl group led to burimamide (Fig. 18.25) which was found to have enhanced activity.

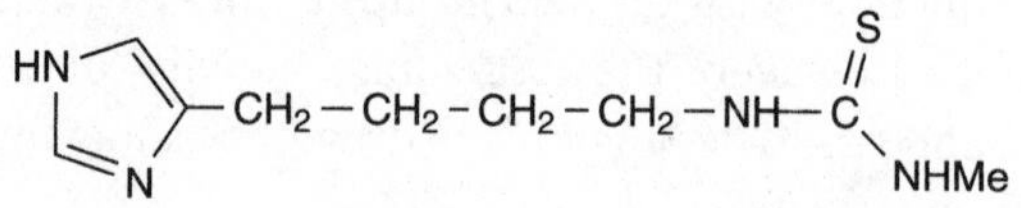

Fig. 18.25 Burimamide.

These results suggest that chain extension has moved the thiourea group closer to the antagonist binding region, and that the addition of the N-methyl group has resulted in a beneficial increase in hydrophobicity. A possible explanation for this latter result will be described in Section 18.12.2.

Burimamide is a highly specific competitive antagonist of histamine at H2-receptors, and is 100 times more potent than N^α-guanylhistamine. Its discovery finally proved the existence of the H2-receptors.

18.9 Development of metiamide

Despite this success, burimamide was not suitable for clinical trials since its antagonist activity was still too low for oral administration. Further developments were needed. Attention was now directed to the imidazole ring of burimamide and, in particular, to the various possible tautomeric forms of this ring. It was argued that if one particular tautomer was preferred for binding with the H2-receptor, then activity might be enhanced by modifying the burimamide structure to favour that tautomer.

At pH 7.4, it is possible for the imidazole ring to equilibrate between the two tautomeric forms (I) and (II) via the protonated intermediate (III) (Fig. 18.26). The necessary proton for this process has to be supplied by water or by an exchangeable proton on a suitable amino acid residue in the binding region. If the exchange is slow, then it is possible that the drug will enter and leave the receptor at a faster rate than the equilibration between the three tautomeric forms. If bonding involves only one of the tautomeric forms, then clearly antagonism would be increased if the structure was varied to prefer that tautomeric form over the others. Our model hypothesis for receptor binding shows that the imidazole ring is important for the binding of both agonists and antagonists. Therefore, it is reasonable to assume that the preferred imidazole tautomer is the same for both agonists and antagonists. If this is so, then the preferred tautomer for a strong agonist such as histamine should also be the preferred tautomer for a strong antagonist.

Figure 18.26 shows that the imidazole ring can exist as one ionized tautomer and two unionized tautomers. Let us first consider whether the preferred tautomer is likely to be ionized or not.

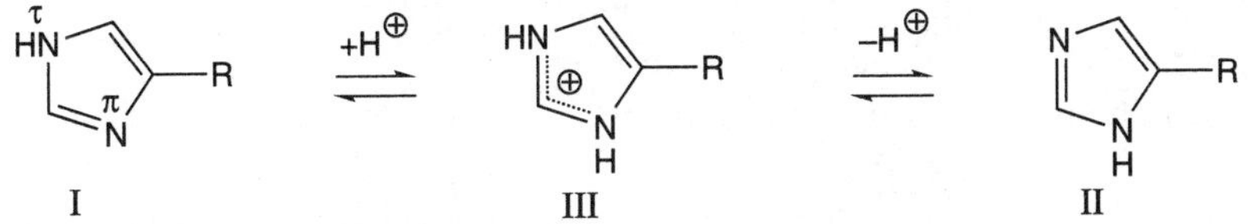

Fig. 18.26 Imidazole ring can equilibriate between tautomeric forms (I and II) via the protonated intermediate (III).

We have already seen that the pK_a for the imidazole ring in histamine is 5.74, meaning that the ring is a weak base and mostly un-ionized. The pK_a value for imidazole itself is 6.80 and for burimamide 7.25. These values show that these imidzaole rings are more basic than histamine and more likely to be ionized. Why should this be so?

The explanation must be that the side-chain has an electronic effect on the imidazole ring. If the side-chain is electron withdrawing or electron donating, then it will affect the basicity of the ring. A measure of the side-chain's electronic effect can be worked out by the Hammett equation (see Chapter 11):

$$pK_{a(R)} = pK_{a(H)} + \rho\sigma_R$$

where $pK_{a(R)}$ is the pK_a of the imidazole ring bearing a side-chain R, $pK_{a(H)}$ is the pK_a of the unsubstituted imidazole ring, ρ is a constant, and σ_R is the Hammett substituent constant for the side-chain R.

From the pK_a values, the value of the Hammett substituent constant can be calculated to show whether the side-chain R is electron withdrawing or electron donating.

In burimamide, the side-chain is calculated to be slightly electron donating (of the same order as a methyl group). Therefore, the imidazole ring in burimamide is more likely to be ionized than in histamine, where the side-chain is electron withdrawing. At pH 7.4, 40 per cent of burimamide is ionized in the imidazole ring compared to approximately 3 per cent of histamine. This represents quite a difference between the two structures and since the binding of the imidazole ring is important for antagonist activity as well as agonist activity, it suggests that a pK_a value closer to that of histamine might lead to better binding and to better antagonist activity.

It was necessary, therefore, to make the side-chain electron withdrawing rather than electron donating. This can be done by inserting an electronegative aom into the side-chain – preferably one which has a minimum disturbance on the rest of the molecule. In other words, an isostere for a methylene group is required – one which has an electronic effect, but which has approximately the same size and properties as the methylene group.

The first isostere to be tried was a sulfur atom. Sulfur is quite a good isostere for the methylene unit in that both groups have similar van der Waals radii and similar bond angles. However, the C-S bond length is slightly longer than a C-C bond, leading to a slight extension (15 per cent) of the structure.

The methylene group replaced was next but one to the imidazole ring. This site was chosen, not for any strategic reasons, but because a synthetic route was readily available to carry out that particular transformation.

As hoped, the resulting compound, thiaburimamide (Fig. 18.27), had a significantly lower pK_a of 6.25 and was found to have enhanced antagonistic activity. This result supported the theory that a reduction in the proportion of ionized tautomer was beneficial to receptor binding and activity.

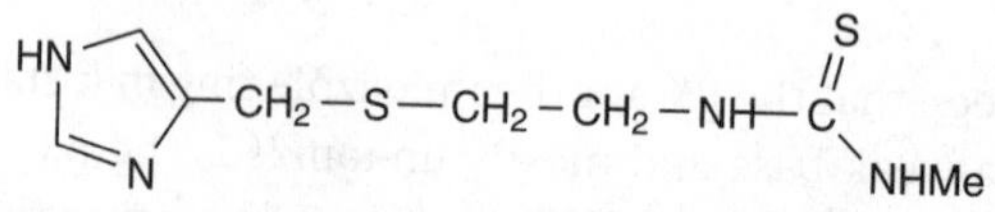

Fig. 18.27 Thiaburimamide.

Thiaburimamide had been synthesized in order to favour the un-ionized imidazole ring over the ionized ring. However, as we have seen, there are two possible un-ionized tautomers. The next question is whether either of these are preferred for receptor binding.

Let us return to histamine. If one of the un-ionized tautomers is preferred over the other in histamine, then it would be reasonable to assume that this is the favoured tautomer for receptor binding. The preferred tautomer for histamine is tautomer I (Fig. 18.26).

Why is tautomer I favoured? The answer lies in the fact that the side-chain on histamine is electron withdrawing. This electron withdrawing effect on the imidazole ring is inductive and therefore the strength of the effect will decrease with distance round the ring. This implies that the nitrogen atom on the imidazole ring closest to the side-chain (Nπ) will experience a greater electron withdrawing effect than the one further away (Nτ). As a result, the closer nitrogen is less basic, which in turn means that it is less likely to bond to hydrogen.

Since the side-chain in thiaburimamide is electron withdrawing, then it too will favour tautomer I.

It was now argued that this tautomer could be further enhanced if an electron donating group was placed at position 4 in the ring. At this position, the inductive effect would be felt most at the neighbouring nitrogen (Nτ), further enhancing its basic character and increasing the population of tautomer I. However, it was important to choose a group which would not interfere with the normal receptor binding interaction. For example, a large substituent might be too bulky and prevent the analogue fitting the receptor. A methyl group was chosen since it was known that 4-methylhistamine was an agonist and also highly selective for the H2-receptor (see Section 18.6.).

The compound obtained was metiamide (Fig. 18.28) which was found to have enhanced activity as an antagonist, supporting the previous theory.

It is interesting to note that the above effect outweighs an undesirable rise in pK_a. By adding an electron donating methyl group, there has been a rise in the pK_a of the imidazole ring to 6.80 compared to 6.25 for thiaburimamide. (Coincidentally, this is the same pK_a as for imidazole itself, which shows that the electronic effects of the methyl group and the side-chain are cancelling each other out as far as pK_a is concerned.) A pK_a of 6.80 means that 20 per cent of metiamide is ionized in the imidazole ring. However, this is still significantly lower than the corresponding 40 per cent for burimamide.

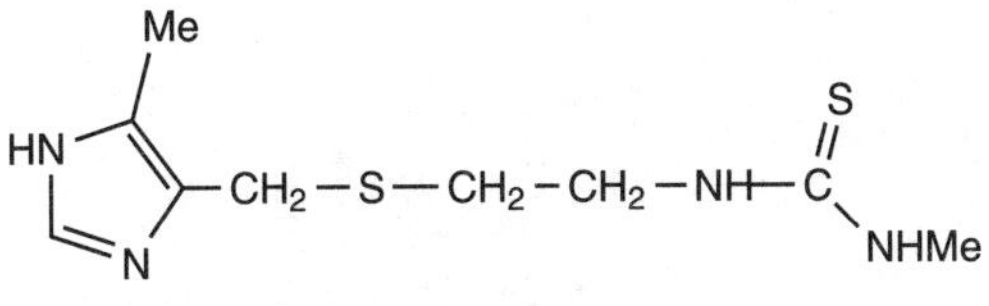

Fig. 18.28 Metiamide.

Compared to burimamide, the percentage of ionized imidazole ring has been lowered in metiamide and the ratio of the two' possible un-ionized imidazole tautomers reversed. The fact that activity is increased with respect to thiaburimamide suggests that the increase in the population of tautomer (I) outweighs the increase in population of the ionized tautomer (III).

4-Methylburimamide (Fig. 18.29) was also synthesized for comparison. Here, the introduction of the 4-methyl group does not lead to an increase in activity. The pK_a is increased to 7.80, resulting in the population of ionized imidazole ring rising to 72 per cent. This demonstrates how important it is to rationalize structural changes. Adding the 4-methyl group to thiaburimamide is advantageous, but adding it to burimamide is not.

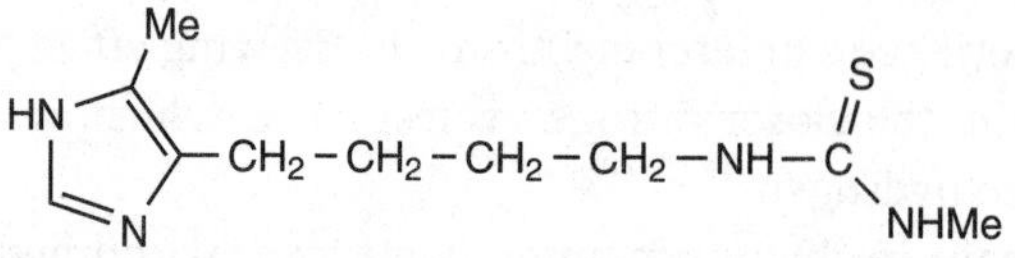

Fig. 18.29 4-Methylburimamide.

The design and synthesis of metiamide followed a rational approach aimed at favouring one specific tautomer. Such a study is known as a dynamic structure-activity analysis.

Strangely enough, it has since transpired that the improvement in antagonism may have resulted from conformational effects. X-ray crystallography studies have indicated that the longer thioether linkage in the chain increases the flexibility of the side-chain and that the 4-methyl substituent in the imidazole ring may help to orientate the imidazole ring correctly for receptor binding. It is significant that the oxygen analogue oxaburimamide (Fig. 18.30) is less potent than burimamide despite the fact that the electron withdrawing effect of the oxygen-containing chain on the ring is similar to the sulfur-containing chain. The bond lengths and angles of the ether link are similar to the methylene unit and in this respect it is a better isostere than sulfur. However, the oxygen atom is substantially smaller. It is also significantly more basic and more hydrophilic than either sulfur or methylene. Oxaburimamide's lower activity might be due to a variety of reasons. For example, the oxygen may not allow the same flexibility permitted by the sulfur atom. Alternatively, the oxygen may be involved in a hydrogen bonding interaction either with the receptor or with its own imidazole ring, resulting in a change in receptor binding interaction.

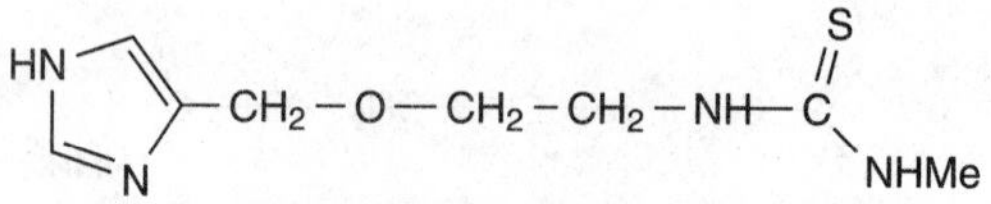

Fig. 18.30 Oxaburimamide.

Metiamide is ten times more active than burimamide and showed promise as an anti-ulcer agent. Unfortunately, a number of patients suffered from kidney damage and granulocytopenia – a condition which results in the reduction of circulating white blood cells and which makes patients susceptible to infection. Further developments were now required to find an improved drug lacking these side-effects.

18.10 Development of cimetidine

It was proposed that metiamide's side-effects were associated with the thiourea group – a group which is not particularly common in the body's biochemistry. Therefore, consideration was given to replacing this group with a group which was similar in property but would be more acceptable in the biochemical context. The urea analogue (Fig. 18.31) was tried, but found to be less active. The guanidine analogue (Fig. 18.32) was also less active, but it was interesting to note that this compound had no agonist activity. This contrasts with the three-carbon bridged guanadine (Fig. 18.18) which we have already seen is a partial agonist. Therefore, the guanidine analogue (Fig. 18.32) was the first example of a guanidine having pure antagonist activity.

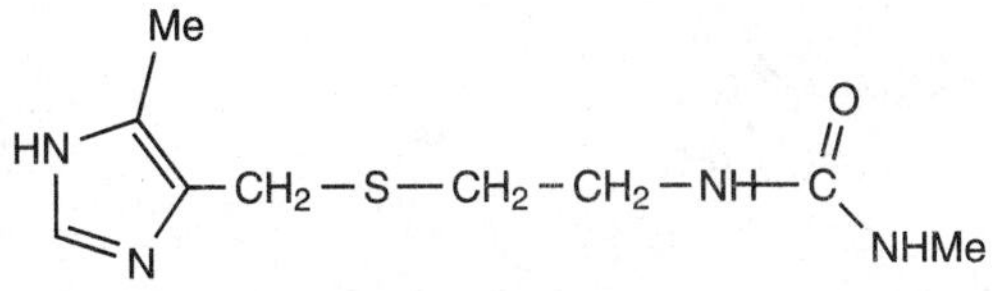

Fig. 18.31 Urea analogue.

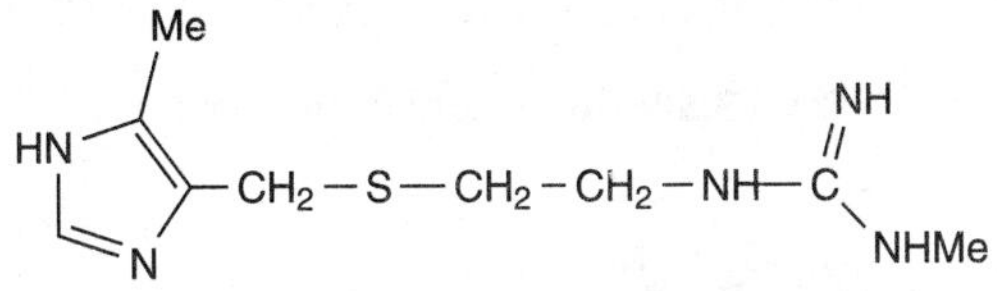

Fig. 18.32 Guanidine analogue.

One possible explanation for this is that the longer four-unit chain extends the guanidine binding group beyond the reach of the agonist binding region (Fig. 18.33), whereas the shorter three-unit chain still allows binding to both agonist and antagonist regions (Fig. 18.34).

The antagonist activity for the guanidine analogue (Fig. 18.32) is weak, but it was decided to look more closely at this compound since it was thought that the guanidine unit would be less likely to have toxic side-effects than the thiourea. This is a reasonable assumption since the guanidine unit is present naturally in the amino acid arginine (Fig. 18.35). The problem now was to retain the guanidine unit, but to increase activity. It seemed likely that the low activity was due to the fact that the basic guanidine group would be ionized at pH7.4. The problem was how to make this group neutral – no easy task, considering that guanidine is one of the strongest bases in organic chemistry.

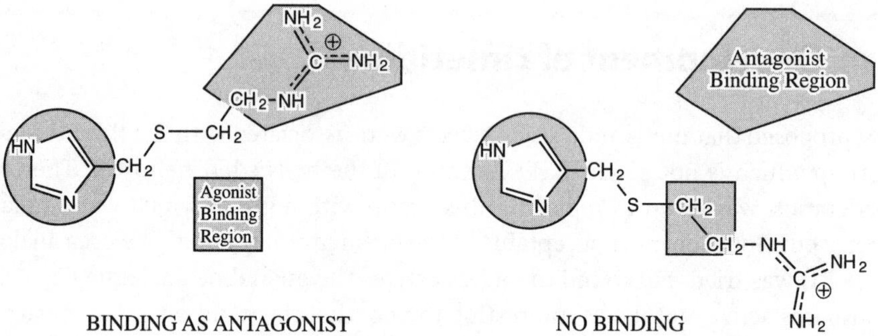

Agonist Binding Region
Antagonist Binding Region
BINDING AS ANTAGONIST
NO BINDING

Fig. 18.33 Four-carbon unit chain.

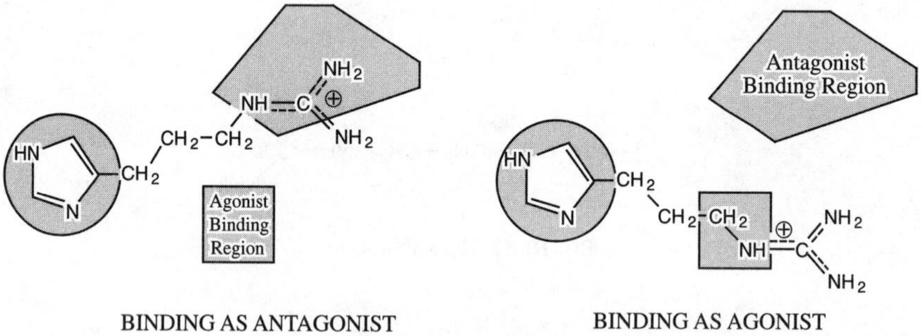

Agonist Binding Region
Antagonist Binding Region
BINDING AS ANTAGONIST
BINDING AS AGONIST

Fig. 18.34 Three-carbon unit chain.

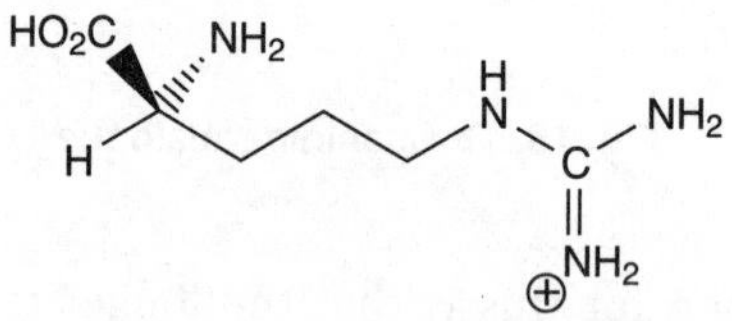

Fig. 18.35 Arginine.

Nevertheless, a search of the literature revealed a useful study on the ionization of monosubstituted guanidines (Fig. 18.36). A comparison of the pK_a values of these compounds with the inductive substitutent constants σ_i for the substituents X gave a straight line as shown in Fig. 18.37, showing that pK_a is inversely proportional to the electron withdrawing power of the substituent. Thus, strongly electron withdrawing substituents make the guanidine group less basic and less ionized. The nitro and cyano groups are particularly strong electron withdrawing groups. The ionization constants for cyanoguanidine and nitroguanidine are 0.4 and 0.9 respectively (Fig. 18.37) – similar values to the ionization constant for thiourea itself (–1.2).

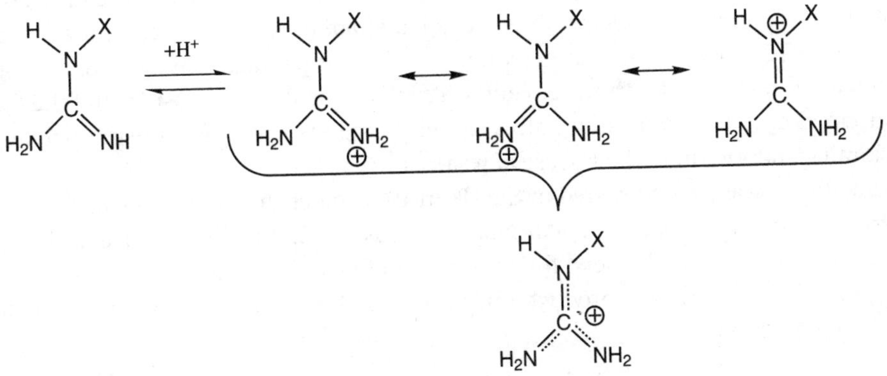

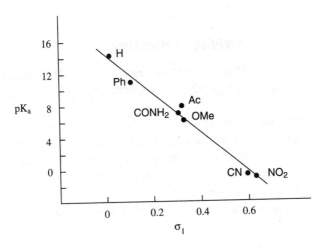

Fig. 18.36 Ionization of monosubstituted guanidines.

Fig. 18.37 pK_a vs. inductive substituent constants (σ_i) for the substituents X in Fig. 18.36.

Both the nitroguanidine and cyanoguanidine analogues of metiamide were synthesized and found to have comparable antagonist activities to metiamide. The cyanoguanidine analogue (cimetidine (Fig. 18.1)) was the more potent analogue and was chosen for clinical studies.

18.11 Cimetidine

18.11.1 Biological activity of cimetidine

Cimetidine inhibits H2-receptors and thus inhibits gastric acid release. The drug does not show the toxic side-effects observed for metiamide and has been shown to be slightly more active. It has also been found to inhibit pentagastrin (Fig. 18.38) from stimulating release of gastric acid. Pentagastrin is an analogue of gastrin and the fact that cimetidine blocks its stimulatory activity suggests some relationship between histamine and gastrin in the release of gastric acid.

Cimetidine was first marketed in the UK in 1976 under the trade name of Tagamet (derived from anTAGonist and ciMETidine). It was the first really effective anti-ulcer drug, doing away with the need for surgery, and for several years, it was the world's biggest selling prescription product, until it was pushed into second place in 1988 by ranitidine (Section 18.13.).

$$N\text{-}t\text{-}BOC\text{-}\beta\text{-}Ala\text{-}Trp\text{-}Met\text{-}Asp\text{-}Phe\text{-}NH_2$$

Fig. 18.38 Pentagastrin.

18.11.2 Structure and activity of cimetidine

The finding that metiamide and cimetidine are both good H2-antagonists of similar activity shows that the cynaoguanidine group is a good biosostere for the thiourea group.

This is despite the fact that three tautomeric forms (Fig. 18.39) are possible for the guanidine group compared to only one for the isothiourea group. In fact, this is more apparent than real, since the imino tautomer (II) is the preferred tautomeric form for the guanidine unit. Tautomer II is favoured since the cyano group has a stronger electron withdrawing effect on the neighbouring nitrogen compared to the two nitrogens further away. This makes the neighbouring nitrogen less basic and therefore less likely to be protonated.

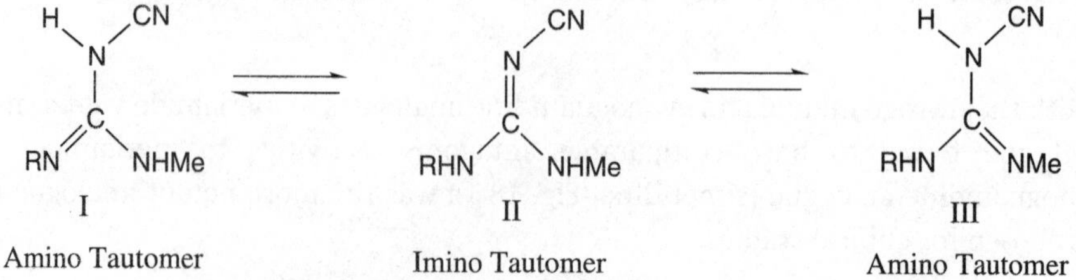

Fig. 18.39 Three tautomeric forms of guanidine unit.

Since tautomer II is favoured, the guanidine group does in fact bear a close structural similarity to the thiourea group. Both groups have a planar π electron system with similar geometries (equal C-N distances and angles). They are polar and hydrophilic with high dipole moments and low partition coefficients. They are weakly basic and also weakly acidic such that they are un-ionized at pH7.4.

18.11.3 Metabolism of cimetidine

It is important to study the metabolism of any new drug in case the metabolites have biological activity in their own right. Any such activity might lead to side-effects. Alternatively, a metabolite might have enhanced activity of the type desired and give clues to further development.

Cimetidine itself is metabolically stable and is excreted largely unchanged. The only metabolites which have been identified are due to oxidation of the sulfur link (Fig. 18.40) or oxidation of the ring methyl group (Fig. 18.41).

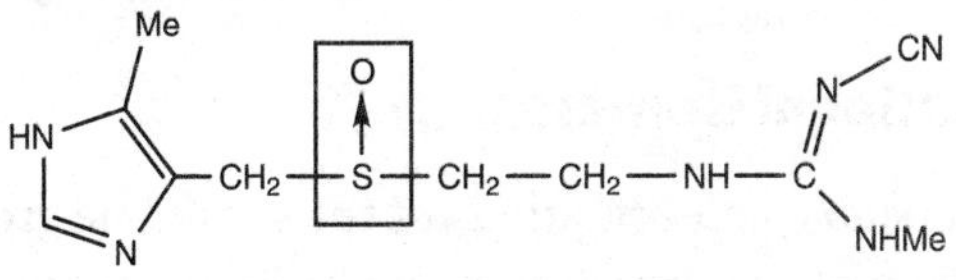

Fig. 18.40 Oxidation of the sulfur link in cimetidine.

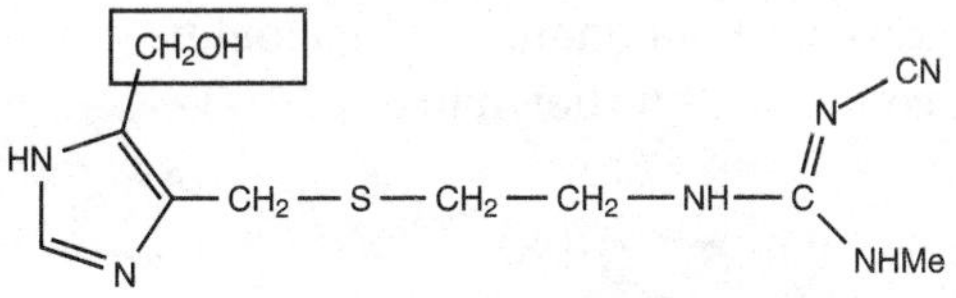

Fig. 18.41 Oxidation of the ring methyl group in cimetidine.

It has been found that cimetidine inhibits the P-450 cytochrome oxidase system in the liver. This is an important enzyme system in the metabolism of drugs and care must be taken if other drugs are being taken at the same time as cimetidine, since cimetidine may inhibit the metabolism of these drugs, leading to higher blood levels and toxic side-effects. In particular, care must be taken when cimetidine is taken with drugs such as diazepam, lidocaine, warfarin, or theophylline.

18.11.4 Synthesis of cimetidine

The synthesis of cimetidine was originally carried out as a four step process as shown in Figure 18.42, where lithium aluminium hydride was used as the reagent for the initial reduction step. However, subsequent research revealed that this reduction could be carried out more cheaply and safely using sodium in liquid ammonia, and so this became the method used in the manufacture of cimetidine.

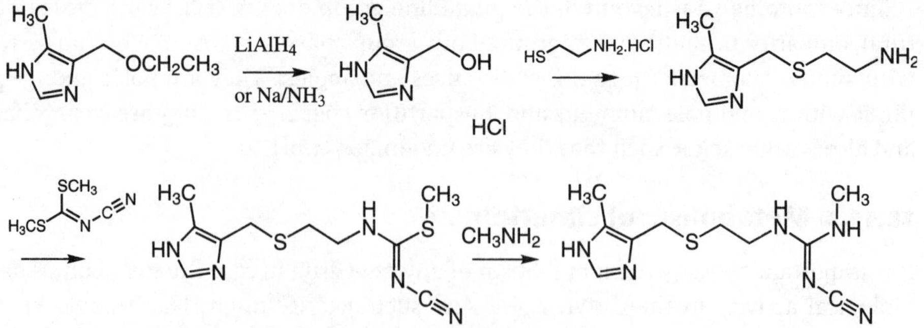

Fig. 18.42 Synthesis of cimetidine.

18.12 Further studies – cimetidine analogues

18.12.1 Conformational isomers

A study of the various stable conformations of the guanidine group in cimetidine led to a rethink of the type of bonding which might be taking place at the antagonist binding region. Up until this point, the favoured theory had been a bidentate hydrogen interaction as shown in Fig. 18.17.

In order to achieve this kind of bonding, the guanidine group in cimetidine would have to adopt the *Z,Z* conformation shown in Fig. 18.43.

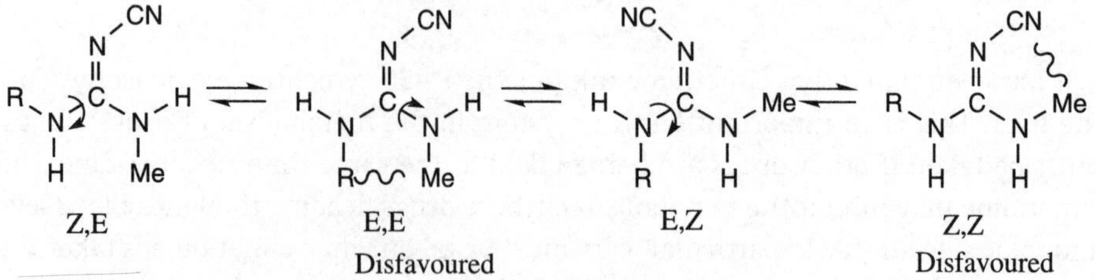

Fig. 18.43 Conformations of the guanidine group in cimetidine.

However, X-ray and NMR studies have shown that cimetidine exists as an equilibrium mixture of the *E,Z* and *Z,E* conformations. The *Z,Z* form is not favoured since the cyano group is forced too close to the *N*-methyl group. If either the *E,Z* or *Z,E* form is the active conformation, then it implies that the chelation type of hydrogen bonding described previously is not essential. An alternative possibility is that the guanidine unit is hydrogen bonding to two distinct hydrogen bonding regions rather than to a single carboxylate group (Fig. 18.44).

Further support for this theory is provided by the weak activity observed for the urea analogue (Fig. 18.31). This compound is known to prefer the Z,Z conformation over the Z,E or E,Z and would therefore be unable to bind to both hydrogen bonding regions.

Clathrate H-bonds not possible   Two separate H-bonds

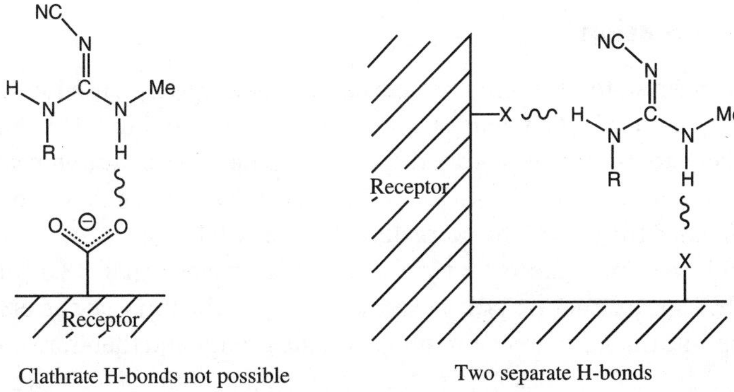

Fig. 18.44 Alternative theory for cimetidine bonding at the agonist region.

If this bonding theory is correct and the active conformation is the E,Z or Z,E form, then restricting the group to adopt one or other of these forms may lead to more active compounds and an identification of the active conformation. This can be achieved by incorporating part of the guanidine unit within a ring – a strategy of rigidification.

For example, the nitropyrrole derivative (Fig. 18.45) has been shown to be the strongest antagonist in the cimetidine series, implying that the E,Z conformation is the active conformation.

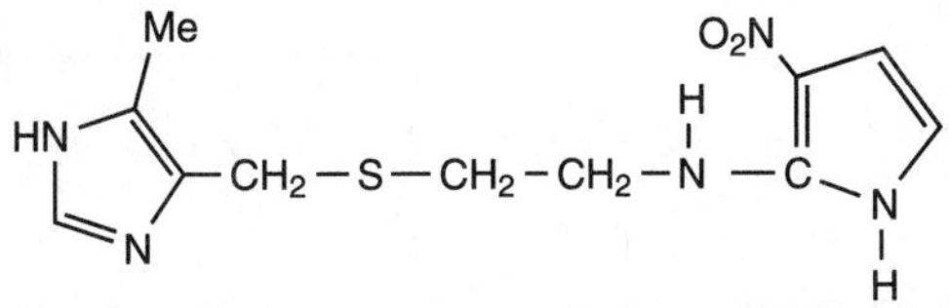

Fig. 18.45 Nitropyrrole derivative of cimetidine.

The isocytosine ring (Fig. 18.46) has also been used to 'lock' the guanidine group, limiting the number of conformations available. The ring allows further substitution and development as seen below (Section 18.12.2.).

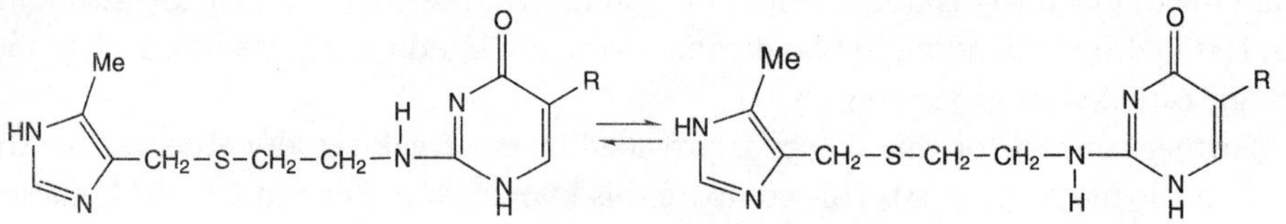

Fig. 18.46 Isocytosine ring.

18.12.2 Desolvation

It has already been stated that the guanidine and thiourea groups, used so successfully in the development of H2-antagonists, are polar and hydrophilic. This implies that they are likely to be highly solvated (i.e. surrounded by a 'water coat'). Before hydrogen bonding can take place to the receptor, this 'water coat' has to be removed. The more solvated the group, the more difficult that will be.

One possible reason for the low activity of the urea derivative (Fig. 18.31) has already been described above. Another possible reason could be the fact that the urea group is more hydrophilic than thiourea or cyanoguanidine groups and therefore more highly solvated. The difficulty in desolvating the urea group might explain why the urea analogue has a lower activity than cimetidine, despite having a lower partition coefficient and greater water solubility.

Leading on from this, if the ease of desolvation is a factor in antagonist activity, then reducing the solvation of the polar group should increase activity. One way of achieving this would be to increase the hydrophobic character of the polar binding group.

A study was carried out on a range of cimetidine analogues containing different planar aminal systems (Z) (Fig. 18.47) to see whether there was any relationship between antagonist activity and the hydrophobic character of the aminal system (HZ).

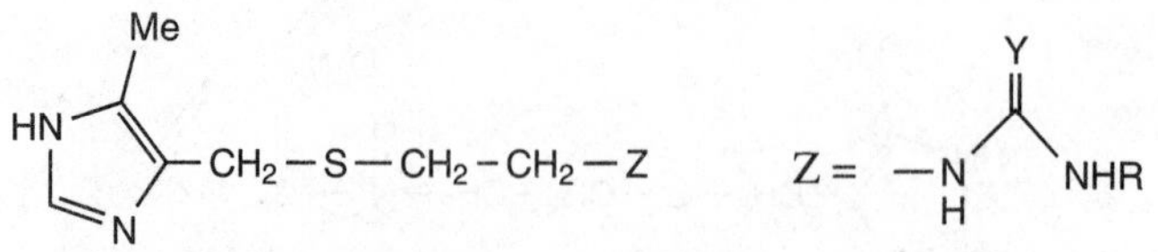

Fig. 18.47 Cimetidine analogue with planar aminal system (Z).

This study showed that antagonist activity was proportional to the hydrophobicity of the animal unit Z (Fig. 18.48) and supported the desolvation theory. The relationship could be quantified as follows:

$$\log(\text{activity}) = 2.0 \log P + 7.4$$

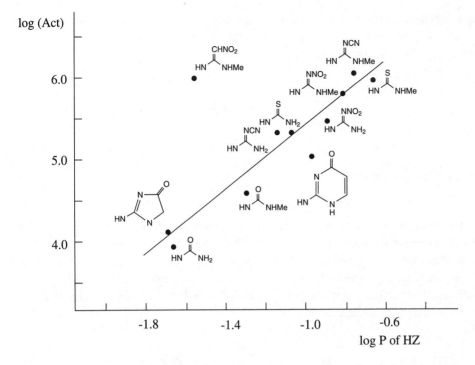

Fig. 18.48 Antagonist activity is proportional to the hydrophobicity of the aminal unit Z.

Further studies on hydrophobicity were carried out by adding hydrophobic substituents to the isocytosine analogue (Fig. 18.46). These studies showed that there was an optimum hydrophobicity for activity corresponding to the equivalent of a butyl or pentyl substituent. A benzyl substituent was particularly good for activity, but proved to have toxic side-effects. These side-effects could be reduced by adding alkoxy substituents to the aromatic ring and this led to the synthesis of oxmetidine (Fig. 18.49), which had enhanced activity over cimetidine. Oxmetidine was considered for clinical use, but was eventually withdrawn since it still retained undesirable side-effects.

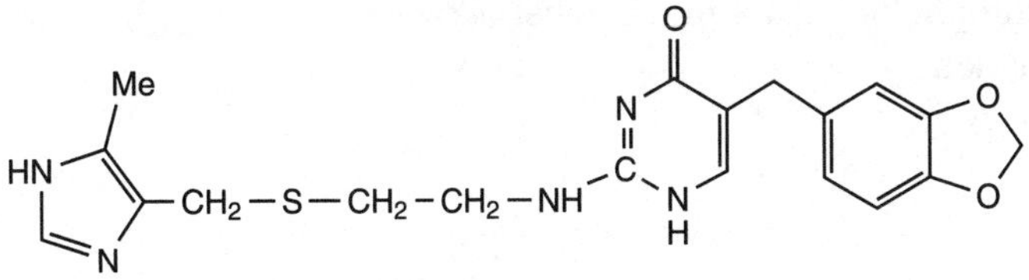

Fig. 18.49 Oxmetidine

The development of the nitroketeneaminal binding group

As we have seen, antagonist activity increases with the hydrophobicity of the polar binding group. It was therefore decided to see what would happen if the polar imino nitrogen of cimetidine was replaced with a non-polar carbon atom. This would result in a keteneaminal group as shown in Fig. 18.50. Unfortunately, keteneaminals are more likely to exist as their amidine tautomers unless a strongly electronegative group (e.g. NO_2) is attached to the carbon atom.

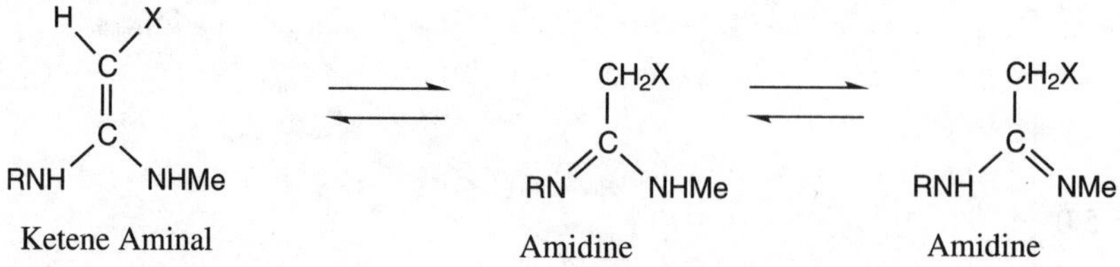

Ketene Aminal   Amidine   Amidine

Fig. 18.50

Therefore, a nitroketeneaminal group was used to give the structure shown in Fig. 18.51. Surprisingly, there was no great improvement in activity, but when the structure was studied in detail, it was discovered that it was far more hydrophilic than expected. This explained why the activity had not increased, but it highlighted a different puzzle. The compound was *too* active. Based on its hydrophilicity, it should have been a weak antagonist (Fig. 18.48).

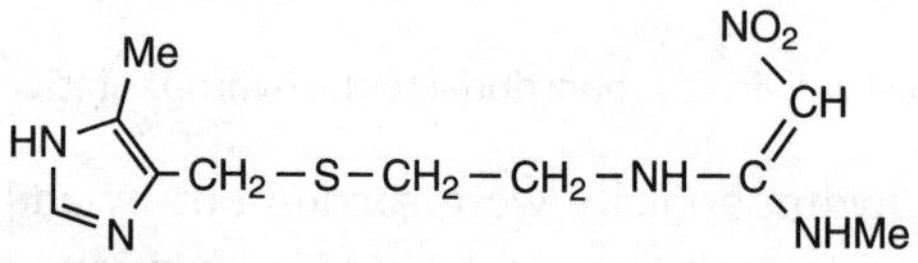

Fig. 18.51 Cimetidine analogue with a nitroketeneaminal group.

It was clear that this compound did not fit the pattern followed by previous compounds since the antagonist activity was 30 times higher than would have been predicted by the equation above. Nor was the nitroketeneaminal the only analogue to deviate from the normal pattern. The imidazolinone analogue (Fig. 18.52), which is relatively hydrophobic, had a much lower activity than would have been predicted from the equation.

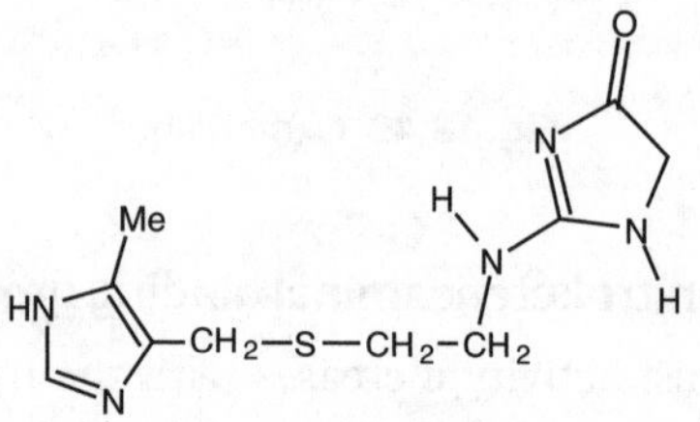

Fig. 18.52 Imidazolinone analogue.

Findings like these are particularly exciting since any deviation from the normal pattern suggests that some other factor is at work which may give a clue to future development.

In this case, it was concluded that the polarity of the group might be important in some way. In particular, the *orientation* of the dipole moment appeared to be crucial. In Fig. 18.53, the orientation of the dipole moment is defined by φ – the angle between the dipole moment and the NR bond. The cyanoguanidine, nitroketeneaminal, and nitropyrrole groups all have high antagonist activity and have dipole moment orientations of 18, 33, and 27° respectively (Fig. 18.54). The isocytosine and imidazolinone groups result in lower activity and have dipole orientations of 2 and –6°, respectively. The strength of the dipole moment (μ) does not appear to be crucial.

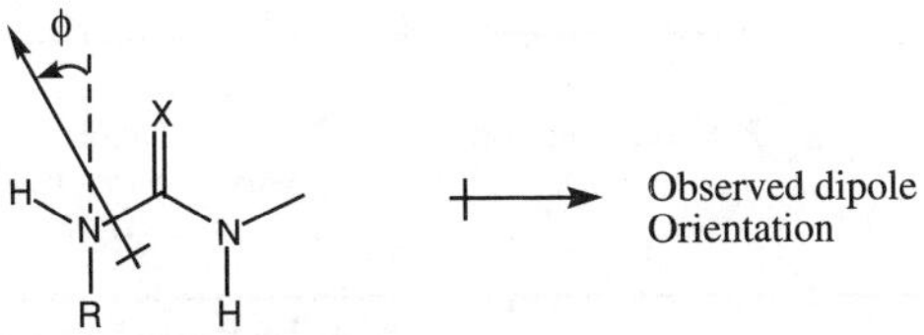

Fig. 18.53 Orientation of dipole moment.

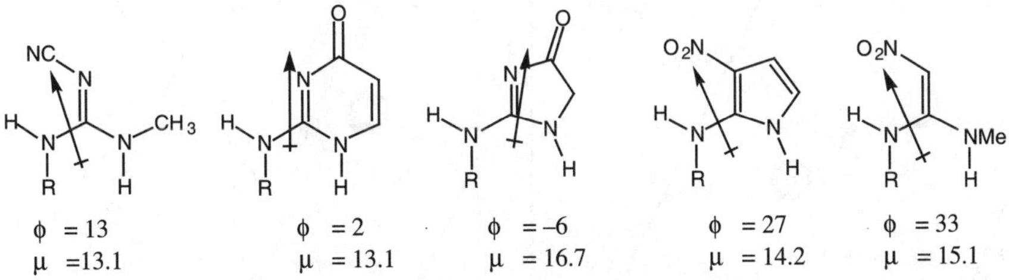

| $\phi = 13$ | $\phi = 2$ | $\phi = -6$ | $\phi = 27$ | $\phi = 33$ |
| $\mu = 13.1$ | $\mu = 13.1$ | $\mu = 16.7$ | $\mu = 14.2$ | $\mu = 15.1$ |

Fig. 18.54 Dipole moments of various antagonistic groups.

Why should the orientation of a dipole moment be important? One possible explanation is as follows. As the drug approaches the receptor, its dipole interacts with a dipole on the receptor surface such that the dipole moments are aligned. This orientates the drug in a specific way before hydrogen bonding takes place and will determine how strong the subsequent hydrogen bonding will be (Fig. 18.55). If the dipole moment is correctly orientated as in the keteneaminal analogue, the group will be correctly positioned for strong hydrogen bonding and high activity will result. If the orientation is wrong as in the imidazolinone analogue, then the bonding is less efficient and activity is lost.

QSAR studies were carried out to determine what the optimum angle φ should be for activity. This resulted in an ideal angle for φ of 30°. A correlation was worked out between the dipole moment orientation, partition coefficient, and activity as follows:

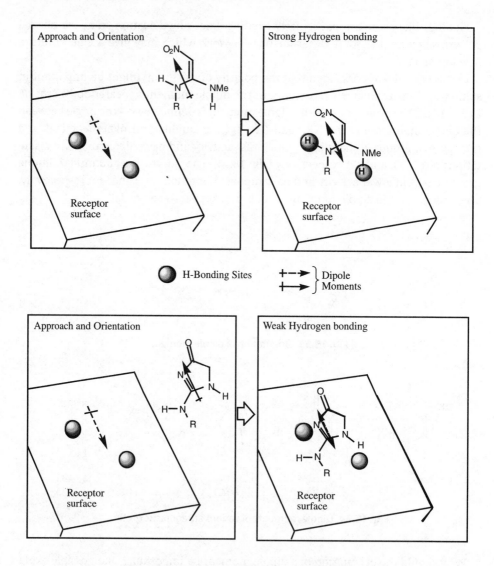

Fig. 18.55 Orientation effects on receptor bonding.

$$\log A = 9.12 \cos \theta + 0.6 \log P - 2.71$$

where A is the antagonist activity, P is the partition coefficient, and θ is the deviation in angle of the dipole moment from the ideal orientation of 30° (Fig. 18.56).

The equation shows that activity increases with increasing hydrophobicity (P). The $\cos \theta$ term shows that activity drops if the orientation of the dipole moment varies from the ideal angle of 30°. At the ideal angle, θ is 0° and $\cos \theta$ is 1. If the orientation of the dipole moment deviates from 30°, then $\cos \theta$ will be a fraction and will lower the calculated activity. The nitroketeneaminal group did not result in a more

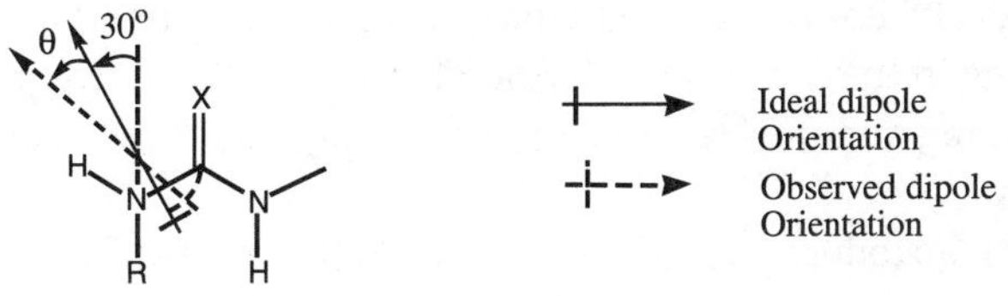

Fig. 18.56

powerful cimetidine analogue, but we shall see it appearing again in ranitidine (Section 18.13.).

18.13 Variation of the imidazole ring – ranitidine

Further studies on cimetidine analogues showed that the imidazole ring could be replaced with other nitrogen-containing heterocyclic rings.

However, Glaxo moved one step further by showing that the imidazole ring could be replaced with a furan ring bearing a nitrogen-containing substituent. This led to the introduction of ranitidine (Zantac) (Fig. 18.57).

Ranitidine has fewer side-effects than cimetidine, lasts longer, and is ten times more active. Structure–activity results for ranitidine include the following:

- The nitroketeneaminal group is optimum for activity, but can be replaced with other planar π systems capable of hydrogen bonding.

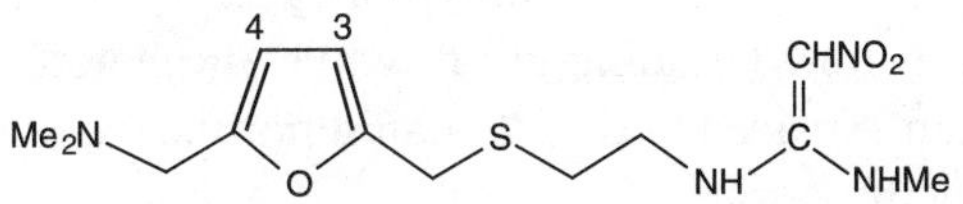

Fig. 18.57 Ranitidine (Zantac).

- Replacing the sulfur atom with a methylene atom leads to a drop in activity.
- Placing the sulfur next to the ring lowers activity.
- Replacing the furan ring with more hydrophobic rings such as phenyl or thiophene reduces activity.
- 2,5-Disubstitution is the best substitution pattern for the furan ring.
- Substitution on the dimethylamino group can be varied, showing that the basicity and hydrophobicity of this group are not crucial to activity.
- Methyl substitution at carbon-3 of the furan ring eliminates activity, whereas the equivalent substitution in the imidazole series increases activity.
- Methyl substitution at carbon-4 of the furan ring increases activity.

The latter two results imply that the heterocyclic rings for cimetidine and ranitidine are not interacting in the same way with the H2-receptor. This is supported by the fact that a corresponding dimethylaminomethylene group attached to cimetidine leads to a drop in activity.

Ranitidine was introduced to the market in 1981 and by 1988 was the world's biggest selling prescription drug.

18.14 Famotidine and nizatidine

During 1985 and 1987 two new antiulcer drugs were introduced to the market – famotidine and nizatidine.

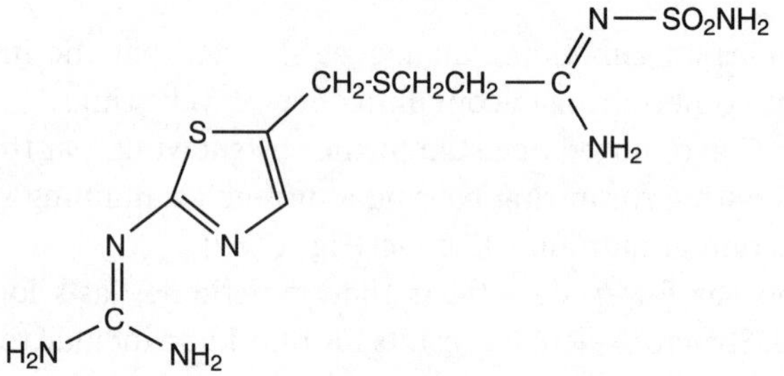

Fig. 18.58 Famotidine.

Famotidine (Pepcid) (Fig. 18.58) is 30 times more active than cimetidine *in vitro*. The side-chain contains a sulfonylamidine group while the heterocyclic imidazole ring of cimetidine has been replaced with a 2-guanidinothiazole ring. Structure–activity studies gave the following results:

• The sulfonylamidine binding group is not essential and can be replaced with a variety of structures as long as they are planar, have a dipole moment, and are capable of interacting with the receptor by hydrogen bonding. A low pK_a is not essential, which allows a larger variety of planar groups to be used than is possible for cimetidine.

• Activity is optimum for a chain length of four or five units.

• Replacement of sulfur with a CH_2 group *increases* activity.

• Modification of the chain is possible with, for example, inclusion of an aromatic ring.

• A methyl substitutent *ortho* to the chain leads to a drop in activity (unlike the cimetidine series).

• Three of the four hydrogens in the two NH_2 groups are required for activity.

There are several results here which are markedly different from cimetidine, implying that famotidine and cimetidine are not interacting in the same way with the

H2-receptor. Further evidence for this is the fact that guanidine substitution at the equivalent position of cimetidine analogues leads to very low activity.

Nizatidine (Fig. 18.59) was introduced into the UK in 1987 by the Lilly Corporation and is equipotent with ranitidine. The furan ring in ranitidine is replaced with a thiazole ring.

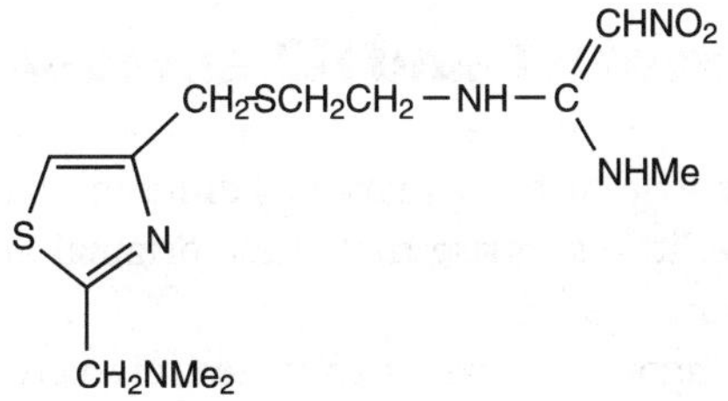

Fig. 18.59 Nizatidine.

18.15 H2-antagonists with prolonged activity

There is presently a need for longer lasting antiulcer agents which require once daily doses. Glaxo carried out further development on ranitidine by placing the oxygen of the furan ring exocyclic to a phenyl ring and replacing the dimethylamino group with a piperidine ring to give a series of novel structures (Fig. 18.60).

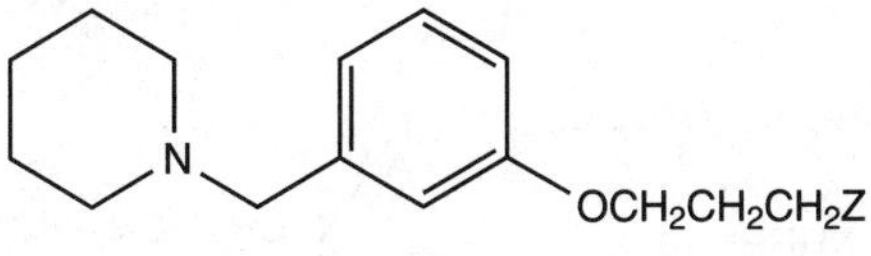

Z= planar and polar H bonding group

Fig. 18.60 Long lasting anti-ulcer agents.

The most promising of these compounds were lamitidine and loxtidine (Fig. 18.61) which were five to ten times more potent than ranitidine and three times longer lasting.

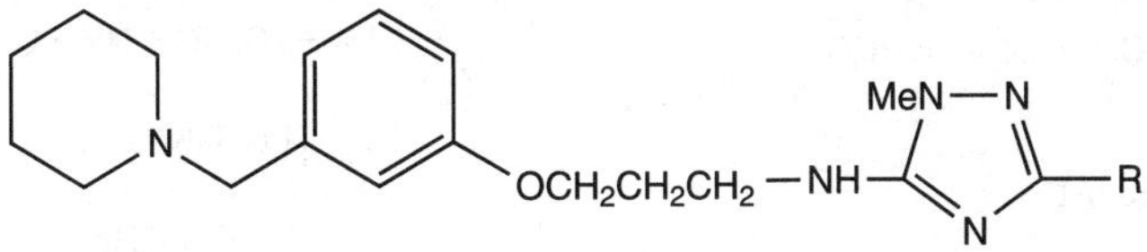

Fig. 18.61 Lamitidine (R = NH$_2$) and loxtidine (R = CH$_2$OH).

Unfortunately, these compounds showed toxicity in long-term animal studies with the possibility that they caused gastric cancer, and they were subsequently withdrawn from clinical study. However, the relevance of these results has been disputed.

18.16 Comparison of H1- and H2-antagonists

The structures of the H2-antagonists are markedly different to the classical H1-antagonists and so there can be little surprise that these original antihistamines failed to antagonize the H2-receptor.

H1-antagonists, like H1-agonists, possess an ionic amino group at the end of a flexible chain. Unlike the agonists, they possess two aryl or heteroaryl rings in place of the imidazole ring. (Fig. 18.62). Because of the aryl rings, H1-antagonists are hydrophobic molecules having high partition coefficients.

In contrast, H2-antagonists are polar, hydrophilic molecules having high dipole moments and low partition coefficients. At the end of the flexible chain they have a polar, π electron system which is weakly amphoteric and un-ionized at pH 7.4. This binding group appears to be the key feature leading to antagonism of H2-receptors (Fig. 18.62). The five-membered heterocycle generally contains a nitrogen atom or, in the case of furan or phenyl, a nitrogen-containing side-chain. The hydrophilic character of H2-antagonists helps to explain why H2-antagonists are less likely to have the CNS side-effects often associated with H1-antagonists.

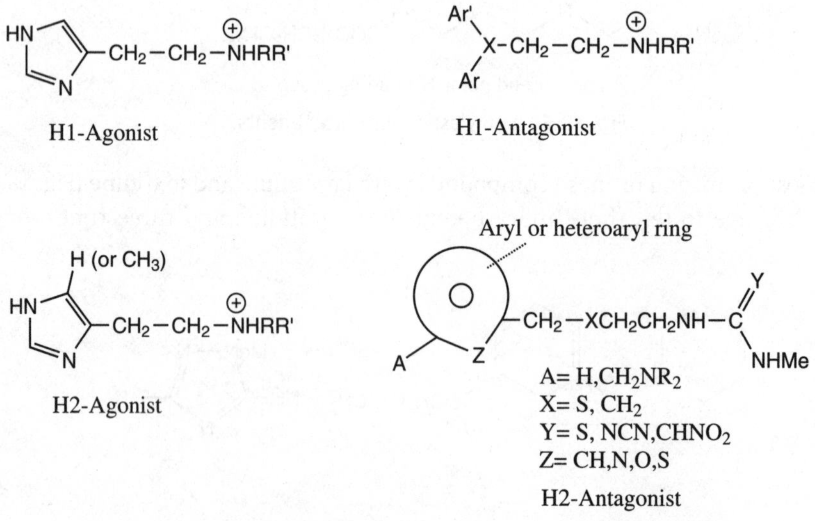

Fig. 18.62 Comparison between H1- and H2-agonists and antagonists.

18.17 The H2-receptor and H2-antagonists

H2-receptors are present in a variety of organs and tissues, but their main role is in acid secretion. As a result, H2-antagonists are remarkably safe and mostly free of side-effects. The four most used agents on the market are cimetidine, ranitidine, famotidine, and nizatidine. They inhibit all aspects of gastric secretion and are rapidly absorbed from the gastrointestinal tract with half-lives of 1–2 h. About 80 per cent of ulcers are healed after 4–6 weeks.

Attention must be given to possible drug interactions when using cimetidine due to inhibition of drug metabolism (Section 18.11.). The other three H2-antagonists mentioned do not inhibit the P-450 cytochrome oxidase system and are less prone to such interactions.

18.18 Recent developments in anti-ulcer therapy

18.18.1 The proton pump inhibitors

Although the H2-antagonists have been remarkably successful in the treatment of ulcers, they are not the only, or even the most effective agents which can be used. In recent years, a class of compounds called the proton pump inhibitors (PPI's) have been found to be superior to the H2-antagonists in the treatment of a variety of ulcers. These agents work by irreversibly inhibiting an enzyme complex called the proton pump (H^+/K^+ ATPase). This enzymic pump is present in the membranes of parietal cells and is responsible for releasing HCl into the stomach. At the same time, protons are exchanged for K^+ ions, and the enzyme ATPase catalyses the hydrolysis of ATP. When PPI's bind to the proton pump, the whole process is inhibited and gastric acid is no longer released. Examples of PPI's which are in clinical use are omeprazole, iansoprazole, pantoprazole and rabeprazole (Fig. 18.63). These are similar in structure, all containing a pyridyl methylsulfinyl benzamidazole skeleton. All these compounds act as prodrugs since they are converted under the acid conditions of the stomach to sulphenamides which then bind irreversibly to exposed cysteine residues of the proton pump.

18.18.2 *Helicobacter pylori*

One of the problems relating to anti-ulcer therapy with both the H2-antagonists and the PPI's is the high rate of ulcer recurrence once the therapy is finished. The reappearance of ulcers has been attributed to the presence of a bacterial strain called *Helicobacter pylori* which is naturally present in the stomachs of many people and which can cause inflammation of the stomach wall. As a result, patients who are found to have *H. pylori* are currently given a combination of three drugs – a PPI in order to reduce gastric acid secretion along with two antibacterial agents (e.g. nitroimidazole, clarithromycin, amoxycillin, or tetracyclin) in order to eradicate the organism.

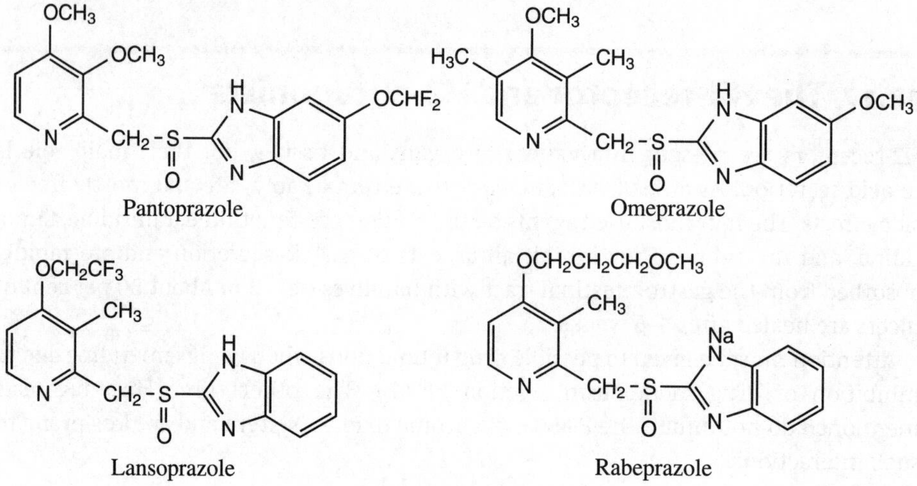

Pantoprazole

Omeprazole

Lansoprazole

Rabeprazole

Fig. 18.63 Proton pump inhibitors.

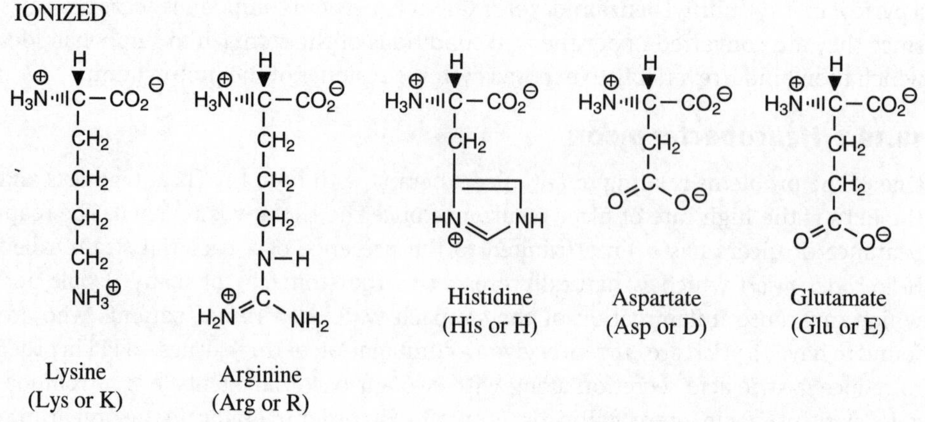

IONIZED

Lysine
(Lys or K)

Arginine
(Arg or R)

Histidine
(His or H)

Aspartate
(Asp or D)

Glutamate
(Glu or E)

Appendix 1
Essential amino acids

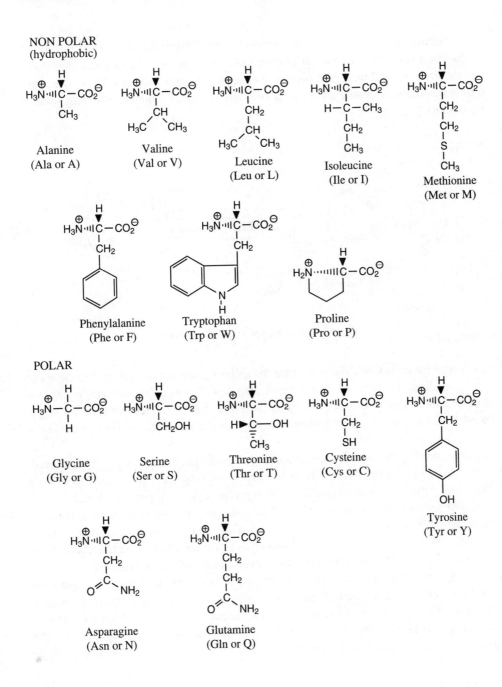

NON POLAR
(hydrophobic)

Alanine
(Ala or A)

Valine
(Val or V)

Leucine
(Leu or L)

Isoleucine
(Ile or I)

Methionine
(Met or M)

Phenylalanine
(Phe or F)

Tryptophan
(Trp or W)

Proline
(Pro or P)

POLAR

Glycine
(Gly or G)

Serine
(Ser or S)

Threonine
(Thr or T)

Cysteine
(Cys or C)

Tyrosine
(Tyr or Y)

Asparagine
(Asn or N)

Glutamine
(Gln or Q)

Appendix 2
The action of nerves

The structure of a typical nerve is shown in Fig. A2.1. The nucleus of the cell is found in the large cell body situated at one end of the nerve cell. Small arms (dendrites) radiate from the cell body and receive messages from other nerves. These messages either stimulate or destimulate the nerve. The cell body 'collects' the sum total of these messages.

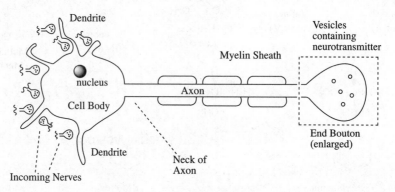

Fig. A2.1 Structure of a typical nerve cell.

It is worth emphasizing the point that the cell body of a nerve receives messages not just from one other nerve, but from a range of different nerves. These pass on different messages (neurotransmitters). Therefore, a message received from a single nerve is unlikely to stimulate a nerve signal by itself unless other nerves are acting in sympathy.

Assuming that the overall stimulation is great enough, an electrical signal is 'fired' down the length of the nerve (the axon). This axon is 'padded' with sheaths of lipid (myelin sheaths) which act to insulate the signal as it passes down the axon.

The axon leads to a knob-shaped swelling (synaptic button) if the nerve is communicating with another nerve. Alternatively, if the nerve is communicating with a muscle cell, the axon leads to what is known as a neuromuscular endplate, where the nerve cell has spread itself like an amoeba over an area of the muscle cell.

Within the synaptic button or neuromuscular endplate there are small globules (vesicles) containing the neurotransmitter chemical. When a signal is received from the axon, the vesicles merge with the cell membrane and release their neurotransmitter into the gap between the nerve and the target cell (synaptic gap). The neurotransmitter binds to the receptor as described in Chapter 5 and passes on its message. Once the message has been received, the neurotransmitter leaves the receptor and is either broken down enzymatically (e.g. acetylcholine) or taken up intact by the nerve cell (e.g. noradrenaline). Either way, the neurotransmitter is removed from the synaptic gap and is unable to bind with its receptor a second time.

To date, we have talked about nerves 'firing' and the generation of 'electrical signals' without really considering the mechanism of these processes. The secret behind nerve transmission lies in the movement of ions across cell membranes, but there is an important difference in what happens in the cell body of a nerve compared to the axon. We shall consider what happens in the cell body first.

All cells contain sodium, potassium, calcium, and chloride ions and it is found that the concentration of these ions is different inside the cell compared to the outside. The concentration of potassium inside the cell is larger than the surrounding medium, whereas the concentration of sodium and chloride ions is smaller. Thus, a concentration gradient exists across the membrane.

Potassium is able to move down its concentration gradient (i.e. out of the cell) since it can pass through the potassium ion channels (Fig. A2.2). However, if potassium can move out of the cell, why does the potassium concentration inside the cell not fall to equal that of the outside? The answer lies in the fact that potassium is a positively charged ion and, as it leaves the cell, an electric potential is set up across the cell membrane. This would not happen if a negatively charged counterion could leave with the potassium ion. However, the counterions in question are large proteins which cannot pass through the cell membrane. As a result, a few potassium ions are able to escape down the ion channels out of the cell and an electric potential builds up across the cell membrane such that the inside of the cell membrane is more negative than the outside. This electric potential (50–80 mV) opposes and eventually prevents the flow of potassium ions.

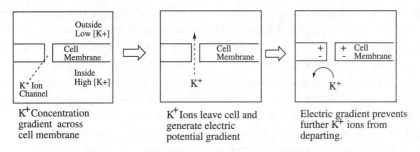

Fig. A2.2 Generation of electric potential across a cell membrane.

But what about the sodium ions? Could they flow into the cell along their concentration gradient to balance the charged potassium ions which are departing? The answer is that they cannot because they are too big for the potassium ion channels. This appears to be a strange argument since sodium ions are smaller than potassium ions. However, it has to be remembered that we are dealing with an aqueous environment where the ions are solvated (i.e. they have a 'coat' of water molecules). Sodium, being a smaller ion than potassium, has a greater localization of charge and is able to bind its solvating water molecules more strongly. As a result, sodium along with its water coat is bigger than a potassium ion with or without its water coat.

Ion channels for sodium do exist and these channels are capable of removing the water coat around sodium and letting it through. However, the sodium ion channels are mostly closed when the nerve is in the resting state. As a result, the flow of sodium ions across the membrane is very small compared to potassium. Nevertheless, the presence of sodium ion channels is crucial to the transmission of a nerve signal.

To conclude, the movement of potassium across the cell membrane sets up an electric potential across the cell membrane which opposes this same flow. Charged protein structures are unable to move across the membrane, while sodium ions cross very slowly, and so an equilibrium is established. The cell is polarized and the electric potential at equilibrium is known as the resting potential.

The number of potassium ions required to establish that potential is of the order of a few million compared to the several hundred billion present in the cell. Therefore the effect on concentration is negligible.

As mentioned above, potassium ions are able to flow out of potassium ion channels. However, not all of these channels are open in the resting state. What would happen if more were to open? The answer is that more potassium ions would flow out of the cell and the electric potential across the cell membrane would become more negative to counter this increased flow. This is known as hyperpolarization and the effect is to destimulate the nerve (Fig. A2.3).

Suppose instead that a few sodium ion channels were to open up. In this case, sodium ions would flow into the cell and as a result the electric potential would become less negative. This is known as depolorization and results in a stimulation of the nerve.

If chloride ion channels are opened, chloride ions flow into the cell and the cell membrane becomes hyperpolarized, destimulating the nerve.

Ion channels do not open or close by chance. They are controlled by the neurotransmitters released by communicating nerves. The neurotransmitters bind with their receptors and lead to the opening or closing of ion channels. For example, acetylcholine controls the sodium ion channel, whereas GABA and glycine control chloride ion channels. The resulting flow of ions leads to a localized hyperpolarization or depolarization in the area of the receptor. The cell body collects and sums all this information such that the neck of the axon experiences an overall depolarization or hyperpolarization depending on the sum total of the various excitatory or inhibitory signals received.

We shall now consider what happens in the axon of the nerve (Fig. A2.4). The cell membrane of the axon also has sodium and potassium ion channels but they are different in character from those in the cell body. The axon ion channels are not controlled by neurotransmitters, but by the electric potential of the cell membrane.

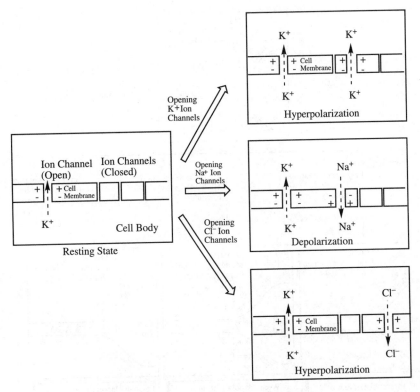

Fig. A2.3 Hyperpolarization and depolarization.

The sodium ion channels located at the junction of the nerve axon with the cell body are the crucial channels since they are the first channels to experience whether the cell body has been depolarized or hyperpolarized.

If the cell body is strongly depolarized then a signal is fired along the nerve. However, a specific threshold value has to be reached before this happens. If the depolarization from the cell body is weak, only a few sodium channels open up and the depolarization at the neck of the axon does not reach that threshold value. The sodium channels then reclose and no signal is sent.

With stronger depolarization, more sodium channels open up until the flow of sodium ions entering the axon becomes greater than the flow of potassium ions leaving it. This results in a rapid increase in depolarization, which in turn opens up more sodium channels, resulting in very strong depolarization at the neck of the axon. The flow of sodium ions into the cell increases dramatically, such that it is far greater than the flow of potassium ions out of the axon, and the electric potential across the membrane is reversed, such that it is positive inside the cell and negative outside the cell. This process lasts less than a millisecond before the sodium channels reclose and sodium permeability returns to its normal state. More potassium channels then open and permeability to potassium ions increases for a while to speed up the return to the resting state.

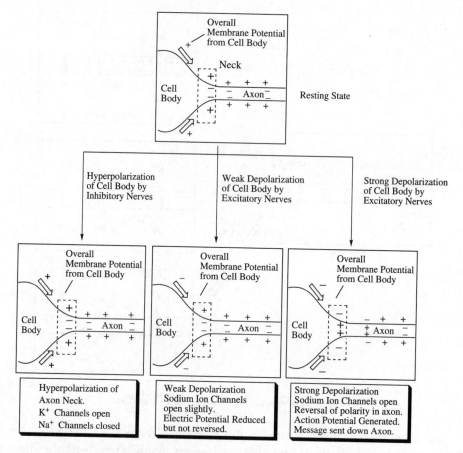

Fig. A2.4 Hyperpolarization and depolarization effects at the neck of the axon.

The process is known as an action potential and can only take place in the axon of the nerve. The cell membrane of the axon is said to be excitable, unlike the membrane of the cell body.

The important point to note is that once an action potential has fired at the neck of the axon, it has reversed the polarity of the membrane at that point. This in turn has an effect on the neighbouring area of the axon and depolarizes it beyond the critical threshold level. it too fires an action potential and so the process continues along the whole length of the axon (Fig. A2.5).

The number of ions involved in this process is minute, such that the concentrations are unaffected.

Once the action potential reaches the synaptic button or the neuromuscular endplate, it causes an influx of calcium ions into the cell and an associated release of neurotransmitter into the synaptic gap. The mechanism of this is not well understood.

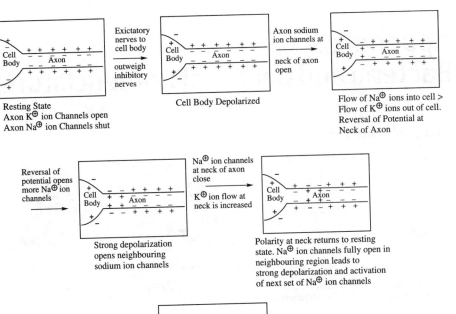

Fig. A2.5 Generation of an action potential.

Appendix 3
Bacteria and bacterial nomenclature

Bacterial nomenclature

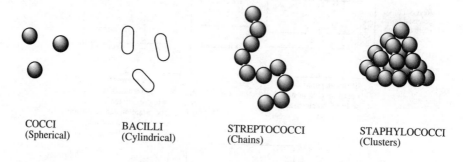

COCCI
(Spherical)

BACILLI
(Cylindrical)

STREPTOCOCCI
(Chains)

STAPHYLOCOCCI
(Clusters)

Fig. A3.1 Bacterial nomenclature.

Some clinically important bacteria

Organism	Gram	Infections
Staphylococcus aureus	Positive	Skin and tissue infections, septicaemia, endocarditis, accounts for about 25% of all hospital infections
Streptococcus	Positive	Several types – commonly cause sore throats, upper respiratory tract infections, and pneumonia
Escherichia coli	Negative	Urinary tract and wound infections, common in the gastrointestinal tract and often causes problems after surgery, accounts for about 25% of hospital infections
Proteus species	Negative	Urinary tract infections
Salmonella species	Negative	Food poisoning and typhoid
Shigella species	Negative	Dysentery
Enterobacter species	Negative	Urinary tract and respiratory tract infections, septicaemia
Pseudomonas aeruginosa	Negative	An 'opportunist' pathogen, can cause very severe infections in burn victims and other compromised patients, i.e. cancer patients, commonly causes chest infections in patients with cystic fibrosis.

Organism	Gram	Infections
Haemophilus influenzae	Negative	Chest and ear infections, occasionally meningitis in young children
Bacteroides fragilis	Negative	Septicaemia following gastrointestinal surgery

The Gram stain

A staining procedure of great value in the identification of bacteria. The procedure is as follows:

1. Stain the cells purple.
2. Decolourize with organic solvent.
3. Stain the cells red.

This test will discriminate between two types of bacterial cell:

(A) Gram-negative bacteria – these bacteria cells are easily decolourized at stage and will therefore be coloured red;

(B) Gram-positive bacteria – these resist the decolourization at stage 2 and will therefore remain purple.

The different result in the Gram test observed between these two types of bacteria is because of differences in their cell wall structure. These differences in cell wall structure have important consequences in the sensitivity of the two types of bacteria to certain types of antibacterial agents. It is believed that Gram-negative bacteria have an extra outer layer.

The cells of Gram-positive bacteria have an outer covering or membrane containing teichoic acids, whereas the walls of Gram-negative bacteria are covered with a smooth, soft lipopolysaccharide which also contains phospholipids, lipoproteins, and proteins. This layer acts a barrier and penicillins have to negotiate a limited number of protein channels to reach the cell.

Glossary

ACTIVE CONFORMATION

The conformation adopted by a compound when it binds to its target binding site.

ACTIVE PRINCIPLE

The single chemical in a mixture of compounds which is chiefly responsible for that mixture's biological activity.

ADDICTION

Addiction can be defined as a habitual form of behaviour. It need not be harmful. For example, one can be addicted to eating chocolate or watching television without suffering more than a bad case of toothache or a surplus of soap operas.

AFFINITY

Affinity is a measure of how strongly a ligand binds to its target binding site.

AGONIST

A drug producing a response at a certain receptor.

ANOMERS

Cyclic stereoisomers of sugars that differ only in their configuration at the hemiacetal (anomeric) carbon.

ANTAGONIST

A drug which interacts with a defined receptor to block an agonist.

ANTIBACTERIAL AGENT

A synthetic or naturally occurring agent which can kill or inhibit the growth of bacterial cells.

ANTIBIOTIC

An antibacterial agent derived from a natural source (e.g. penicillin from penicillium mould).

ANTISENSE THERAPY

The design of molecules which will bind to specific regions of m-RNA and thus prevent them being used to code for protein synthesis.

BACTERIOSTATIC

Bacteriostatic drugs inhibit the growth and multiplication of bacteria, but do not directly kill them (e.g. sulfonamides, tetracyclines, chloramphenicol).

BACTERICIDAL

Bactericidal drugs irreversibly damage and kill bacteria, usually by attacking the cell wall or plasma membrane (e.g. penicillins, cephalosporins, polymyxins).

BIOISOSTERE

A chemical group which can replace another chemical group in a drug without affecting the biological activity of the drug.

CASPASES

Enzymes which have important roles to play in the ageing process of cells.

CHEMOTHERAPEUTIC INDEX

A comparison of the minimum effective dose of a drug with the maximum dose which can be tolerated by the host.

CO-ENZYME

The name given to small organic molecules acting as cofactors.

COFACTOR

Cofactors are ions or small organic molecules (other than the substrate) which are bound to the active site of an enzyme and take part in the enzyme catalysed reaction.

DATABASE MINING

The use of computers to automatically search databases of compounds for structures containing specified pharmacophores.

DEPENDENCE

A compulsive urge to take a drug for psychological or physical needs. The psychological need is usually why the drug was taken in the first place (to change one's mood) but physical needs are often associated with this. This shows up when the drug is no longer taken, leading to psychological withdrawal symptoms (feeling miserable) and physical withdrawal symptoms (headaches, shivering, etc.) Dependence need not be a serious matter if it is mild and the drug is non-toxic (e.g. dependence on coffee). However, it is a serious matter if the drug is toxic and/or shows tolerance. Examples: opiates, alcohol, barbiturates, diazepams.

ED_{50}

The ED_{50} is the mean effective dose of a drug necessary to produce a therapeutic effect in 50% of the test sample.

EFFICACY

Efficacy is a measure of how effectively an agonist activates a receptor. It is possible for a drug to have high affinity for a receptor (i.e. have strong binding interaction) but have low efficacy.

FIRST PASS EFFECT

The first pass effect refers to the extent to which an orally administered drug is metabolized during its first passage through the gut wall and the liver.

GATING

The mechanism by which ion channels are opened or closed.

INVERSE AGONIST

A compound which acts as an antagonist, but which also decreases the 'resting' activity of target receptors (i.e. those receptors which are active in the absence of agonist).

ISOSTERE

A chemical group which can be considered to be equivalent in size and behaviour to another chemical group; for example, replacing a methylene group with an ether bridge (CH_2 for O).

KINASES

Enzymes which catalyse the phosphorylation of alcoholic or phenolic groups present in the substrate.

LD_{50}

The LD_{50} is the mean lethal dose of a drug required to kill 50 per cent of the test sample.

LEAD COMPOUND

A compound showing a desired pharmacological property which can be used to initiate a medicinal chemistry project.

METHYLENE SHUFFLE

A strategy used to alter the hydrophobicity of a molecule. One alkyl chain is shortened by one carbon unit, while another is lengthened by a one carbon unit.

MIX AND SPLIT

The procedure involved when synthesizing mixtures of compounds by combinatorial synthesis.

PARTIAL AGONIST

A drug which acts like an antagonist by blocking an agonist, but retains some agonist activity of itself.

PEPTOIDS

Peptides which are partly or wholly made up of non naturally occurring amino acids. As such, they may no longer be recognized as peptides by the body's protease enzymes.

PHARMACOPHORE

Defines the atoms and functional groups required for a specific pharmacological activity, and their relative positions in space.

PINOCYTOSIS

A method by which molecules can enter cells without passing through cell membranes. The molecule is 'engulfed' by the cell membrane and taken into the cell in a membrane bound vesicle.

PROSTHETIC GROUP

A cofactor which is covalently linked to the active site of an enzyme.

PROSTAGLANDINS

Natural chemical messengers in the body with a large variety of biological functions.

PROTEASES

Enzymes which hydrolyse peptide bonds.

QUANTITATIVE STRUCTURE ACTIVITY RELATIONSHIPS

Studies which relate the physicochemical properties of compounds with their pharmacological activity.

RECEPTOR

A protein in the cell membrane of a nerve or target organ with which a transmitter substance or drug can interact to produce a biological response. Example: cholinergic receptors at nerve synapses.

RECURSIVE DECONVOLUTION

A method for identifying the constituents in a combinatorial synthetic mixture. The method requires the storage of intermediate mixtures.

SECONDARY MESSENGER

A secondary messenger is a natural chemical (e.g. cyclic AMP) which is produced by the cell as a result of receptor activation, and which carries the chemical message from the cell membrane to the cycloplasm.

SIGNAL TRANSDUCTION

The mechanism by which an activated receptor transmits a message into the cell resulting in a cellular response.

STRUCTURE ACTIVITY RELATIONSHIPS (SAR)

Studies carried out to determine those atoms or functional groups which are important to a drug's activity.

SUBSTRATE

A chemical which undergoes a reaction catalysed by an enzyme.

SUICIDE SUBSTRATES

Enzyme inhibitors which have been designed to be activated by an enzyme catalysed reaction, and which will bind irreversibly to the active site as a result.

TAGGING

A method of identifying what structures are being synthesized on any one resin bead during a combinatorial synthesis. The tag is a peptide or nucleotide sequence which is constructed in parallel with the synthesis.

THERAPEUTIC INDEX (OR RATIO)

The therapeutic index is the ratio of a drug's undesirable effects with respect to its desirable effects and is therefore a measure of how safe that drug is. Usually this involves comparing the dose levels leading to a toxic effect with respect to the dose levels leading to a therapeutic effect. The larger the therapeutic index, the safer the drug.

To be more precise, the therapeutic index compares the drug dose levels which lead to toxic effects in 50 per cent of cases studied, with respect to the dose levels leading to maximum therapeutic effects in 50 per cent of cases studied. This is a more reliable method of measuring the index since it eliminates any peculiar individual results.

TOLERANCE

Repeat doses of a drug may result in smaller biological results. The drug may block or antagonize its own action and larger doses are needed for the same pharmacological effect.

Alternatively, the body may 'learn' how to metabolize the drug more efficiently. Again, larger doses are needed for the same pharmacological effect, increasing the chances of toxic side-effects.

Examples: morphine, hexamethonium.

TRANSITION STATE ANALOGUES

Enzyme inhibitors which have been designed to mimic the transition state of an enzyme catalysed reaction.

Further reading

Albert, A (1987). *Xenobiosis*. Chapman and Hall

Bowman, W.C. and Rand, M.J. (1980). *Textbook of Pharmacology*. Blackwell Scientific.

Ganellin, C.R. and Roberts. S.M. (1993). *Medicinal chemistry – the role of organic chemistry in drug research*, Academic Press.

King, F.D. (1994). *Medicinal chemistry – principles and practice*. The Royal Society of Chemistry.

Mann, J. (1992). *Murder, magic and medicine*. Oxford University Press.

Nogrady, T. (1988). *Medicinal chemistry: a biochemical approach*. Oxford University Press, New York.

Rang, H.P., Dale, M.M. and Ritter, J.M. (1999). *Pharmacology (4th edition)*. Churchill Livingstone.

Roberts, S.M. and Price, B.J. (ed) (1985). *Medicinal chemistry – the role of organic chemistry in drug research*. Academic Press.

Sammes, P.G. (ed) (1990). *Comprehensive medicinal chemistry*. Pergamon Press.

Silverman, R. (1992). *The organic chemistry of drug design and drug action*. Academic Press.

Smith, C.M. and Reynard, A.M. (1992). *Textbook of pharmacology*. W.B. Saunders.

Sneader, W. (1985). *Drug discovery: the evolution of modern medicine*. Wiley.

Stenlake, J.B. (1979). *Foundations of molecular pharmacology*. Volumes 1 and 3. Athlone Press.

Suckling, K.E. and Suckling, C.J. (1980). *Biological chemistry*. Cambridge University Press.

Wermuth, C.G. (1996). *The practice of medicinal chemistry*. Academic Press.

Wolff, M.E. *Burger's medicinal chemistry and drug discovery (5th edition)*. (1995). Volumes 1–5. Wiley.

Index